UNCOVERING EARLY GALAXY EVOLUTION
IN THE ALMA AND JWST ERA

IAU SYMPOSIUM 352

COVER ILLUSTRATION:

A multi-wavelength view of the deep Universe in the Hubble Ultra Deep Field. In blue, the optical/near-infrared emission of galaxies imaged using the Hubble Space Telescope is shown. Orange shows the carbon monoxide emission (indicating molecular gas out of which new stars can form) imaged by ALMA. This is an example of multi-wavelength images that will be obtained in the future by combining deep JWST and ALMA observations. This image is based on the very deep ALMA survey by the ALMA Spectroscopic Survey in the HUDF (ASPECS) collaboration (see also ESO Science Release #1633).

Image credit: B. Saxton (NRAO/AUI/NSF); ALMA (ESO/NAOJ/NRAO); NASA/ESA Hubble

Over the image we show an outline of the 16th-century Portuguese fortress that served as the venue for this IAU Symposium, the Fort of Santiago da Barra in Viana do Castelo.

IAU SYMPOSIUM PROCEEDINGS SERIES

INTERNATIONAL ASTRONOMICAL UNION

UNION ASTRONOMIQUE INTERNATIONALE

UNCOVERING EARLY GALAXY EVOLUTION IN THE ALMA AND JWST ERA

PROCEEDINGS OF THE 352nd SYMPOSIUM OF THE INTERNATIONAL ASTRONOMICAL UNION HELD IN VIANA DO CASTELO, PORTUGAL 3–7 JUNE, 2019

Edited by

ELISABETE DA CUNHA
Australian National University & University of Western Australia, Australia

JACQUELINE HODGE
Leiden University, The Netherlands

JOSÉ AFONSO
Instituto de Astrofísica / University of Lisbon, Portugal

LAURA PENTERICCI
INAF Roma, Italy

and

DAVID SOBRAL
Lancaster University, United Kingdom

CAMBRIDGE
UNIVERSITY PRESS

University Printing House, Cambridge CB2 8BS, United Kingdom

One Liberty Plaza, 20th Floor, New York, NY 10006, USA

477 Williamstown Road, Port Melbourne, VIC 3207, Australia

314-321, 3rd Floor, Plot 3, Splendor Forum, Jasola District Centre, New Delhi - 110025, India

79 Anson Road, #06-04/06, Singapore 079906

Cambridge University Press is part of the University of Cambridge.

It furthers the University's mission by disseminating knowledge in the pursuit of education, learning and research at the highest international levels of excellence.

www.cambridge.org
Information on this title: www.cambridge.org/9781108492133

First published 2020

A catalogue record for this publication is available from the British Library

ISBN 978-1-108-49213-3 Hardback

Table of Contents

SESSION 1: Epoch of reionization

SESSION 2: Theoretical models and simulations

SESSION 3: Spectral energy distribution models

SESSION 5: The interstellar medium of high-redshift galaxies

SESSION 6: Spatially-resolved analyses of $z > 2$ galaxies

SESSION 7: Lessons from local galaxies and high-z analogues

SESSION 8: Synergies with other facilities & future outlook

Preface

The 352nd Symposium of the International Astronomical Union, on *Uncovering early galaxy evolution in the ALMA and JWST era,* happened in the Portuguese coastal town of Viana do Castelo from June 2nd to 7th, 2019. It is noteworthy that this was only the second IAU symposium ever hosted in Portugal. The main goal of this IAU Symposium was to bring together the international community of observational and theoretical astronomers in the field of early galaxy evolution, with a special focus on new results from the Atacama Large Millimetre Array (ALMA) and preparation for new research and synergies with the soon-to-be-launched James Webb Space Telescope (JWST). We consider this symposium to have been a big success! The symposium brought together 172 participants from 24 countries in a week of very high quality science talks and discussions, as well as some important collateral activities that had the goal of involving the local community through various teaching and outreach events.

Thanks to deep observations in the last few decades with the Hubble Space Telescope, the Spitzer Space Telescope, and ground-based 8-10-metre class telescopes, we know more about the young Universe than ever before, having reached tantalisingly close to the dark ages and the formation of the first stars and galaxies. It is now well established that the rate of cosmic star formation rose rapidly from the epoch of reionization to a maximum at $z \sim 2$. The first three billion years of cosmic time were therefore the prime epoch of galaxy formation. Characterising galaxies at this epoch, both observationally and theoretically, is thus crucial to achieve a major goal of modern astrophysics: to understand how galaxies such as our Milky Way emerged from the primordial density fluctuations in the early Universe and evolved through cosmic time.

Many questions have to be addressed with the next generation of observing facilities and theoretical models. For example, what physical processes drove the rise in star formation rate in the first three billion years? How is the formation and evolution of galaxies determined by their dark matter haloes and large-scale environment? How did black hole growth follow this rise, and how important is the galaxy-AGN connection at early cosmic times? Which star-forming galaxies are responsible for re-ionising the Universe (a major event in cosmic history), and how important is the contribution by early quasars? How do the gas, metals and dust in the interstellar medium of early galaxies evolve? What regulates star formation in galaxies, and what are the physical drivers behind the close correlation between stellar mass and star formation rate (the so-called 'star-forming main sequence')? Are there different star formation modes associated with secular and interaction-driven starbursts, and how important were those processes in shaping the general galaxy population? What dynamical processes established the morphologies of galaxies we observe today? Recent major international investments in two major facilities, the Atacama Large Millimetre Array (ALMA) and the James Webb Space Telescope (JWST), promise to shed light on these questions and uncover the rise of galaxies from the cosmic dark ages. ALMA has been operating since 2011 and has already started changing our view of the distant Universe by detecting dust heated by young star formation and cold molecular gas i.e. the fuel for new star formation, with unprecedented sensitivity and spatial resolution, reaching all the way to the epoch of reionization $(z > 6)$. ALMA gives us an exquisite view into the physical state of the dense interstellar medium in the young Universe, which is determinant in understanding star formation and feedback processes. The soon-to-be-launched JWST will bring a necessary and complementary view of the stellar populations and ionised interstellar medium in galaxies at those epochs. It will directly observe young stars radiating in the rest-frame ultraviolet as well as more evolved stars emitting mostly in the optical and near-infrared (which comprise most of the total stellar mass), and it will access the nebular emission from gas Ionised by the young stars, AGN, and shocks. Combining ALMA and JWST will be crucial to go beyond simply detecting large samples of galaxies in the young Universe, but also characterising in detail the physical processes governing their evolution.

The main science driver of the symposium was the need for the extragalactic community to be prepared to maximise what we can learn from having those facilities simultaneously available during the lifetime of JWST. We brought together the community of theoretical and observational experts to discuss and strategise on how we can make the most of ALMA and JWST synergies in advancing our understanding of galaxy evolution in the young Universe during the next decade. The goal was to formulate the key questions that will be answered with ALMA+JWST, and discuss what observations, diagnostics, and theoretical models/simulations will need to be developed to address them. To achieve this goal, the symposium included an overview of the state-of-the-art in observations and theoretical models of high-redshift galaxies, identifying strategic areas where the overlap between ALMA and JWST will be crucial, and fostering exchanges and international collaboration between theorists and observers, as well as astronomers traditionally observing in different spectral regions.

A major challenge in the JWST era will be how to optimally and reliably measure the physical properties of galaxies from observations in order to compare with theoretical predictions and build a more robust physical understanding of galaxy evolution at early cosmic times. To do this, we need spectral models of galaxies that reliably account for the evolution of young stellar populations, as well as the transfer of starlight through gas and dust in their interstellar medium. For example, the evolution of low-metallicity massive stars is a key ingredient to characterising galaxy spectra in the rest-frame ultraviolet, and importantly the production of ionising photons in galaxies, and has recently arisen as a major source of discrepancy between current models, especially when accounting for the effects of stellar rotation and binaries. Another major source of uncertainty is the potential evolution of the properties of interstellar dust towards early cosmic epochs, which will have to be addressed observationally by combining JWST and ALMA observations to fully characterise both dust attenuation and emission. As spectral models are becoming increasingly complex, we must use advanced statistical techniques to compare models and observations. The symposium also addressed how to put the wealth of new observational information in the context of theoretical models and the variety of increasingly more detailed numerical simulations of galaxy formation. As observations improve in spatial resolution and sensitivity, they present a further challenge to simulations that aim to reproduce both the statistical properties of the galaxy population and the detailed physical processes occurring in galaxies, such as star formation and AGN feedback, or gas accretion and outflows. Only by bringing together the observers and simulators can we as a community build a framework where observations can be optimally used to test and further develop our theoretical physical understanding of the processes that affect galaxy evolution. We also made important links to the local/low-redshift galaxy community to understand what can be learnt from detailed observations of nearby galaxies at resolutions that cannot be achieved in the high-redshift Universe even with ALMA and JWST. For example, studies of resolved stellar populations in nearby dwarf galaxies and other analogues can reveal the star formation history of these objects far into the past, all the way to the epoch of reionization when they might have played a major role as a source of ionizing photons. Also, studies of resolved star clusters and star-forming regions in nearby galaxies give us a detailed insight into the physics of star formation, as well as providing calibrations for diagnostics of, e.g., the star formation rate and gas-phase metallicity. Finally, detailed, high-resolution studies of the interstellar medium of local galaxies give us a detailed picture of the physical processes happening between stars and the surrounding gas and dust.

Bringing together these and other science topics under a common theme, and with the particular focus on synergies between ALMA and JWST for studies of high-redshift galaxy evolution made this symposium unique and very timely. The field was ripe for a meeting that brought together both users of ALMA and JWST as well as theorists, to strategise for the next decade when both these facilities will be available. This will be a transformative time when not only we will detect the first galaxies, but we will start to understand the underlying physics of early galaxy evolution and the sources that drove one of the most important events in cosmic history, the reionization of the Universe.

We would like to emphasize the lively, collegial, and inclusive atmosphere of the symposium, which we believe was a major factor in the fruitful scientific interactions and discussions. The science organizing committee was very active in diversity and equity efforts, from ensuring gender, seniority, and geographical balance of invited and contributed speakers, as well as session chairs, to ensuring poster presenters (especially early-career researchers) had a high visibility throughout the symposium (through poster sparkler sessions, poster viewing sessions, as well as a 'best poster' competition and a 'poster quiz' for participants). We also actively encouraged family caregivers to attend the symposium (who are still women in the majority of cases), by providing an on-site babysitter service free of charge to the participants. This was a very successful initiative in the sense that the participants who used the service were very positively impacted. We view this as a constructive initiative, and encourage future conference and symposium organizers to consider offering this service.

Along with the scientific programme, we also had various collateral for the local community of Viana do Castelo, which had a very positive impact in the community's engagement with astronomy in general, and in the visibility of the IAU in particular. This was achieved thanks to the excellent collaboration with the local authorities, in particular the Viana do Castelo City Council and Mayor, and the important activities co-organized between the LOC and the University of Porto. The connection with the local community and sense of contributing to the local scientific culture were no doubt one of the most rewarding parts of organizing this symposium. We strongly encourage future IAU symposium organizers to think about how they can use the opportunity of organizing a symposium to not only produce a stimulating meeting for professional astronomers and astronomy students, but also to enrich and build ties with local communities that do not always have access to science events. We included a fully accredited teacher training workshop, co-organized by the University of Porto and the 'Ciencia Viva' organization, with the topic *New tools for Astronomy in secondary school.* This workshop was taught by two professors of the University of Porto, Prof Carlos Martins (LOC co-chair) and Prof Paulo Mauricio, for a total of 25 hours between March and June, 2019. This was a unique opportunity for local high-school teachers to obtain fully accredited training in astronomy, free of charge and without having to travel outside of Viana do Castelo.

We also had two events organized with the Viana do Castelo City Council: a welcome reception and a press conference. The welcome reception was a chance for the conference participants to get to know each other and also to interact with the local authorities, including the Mayor. The press conference increased the visibility of the symposium (and of the IAU) in the local press, and was a valuable opportunity to talk about the science topics of the symposium to a broad audience, and to advertise the public events that happened later in the week. One of these events was a free evening of public talks (in Portuguese) in the beautiful Sá de Miranda Municipal Theatre. The event was presented by Miguel Gonçalves, a well-known Portuguese science communicator, and included talks about the IAU (by pioneer Portuguese astronomer and IAU General Secretary Prof Teresa

Lago), the European Space Agency and the JWST (by symposium participant and Viana native Dr Catarina Alves de Oliveira), and early galaxies (by SOC member Dr David Sobral). The talks were followed by a panel Q&A including all the speakers and also Dr Elisabete da Cunha, symposium organizer. This event was a success, very well attended by the locals, and with great visibility in the press. The symposium week ended with a stargazing night for the public at the city beach, organized in collaboration with the Planetarium of Porto and City Council. The attendance exceeded expectations, with about 300 locals of all ages coming to the beach despite the windy evening. The main attractions were the Moon craters and Jupiter. This event was such a success that it motivated City Council to organize more regular stargazing events in Viana do Castelo in the future. We believe this has a lot of potential to become a lasting positive impact of the symposium in the local community.

As a final note we would like to thank the IAU for the support in organizing this conference, and specifically the support of Division J (Galaxies and Cosmology), our co-ordinating division, as well as Division H (Interstellar Matter and Local Universe), Division G (Stars and Stellar Physics), and Division D (Education, Outreach, and Heritage). We are also immensely grateful to all the members of the scientific and local organizing committees without whom this symposium could not have happened (in particular the Leiden members for their tireless support with registration and finances). Many thanks as well to our invited speakers, discussion leads, contributed speakers, and poster presenters for contributing to the high level of scientific talks and discussions, as well as all participants who enthusiastically asked questions, engaged in discussions, and contributed to the overall excellent atmosphere of the symposium, inside and outside of the seminar room. We thank Richard Ellis and Gabriella de Lucia in particular for having accepted the challenge of giving the summary talks. We also thank Carlos Martins and Paulo Mauricio for their efforts in leading the teacher training workshop, and Planetário do Porto for their major contribution to the stargazing event. A huge thanks to Prof Teresa Lago, Miguel Gonçalves, Catarina Alves de Oliveira, and David Sobral, for having accepted our invitation to be the stars of our public talk event, and for making it such an exciting success. Also, additional thanks to Catarina for organizing to send the JWST promotional materials for the participants through ESA. Finally, a very special thanks to the Camara Municipal de Viana do Castelo, in particular Ms Manuela Passos Silva, Dr Ricardo Carvalhido, and Mayor Jose Maria Costa, for their enthusiastic and invaluable support in all aspects of the symposium, from providing gifts for the participants, to buses, to the welcome reception and press conference, and venues for the collateral events, as well as Mr Vítor Ramalhete for his prompt help in dealing with technical difficulties. The city truly embraced the event and welcomed everyone with open arms in true northern Portuguese fashion.

Organizing this symposium was a truly rewarding experience for both of us, scientifically and personally, and we hope everyone who was involved has as many great memories of the event as we do.

Elisabete da Cunha and Jacqueline Hodge, SOC co-chairs
January 2020

Editors

Elisabete Da Cunha
National University & University of Western Australia, Australia

Jacqueline Hodge
Leiden University, The Netherlands

José Afonso
Instituto de Astrofísica / University of Lisbon, Portugal

Laura Pentericci
INAF Roma, Italy

David Sobral
Lancaster University, United Kingdom

Organizing Committees

Scientific Organizing Committee

José Afonso	(IA/U. Lisbon, Portugal)
Franz Bauer	(PUC, Chile)
Gustavo Bruzual	(UNAM, Mexico)
Karina Caputi	(Groningen, Netherlands)
Elisabete da Cunha	(*co-chair,* ANU, Australia)
Romeel Davé	(ROE, United Kingdom)
Jacqueline Hodge	(*co-chair,* U. Leiden, The Netherlands)
Nicole Nesvadba	(IAS, France)
Masami Ouchi	(U. Tokyo, Japan)
Roderik Overzier	(Obs. Nacional, Brazil)
Laura Pentericci	(INAF Roma, Italy)
Alex Pope	(UMass Armherst, USA)
David Sobral	(Lancaster U., United Kingdom)
Rachel Somerville	(Flatiron Institute, USA)
Kim-Vy Tran	(UNSW, Australia, Texas A&M, USA)
Fabian Walter	(MPIA, Germany)

Local Organizing Committee

Ana Afonso	(MPE, Germany)
Iris Breda	(IA/U. Porto, Portugal)
Leandro Cardoso	(IA/U. Porto, Portugal)
Elisabete da Cunha	(*co-chair*, ANU, Australia)
Erik Deul	(U. Leiden, The Netherlands)
Evelijn Gerstel	(U. Leiden, The Netherlands)
Els Heijsman	(U. Leiden, The Netherlands)
Carlos Martins	(*co-chair*, IA/U. Porto, Portugal)
Sandra Reis	(IA/U. Lisbon, Portugal)
Ana Rita Silva	(IA/U. Porto, Portugal)
Elsa Silva	(CAUP, Portugal)
Léo Vacher	(CAUP, Portugal)

Partners & Sponsors

The International Astronomical Union
The University of Leiden
The Australian National University
Instituto de Astrofísica/Universidade do Porto
Centro de Astrofísica da Universidade do Porto
Planetário do Porto
Câmara Municipal de Viana do Castelo
Ciência Viva
TAP Air Portugal
Turismo do Porto e Norte de Portugal

Participants

José Afonso (Portugal)
Ana Afonso (Portugal)
Stacey Alberts (USA)
Hiddo Algera (Netherlands)
Javier Alvarez-Marquez (Spain)
Catarina Alves de Oliveira (USA)
Stergios Amarantidis (Portugal)
Ricardo Amorin (Chile)
Shohei Arata (Japan)
Manuel Aravena (Chile)
Hakim Atek (France)
Roland Bacon (France)
Tom Bakx (Japan)
Eduardo Bañados (Germany)
Andrew Battisti (Australia)
Rachel Bezanson (USA)
Fuyan Bian (Chile)
Manuela Bischetti (Italy)
Leindert Boogaard (Netherlands)
Rychard Bouwens (Netherlands)
Rebecca Bowler (UK)
Iris Breda (Portugal)
Joanna Bridge (USA)
Gustavo Bruzual (Mexico)
Andrew Bunker (UK)
Kirsty Butler (Netherlands)
Leandro Cardoso (Portugal)
Stefano Carniani (Italy)
Caitlin Casey (USA)
Onur Catmabacak (Switzerland)
Lana Ceraj (Croatia)
Daniel Ceverino (Denmark)
Jaclyn Champagne (USA)
Stéphane Charlot (France)
Jason Chu (USA)
Rachel Cochrane (UK)
Luis Colina (Spain)
Emma Curtis Lake (UK)
Elisabete da Cunha (Australia)
Daniel Dale (USA)
Romeel Davé (UK)
Pratika Dayal (Netherlands)
Gabriella De Lucia (Italy)
Miroslava Dessauges (Switzerland)
Gustavo Dopcke (Brazil)
Quirino D'Amato (Italy)
Richard Ellis (UK)
Edith Falgarone (France)
Xiaohui Fan (USA)
Carl Ferkinhoff (USA)
Steven Finkelstein (USA)
Deanne Fisher (Australia)
Cristina Garcia-Vergara (Netherlands)
Karl Glazebrook (Australia)
Jean Michel Gomes (Portugal)
Thiago Gonçalves (Brazil)
Valentino Gonzalez (Chile)
Kathryn Grasha (Australia)
Takuya Hashimoto (Japan)
Bunyo Hatsukade (Japan)
Ryley Hill (Canada)
Michaela Hirschmann (Denmark)
Jacqueline Hodge (Netherlands)
Andrew Hopkins (Australia)
Anne Hutter (Netherlands)
Raphael Hviding (USA)
Takuma Izumi (Japan)
Bethan James (USA)
Katie Jameson (Australia)
Anne Jaskot (USA)
Zhiyuan Ji (USA)
Yifei Jin (Australia)
Jean-Baptiste Jolly (Sweden)
Logan Jones (USA)
Gareth Jones (UK)
Stéphanie Juneau (USA)
Melanie Kaasinen (Germany)
Susan Kassin (USA)
Chiaki Kobayashi (UK)
Tlmea Orsola Kovács (Hungary)
Ivo Labbé (Australia)
Phillip Lang (Germany)
Andrea Lapi (Italy)
Rebecca Larson (USA)
Olivier Le Fevre (France)
Andrew Lehmann (France)
Joel Leja (USA)
Tsz Kuk Daisy Leung (USA)
Qi Li (USA)
Daizhong Liu (Germany)
Xiangcheng Ma (USA)
Andrea Macciò (United Arab Emirates)
Georgios Magdis (Denmark)
Allison Man (Canada)
Sinclaire Manning (USA)
Rui Marques-Chaves (Spain)
Carlos Martins (Portugal)
Michael Maseda (Netherlands)
Jorryt Matthee (Switzerland)
Israel Matute (Portugal)
Jed McKinney (USA)
Karin Menendez-Delmestre (Brazil)
Hugo Messias (Chile)
Uros Mestric (Australia)
Simon Mutch (Australia)
Rohan Potham Naidu (USA)
Themiya Nanayakkara (Netherlands)
Desika Narayanan (USA)
Nicole Nesvadba (France)
Pascal Oesch (Switzerland)
Masami Ouchi (Japan)
Roderik Overzier (Brazil)
Lara Pantoni (Italy)
Cirino Pappalardo (Portugal)

Antonio Pensabene (Italy)
Nor Pirzkal (USA)
Adele Plat (France)
Annagrazia Puglisi (France)
Timothy Rawle (USA)
Sandra Reis (Portugal)
Alvio Renzini (Italy)
George Rieke (USA)
Marcia Rieke (USA)
Elisa Ritondale (Germany)
Fernanda Roman de Oliveira (Brazil)
Donatella Romano (Italy)
Wiphu Rujopakarn (Thailand)
Matus Rybak (Netherlands)
Paola Santini (Italy)
Daniel Schaerer (Switzerland)
Jan-Torge Schindler (Germany)
Allan Schnorr-Müller (Brazil)
Sander Schouws (Netherlands)
Peter Senchyna (USA)
Chelsea Sharon (Singapore)
Prajval Shastri (India)
Sydney Sherman (USA)
Irene Shivaei (USA)
Ana Rita Silva (Portugal)
Renske Smit (UK)
David Sobral (UK)
Alyssa Sokol (USA)
Mimi Song (USA)
Justin Spilker (USA)
Flora Stanley (Sweden)
Elizabeth Stanway (UK)
Mauro Stefanon (Netherlands)
Katherine Suess (USA)
Ken-ichi Tadaki (Japan)
Mengtao Tang (USA)
Anthony Taylor (USA)
Mônica Tergolina (Brazil)
Maxime Trebitsch (France)
Hideki Umehata (Japan)
Tanya Urrutia (Germany)
Léo Vacher (France)
Dieuwertje van der Vlugt (Netherlands)
Bram Venemans (Germany)
Alba Vidal García (France)
Ilhuiyolitzin Villicana Pedraza (USA)
Fabian Walter (Germany)
Daniel Weisz (USA)
Christina Williams (USA)
Christopher Willmer (USA)
Aida Wofford (Mexico)
Charity Woodrum (USA)
Stijn Wuyts (UK)
Chentao Yang (Chile)
L. Y. Aaron Yung (USA)
Jorge Zavala (USA)
Luwenjia Zhou (France)
Tom Zick (USA)

Conference Photographs

Participants of the symposium. *Credit:* Paulo Carmo

Participants chatting with Viana do Castelo Mayor José Maria Costa and his assessor Ms Manuela Passos at the welcome reception. *Credit*: Armando Belo

Press conference with the local press, including SOC co-chair Dr Elisabete da Cunha and Viana Councilman for Science Dr Ricardo Carvalhido. *Credit*: Armando Belo

Session 4 discussion session, including Xiaohui Fan, Stacey Alberts, Takuma Izumi, Hideki Umehata, Bram Venemans, Chelsea Sharon, and Stephanie Juneau. *Credit*: Elisabete da Cunha

SOC co-chairs Prof Jacqueline Hodge and Dr Elisabete da Cunha. *Credit*: Elisabete da Cunha

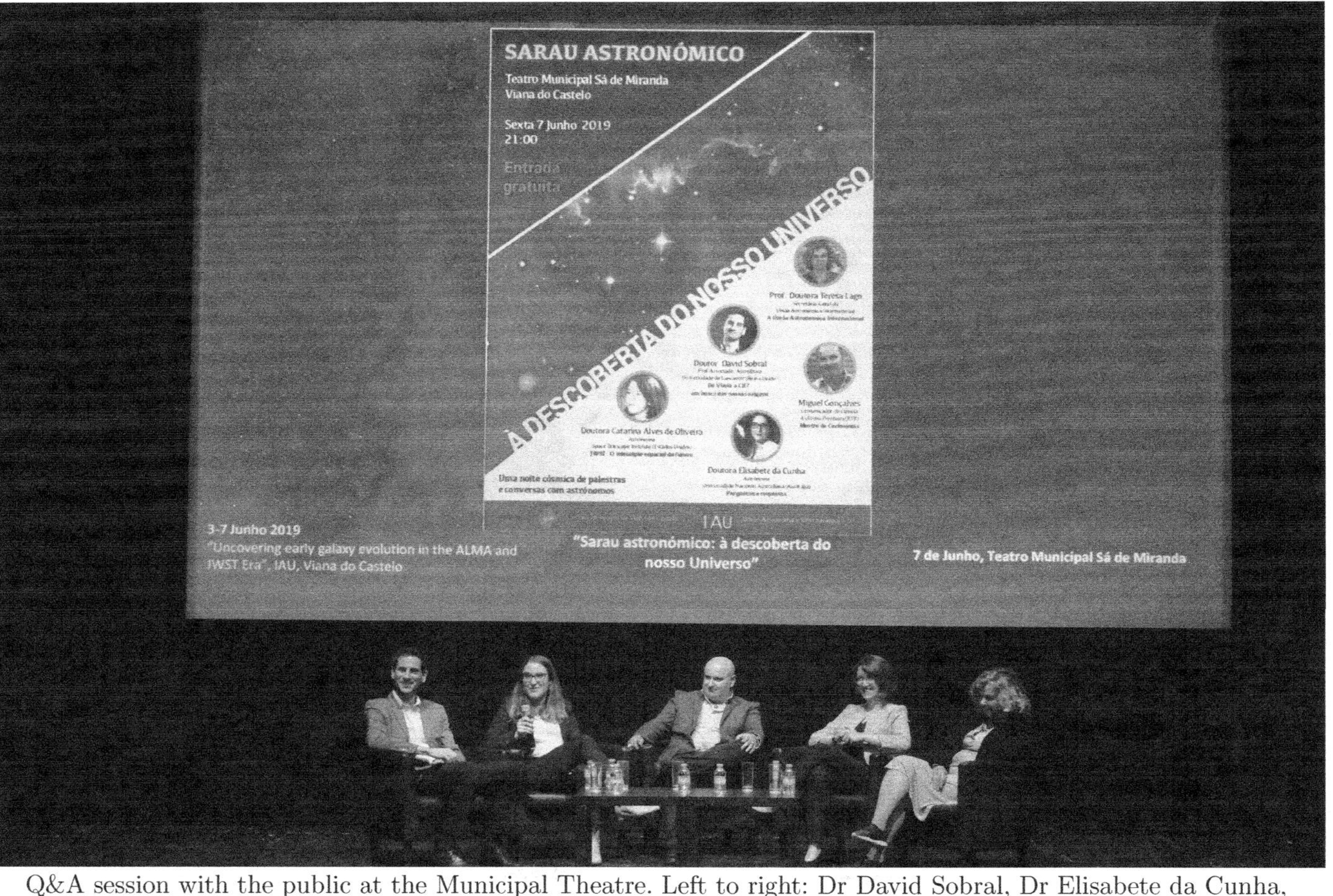

Q&A session with the public at the Municipal Theatre. Left to right: Dr David Sobral, Dr Elisabete da Cunha, Mr Miguel Gonçalves, Dr Catarina Alves de Oliveira, Prof Teresa Lago. *Credit*: Armando Belo

Local boy looking through telescope at the public stargazing event.
Credit: Elisabete da Cunha

Science highlights

The science programme started with several interesting talks about the current state of the art of detections of the faintest, most distant early galaxies (Smit, Finkelstein, Bouwens, Oesch, Atek). Thanks to HST and Spitzer, often aided by gravitational lensing, we can go deep into the low end of luminosity (and mass) functions of early galaxies, getting close to recovering all the sources responsible for reionization. However in order to really get there, we will need JWST (Oesch, Bouwens, Renzini talks). Efforts are now focusing on not only detecting the faintest, most distant galaxies, but also characterizing their physical properties. ALMA is providing exciting new results in this area, as shown by Smit, Hashimoto, Bowler, Carniani. Mostly [CII] and [OIII] line detections at $z > 6$ with ALMA are providing spectroscopic redshifts for the sources, as well as in some cases dynamical masses and information on the ionized gas. To measure the dust continuum in these sources remains challenging, with some tantalizing observations suggesting that dust grains might be very different than what we observe in the local Universe. A current open question is what is the dust temperature and luminosity in these sources. Answering this will require multi-band ALMA measurement.

In the theory session we had many interesting updates on the state of theoretical early galaxy formation models (Dayal, Arata, Ceverino, Ma, Hutter, Naidu). These models are making predictions on the properties of galaxies in the epoch of reionization that will be directly in the future tested by JWST observations of those galaxies, and also by Square Kilometre Array observations of the 21cm neutral hydrogen signal at reionization. We also highlight the talk by Narayanan, who presented a new model for interstellar dust formation, growth, and destruction that takes into account all known physical processes affecting the evolution of dust grains, and is included in a large-scale cosmological simulation, allowing us to follow the total dust content of galaxies in a way that can be compared with future ALMA and JWST observations.

Spectral energy distribution (SED) models of galaxies are needed to translate our multi-wavelength observations into physical parameters that can be compared to theoretical simulations and models. Currently SED model developers are getting ready to interpret the future vast amount of data on high-redshift sources that will be enabled by the JWST (e.g., talks by Leja, Gomes). A critical step is to produce realistic spectra, including both stellar continuum and nebular emission, especially in the rest-frame UV and optical (talks by Charlot, Hirschmann). In order to do that new stellar evolution ingredients that are still uncertain are needed (for example, stellar binaries and rotation; Nanayakkara, Schaerer, Stanway), as well as robust photoionization models. We had an interesting discussion session about the challenges in producing SED models for the future, when many of their ingredients (stellar evolution, stellar spectra, initial mass function) are very uncertain at the early evolutionary stages when the metallicity was low.

We had several talks about the detection of quasars at high-redshift (Bañados, Venemans, Fan), which trace high density environments at high redshift ($z > 6$) where galaxies and massive central black holes are rapidly growing. These are ideal environments to test the galaxy-AGN connection (e.g., Alberts, Bischetti, Juneau, Humehata talks). ALMA is enabling great strides in this kind of science by observing the interstellar medium of these objects, both in dust continuum, [CII], and CO (e.g., Venemans, Izumi, Sharon talks).

In the session dedicated to the interstellar medium, we had updates on the ALMA large programmes ASPECS (Aravena), and ALPINE (Le Fevre), as well as additional contributed talks reporting on ALMA studies of the molecular gas content of high- redshift star-forming galaxies with ALMA (Williams, Suess, Magdis, Liu). These studies are unlocking a key component in our understanding of how star formation is fuelled at cosmic dawn and beyond. Another avenue worth highlighting are detailed studies of strongly lensed, bright dusty star-forming galaxies, which allow us to dissect the interstellar medium down to molecular cloud scales and study small-scale physics effects (such as turbulence) on star formation (Spilker, Falgarone, and the talks by Rujopakarn, Dessauges, and Tadaki in the following session). Another highlight is the use of CO isotopes in the ISM, which can be observed with ALMA at high-redshift, to trace variations of the stellar initial mass function (Romano talk).

Another approach that holds a lot of promise for future combined studies with ALMA and JWST is to obtain resolved, multi-wavelength observations of star-forming galaxies so we can trace both their stellar component (past and current star formation), and their interstellar medium (fuel for future star formation), at similar scales. ALMA and ground-/space-based large telescopes (VLT, Keck and HST now, ELT and JWST in the future) are ideally matched for this purpose (talks by Shivaei, Wuyts, Bezanson, James, Lang, Ritondale, Cochrane). A common result from current studies seems to be that there is often a mismatch between the extent and/or location of the rest-frame UV/optical emission produced by stars, and the infrared emission produced by dust. This can tell us something about the morphology and structure of early galaxies, and how the stellar component is growing, but highlights that caution must be taken when correcting the stellar emission for dust attenuation. Some ambiguities remain due to the fact that high spatial resolutions cannot be achieved in the rest-frame near-infrared with current facilities, which would be closer to a pure stellar distribution (i.e., unaffected by dust); JWST will help resolve this issue.

We had a very interesting session on what can be learnt from local analogues to high-z galaxies; this demonstrated that there is indeed a lot of potential for the two communities of local and distant Universe studies to collaborate. A promising approach is to select low-mass, low-metallicity, high star formation galaxies locally that presumably resemble the first galaxies that contributed to reionization, and study them in the exquisite detail enabled by their relative closeness (e.g., talks by Amorin, Gonalves, Senchyna, Fisher). Such studies focus, for example, how calibrating observational diagnostics for star formation and metallicity, or measuring quantities like the escape fraction of ionizing photons. The main challenges are still how to define exactly what constitutes a low-redshift analogue, and whether true analogues of young galaxies can be found in our evolved Universe.

In our last session, we had several talks about new observations with the MUSE integral field spectrograph on the ELT (Bacon, Maseda, Boogaard), which give us a preview of the potential of combining this type of observations with ALMA. We highlight the discovery enabled by MUSE of ultra-faint Lyman-alpha emitters at high-redshift, which are very young primeval galaxies with very low metallicities and high star-formation rates, and presumably would contribute significantly to reionization (Maseda); these will be prime targets for JWST in the future. We also had exciting previews of the capabilities of the JWST instruments for high-redshift galaxy evolution science, and descriptions of the planned GTO programmes (Alves de Oliveira, Rieke, Bunker).

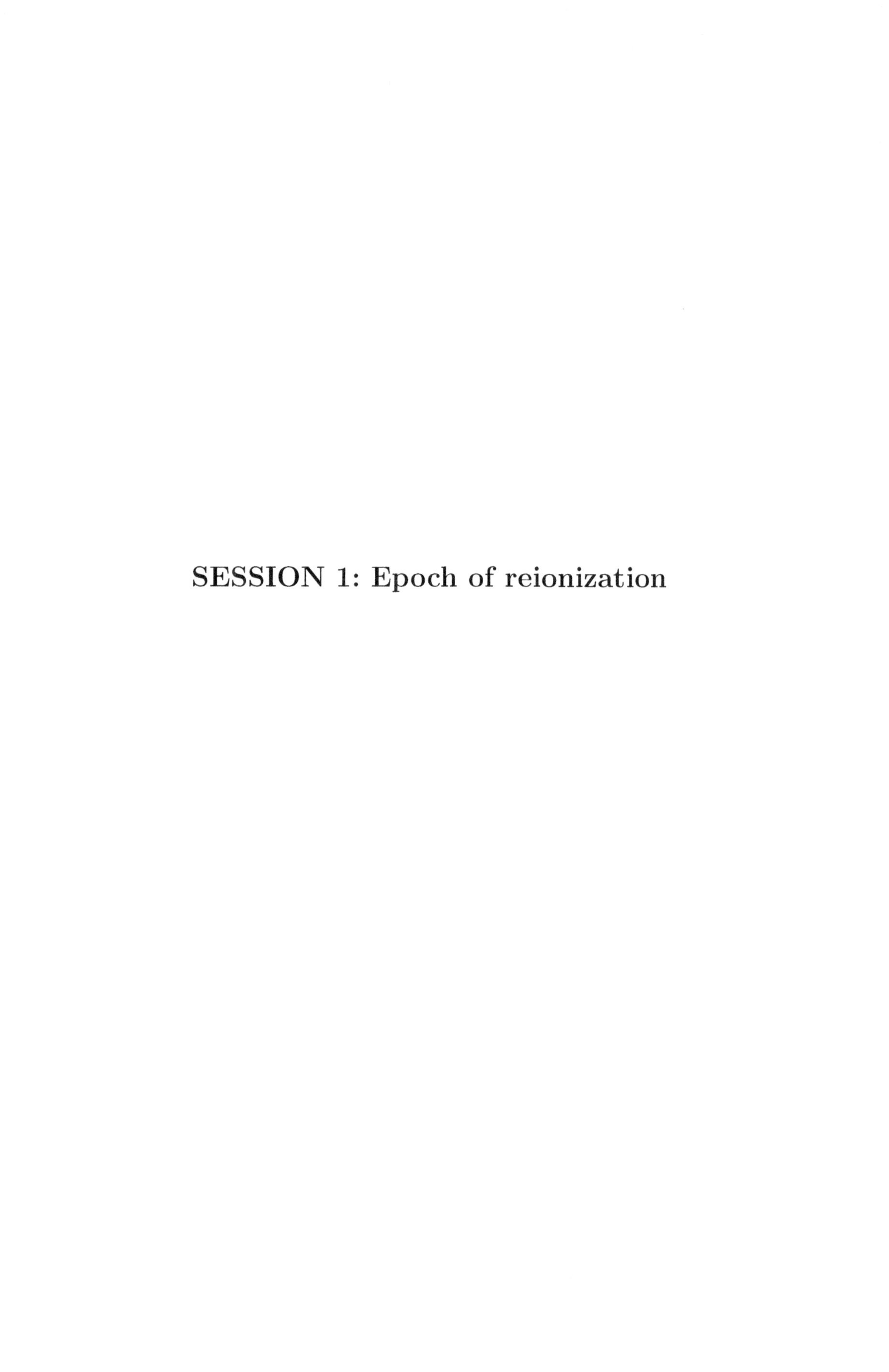

SESSION 1: Epoch of reionization

Uncovering Early Galaxy Evolution in the ALMA and JWST Era
Proceedings IAU Symposium No. 352, 2019
E. da Cunha, J. Hodge, J. Afonso, L. Pentericci & D. Sobral, eds.

doi:10.1017/S1743921320001052

An ALMA view of galaxies in the Epoch of Reionisation

Renske Smit

University of Cambridge, United Kingdom

Abstract. In the past decade hundreds of galaxies have been identified in the Epoch of Reionisation, selected from their rest-frame UV light. Only a handful of these sources, however, have spectroscopic redshift determinations and we have limited understanding of their physical properties. ALMA is currently transforming this field by providing the first view of the dust obscured star- formation, the kinematics of these sources, the cool gas traced by [CII] and highly ionised gas traced by [OIII]. In this talk I will discuss new and recent results on the UV-bright galaxy population during the first billion years of cosmic time and what they imply for their observational and physical properties.

Uncovering Early Galaxy Evolution in the ALMA and JWST Era
Proceedings IAU Symposium No. 352, 2019
E. da Cunha, J. Hodge, J. Afonso, L. Pentericci & D. Sobral, eds.

doi:10.1017/S1743921319008901

Discovery of the most distant star-forming and quenched galaxies in the universe

Steven L. Finkelstein

The University of Texas at Austin, Austin, TX, 78712, USA
email: stevenf@astro.as.utexas.edu

Abstract. While the high-redshift component of the CANDELS survey was designed with the $z \sim 6$–8 era in mind, these data do probe the far-UV of galaxies at even higher redshift. A few studies have ventured this far out, and have published conflicting results - some continue to find significant star-formation, while others conclude there is a steep decline in this quantity. Here I report on a new search for $z = 9$–10 galaxies, making significant use of the Spitzer/IRAC data in the CANDELS fields. We have discovered a larger number of galaxies in this epoch than previous works, implying the UV luminosity function, and thus the SFR density, may not evolve as steeply as previously thought. This implies that star-formation begins early in the universe. I will also report on a new study searching for the earliest quenched galaxies at $3 < z < 5$, which are not predicted by models, yet may exist if galaxies form very early, and thus can approach their quenching phase quicker.

Keywords. galaxies: formation — galaxies: high-redshift

1. Introduction

While the past decade, since the addition of the infrared-sensitive Wide Field Camera 3 (WFC3) to the *Hubble Space Telescope* (*HST*), has seen the solution to a number of pressing questions, more have since been opened. How rapidly does the process of the reionization of the intergalactic medium (IGM) end, and what is the dominant driver? When do the first galaxies turn on, and how rapidly do they build up? And, how fast do these early galaxies begin the process of shutting down their star formation, producing quiescent galaxies? In this proceeding, I summarize progress in these three areas.

2. How Rapidly Does Reionization End?

Typical models of reionization from earlier in this decade were able to show that high-redshift galaxies were capable of completing reionization by $z \sim 6$ with a few assumptions, most importantly that all galaxies have ionizing photon escape fractions of ~10–20% (Finkelstein *et al.* 2012; Robertson *et al.* 2013; Finkelstein *et al.* 2015; Robertson *et al.* 2015; Bouwens *et al.* 2015). However, such high escape fractions are rarely seen, and most studies of the average galaxy populations at $z \sim 2$–4 find upper limits which are much lower (<5%; e.g., Siana *et al.* 2010; Grazian *et al.* 2017). Simulations frequently predict that is the lowest-mass galaxies which have the highest escape fractions (e.g., Paardekooper *et al.* 2015; Xu *et al.* 2016). This was recently explored by Finkelstein *et al.* (2019b), who found that if one assumes the escape fraction is halo-mass dependent (following Paardekooper *et al.* (2015)), one can still complete reionization "on time" by $z \sim 6$, but reionization starts earlier and proceeds more smoothly, due to the dependence on the faintest galaxies. These two models differ strongly at $z \sim 7$–10. The first model predicts an IGM which is ~50% neutral and $z = 7$, while the second model predicts ~20% neutral. This is summarized in Figure 1.

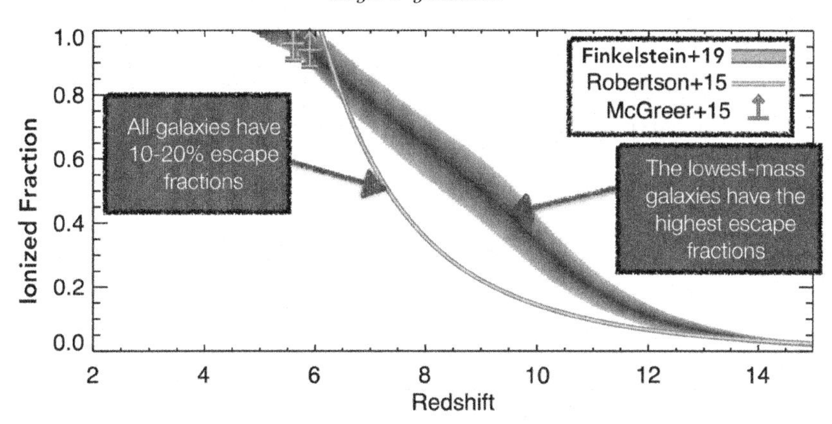

Figure 1. Adapted from Finkelstein *et al.* (2019b). The gray curve shows the reionization history predicted for a model where all galaxies contribute, with a constant escape fraction. The blue curve shows that from a model where only the faintest galaxies have significant escape fractions.

Additional observations of the neutral fraction in the epoch of reionization can help distinguish between these (and other) models. The most promising presently available method to do so is via Lyα emission. Lyα photons are resonantly scattered by neutral hydrogen, thus an increasing IGM neutral fraction should be traced by a decrease in detectable Lyα emission (e.g., Miralda-Escudé & Rees 1998; Malhotra & Rhoads 2004). In spite of some detections (e.g., Shibuya *et al.* 2012; Finkelstein *et al.* 2013; Oesch *et al.* 2015; Zitrin *et al.* 2015; Stark *et al.* 2016; Hoag *et al.* 2017; Laporte *et al.* 2017; Larson *et al.* 2018), the majority of $z > 7$ galaxies go undetected with spectroscopic followup. Based on these results, the most recent studies infer $X_{HI} = 60$–90% at z=7-8 (Mason *et al.* 2018, 2019; Hoag *et al.* 2019), due to just a few detections, and often none. However, converting observed spectra to constraints on the neutral fraction has lots of assumptions and potential for systematic uncertainties, including: incomplete spectral coverage of the full $P(z)$, copious telluric emission lines, uncertainty in spectroscopic depth of non-detections due to uncertain Lyα line width (and sometimes shallow depth), and sample contamination.

In the thesis work of newly minted Dr. Intae Jung (published in Jung *et al.* (2018, 2019) and Jung *et al.* (2019b)), we used data taken over 18 nights from Keck with DEIMOS and MOSFIRE to try to overcome some of these systematics, specifically:

- Depth: Our MOSFIRE integrations range from 5-20 hours.
- Sample selection: Significant effort to improve the photo-z's (and minimize sample contamination) of the observed galaxies.
- Wavelength: ~20 of our sources are covered by both DEIMOS and MOSFIRE.

Out of our full MOSFIRE sample of 72 galaxies in the GOODS-N field observed with exposure times greater than 5 hours, we found nine Lyα lines at $7.1 < z < 7.9$ detected at $\geqslant 4\sigma$ significance (including five at $\geqslant 5\sigma$). We used our observations to model the Lyα equivalent width (EW) distribution, accounting for all sources of incompleteness and uncertainty by implementing via MCMC (similar to Jung *et al.* (2018)). We constrain the characteristic scale-length of this distribution, predicting for a given value the number of lines we should detect at a given S/N in our data. While the scale length of Lyα emission is typically ~70Å at $z < 6$, we find 30 ± 10 Å at $z \sim 7.5$. We conclude that in contrast to previous results, Lyα is very detectable in this epoch. Our measured scale length implies a neutral fraction which is non-zero, but likely not as high as previously published.

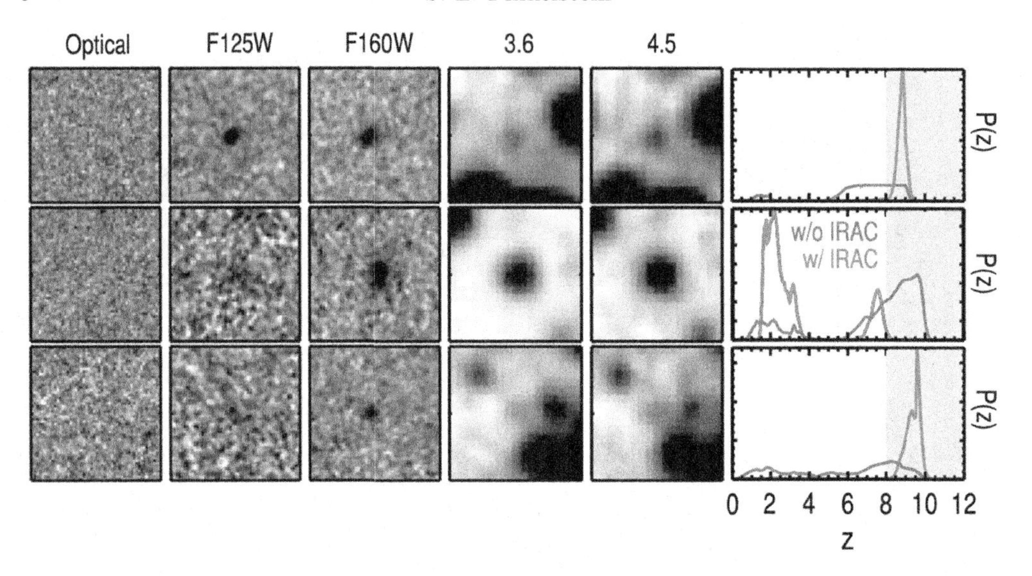

Figure 2. Adapted from Finkelstein *et al.* (2019c), in prep. Each row shows one object, with the right-hand panel showing the $P(z)$ without and with the inclusion of IRAC data. The first source is faint yet detectable in IRAC, with a red [3.6]−[4.5] color indicative of strong [O III] emission in the redder channel, consistent with $z \sim$ 8.5–9. the second source is very bright in IRAC, showing a very red slope consistent with a low-redshift solution. The final source is not strongly detected in IRAC, again consistent with a high redshift nature.

3. When do the First Massive Galaxies Form?

While studies of the rest-ultraviolet (UV) luminosity function, and by extension the cosmic star-formation rate (SFR) density, agree nicely that there is smooth evolution from $z = 4$ to $z = 8$ (Finkelstein 2016), there is significant tension at higher redshifts. While studies by Coe *et al.* (2013) and McLeod *et al.* (2015) imply that the smooth evolution continues to $z \geqslant 9$, other studies by Bouwens *et al.* (2016) and Oesch *et al.* (2018) find an accelerated evolution downward. This discrepancy is understandable, as at $z > 9$, galaxies are only robustly detected in a single *HST* filter (F160W), and are just barely detectable with this 2.4m telescope. This is thus exceedingly difficult, and previous work pushing previous generations of *HST* instruments to their limits have shown that it is possible to underestimate the true abundance of such difficult-to-find galaxies (e.g. Bouwens *et al.* 2004). The hopefully imminent launch of the *James Webb Space Telescope* (*JWST*) will rapidly clear this up. The approved Cosmic Evolution Early Release Science (CEERS) Survey (Finkelstein *et al.* 2017) is designed to robustly measure the SFR density at $z \sim 10$, going ∼two magnitudes deeper than CANDELS, and detecting $z \sim 10$ galaxies in 5–8 filters.

However, that is still a few years away. We can make progress now by using a telescope even smaller than *Hubble* – the *Spitzer Space Telescope*. While the sensitivity of *Spitzer*'s IRAC camera is less than WFC3, deep surveys such as S-CANDELS (PI Fazio) and GREATS (PI Labbé) can detect rest-frame optical emission from galaxies at $z \sim$ 9–10. As shown in Figure 2, the inclusion of *Spitzer*/IRAC photometry can significantly shrink the uncertainties on the photometric redshift probability distribution function, allowing the selection of a robust $z \sim$ 9–10 galaxy candidate whose nature remained muddle with *HST* data only.

Inspired by this, we have embarked on a search for $z > 8$ galaxies in the CANDELS fields including *Spitzer*/IRAC photometry in the photometric redshift selection, to try

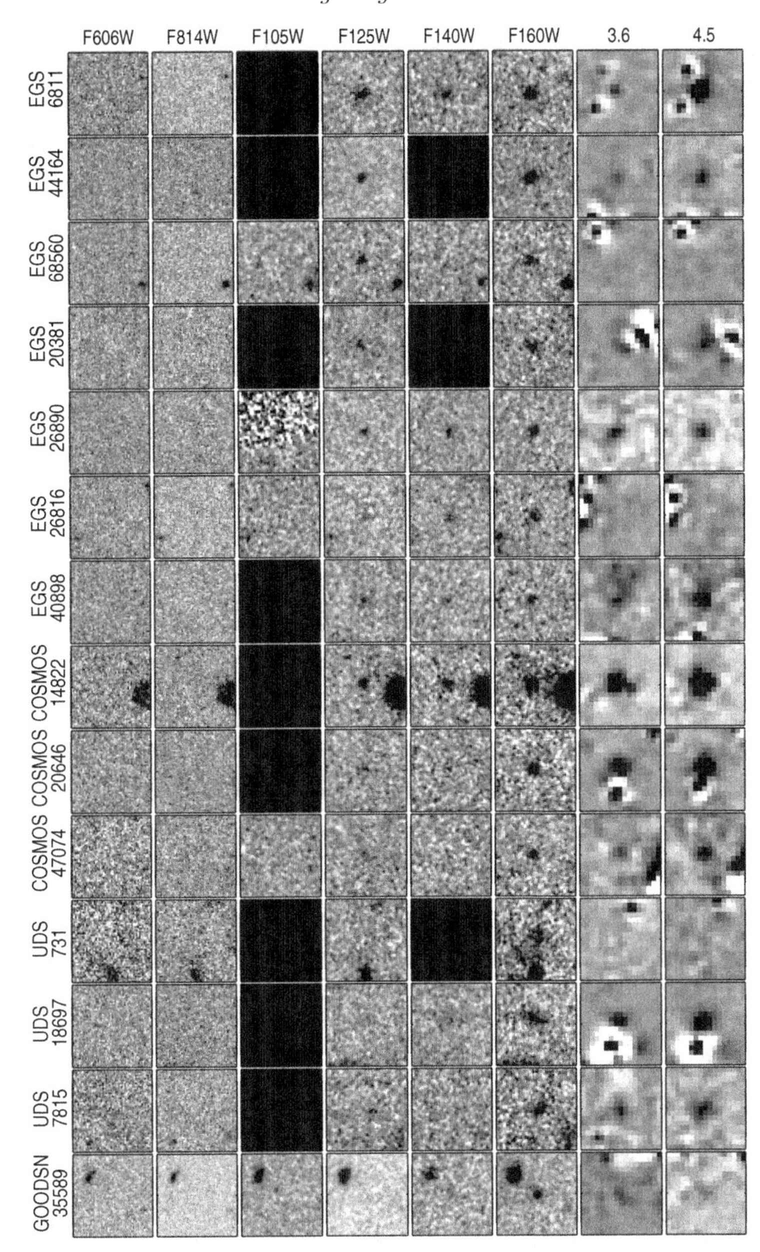

Figure 3. Adapted from Finkelstein *et al.* (2019c), in prep. Each row shows one candidate $z > 8.5$ galaxy from the CANDELS fields, selected with both *HST* and *Spitzer*/IRAC imaging. Each column shows one filter, with the last two showing the two IRAC bands after subtraction of neighbors with TPHOT. Further inclusion of ground-based imaging show that the sources COSMOS 14822 and UDS 731 are likely low-redshift interlopers.

to better constrain the abundance of galaxies in this epoch. Over the five CANDELS fields, we have discovered 14 bright $H < 26.5$ galaxies at $z > 8.5$, larger than previously published samples. After an extensive vetting process including all available ground and space-based data (described in Finkelstein *et al.* (2019c), in prep), we find 12 of these sources remain very likely high-redshift candidates. This analysis is ongoing, but the presence of these 12 promising candidates results in a volume density of bright sources

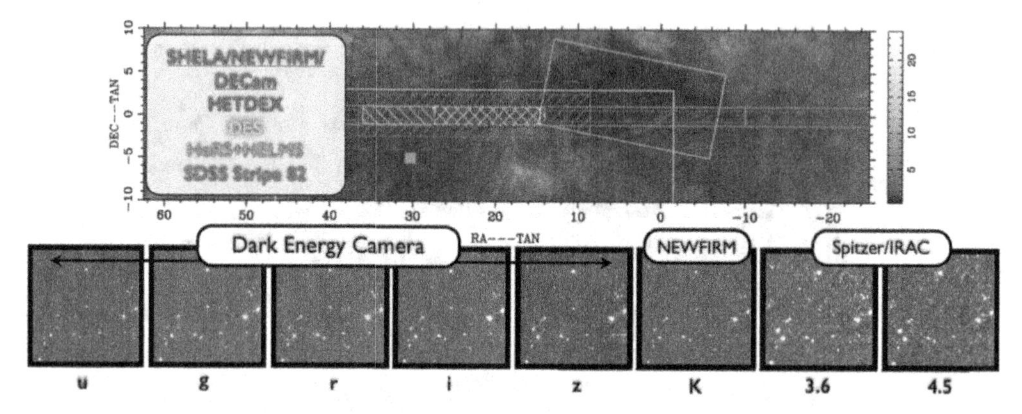

Figure 4. A schematic diagram of the 20 deg^2 SHELA field, highlighting the datasets available. We use these data, led by our new K_s-band imaging, to select candidate massive quiescent galaxies at $3 < z < 5$.

which is essentially unchanged from $z \sim 6$ to $z \sim 9$, presenting a challenge to models of early universe galaxy formation.

What could this mean? Maybe we are just bad at weeding out contaminants? That would be unsatisfying as we all work very hard at this, but we need spectroscopic confirmation. We are attempting to confirm the redshifts of these galaxies via Lyα, with a MOSFIRE J-band program being led by Rebecca Larson, as well as [O III] 88 μm, with an ALMA program led by Takuya Hashimoto. If we confirm these sources to be real, bright, distant galaxies must be better at making stars, possibly due to a steepening of the star-formation law at high gas densities (Yung *et al.* (2019) shows this is needed to match high-redshift rest-UV luminosity functions).

4. How early do galaxies begin to shut down their star formation?

If the universe is making bright and massive galaxies earlier than predicted, might it also produce massive quiescent galaxies early as well? Models currently predict that *all* massive galaxies should be star-forming at $z > 3$ (e.g., Brennan *et al.* 2015). Observational studies in this epoch show that such quiescent galaxies do exist at $z > 3$ (e.g., Glazebrook *et al.* 2017), but the fraction of massive galaxies which are quiescent has been observed to range from ~10–50%. This spread is largely driven by the very small volumes probed by many studies.

To make progress, we are using the 20 deg^2 Spitzer-HETDEX Exploratory Large Area (SHELA; Figure 4) survey to perform a search for massive quiescent galaxies at $3 < z < 5$. This field contains fairly deep CTIO/DECam optical (Wold *et al.* 2019) and *Spitzer*/IRAC (Papovich *et al.* 2016) imaging. To this we have added K_s, AB =23 imaging from our 93 night KPNO/NEWFIRM survey program (PI Finkelstein; Stevans *et al.* in prep).

Our K_s-selected catalog of all 1.6 million sources in this field is presented in Stevans *et al.*, in prep. We selected a sample of ~4000 massive (log $[M/M_\odot] > 11$) galaxies with photometric redshifts between $3 < z < 5$ using EAZY-py. Using the measured EAZY-py SFRs, we then classified galaxies as quiescent if they had specific SFRs $< 10^{-11}$ yr^{-1}, finding ~500 such galaxies. While this selection is difficult due to the limited number of photometric bands in this field, we can make two main conclusions at this point. First, this number is decidedly non-zero, therefore *all* models which predict it should be zero (which is all models presently) need revision. Simply increasing feedback may not be enough, as that will simply delay the formation of such massive galaxies. Second,

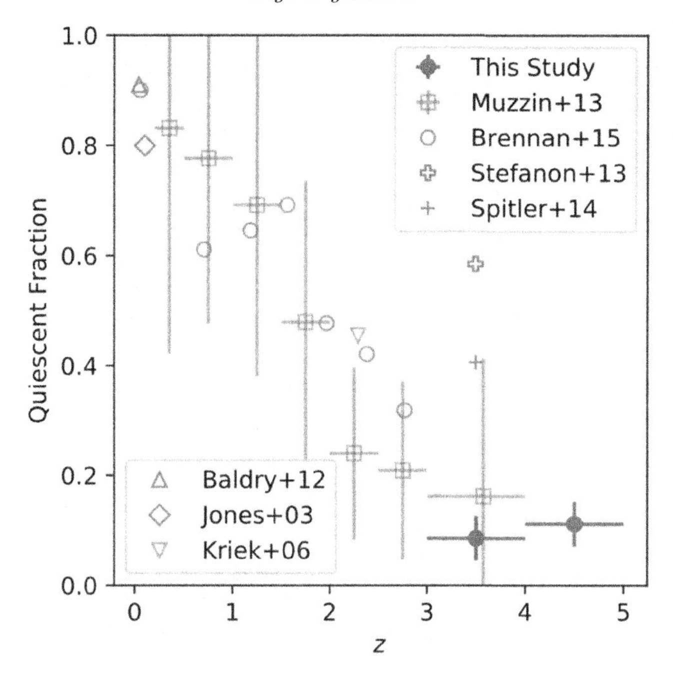

Figure 5. Adapted from Stevans *et al.* (2019). The quiescent fraction of massive galaxies (log $[M/M_\odot] > 11$) as a function of redshift. The results from this study are shown in red, compared to a large number of studies from the literature. Our measure is at the lower end of previous observations, with significantly reduced uncertainties.

our measured quiescent fraction of $\sim$12 $\pm$ 4% (Figure 5) is at the low-end of previous observational estimates, but with a much smaller uncertainty, and a negligible cosmic variance uncertainty. These observations tell us that whatever mechanism is responsible for quenching, it must be able to manifest rapidly, as these galaxies are quenched only $\sim$1.5 Gyr after the Big Bang.

5. Lingering Questions

In these proceedings, I have discussed new results on studies of reionization, and the formation and quenching of massive galaxies at high redshift. These results, while interesting, are far from conclusive. Here I discuss some lingering questions.

• What is the true reionization history? Our new Lyα observations imply a fairly ionized IGM at $z \sim 7.5$, although we have not yet estimated the neutral fraction from our dataset. If this is confirmed, how is it reconciled with other studies, which infer a much larger neutral fraction? Reionization cosmic variance could be to blame, though it could also highlight the difficulty of working at faint flux levels with ground-based near-IR spectroscopy. Some progress may be made with deeper, telluric-free, observations from *JWST*, though the spectroscopic field-of-view is very small. Longer term, wide-field spectroscopic surveys with the Giant Magellan Telescope offer a promising future for Lyα-based studies of reionization (more information on this can be found in the Astro2020 white paper by Finkelstein *et al.* (2019a)).

• Are the bright $z \sim$ 9–10 candidates we've found real? Spectroscopic confirmation is needed to know for sure. ALMA observations of the [O III] 88 μm line are likely the most efficient method in the pre-*JWST* era (Hashimoto *et al.* 2018). However, the majority of our sources are too far north for ALMA observations. NOEMA in principle can do this, but this requires NOEMA's Band 4, which has not yet been installed. Lyα remains a possibility, though this requires pushing current 10m-class telescopes to their limits, and is risky. Ultimately, rest-UV observations with *JWST* will likely be needed (or perhaps single-slit rest-optical observations with MIRI).

• If these candidates are real, why is the bright-end of the LF evolving so slowly? Is it due to a change in the physics regulating star-formation, or a reduction in dust attenuation? If the latter, once these galaxies hit dust-free, luminosity function evolution should continue, thus *JWST* observations pushing to $z > 10$ should prove illuminating. Cosmic variance is also certainly playing a role, as can be seen by the variance in number of galaxies between our five fields (7, 2, 2, 1 and 0 in the five CANDELS fields). Out to $z \sim 10$, *WFIRST* surveys will negate these uncertainties, improving these abundance estimates.

• What are the implications of massive quiescent galaxies forming at $z > 3$? The presence must certainly inform the implementation of various forms of feedback in modern cosmological simulations. However, before we assume the model physics need revision, we need to make sure the observations are right. First, we need to test whether any of our candidate quiescent galaxies are dusty interlopers. This is easily done with minutes of ALMA observations, and we have a pilot program for 100 sources approved for Cycle 7 (PI Finkelstein). Once we have a clean sample of quiescent candidates, we can then pursue near-IR spectroscopic confirmation of the 4000 Å break. Due to their bright magnitudes ($K_{AB} \sim$ 20–22), this should also be fairly quick, perhaps a few hours with Keck/MOSFIRE. Combining these results will provide us with a more reliable estimate of the quiescent fraction, which can then be used to constrain future simulations.

References

Bouwens, R. J., Illingworth, G. D., Oesch, P. A., *et al.* 2015, *ApJ*, 811, 140
Bouwens, R. J., Thompson, R. I., Illingworth, G. D., *et al.* 2004, *ApJ*, 616, L79
Bouwens, R. J., Oesch, P. A., Labbé, I., *et al.* 2016, *ApJ*, 830, 67
Brennan, R., Pandya, V., Somerville, R. S., *et al.* 2015, *MNRAS*, 451, 2933
Coe, D., Zitrin, A., Carrasco, M., *et al.* 2013, *ApJ*, 762, 32
Finkelstein, S., Dickinson, M., Ferguson, H., *et al.* 2017, The Cosmic Evolution Early Release Science (CEERS) Survey, JWST Proposal ID 1345. Cycle 0 Early Release Scienc
Finkelstein, S., Bradac, M., Casey, C., *et al.* 2019a, *BAAS*, 51, 221
Finkelstein, S. L. 2016, *PASA*, 33, e037
Finkelstein, S. L., Papovich, C., Ryan, R. E., *et al.* 2012, *ApJ*, 758, 93
Finkelstein, S. L., Papovich, C., Dickinson, M., *et al.* 2013, *Nature*, 502, 524
Finkelstein, S. L., Ryan, Jr., R. E., Papovich, C., *et al.* 2015, *ApJ*, 810, 71
Finkelstein, S. L., D'Aloisio, A., Paardekooper, J.-P., *et al.* 2019b, arXiv e-prints
Glazebrook, K., Schreiber, C., Labbé, I., *et al.* 2017, *Nature*, 544, 71
Grazian, A., Giallongo, E., Paris, D., *et al.* 2017, *A&A*, 602, A18
Hashimoto, T., Laporte, N., Mawatari, K., *et al.* 2018, *Nature*, 557, 392
Hoag, A., Bradač, M., Trenti, M., *et al.* 2017, *Nature Astronomy*, 1, 0091
Hoag, A., Bradač, M., Huang, K., *et al.* 2019, *ApJ*, 878, 12
Jung, I., Finkelstein, S. L., Livermore, R. C., *et al.* 2018, *ApJ*, 864, 103
Jung, I., Finkelstein, S. L., Dickinson, M., *et al.* 2019, arXiv e-prints, arXiv:1901.05967
Laporte, N., Nakajima, K., Ellis, R. S., *et al.* 2017, *ApJ*, 851, 40
Larson, R. L., Finkelstein, S. L., Pirzkal, N., *et al.* 2018, *ApJ*, 858, 94
Malhotra, S. & Rhoads, J. E. 2004, *ApJL*, 617, L5
Mason, C. A., Treu, T., Dijkstra, M., *et al.* 2018, *ApJ*, 856, 2
Mason, C. A., Fontana, A., Treu, T., *et al.* 2019, *MNRAS*, 485, 3947
McLeod, D. J., McLure, R. J., Dunlop, J. S., *et al.* 2015, *MNRAS*, 450, 3032
Miralda-Escudé, J. & Rees, M. J. 1998, *ApJ*, 497, 21
Oesch, P. A., Bouwens, R. J., Illingworth, G. D., Labbé, I., & Stefanon, M. 2018, *ApJ*, 855, 105

Oesch, P. A., van Dokkum, P. G., Illingworth, G. D., *et al.* 2015, *ApJL*, 804, L30
Paardekooper, J.-P., Khochfar, S., & Dalla Vecchia, C. 2015, *MNRAS*, 451, 2544
Papovich, C., Shipley, H. V., Mehrtens, N., *et al.* 2016, *ApJS*, 224, 28
Robertson, B. E., Ellis, R. S., Furlanetto, S. R., & Dunlop, J. S. 2015, *ApJL*, 802, L19
Robertson, B. E., Furlanetto, S. R., Schneider, E., *et al.* 2013, *ApJ*, 768, 71
Shibuya, T., Kashikawa, N., Ota, K., *et al.* 2012, *ApJ*, 752, 114
Siana, B., Teplitz, H. I., Ferguson, H. C., *et al.* 2010, *ApJ*, 723, 241
Stark, D. P., Ellis, R. S., Charlot, S., *et al.* 2016, *MNRAS*
Wold, I. G. B., Kawinwanichakij, L., Stevans, M. L., *et al.* 2019, *ApJS*, 240, 5
Xu, H., Wise, J. H., Norman, M. L., Ahn, K., & O'Shea, B. W. 2016, *ApJ*, 833, 84
Yung, L. Y. A., Somerville, R. S., Finkelstein, S. L., Popping, G., & Davé, R. 2019, *MNRAS*, 483, 2983
Zitrin, A., Labbé, I., Belli, S., *et al.* 2015, *ApJL*, 810, L12

Uncovering Early Galaxy Evolution in the ALMA and JWST Era
Proceedings IAU Symposium No. 352, 2019
E. da Cunha, J. Hodge, J. Afonso, L. Pentericci & D. Sobral, eds.

doi:10.1017/S1743921320001064

Galaxy build-up at cosmic dawn: Insights from deep observations with Hubble, Spitzer, and ALMA

Pascal Oesch

University of Geneva, Switzerland

Abstract. Over the last few years, great progress has been made in understanding the build-up of the first generations of galaxies based on deep optical and near-infrared imaging from the Hubble Space Telescope. However, HST only samples the rest-frame UV light of galaxies at $z \geqslant 4$, providing only limited information on the dust obscuration and on stellar masses of these sources. Fortunately, several Spitzer/IRAC programs have complemented the extragalactic HST fields with ultra-deep imaging data, allowing for a rest-frame optical view on early galaxies. Together with first ALMA/ NOEMA (sub)mm observations on distant galaxies, we are starting to gain a more and more complete picture of galaxy star-formation and mass build-up in the early universe. In this talk, I will present an overview of our current understanding of normal star-forming galaxies at $z > 3$ based the combination of HST+Spitzer+ALMA/NOEMA data. In particular, I will show how HST as already pushed into JWST territory with the discovery and spectroscopic confirmation of a galaxy at $z = 11.1 \pm 0.1$, only ~ 400 Myr after the Big Bang. I will also highlight some of the exciting possibilities that lie ahead with JWST to push the spectroscopic frontier to the cosmic dawn and to finally probe the physics of early galaxies.

Uncovering Early Galaxy Evolution in the ALMA and JWST Era
Proceedings IAU Symposium No. 352, 2019
E. da Cunha, J. Hodge, J. Afonso, L. Pentericci & D. Sobral, eds.

doi:10.1017/S1743921319008962

Properties of galaxies at $z \approx 7-9$ revealed by ALMA

Takuya Hashimoto[1,2,3]

[1]Faculty of Science and Engineering, Waseda University, 3-4-1 Okubo, Shinjuku, Tokyo 169-8555, Japan

[2]Department of Environmental Science and Technology, Faculty of Design Technology, Osaka Sangyo University, 3-1-1, Nagaito, Daito, Osaka 574-8530, Japan

[3]National Astronomical Observatory of Japan, 2-21-1 Osawa, Mitaka, Tokyo 181-8588, Japan
email: thashimoto@obsap.phys.waseda.ac.jp

Abstract. Understanding properties of galaxies in the epoch of reionization (EoR) is a frontier in the modern astronomy. With the advent of ALMA, it has become possible to detect far-infrared fine structure lines (e.g. [CII] 158 μm and [OIII] 88 μm) and dust continuum emission in star-forming galaxies in the EoR. Among these lines, our team is focusing on [OIII] 88 μm observations in high-z galaxies. After the first detection of [OIII] in the epoch of reionization (EoR) in 2016 from our team at $z = 7.21$, there are now more than ten [OIII] detections at $z > 6$ up to $z = 9.11$. Interestingly, high-z galaxies typically have very high [OIII]-to-[CII] luminosity ratio ranging from 3 to 12 or higher, demonstrating [OIII] is a powerful tracer at high-z. The high luminosity ratios may imply that high-z galaxies have low gas-phase metallicity and/or high ionization states.

Keywords. galaxies: formation — galaxies: high-redshift — galaxies: ISM

---

1. Introduction

1.1. *Far-infrared oxygen line as a new probe of galaxies in the reionization epoch*

Understanding properties of galaxies during reionization, at redshift $z \gtrsim 6-7$, is important. While a large number of galaxy candidates are selected with a dropout technique at $z \gtrsim 7$ owing to the Hubble Space Telescope (*HST*) and *Spitzer*/IRAC data (e.g., Ellis *et al.* 2013), the spectroscopic identifications at $z \gtrsim 7$ remain difficult (e.g., Stark *et al.* 2017 and references therein). This is mainly due to the fact that the most prominent hydrogen Lyα line is significantly attenuated by the intergalactic medium (IGM).

With the advent of the Atacama Large Millimeter/Submillimeter Array (ALMA) telescope, it has become possible to detect rest-frame far-infrared (FIR) fine structure lines in star-forming galaxies at $z > 5$ (e.g., Capak *et al.* 2015). A most commonly used line is [CII] 158 μm, which is one of the brightest lines in local galaxies (e.g., Brauher *et al.* 1998). To date, more than 21 [CII] detections are reported at $5 \lesssim z \lesssim 7$ (Carniani *et al.* 2018 and references therein).

However, based on a compiled sample with [CII] observations at $z \gtrsim 5$, Harikane *et al.* (2018) and Carniani *et al.* (2018) have revealed that [CII] may be weak for galaxies with strong Lyα emission, so-called Lyα emitters (LAEs; rest-frame Lyα equivalent widths $EW_0(Ly\alpha) \gtrsim 20-30$ Å). Harikane *et al.* (2018) have interpreted the trend with photoionization models of CLOUDY (Ferland *et al.* 2013) implemented in spectral energy distribution (SED) models of BEAGLE (Chevallard *et al.* 2016). The authors show that low metallicity or high ionization states in LAEs lead to weak [CII]. Theoretical studies also show that such ISM conditions lead to the decrease in the [CII] luminosity (e.g., Lagache *et al.* 2018). If we assume that $z \gtrsim 7$ galaxies in general have low metallicity or high ionization states, [CII] may not be the best line to spectroscopically confirm $z \gtrsim 7$ galaxies. Indeed, a number of null-detections of [CII] are reported at $z \gtrsim 7$ (e.g., Ota *et al.* 2014).

In fact, based on *Herschel* spectroscopy for local dwarf galaxies, Cormier *et al.* (2015) have demonstrated that [OIII] 88 μm becomes brighter than [CII] at low metallicity. Based on calculations of CLOUDY, Inoue *et al.* (2014) also theoretically predict that the [OIII] line at high-z should be bright enough to be detected with ALMA.

Motivated by these backgrounds, we are conducting follow-up observations of the [OIII] 88 μm line for $z > 7$ galaxies with ALMA. After the first detection of [OIII] in the reionization epoch in Inoue *et al.* (2016) at $z = 7.21$ by our team, the number of [OIII] detections is rapidly increasing. There are currently ten objects with [OIII] detections at $z \approx 6-9$ (Carniani *et al.* 2017; Laporte *et al.* 2017; Marrone *et al.* 2018; Hashimoto *et al.* 2018; Tamura *et al.* 2018; Hashimoto *et al.* 2019; Walter *et al.* 2018). Remarkably, Hashimoto *et al.* (2018) have detected [OIII] in a $z = 9.11$ galaxy with a high significance level of 7.4σ. These results clearly demonstrate that [OIII] is a powerful tool to confirm $z > 6-7$ galaxies.

1.2. *Far-infrared line luminosity ratios to diagnose properties of galaxies at $z > 7$*

Inoue *et al.* (2016) have also investigated the FIR line ratio at $z > 7$. In a combination with the null detection of [CII], the authors have shown that their $z = 7.21$ LAE has a line ratio of [OIII]/[CII] > 12 (3σ). The line ratio would give us invaluable information on properties of the interstellar medium (ISM). Given the fact that [OIII] originates only from HII regions whereas [CII] originates both from HII regions and photo-dissociated regions (PDRs), Inoue *et al.* (2016) have interpreted the high line ratio as the $z = 7.21$ LAE having highly ionized HII regions but less PDRs. Such properties would lead to a high escape fraction of ionizing photons, which is a key parameter to understand reionization. It is therefore of interest to test whether high FIR line luminosity ratios are ubiquitous for galaxies in the reionization epoch.

In this article, we will first show individual [OIII] detections. We will then discuss ubiquitously large line luminosity ratios of [OIII]/[CII] at $z = 6-9$ and its implications on reionization.

2. An example of our ALMA observations: MACS1149-JD1

In this section, we first show an example of our ALMA observations for MACS1149-JD1 at $z = 9.11$ (Hashimoto *et al.* 2018). For detailed observations on [OIII] and [CII], we refer the reader to Hashimoto *et al.* (2018) and Laporte *et al.* (2019), respectively.

2.1. *[OIII] 88 μm observations*

We observed MACS1149-JD1 with ALMA in Band 7 with a configuration C40-3 (ID 2015.1.00428.S, PI: A. K. Inoue). To cover the uncertainty derived from the photometric redshift analysis (e.g., Zheng *et al.* 2012), we used four setups with contiguous frequencies labeled as T3, T4, T5, and T6 encompassing the frequency range 314.4 – 340.5 GHz and the redshift range $z = 9.0 - 9.8$. In each setup, a total bandwidth of 7.5 GHz was used, split into four spectral windows (SPWs) each with a 1.875 GHz bandwidth in the Frequency Division Mode (FDM). Each SPW has a 7.8125 MHz resolution, corresponding to a velocity resolution of ~ 7 km s^{-1}. The total on-source exposure times are 75.6, 35.3, 119.4, and 42.3 minutes, for T3, T4, T5, and T6, respectively. The T3, T4, T5, and T6 data were reduced using the CASA pipeline version 4.7.0, 4.5.2, 4.7.2, and 4.6.0,

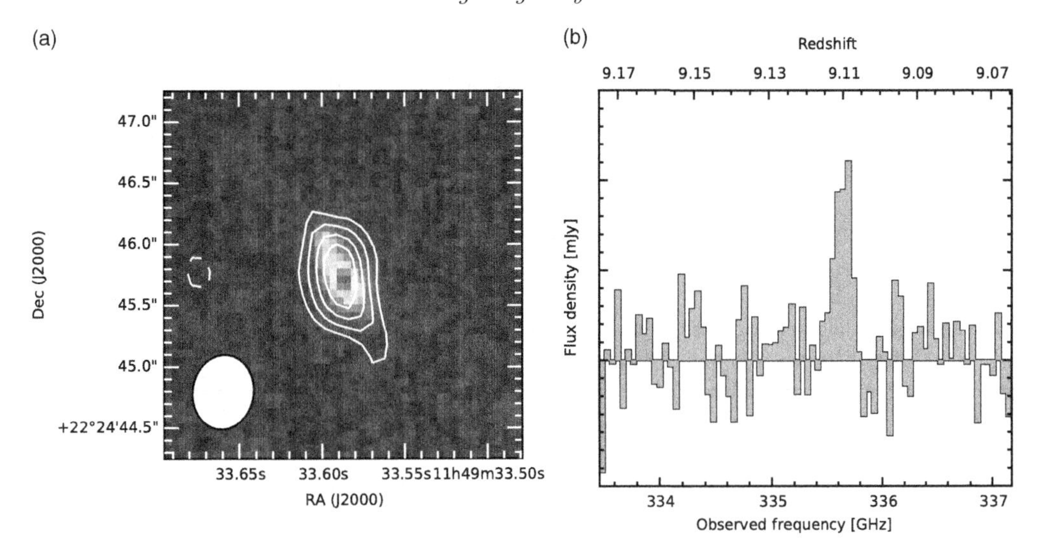

Figure 1. ALMA [OIII] contours and spectrum of MACS1149-JD1. (a) Zoom on an *HST* image (F160W) with the ALMA [OIII] contours overlaid. Contours are drawn at 1σ intervals from ± 3 to $+6\sigma$ where $\sigma = 17.5$ mJy km s^{-1} beam^{-1}. Negative contours are shown by the dashed line. Ellipse at the lower left corner indicates the synthesized beam size of ALMA. (b) ALMA [OIII] 88 μm spectrum in frequency space with a resolution of ~ 42 km s^{-1}.

respectively, with a standard calibration script provided by the ALMA observatory. We then produced final images and cubes with the `CLEAN` task using natural weighting to maximize point-source sensitivity. The spatial resolution is $0''.62 \times 0''.52$ (FWHM) and the beam position angle was PA $= -8.9°$. A quasar, J1229+0203, was used for bandpass and flux calibrations, for which a flux uncertainty is estimated to be $\lesssim 10\%$.

To search for a line, we created a data cube, six native channels of which are binned, resulting in a velocity resolution of ~ 42 km s^{-1}. In the T5 setup at ~ 335 GHz, we found a $> 3.0\sigma$ signal in five continuous binned-channels, where 1σ is the local noise estimated with the CASA task `imstat`. This frequency region is free from atmospheric absorption features. We then created a velocity-integrated intensity image between 335.5 and 335.8 GHz. The peak intensity of MACS1149-JD1 is 129.8 ± 17.5 mJy km s^{-1} beam^{-1} corresponding to a significance level of 7.4σ, where 1σ error values in this paper denote the 1σ rms or standard deviation unless otherwise specified.

The spatial centroid of the emission line is in good positional agreement with that of the UV continuum emission observed by *HST* (Figure 1, left). Both images are similarly elongated along the gravitational lensing shear. We measured the integrated line flux using the CASA task `imfit` to be 0.229 ± 0.048 Jy km s^{-1}. To obtain the redshift, we extracted the 1D-spectrum from the region with $> 3\sigma$ signals in the velocity-integrated intensity image (Figure 1, right). As can be seen, the [OIII] line is detected at around 335.6 GHz (or 893.2 μm) in the Solar system barycentric frame. Applying a Gaussian fit to the line, and with a rest-frame [OIII] frequency 3393.006244 GHz, we obtain a redshift $z = 9.1096 \pm 0.0006$ and FWHM of 154 ± 39 km s^{-1}, which is reasonable for a low mass galaxy. The integrated flux and redshift leads to an observed luminosity of $(7.4 \pm 1.6) \times (10/\mu) \times 10^7$ $L_\odot$, where μ is the magnification factor. We adopt $\mu = 10$ as our fiducial lensing magnification factor.

With the CASA task `imfit`, we obtained the deconvolved size of $(0''.82 \pm 0''.25) \times (0''.30 \pm 0''.14)$. Assuming that lensing effects are equal for the major and minor axes, the intrinsic size is $(3.7 \pm 1.1)/\sqrt{\mu}$ [kpc] $\times$ $(1.4 \pm 0.9)/\sqrt{\mu}$ [kpc].

2.2. *[CII] 158 μm observations*

After the [OIII] detection, [CII] observations were also carried out in band 5 during ALMA Cycle 6 under DDTs (2017.A.00026 and 2018.A.0004 - PI: N. Laporte). We covered a redshift range $8.96 \leqslant z \leqslant 9.16$ and a total exposure time was 6.2hrs. We used the configuration C43-4 to achieve a beam size of 0.75" × 0.63", which is similar to those used in [OIII] observations. Data were reduced using the version 5.4.0 of the CASA pipeline. We searched for line emission in a 1.5" radius circle around the UV-rest frame position (corresponding to a physical size of 13.2 kpc) and allowing a velocity offset respective to the [OIII]88μm redshift ranging from -500 km/s to 500km/s. We rebinned the data assuming a FWHM of 100km/s for [CII]158μm. No emission is detected with a 3σ upper limit on the [CII]158μm luminosity $< 3.98 \times 10^6 \times (10/\mu) L_\odot$, assuming a FWHM=100km/s, with the rms measured in several beam size apertures (with $\theta_{min} = 0.63$" and $\theta_{maj} = 0.75$") distributed in a 1.5" radius circle around the UV restframe position and taking into account the best magnification($\mu = 10$).

With these observations, we obtain a luminosity ratio of $> 18.5(3\sigma)$, which is the highest among so far reported. A similar non-detection of [CII]158μm was reported by Inoue *et al.* (2016) for a Lyman-α emitter at $z = 7.2$ with [OIII]88μm emission, as well as for a $z = 8.38$ LBG with [OIII] detection.

These results clearly demonstrate that [OIII] 88 μm is a powerful tool to identify galaxies in the reionization epoch.

3. High [OIII]/[CII] luminosity ratios and implications

The line luminosity ratio, [OIII]/[CII], would give us invaluable information on chemical and ionization properties of galaxies (e.g., Inoue *et al.* 2016). For example, in local galaxies, a number of studies have examined the line ratio (e.g., Brauher *et al.* 1998). These studies have shown that dwarf metal-poor galaxies have high line ratios, [OIII]/[CII] $\approx 2 - 10$, whereas metal-rich galaxies have low line ratios, [OIII]/[CII] ≈ 0.5. Alternatively, if the ISM of galaxies is highly ionized, the [CII] luminosity would be weak because [CII] emission is predominantly emitted from the PDR (e.g., Katz *et al.* 2017).

Figure 2 summarizes the FIR line luminosity ratios of high-z galaxies (see also Laporte *et al.* 2019 for the updated data points): three $z \approx 7$ star-forming galaxies and two $z \approx 6 - 7$ sub-millimeter galaxies (SMGs). Inoue *et al.* 2016 have detected [OIII] from a $z = 7.21$ LAE with the $EW_0(Ly\alpha)$ value of 33 Å (SXDF-NB1006-2). With the null detection of [CII], the authors have shown that SXDF-NB1006-2 has a total line luminosity ratio of [OIII]/[CII] > 12 (3σ). Carniani *et al.* (2017) have reported detections of [OIII] and [CII] in a galaxy at $z = 7.11$ (BDF-3299). BDF-3299 has a large $EW_0(Ly\alpha)$ $= 50$ Å and thus can be categorized into LAEs. The galaxy has spatial offsets between [OIII], [CII], and UV emission. Under the assumption that both [CII] and [OIII] are associated with the UV emission, we obtain the total line ratio of 3.7 ± 0.6 using the [CII] luminosity ($4.9 \pm 0.6 \times 10^8$ $L_\odot$) and the [OIII] luminosity ($18 \pm 2 \times 10^8$ $L_\odot$) Hashimoto *et al.* (2018) have detected all of [OIII], [CII], and dust continuum in the two ALMA bands in a remarkably luminous LBG at $z = 7.15$. The galaxy is comprised of two UV clumps, suggesting a presence of a merger. In the whole system, the FIR luminosity ratio is 3.1 ± 0.6. Recently, Marrone *et al.* (2018) have detected both [OIII] and [CII] from a lensed SMG at $z = 6.90$ comprised of two galaxies (SPT0311-058E and SPT0311-058W). The total line luminosity ratio is 1.27 ± 0.18 and 0.56 ± 0.17 for SPT0311-058E and SPT0311-058W, respectively, where the 1σ values take the uncertainties on magnification factors into account. Finally, Walter *et al.* (2018) have detected [OIII] in an SMG at $z = 6.08$ located at the projected distance of ≈ 61 kpc from a quasar at the same redshift. In the SMG, J2100-SB, the authors have presented the line luminosity ratio of 1.58 ± 0.24 combining the previous [CII] detection.

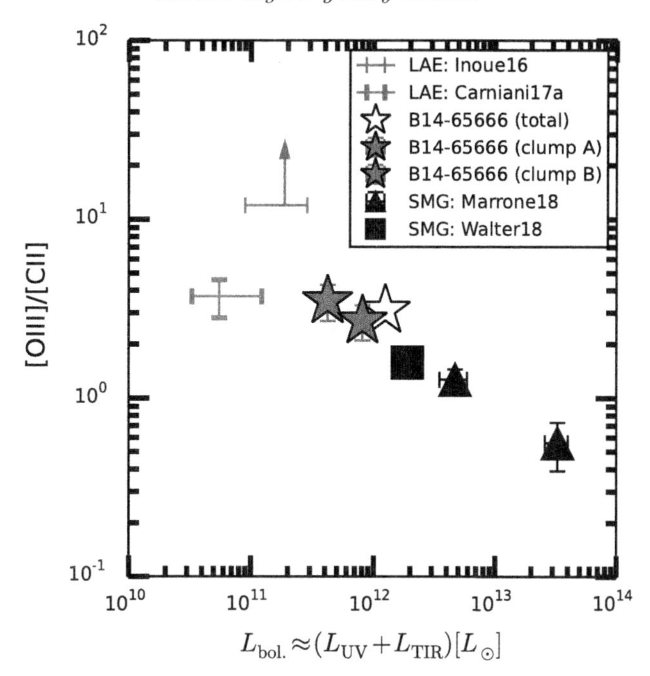

Figure 2. [OIII]-to-[CII] line luminosity ratio plotted against the bolometric luminosity estimated as the summation of the UV and IR luminosities for $z \approx 6-7$ objects. The red arrow represents the 3σ lower limit of the line luminosity ratio in the LAE of Inoue *et al.* (2016) For the two LAEs without L_{TIR} measurements, the upper limits of $L_{\mathrm{bol.}}$ are estimated as the summation of L_{UV} and the 3σ upper limits on L_{TIR}, where we assume $T_{\mathrm{d}} = 50$ K and $\beta_{\mathrm{d}} = 1.5$. The lower limits of $L_{\mathrm{bol.}}$ for the two LAEs correspond to L_{UV}.

Figure 2 shows a clear anti-correlation, although a larger number of galaxies are needed for a definitive conclusion. Given that the bolometric luminosity traces the mass scale of a galaxy (i.e., the stellar and dark matter halo masses and/or the SFR), the possible trend implies that lower mass galaxies having higher luminosity ratios. These would in turn indicate that lower mass galaxies have lower metallicity and/or higher ionization states (cf., Laporte *et al.* 2019).

References

Brauher, J. R., Dale, D. A., & Helou, G. 2008, *ApJS*, 178, 280

Capak, P. L., Carilli, C., Jones, G., Casey, C. M., Riechers, D., Sheth, K., Carollo, C. M., Ilbert, O., *et al.* 2015, *Nature*, 522, 455

Carniani, S., Maiolino, R., Pallottini, A., Vallini, L., Pentericci, L., Ferrara, A., Castellano, M., Vanzella, E., *et al.* 2017, *A&A*, 605, A42

Carniani, S., Maiolino, R., Amorin, R., Pentericci, L., Pallottini, A., Ferrara, A., Willott, C. J., Smit, R., *et al.* 2018, *MNRAS*, 478, 1170

Chevallard, J. & Charlot, S. 2016, *MNRAS*, 462, 1415

Cormier, D., Madden, S. C., Lebouteiller, V., Abel, N., Hony, S., Galliano, F., Rémy-Ruyer, A., Bigiel, F., *et al.* 2015, *A&A*, 578, A53

da Cunha, E., Groves, B., Walter, F., Decarli, R., Weiss, A., Bertoldi, F., Carilli, C., Daddi, E., *et al.* 2013, *ApJ*, 766, 13

Ellis, R. S., McLure, R. J., Dunlop, J. S., Robertson, B. E., Ono, Y., Schenker, M. A., Koekemoer, A., Bowler, R. A. A., *et al.* 2013, *ApJL*, 763, L7

Ferland, G. J., Porter, R. L., van Hoof, P. A. M., Williams, R. J. R., Abel, N. P., Lykins, M. L., Shaw, G., Henney, W. J., *et al.* 2013, *RMxAA*, 49, 137

Harikane, Y., Ouchi M., Shibuya, T., Kojima, T., Zhang H., Itoh, R., Ono, Y., Higuchi, R., *et al.* 2018, *ApJ*, 859, 84

Hashimoto, T., Laporte, N., Mawatari, K., Ellis, R. S., Inoue, A. K., Zackrisson, E., Roberts-Borsani, G., Zheng, W., *et al.* 2018, *Nature*, 557, 392

Hashimoto, T., Inoue, A. K., Mawatari, K., Tamura, Y., Matsuo, H., Furusawa, H., Harikane, Y., Shibuya, T., *et al.* 2019, *PASJ* in press (arXiv:1806.00486)

Hildebrand, R. H. 1983, *QJRAS*, 24, 267

Inoue, A. K., Shimizu, I., Tamura, Y., Matsuo, H., Okamoto, T., & Yoshida, N. 2014, *ApJL*, 780, L18

Inoue, A. K., Tamura,Y., Matsuo, H., Mawatari, K., Shimizu, I., Shibuya, T., Ota, K., Yoshida, N., *et al.* 2016, *Science*, 352, 1559

Knudsen, K. K., Watson, D., Frayer, D., Christensen, L., Gallazzi, A., Michałowski, M. J., Richard, J., & Zavala, J. 2017, *MNRAS*, 466, 138

Katz, H., Kimm, T., Sijacki, D., & Haehnelt, M. G. 2017, *MNRAS*, 468, 4831

Lagache, G., Cousin, M., & Chatzikos, M. 2018, *A&A*, 609, A130

Laporte, N., Ellis, R. S., Boone, F., Bauer, F. E., Quénard, D., Roberts-Borsani, G. W., Pelló, R., Pérez-Fournon I., *et al.* 2017, *ApJL*, 837, L21

Laporte, N., Katz, H., Ellis, R. S., Lagache, G., Bauer, F. E., Boone, F., Inoue, A. K., Hashimoto, T., *et al.* 2019, *ApJL*, 487, L81

Marrone, D. P., Spilker, J. S., Hayward, C. C., Vieira, J. D., Aravena, M., Ashby, M. L. N., Bayliss, M. B., Béthermin, M., *et al.* 2018, *Nature*, 553, 51

Ota, K., Walter, F., Ohta, K., Hatsukade, B., Carilli, C. L., da Cunha, E., González-López, J., Decarli, R., *et al.* 2014, *ApJ*, 792, 34

Stark, D. P., Ellis, R. S., Charlot, S., Chevallard, J., Tang, M., Belli, S., Zitrin, A., Mainali, R., *et al.* 2017, *MNRAS*, 464, 469

Tamura, Y., Mawatari, K., Hashimoto, T., Inoue, A. K., Zackrisson, E., Christensen, L., Binggeli, C., Matsuda, Y., *et al.* 2019, *ApJ*, 874, 27

Walter, F., Riechers, D., Novak, M., Decarli, R., Ferkinhoff, C., Venemans, B., Bañados, E., Bertoldi, F., *et al.* 2018, *ApJ*, 869L, 22

Zheng, W., Postman, M., Zitrin, A., Moustakas, J., Shu, X., Jouvel, S., Host, O., Molino, A., *et al.* 2012, *Nature*, 489, 406

Discussion

A: In [CII] observations for MACS1149-JD1, did you use similar beam sizes to those used in [OIII] observations?

HASHIMOTO: Yes, we used almost the same beam sizes in the two observations. Therefore, it is unlikely that [CII] emission is partly resolved out.

SMIT: Do you see any spatial offsets between UV and dust continuum emission?

HASHIMOTO: In the case of B14-65666, an LBG at $z = 7.15$, we do not see any statistically significant spatial offset.

C: You said that some theoretical models cannot reproduce the large [OIII]-to-[CII] luminosity ratios. In those simulations, what did they assume?

HASHIMOTO: I am not sure about the details on the simulations. However, these simulated galaxies have similar SFRs and stellar masses to those of our observations. Therefore, it is likely that something is missing in these simulations.

D: Is it possible that AGN activity enhances the luminosity ratios?

HASHIMOTO: It is unlikely because, in the local Universe, the luminosity ratios are similar for AGN-dominated galaxies and normal star-forming galaxies.

Uncovering Early Galaxy Evolution in the ALMA and JWST Era
Proceedings IAU Symposium No. 352, 2019
E. da Cunha, J. Hodge, J. Afonso, L. Pentericci & D. Sobral, eds.

doi:10.1017/S1743921320001076

The prevalence and physical properties of extremely low-luminosity galaxies in the early universe

Rychard Bouwens

Leiden University, The Netherlands

Abstract. Gravitational lensing from galaxy clusters has great potential for deriving the prevalence and physical properties of ultra-faint galaxies at early times, with recent very impressive results from the Hubble Frontier Fields program. Important issues in deriving the most accurate results are accurate constraints on source sizes and a robust treatment of uncertainties in the magnification models. Using >3300 $z = 2 - 10$ galaxies behind the 6 Hubble Frontier Fields clusters and a forwards modeling approach, I describe the efforts of my collaborators and me to map out the galaxy luminosity functions at ~ -13 mag from $z \sim 9$ to $z \sim 2$, i.e, a factor of 1000 below L^* and to the typical luminosity of galaxies suspected to drive cosmic reionization. Additionally, I discuss the constraints we can obtain on the properties of faint sources, in particular their stellar masses, mass-to-light ratios, colors, and stellar population ages. I conclude with a prospective on using cluster lenses to study the distant universe with the James Webb Space Telescope.

Uncovering Early Galaxy Evolution in the ALMA and JWST Era
Proceedings IAU Symposium No. 352, 2019
E. da Cunha, J. Hodge, J. Afonso, L. Pentericci & D. Sobral, eds.

doi:10.1017/S1743921320001088

Unveiling the nature of the brightest $z > 6$ galaxies with ALMA and JWST

Rebecca Bowler

University of Oxford, United Kingdom

Abstract. The very brightest $z > 6$ galaxies are ideal laboratories for studying the physical properties of star-forming objects into the epoch of reionization. Selected from degree-scale, ground-based fields, these rare objects provide a key insight into early dust production and may harbour faint AGN. Targeted follow-up of small samples have unexpectedly shown both Lyman-alpha emission and other rest-frame UV lines (e.g CIV and HeII), suggesting unique star-formation conditions (or AGN) at early times. Furthermore, ALMA observations have revealed that 75% of the star-formation in these galaxies may be obscured. I will talk about HST/ALMA follow-up of bright $z \sim 7$ LBGs in COSMOS and present new results from even brighter samples from $z = 6 - 9$ selected over ~ 5 deg^2. The power of both ALMA and JWST, coupled with the intrinsic luminosity of these sources, will provide a unique insight into the formation and evolution of vigorously star-forming galaxies in the first billion years.

Uncovering Early Galaxy Evolution in the ALMA and JWST Era
Proceedings IAU Symposium No. 352, 2019
E. da Cunha, J. Hodge, J. Afonso, L. Pentericci & D. Sobral, eds.

doi:10.1017/S1743921319009451

Unveiling the most luminous Lyman-α emitters in the epoch of reionisation

Jorryt Matthee[1] and David Sobral[2]

[1]Department of Physiscs, ETH Zürich, Wolfgang-Pauli-Strasse 27, 8093 Zürich, Switzerland
email: `mattheej@phys.ethz.ch`

[2]Department of Physics, Lancaster University, Lancaster, LA1 4YB, United Kingdom

Abstract. Distant luminous Lyman-α emitters are excellent targets for detailed observations of galaxies in the epoch of reionisation. Spatially resolved observations of these galaxies allow us to simultaneously probe the emission from young stars, partially ionised gas in the interstellar medium and to constrain the properties of the surrounding hydrogen in the circumgalactic medium. We review recent results from (spectroscopic) follow-up studies of the rest-frame UV, Lyman-α and [CII] emission in luminous galaxies observed $\sim$ 500 Myr after the Big Bang with ALMA, HST/WFC3 and VLT/X-SHOOTER. These galaxies likely reside in early ionised bubbles and are complex systems, consisting of multiple well separated and resolved components where traces of metals are already present.

Keywords. galaxies: formation, galaxies: evolution, galaxies: high-redshift

1. Introduction

Thanks to its rest-frame UV wavelength and intrinsic brightness in star-forming regions, the Lyman-α (Lyα, $\lambda_0 = 1215.67$ Å) emission line is extremely useful for identifying galaxies in the early Universe ($z > 2$). In recent years, significant progress has been made in enlarging the dynamical range in the luminosities of known Lyman$-\alpha$ emitters (LAEs): either by identifying samples of very luminous LAEs using wide-field narrow-band surveys (e.g. Matthee *et al.* 2015; Zheng *et al.* 2017; Konno *et al.* 2018; Sobral *et al.* 2018) or by identifying the faintest LAEs using sensitive integral field spectrographs such as MUSE (e.g. Bacon *et al.* 2017).

Due to the high scattering cross-section at line-centre, Lyα photons are subject to radiative transfer effects in the presence of neutral hydrogen, which results in an uncertain Lyα escape fraction. This significantly challenges the interpretation of observed Lyα light. However, the sensitivity of Lyα to the HI column density also provides an opportunity to study the HI structure in and around galaxies. This is of particular relevance for the escape of ionising photons from the ISM (e.g. Verhamme *et al.* 2015) and for using Lyα emission as a tracer of reionisation (e.g. Mason *et al.* 2018).

Here we present highlights on results obtained using spectroscopic observations of luminous LAEs at $z \approx 6.5$ (selected in Matthee *et al.* 2015; Sobral *et al.* 2015). While all galaxies in the sample have a high Lyα luminosity, they span a range in UV continuum luminosity (1-5 $M^{\star}_{1500}$) and a range in Lyα equivalent width (EW_0=30-200 Å). Current facilities such as ALMA and VLT/X-SHOOTER can be used to obtain a first glimpse of the properties of the ISM in these galaxies in the first billion year of cosmic time.

In §2 we focus on recent results on the LAE 'COLA1' (Hu *et al.* 2016; Matthee *et al.* 2018), a unique double peaked Lyα emitter at $z = 6.59$ that is of particular interest for the study of the epoch of reionization. In §3 we review results of recent ALMA follow-up

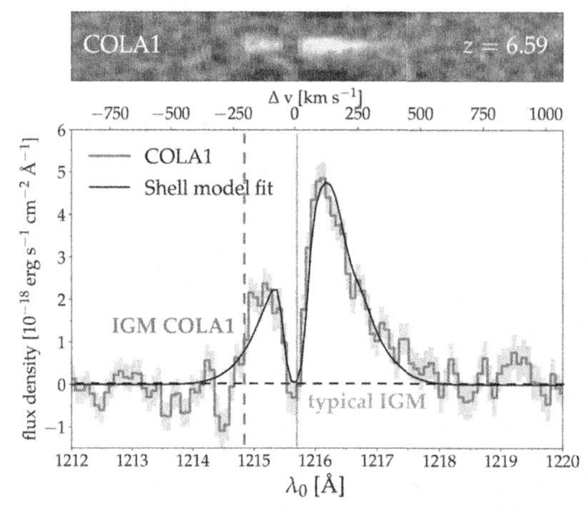

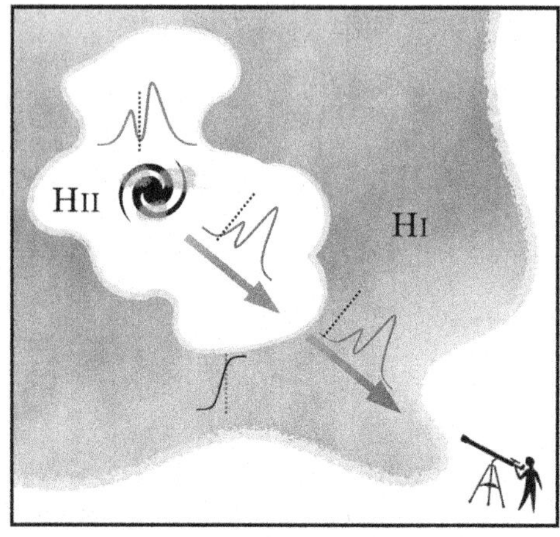

Figure 1. Left: VLT/X-SHOOTER spectrum of the double peaked Lyα line in COLA1 at $z = 6.591$ (adapted from Matthee *et al.* 2018).The red line indicates the typical velocity at which the Lyα spectrum in typical LAEs is truncated, likely due to low IGM transmission. The blue dashed line indicates the velocity at which this seems to occur for COLA1. Right: Illustration of a scenario in which COLA1 resides in a large ionised bubble. Lyα redshift out of the resonant frequency on the Hubble flow before encountering significant amounts of neutral hydrogen. This allows the blue peak to be observed.

observations of Lyα-selected galaxies at $z \approx 6-7$. We use a ΛCDM cosmology with $\Omega_\Lambda = 0.70$, $\Omega_M = 0.30$ and $H_0 = 70$ km s^{-1} Mpc^{-1} and a Salpeter (1995) IMF.

2. COLA1: a double peaked Lyα emitter at $z = 6.59$

At high redshift ($z > 6$) virtually all LAEs show a single red asymmetric peak (e.g. Hu *et al.* 2010; Matthee *et al.* 2017b), likely because the increasing hydrogen density around galaxies results in a low transmission on the blue side of line-center (e.g. Laursen *et al.* 2011). The discovery of double peaked Lyα emission at $z = 6.59$ in 'COLA1' by Hu *et al.* (2016) challenged this picture.

Initially, the reality of COLA1 being a LAE at $z = 6.59$ was uncertain. The peak separation is consistent with the line being the [OII] doublet at $z = 1.47$ and tentative optical flux has been observed in bands blue-wards of the Lyman-break at $z = 6.6$ (Matthee *et al.* 2018). However, new higher resolution, deeper VLT/X-SHOOTER observations (left panel of Fig. 1; Matthee *et al.* 2018) rule out that COLA1 is an [OII] emitter at $z = 1.47$ based on 1) the asymmetry of the red line, 2) the zero-flux between the lines, 3) the ratio of the blue-to-red line, 4) the non-detection of Hα at $z = 1.47$ and 5) the extreme observed EW. Moreover, the spectrum also revealed foreground Lyα and [OIII]$_{5008}$ emission at $z = 2.14$ at a close ($< 1''$) separation that explains the flux in blue optical bands. Therefore, COLA1 is confirmed as a double peaked LAE at $z = 6.59$.

How is it possible that double peaked Lyα emission is seen in COLA1? The answer likely lies in the narrowness of the Lyα peak separation (220 km s^{-1}). Using UV observations of green pea galaxies at $z \approx 0.3$ with *HST*, Izotov *et al.* (2018) recently found a strong anti-correlation between the Lyman-continuum escape fraction ($f_{esc,LyC}$) and the peak separation of the double peaked Lyα line. The physical explanation is that LAEs with narrow peak separation have an ISM with low HI column density channels through which ionising photons escape (Verhamme *et al.* 2015). Following the trend from Izotov *et al.* 2018), the peak separation in COLA1 corresponds to a high $f_{esc,LyC} \approx 30$ %.

In order to observe double peaked Lyα emission at $z > 6$, galaxies need to reside in ionised regions that are large enough to allow Lyα photons to redshift out of the resonance

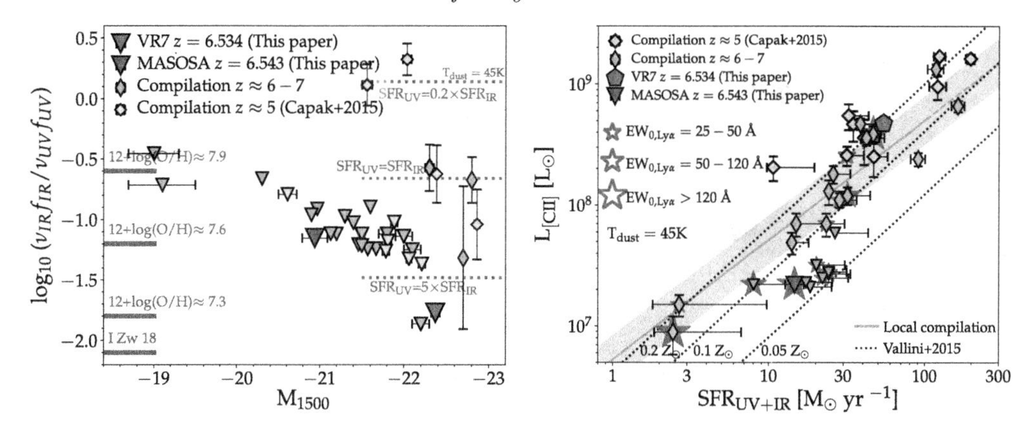

Figure 2. Left: Observed rest-frame UV ($\lambda_0 = 1500$Å) to IR ($\lambda_0 = 160\mu$m) flux ratio as a function of UV luminosity for a compilation of UV and Lyα-selected galaxies at $z \approx 5-7$. The gas-phase metallicity corresponding to galaxies in the local Universe with similar UV to IR flux ratios (Maiolino *et al.* 2015) are illustrated. We also illustrate the obscured to unobscured SFR ratio when converting IR flux to IR luminosity assuming a dust temperature of 45 K. Right: [CII] luminosity versus SFR$_{\rm UV+IR}$ for a compilation of galaxies at $z \approx 5-7$ compared to galaxies in the local Universe (blue band). Above SFR> 30 M$_\odot$ yr^{-1}, most high-redshift galaxies are relatively luminous in [CII], while low [CII] luminosities (and relatively deep upper limits on the [CII] luminosity) are found in fainter galaxies. Both panels adapted from Matthee *et al.* (2019).

wavelength (right panel of Fig. 1). Smaller regions are thus required for galaxies with smaller Lyα peak separations. Since the escape fraction is anti-correlated with peak separation, galaxies with smaller peak separation will more easily reside in large ionised regions. Such galaxies are therefore likely more easily observed at high redshift.

Assuming that the systemic redshift of COLA1 lies between the two Lyα lines (as in lower-redshift double peaked LAEs) and that the blue line needs to redshift on the Hubble flow by 220 km s^{-1} before encountering neutral hydrogen, a Strömgren sphere with radius 300 proper kpc is required (Matthee *et al.* 2018). Assuming an ionising efficiency $\xi_{ion} = 10^{25.4}$ Hz erg^{-1}, an escape fraction of 15 % (conservative, compared to inferred 30 %) and a SFR of 30 M$_\odot$ yr^{-1} (based on the UV luminosity), COLA1 can plausibly ionise the required region on its own in 10^7 yr.

COLA1 is the first galaxy that can be used as both a tracer and an agent of reionisation. Future observations can assess the properties of the ISM and stellar populations present in COLA1 and investigate whether COLA1 resides in a large over-density. Additionally, detailed statistical investigation of the presence of double peaked Lyα lines (and their peak separations) in galaxies at different UV luminosities and redshifts may provide useful indirect information on which (star-forming) galaxies contribute to the global ionising background throughout the evolution of the Universe.

3. ALMA results on luminous LAEs at $z \approx 6-7$

Since the advent of the Atacama Large (Sub)Millimetre Array (ALMA), observations have been performed to study the ISM and dust properties of high-redshift galaxies using far-infrared emission lines such as [CII]158μm and the FIR continuum (IR hereafter).

As compiled in Matthee *et al.* (2019), the number of UV and Lyα selected galaxies with dust continuum detections at $z \approx 5-7$ is low. Furthermore, detections/upper limits in a single IR continuum frequency need to assume a dust temperature distribution in order to infer obscured SFR or dust mass. The left panel of Fig. 2 shows the observed UV to IR (at $\lambda_0 = 160\mu$m) flux ratio as a function of UV luminosity for UV or Lyα selected

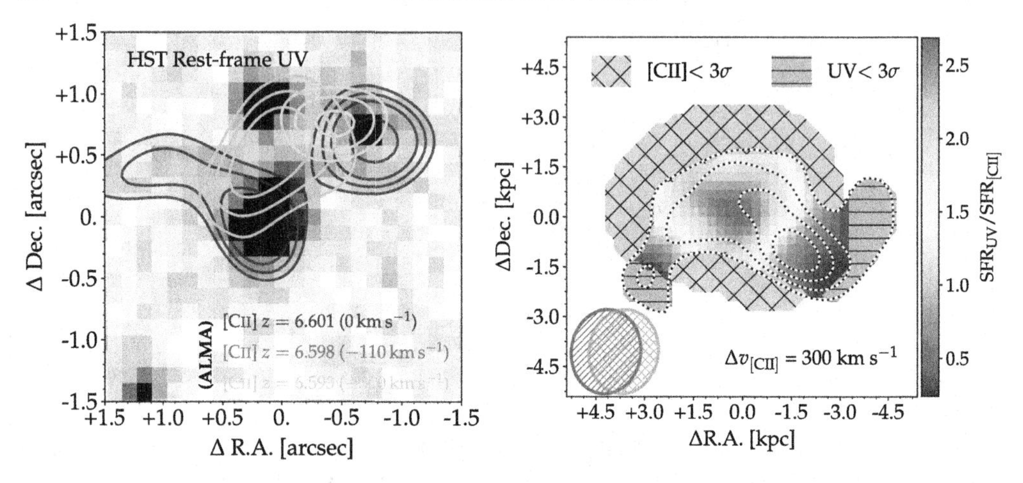

Figure 3. Left: Resolved rest-frame UV and [CII] image of luminous LAE CR7 (Sobral *et al.* 2015, 2019) for which three UV and four [CII] components are identified. Right: Resolved UV to [CII] ratio of luminous LAE VR7 (Matthee *et al.* 2019). Variations in the UV-[CII] ratio of a factor 5 are observed on ~ 2 kpc scales.

galaxies at $z \approx 5-7$. The majority of measurements are upper limits and it is clear that dust continuum detections are only found in systems that are more UV-luminous. This likely points to a relatively monotonic relation between UV luminosity and (dust) mass. The most UV-luminous systems with the strongest constraints on UV-IR flux ratio are two luminous LAEs (CR7 and VR7). This indicates that those systems either have the least dust (at fixed UV luminosity), and/or the highest dust temperatures (at these frequencies, a higher dust temperature would decrease the UV-IR ratio for a fixed integrated IR luminosity). Such conditions are expected in LAEs, where high ISM ionisation states and little dust reddening are typical at $z \sim 2$ (e.g. Trainor *et al.* 2016).

Early [CII] observations of LAEs yielded contradictory results: studies found relatively high [CII] luminosities (Capak *et al.* 2015) or only strong upper limits on the [CII] luminosity (e.g. Ouchi *et al.* 2013), well below expectations from the local Universe. Other observations of UV-selected galaxies (but with known Lyα redshift) indicated mildly low [CII] luminosities (e.g. Pentericci *et al.* 2016). Weak [CII] emission in high-redshift galaxies severely challenged the promise of using ALMA as a 'redshift-machine' and it was speculated that this was due to the strong Lyα emission (because of the pre-requisite of a known spectroscopic redshift) biasing follow-up observations towards galaxies with low metallicities and/or high ionisation states (both decreasing [CII] luminosity).

More recently however, relatively strong [CII] emission has been detected in galaxies with strong Lyα emission (Matthee *et al.* 2017a; Carniani *et al.* 2018) and [CII] has also been detected in galaxies without strong Lyα emission (Smit *et al.* 2018). The right panel of Fig. 2 shows that while there is a large range in [CII]-UV ratios in galaxies at $z \approx 5-7$, relatively low [CII] luminosities are mostly found in faint galaxies (SFR< 25 M$_{\odot}$ yr^{-1}), while [CII] is relatively strong in galaxies with higher SFRs. This indicates that either metallicity and/or ionisiation state are relatively strong functions of UV luminosity. As most LAEs are found in relatively faint galaxies, this explains why initial results on strong LAEs mostly found low [CII] luminosities, together with a better understanding of ALMA data reduction. Future deep observations still need to confirm these low [CII]/UV ratios.

Another important result is that [CII] emission in luminous LAEs typically consist of multiple components that are resolved in observations with ~ 2kpc resolution (e.g. Matthee *et al.* 2017a; Carniani *et al.* 2018). Such components are seen with separations up to ~ 5kpc and ~ 200 km s^{-1} and overlap (in most cases) with components seen in high-resolution rest-frame UV observations from *HST*, see Fig. 3. On small scales, [CII]-UV ratios can vary in the same galaxy, with differences up to factor 5 on $\sim$ 2kpc scales (Fig. 3; Matthee *et al.* 2019). These observations indicate that luminous galaxies build-up through the relative complex assembly of different components.

4. Conclusions & Outlook

We have highlighted recent results on deep, resolved spectroscopic observations of luminous LAEs at $z \sim 6-7$. These reveal the first glimpse on the detailed properties of galaxies that reside in large ionised regions at the end of reionization. In the near future studies will likely increase the sample of observed galaxies and span the parameter space towards both fainter and brighter galaxies. Significant progress is anticipated when multiple emission lines in the same systems can be observed (for example with ALMA; e.g. Hashimoto *et al.* 2019), but particularly with sensitive resolved infrared spectroscopy with *JWST*. In addition, the results on COLA1 show that deep, high resolution spectroscopy of the Lyα line is incredibly valuable, even in the era when spectroscopic confirmations can and (finally) will be provided using alternative emission lines. Lyα observations of high-redshift galaxies should therefore start to be seen even more as a physical tool in addition to be the means for spectroscopic redshift determination.

References

Bacon, R., Conseil, S., Mary, D., *et al.* 2017, *A&A*, 608, 1
Capak, P., Carilli, C., Jones, G., *et al.* 2015, *Nature*, 522, 455
Carniani, S., Maiolino, R., Amorin, R., *et al.* 2018, *MNRAS*, 478, 1170
Hashimoto, T., Inoue, A., Mawatari, K., *et al.* 2019, *PASJ*, 70
Hu, E., Cowie, L., Barger, A., *et al.* 2010, *ApJ*, 725, 394
Hu, E., Cowie, L., Songaila, A., *et al.* 2016, *ApJ*, 825, 7
Izotov, Y., Worseck, G., Schaerer, D., *et al.* 2018, *MNRAS*, 478, 4851
Konno, A., Ouchi, M., Shibuya, T., *et al.* 2018, *PASJ*, 70, 16
Laursen, P., Sommer-Larsen, J., & Razoumov, A. 2011, *ApJ*, 728, 52
Maiolino, R., Carniani, S., Fontana, A., *et al.* 2015, *MNRAS*, 452, 54
Mason, C., Treu, T., Dijkstra, M., *et al.* 2018, *ApJ*, 857, 11
Matthee, J., Sobral, D., Santos, S., *et al.* 2015, *MNRAS*, 451, 400
Matthee, J., Sobral, D., Boone, F., *et al.* 2017, *ApJ*, 851, 145
Matthee, J., Sobral, D., Darvish, B., *et al.* 2017, *MNRAS*, 472, 772
Matthee, J., Sobral, D., Gronke, M., *et al.* 2018, *A&A*, 619, 136
Matthee, J., Sobral, D., Boogaard, L., *et al.* 2019, *ApJ*, arXiv:1903.08171
Ouchi, M., Ellis, R., Ono, Y., *et al.* 2013, *ApJ*, 778, 102
Pentericci, L., Carniani, S., Castellano, M., *et al.* 2016, *ApJ*, 829, 11
Salpeter, E. 1995, *ApJ*, 121, 161
Smit, R., Bouwens, R., Carniani, S., *et al.* 2018, *Nature*, 553, 178
Sobral, D., Matthee, J., Darvish, B., *et al.* 2015, *ApJ*, 808, 139
Sobral, D., Santos, S., Matthee, J., *et al.* 2018, *MNRAS*, 476, 4725
Sobral, D., Matthee, J., Brammer, G., *et al.* 2019, *MNRAS*, 482, 2422
Trainor, R., Strom, A., Steidel, C., *et al.* 2016, *ApJ*, 832, 171
Verhamme, A., Orlitova, I., Schaerer, D., *et al.* 2015, *A&A*, 578, 7
Zheng, Z., Wang, J., Rhoads, J., *et al.* 2017, *ApJ*, 842, 22

Uncovering Early Galaxy Evolution in the ALMA and JWST Era
Proceedings IAU Symposium No. 352, 2019
E. da Cunha, J. Hodge, J. Afonso, L. Pentericci & D. Sobral, eds.

doi:10.1017/S174392132000109X

Probing the faintest galaxy population at the epoch of reionization with gravitational lensing

Hakim Atek

Institut d'Astrophysique de Paris, France

Abstract. Ultra-deep observations of blank fields with the Hubble Space Telescope have made important inroads in characterizing galaxy populations at redshift $z = 6 - 10$. Gravitational lensing by massive galaxy clusters offers a new route to identify the faintest sources at the epoch of reionization. In particular, thanks to the Hubble Frontier Fields program, we robustly pushed the detection limit down to $M_{\rm AB} = -15$ mag at $z \sim 6$. I will present the latest results based on the complete dataset of the HFF clusters and parallel fields, and their implications on the ability of galaxies to reionize the Universe. I will also discuss the results of a comprehensive end-to-end modeling effort towards constraining the systematic uncertainties of the lens models, which are currently the last hurdle before extending the UV LF to fainter luminosities. Finally, I will discuss the great discoveries awaiting combination of such cosmic lenses with the upcoming James Webb Space Telescope and the exciting opportunity to probe the turnover of the UV LF, hence the limit of the star formation process at those early epochs.

Uncovering Early Galaxy Evolution in the ALMA and JWST Era
Proceedings IAU Symposium No. 352, 2019
E. da Cunha, J. Hodge, J. Afonso, L. Pentericci & D. Sobral, eds.

doi:10.1017/S1743921319009098

ALMA witnesses the assembly of first galaxies

Stefano Carniani

Scuola Normale Superiore, Piazza dei Cavalieri 7, I-56126 Pisa, Italy
email: `stefano.carniani@sns.it`

Abstract. Characterising primeval galaxies entails the challenging goal of observing galaxies with modest star formation rates (SFR < 100 M$_\odot$ yr^{-1}) and approaching the beginning of the reionisation epoch ($z > 6$). To date a large number of primeval galaxies have been identified thanks to deep near-infrared surveys. However, to further our understanding on the formation and evolution of such primeval objects, we must investigate their nature and physical properties through multi-band spectroscopic observations. Information on dust content, metallicity, interactions with the surrounding environment, and outflows can be obtained with ALMA observations of far-infrared (FIR) lines such as the [CII] at 158 μm and [OIII] at 88 μm. Here, we, thus, discuss the recent results unveiled by ALMA observations and present new [CII] observations of BDF-3299, a star-forming galaxy at $z = 7.1$ showing a spatial and spectral offset between the rest-frame UV and the FIR lines emission.

Keywords. galaxies: high-redshift, galaxies: ISM, infrared: ISM

---

1. Introduction

In the last decade deep optical and near-infrared imaging surveys carried out with the *Hubble Space Telescope* (*HST*) have identified a large number of galaxies within the Epoch of Reionization ($z > 6$). Such observations have shown that primeval galaxies are dramatically different from their low-z counterparts being more compact, metal poor, and very clumpy likely due to intense merging activity (see review by Dayal & Ferrara 2018). However, despite the remarkable progress, *HST* observations map only the rest-frame UV light powered by stellar population providing only partial information on galaxy assembly and evolution. It is thus fundamental to combine *HST* images with those obtained with the Atacama Large Millimetre/submillimetre array (ALMA) probing the interstellar medium (ISM), the main fuel of star formation activity in galaxies.

To date more than 40 star-forming galaxies with SFR < 100 M$_\odot$ yr^{-1}, which represent the typical galaxy population in the early Universe (e.g. Carniani *et al.* 2015; Robertson *et al.* 2015), have been studied with ALMA through the emission of the fine structure far-infrared (FIR) lines of [CII] at 158μm and [OIII] at 88μm (e.g., Maiolino *et al.* 2015; Pentericci *et al.* 2016; Carniani *et al.* 2017, 2018a,b; Matthee *et al.* 2018; Hashimoto *et al.* 2019). The [CII] line is emitted mainly from the neutral diffuse and partially ionised gas of the ISM, while the [OIII] arises from the ionised gas excited by star formation. ALMA observations have revealed that FIR lines are perfect tools to characterise ISM conditions, obtain precise redshift, and investigate gas kinematics. ALMA extended array configurations have also enabled to obtain images with angular resolution as high as 0.3$''$ (Figure 1), which is comparable to the size of high-z galaxies, and thus asses the galaxy morphology. For a large fraction of high-z galaxies, [CII] emission breaks into multiple components associated to on-going merger processes (Carniani *et al.* 2018).

The high-resolution ALMA observations have also unveiled another puzzling scenario. For some galaxies the [CII] (or [OIII]) emission appears to be significantly offset both

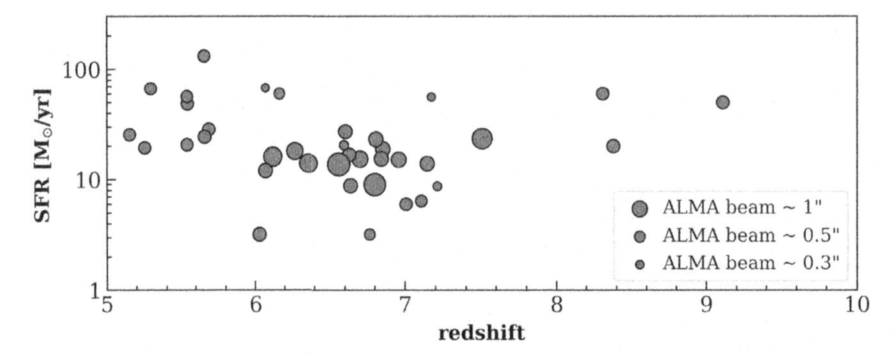

Figure 1. The redshift and SFR of high-z galaxies observed with ALMA in either [CII] or [OIII]. The size of the circle marks denotes the angular resolution (ALMA beam) of the observations.

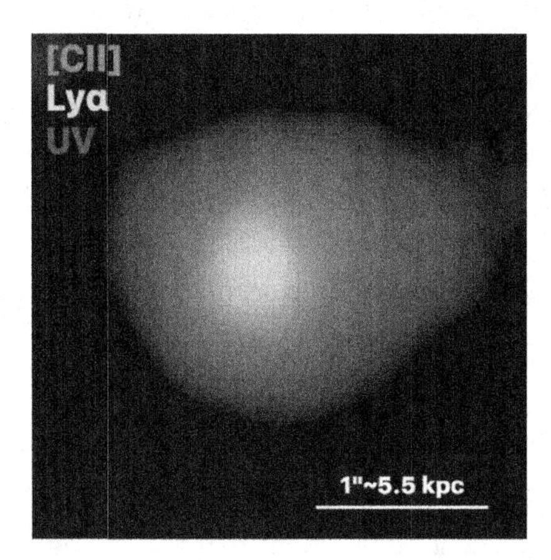

Figure 2. RGB image of Himiko, a star-forming galaxy at $z \sim 7$. Red, green, and blue images correspond to [CII], Lyα, an d rest-frame UV.

in spatial position and in velocity relative to the Lyα and rest-frame UV emission. An example of displaced [CII] emission is illustrated in Figure 2 showing an RGB image of a Lyα emitter (LAE) a $z = 6.6$, Himiko, **comprised** of three rest-frame UV bright clumps embedded in a extended Lyα nebula. The [CII] emission peaks in a region not detected in the rest-frame UV observations. **According to** current models, the [CII] deficit emission co-spatial to rest-frame UV region could be associated to intense stellar feedback removing or destroying molecular clouds and, thus regulating [CII] luminosity (Vallini *et al.* 2015; Ferrara *et al.* 2019). On the other hand, the displaced [CII] line may trace either outflowing or accreting gas (Maiolino *et al.* 2015; Vallini *et al.* 2015; Pallottini *et al.* 2017). In the context of the origin of spatial and spectral offset of FIR lines, a key result has been provided by the ALMA observations of the star forming galaxy BDF-3299 at $z = 7.1$ (Maiolino *et al.* 2015; Carniani *et al.* 2017) showing for the first time a clear evidence of displaced FIR emission. Here, we briefly review previous observations and new ALMA data of BDF-3299 confirming the detection of the spatially offset [CII] emission.

2. BDF-3299

BDF-3299 is a spectroscopically confirmed LAE at $z = 7.109$ with SFR ≈ 6 M$_\odot$ yr^{-1} (Vanzella *et al.* 2011). The galaxy is located in the Bremer Deep Field (BDF) where two additional LAEs, within a projected distance of ~ 2 Mpc from BDF-3299, have been recently identified indicating that the BDF area is an overdense, reionized region in the

Table 1. Properties of ALMA [CII] observations

Obs. (a)	$\sigma_{\rm cont}$ (b)	$\sigma_{\rm line}$ (c)	beam (d)
Cycle 1	8	62	$0.8'' \times 0.6''$
Cycle 2	11	90	$0.3'' \times 0.2''$
Cycle 1+2	7	58	$0.6'' \times 0.5''$
Cycle 4	11	70	$0.6'' \times 0.5''$

Note: (a) ALMA Cycle; (b) sensitivity on continuum image in unit of μJy beam^{-1}; (c) sensitivity over a spectral channel of 100 km s^{-1} in unit of μJy beam^{-1}; (d) angular resolution.

$z \sim 7$ Universe (Castellano *et al.* 2016, 2018). Among the three LAEs, BDF-3299 is the faintest in Lyα with a flux of 1.2×10^{-17} erg s^{-1} cm^{-2} and EW(Lyα) = 50 Å.

During ALMA Cycle 1 we led an **observational** program to investigate the strength of the [CII] line of BDF-3299 (Maiolino *et al.* 2015). The deep observations ($\sim$ 8h on source) have revealed a carbon emission line consistent with the redshift of the galaxy. The [CII] line is slightly blueshifted by 64 km s^{-1} relative to Lyα redshift, which is expected since the blue side of Lyα is heavily absorbed by intervening intergalactic medium, hence artificially redshifted relative to the actual rest frame. However, the detection is offset by $0.7''$ ($\sim$4 kpc) relative to the rest UV counterpart. This has been the first detection of displaced [CII] emission in the distant Universe and **such an offset** is not ascribed to astrometric uncertainties, as astrometry has been checked through serendipitous sources. These results seem to support model predictions expecting that molecular clouds in the central parts of primordial galaxies are rapidly disrupted by stellar feedback. As a result, [CII] emission mostly arises from more accreting/satellite clumps of neutral gas (Vallini *et al.* 2015).

In Carniani *et al.* (2017) we have reported new follow-up [CII] observations carried out with ALMA in Cycle 2 with the goal of obtaining a [CII] map at a resolution ~ 3 times higher than previous observations. Unfortunately, the observations were affected by various instrumental issues (differential timing among the different antennae, resulting in phase shifts across the field of view) and the sensitivity of the delivered data was worse than requested (Table 1). Given the low sensitivity and the high angular resolution of observations ($\sim 0.3''$, Table 1), most of the diffuse [CII] emission is resolved out and only $\sim 20\%$ of the total emission observed by Maiolino *et al.* (2015) is detected in the new dataset, indicating that the gas is extended on scales larger than 1 kpc. By combining the Cycle 1 and 2 datasets we have obtained a deeper [CII] image at a intermediate angular resolution with respect to the two separate datasets (Table 1). **The combined image revealed** that the displaced [CII] emission has a clumpy structure, likely associated to a interacting system and/or a result of a feedback process (Carniani *et al.* 2017).

In ALMA Cycle 2 we also proposed [OIII] observations of BDF-3299. [OIII] emission was detected at high significance ($\sim 9\sigma$) and spatially offset relative to the optical counterpart in the same direction as the [CII] clump, but not completely overlapping with the bulk of the [CII] emission (see Fig. 4 by Carniani *et al.* 2017). The central velocity of the [OIII] is redshifted by about 440 km s^{-1} with respect to the Lyα peak and about 500 km s^{-1} relative to the [CII] emission, indicating that the two emission FIR lines, [OIII] and [CII] are tracing two different clumps next to BDF-3299. By comparing our data with models by Vallini *et al.* (2017) we have concluded that the spatial and spectral offset [OIII] line cannot be excited by the UV radiation coming from BDF-3299, but the oxygen line must be excited by in-situ star formation that is not observable in current rest-frame UV images due to dust extinction (Carniani *et al.* 2017).

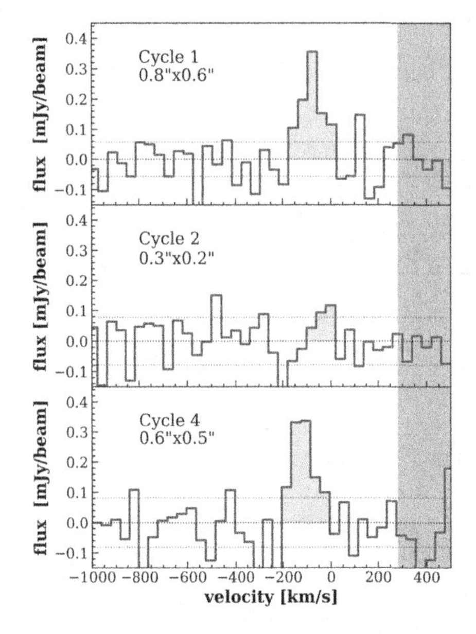

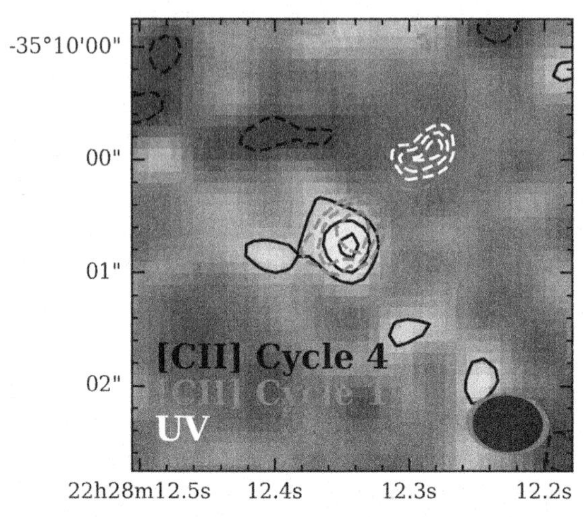

Figure 3. *Left:*. [CII] spectra of BDF-3299 extracted from the three individual datasets The rms noise levels are shown by the black dotted lines. The grey shaded region indicates the part of the spectrum affected by higher noise because of atmospheric absorption. *Right:* [CII] map of BDF-3299 from new Cycle 4 observations (black contours are in steps of 1σ, starting at 2σ). Blue contours trace the [CII] surface brightness from the Cycle 1 data (Maiolino *et al.* 2015) and contours are at levels 2, 3 and 4 times noise per beam. The white contours trace the Y-band emission (UV-rest frame). The ALMA synthesised beam of Cycle 1 and 4 observations are shown in red and blue in the bottom-right corner, respectively.

The complex scenario revealed by ALMA in the $z = 7.1$ galaxy is consistent with expectations by recent models and cosmological simulations (Vallini *et al.* 2015; Katz *et al.* 2017; Pallottini *et al.* 2017, 2019). The observational properties can be interpreted as a primeval system in the process of being assembled, where different emission lines trace distinct components, each possibly characterised by a different metallicity and/or excitation (ionisation parameter). In BDF-3299 the observable UV light is associated with the least obscured region, either because recently accreted, hence with low metallicity, or because strong feedback has removed the bulk of dust and gas content; both scenarios would account for the weakness of FIR line emission at the location of the optical image.

Cycle 1 and 2 ALMA observations are clearly tracing the process of assembly of a primeval galaxy and supporting the expectation of theoretical models. However, our capability of constraing galaxy formation models and properties of the intergalactic and circumgalactic medium is still limited by the sensitivity of our observations. In order to overcome this, we proposed deeper observations in [CII] and [OIII] of BDF-3200 in Cycle 4 aimed at achieving sensitivities twice higher than previous ones and studying the gas kinematics. Unfortunately [OIII] observations were not performed and **the** [CII] program was observed only for half of the requested time reaching a sensitivity comparable to the other previous datasets (see Table 1).

The displaced [CII] emission is detected in the new individual dataset with a level of significance $> 4.5\sigma$ and its location is in agreement with previous observations (Figure 3). The line width and the centroid are slightly different from those reported by Maiolino *et al.* (2015) and Carniani *et al.* (2017), but they are consistent within the uncertainties. We note that the angular resolution of Cycle 4 observations is 1.6 times better than in Cycle 1 and some of the diffuse emission could be resolved out in the new observations resulting in a slightly different line profile. At the location of the [OIII] clump any emission line has been detected with a level of significance higher than 2σ. A detailed analysis of the new data and the results obtained by combining all ALMA observations from the various cycles are discussed in a **separate** (Carniani *et al.* in prep.). The new observations of BDF-3299 confirm our previous results indicating the [CII] emission line **traces** the diffuse and accreting gas around high-z galaxies. Intense feedback mechanisms may destroy molecular gas from the central galaxy and explain the deficit of [CII] observed in several primeval galaxies. We note that the detection of displaced [CII] emission is not limited to this galaxy, but this scenario is observable **in several other** $z > 6$ star-forming galaxies. By taking into account that in some cases the [CII] line is not associated to the UV counterpart, we have recently found that the $L_{[CII]} - SFR$ relation at early epochs is fully consistent with the local relation (Carniani *et al.* 2018)

These preliminary results **suggest** that future deep ALMA observations in both [OIII] and [CII] applied to a large sample will be crucial to put stringent constrains on ISM properties and galaxy formation models.

3. Summary and conclusions

We have briefly summarised recent stat-of-art of ALMA observations targeting $z > 6$ star-forming galaxies. The ALMA extended array configurations have enabled us to reach high-angular resolution and exploit FIR fine structure emission lines, such as [CII] and [OIII], as powerful diagnostics to assess the morphology of primeval galaxies. In particular ALMA observations have revealed that in several galaxies the [CII] emission is not associated with any UV counterpart at the current limits. In this context, we have reviewed ALMA observations of BDF-3299 **which** was the first high-z star-forming galaxy showing a spatial offset between the FIR lines and UV emission. Finally we have discussed the preliminary results obtained by new [CII] observations of BDF-3299 confirming that the displaced [CII] emission is not due a glitch in the observation, but is a real detection, tracing circumgalactic gas in accretion or ejected by the main galaxy.

Acknowledgements

The results presented in this paper were obtained in collaboration with several other astronomers, and in particular: A. Ferrara, R. Maiolino, A. Pallottini, L. Vallini, S. Gallerani, and L. Pentericci. This work is supported by the ERC Advanced Grant INTERSTELLAR H2020/740120. This work reflects only the author's view and the European Research Commission is not responsible for information it contains.

References

Dayal & Ferrara, 2018, *PhR*, 780, 1D
Carniani, S., Maiolino, R., De Zotti, G., *et al.* 2015, *A&A*, 584A, 78C
Carniani, S., Maiolino, R., Pallottini, A., *et al.* 2017, *A&A*, 605A, 42C
Carniani, S., Maiolino, R., Smit, R., *et al.* 2018a, *ApJ*, 854L, 7C
Carniani, S., Maiolino, R., Amorin, R., *et al.* 2018b, *MNRAS*, 478, 1170C
Castellano, M., Dayal, P., Pentericci, L., *et al.* 2016, *ApJ*, 818L,3C
Castellano, M., Pentericci, L., Vanzella, E., *et al.* 2018, *ApJ*, 863L, 3C
Ferrara, A., Vallini, L., Pallottini, A., *et al.* 2019, *MNRAS*, 489, 1F
Hashimoto, Takuya, Inoue, Akio K., Mawatari, Ken., *et al.* 2019, *PASJ*, tmp, 70H
Kohandel, M., Pallottini, A., Ferrara, A., *et al.* 2019, *MNRAS*, 487, 3007K
Maiolino, R., Carniani, S., Fontana, A., *et al.* 2015, *MNRAS*, 452, 54M
Matthee, J., Sobral, D., Boone, F., *et al.* 2017, *ApJ*, 851, 145M
Pallottini, A., Ferrara, A., Gallerani, S., *et al.* 2017, *MNRAS*, 465, 2540P
Pallottini, A., Ferrara, A., Decataldo, D., *et al.* 2019, *MNRAS*, 487, 1689P

Pentericci, L., Carniani, S., Castellano, M., *et al.* 2016, *ApJ*, 829L, 11P
Robertson, B. E., Ellis, R. S., Furlanetto, S. R., & Dunlop, J. S. 2015, *ApJ*, 802, L1
Vallini, L., Gallerani, S., Ferrara, A., *et al.* 2015, *ApJ*, 813, 36V
Vallini, L., Ferrara, A., Pallottini, A., *et al.* 2017, *MNRAS*, 467, 1300V
Vanzella, E., Pentericci, L., Fontana, A., *et al.* 2011, *ApJ*, 730L, 35V

Uncovering Early Galaxy Evolution in the ALMA and JWST Era
Proceedings IAU Symposium No. 352, 2019
E. da Cunha, J. Hodge, J. Afonso, L. Pentericci & D. Sobral, eds.

doi:10.1017/S1743921319008846

Witnessing globular cluster formation at z ∼ 3−10 with JWST and ELT

Alvio Renzini

INAF – Osservatorio Astronomico di Padova, Vicolo dell'Osservatorio 5, 35122, Padova, Italy
email: alvio.renzini@inaf.it

Abstract. The opportunities offered by JWST and the ELT for the detection and study of forming/just formed globular clusters at high redshifts are illustrated, also alluding at the unique insight we may get on the very early stages of galaxy formation.

Keywords. globular clusters: general, galaxies: high-redshift, galaxies: evolution, galaxies: formation, galaxies: star clusters.

1. Introduction

The possibility of seeing globular clusters (GC) in formation at high redshift and their possible role in cosmic reionization has been occasionally discussed in the literature, yet with rapidly growing interest as the operation of new major facilities (notably JWST and ELT) is approaching (see Carlberg 2002; Ricotti 2002; Schraerer & Charbonnel 2011; Katz & Ricotti 2013, 2014; Trenti *et al.* 2015; Renzini 2017; Zick *et al.* 2018; Boylan-Kolchin 2018). First observational hints at objects that may be them can be found in Vanzella *et al.* (2016, 2017a,b, 2019), Elmegreen (2017) and Bouwens *et al.* (2017, 2018).

Figure 1 shows what is perhaps the best example of a just formed GC at $z = 6.145$ from Vanzella *et al.* (2019). What we see is a galaxy highly magnified and stretched by cluster gravitational lensing, with object D1(core) appearing perfectly round in spite of the stretching, which sets an upper limit of only 13 pc to its half-light radius. Vanzella *et al.* estimate for it a mass of $\sim 7 \times 10^5\ M_\odot$ and an age of 10 Myr, thus redshift, size, mass and age appear to be what indeed we would expect a just formed GC should look like.

Most Milky Way GCs formed $\sim 12.5 \pm 1$ Gyr ago (e.g., Leaman, VandenBerg & Mendel 2013), hence corresponding to redshifts beyond 3, when only ∼6% of the present day stellar mass was already formed (Madau & Dickinson 2014), which means that the bulk GC formation precedes the build up of massive galaxies, hence the observation of forming GCs can provide crucial information on the very early stages of galaxy formation.

In this contributed paper I build on the results of a recent attempt to calculate the expected number counts of GC progenitors (i.e., young GC at $z > 3$), specifically for the early observations with JWST (Pozzetti, Maraston & Renzini 2019), then mentioning some of the opportunities offered by a follow-up with the ELT and its early instrumentation.

2. Number Counts

The number counts of any kind of astronomical precursor at high redshift is given by the sampled comoving volume times the local number density of the specific kind of astronomical objects. Examples are given in Table 1, where for a standard cosmology ($\Omega_\Lambda = 0.7$, $\Omega_{\rm M} = 0.3$ and $H_\circ = 70$) the comoving volume between $z = 3$ and 10 sampled by the ∼10 arcmin^2 field of view of the NIRCam camera on board JWST will be

Table 1. What to find at $z > 3$ in a NIRCam frame ($\sim$10 arcmin^2)

Object	Local Number Density [Mpc^{-3}]	Number of Precursors in a Frame
Galaxy Clusters & BCGs	10^{-5}	$\sim$1
$M_* > 10^{11} M_\odot$ Galaxies	2×10^{-3}	$\sim$200
Globular Clusters	1.5	$\sim$200, 000
GCPs within 10 Myr	from peak luminosity	$\sim$1, 500

Note: The number of precursors within one NIRCam frame is given by the sampled comoving volume between $z = 3$ and 10 ($\sim$130, 000 Mpc^3) times the local number density of their progeny.

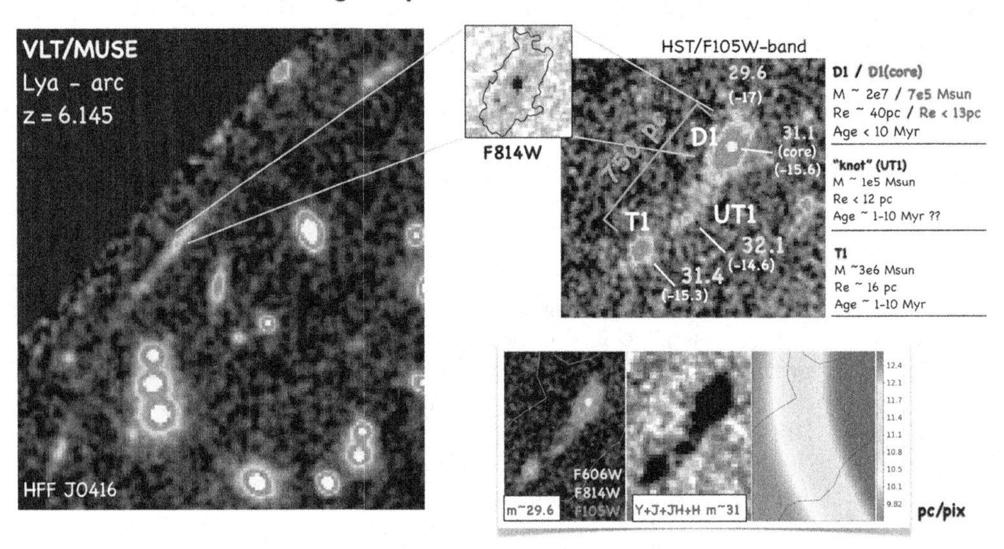

Figure 1. Zoomed-in images of a highly magnified and stretched galaxy at $z = 6.145$ with its point-like core, less than 13 pc in size, offering a very plausible example of a just formed globular cluster. (Courtesy of Eros Vanzella).

$\sim$130, 000 Mpc^3. Thus, given the local number density of GCs ($\sim$1.5 Mpc^{-3}) each NIRCam 10 arcmin^2 frame will include the precursors of some 200,000 GCs. Of course, the vast majority of them will be too faint for being detected, either because they had not yet started to form stars, or because the young GC progenitors (GCP) may have already faded below detectability. We estimate that only $\sim$1, 500 among them will be caught within 10 Myr from their peak luminosity.

In this approach, in Pozzetti *et al.* we have calculated the expected number counts of GCPs starting from the following assumptions:

• The MW is not atypical, so most GCs in the Universe formed within the first $\sim$2 Gyr from the Big Bang, i.e., at $z \gtrsim 3$.

• The mass function of forming GCs in the early Universe is either 1) identical to the local GC mass function, i.e., a Gaussian with log $M* = 5.3$ and $\sigma = 0.52$ dex (Harris *et al.* 2014), or 2) a scaled-up version of it with log $M* = 6.3$ (i.e., $10\times$ the local $M*$). The first option assumes that the stellar mass of GCs is constant in time (no mass loss of any kind), the second option assumes that GCs were 10 times more massive at formation. With this *mass budget factor* (MBF) varying from 1 to 10 we hope to bracket reality.

• GCPs are approximated as unreddened simple stellar populations (SSP) ignoring the possible contribution of emission lines to broad-band photometry.

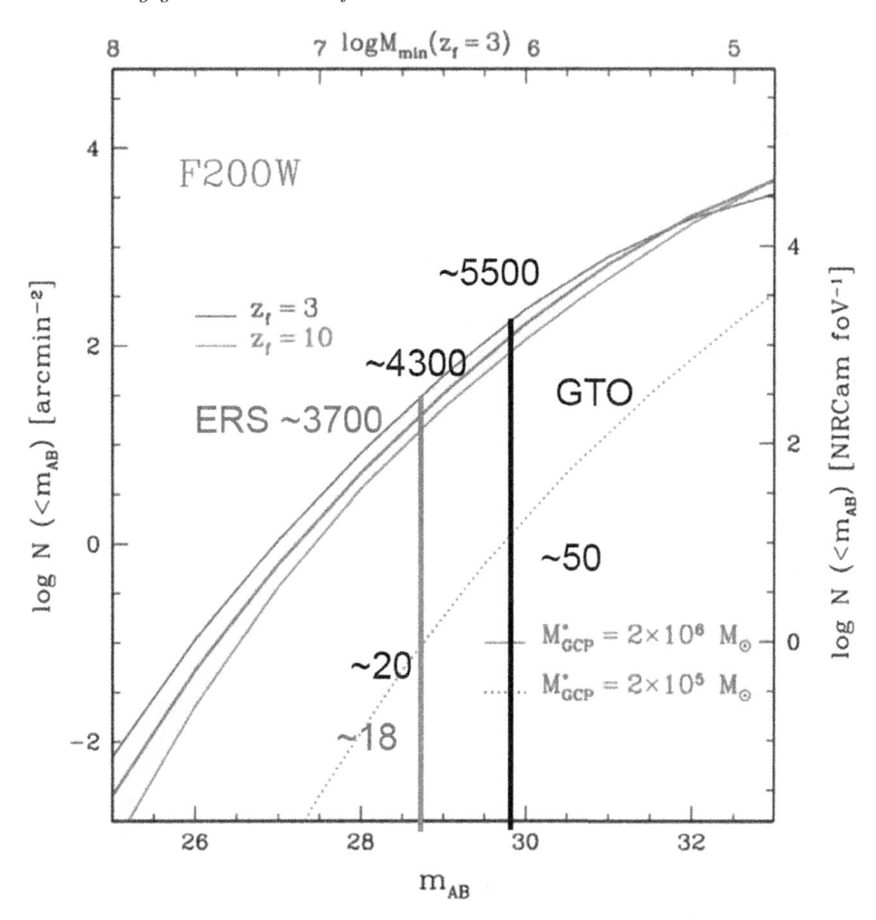

Figure 2. The expected number counts of globular cluster progenitors for the F200W filter of the NIRCam camera on JWST as from Pozzetti, Maraston & Renzini (2019). The upper scale gives the minimum mass a GCP should have in order to reach at its maximum luminosity the magnitude given in the lower scale. The dotted line correspond to assume a mass budget factor MBF=1, whereas the yellow band corresponds to MBF=10. The lower (magenta) bound of the yellow band refers to all GCPs forming at $z = 10$ whereas the upper (blue) bound refers to all forming at $z = 3$. The red line in the middle refers to the case in which GCPs forms at constant rate in time, between $z = 10$ and $z = 3$. The vertical lines show the limiting magnitudes that will be reached by the ERS and GTO observations with the NIRCam camera, with the corresponding numbers of GCP detections, respectively in red and black for ERS and GTO.

In Pozzetti *et al.* we extensively discuss justifications and limits for this set of assumptions, for example ignoring reddening may be reasonable for the metal poor half of the GCP population, but will certainly fail for the metal rich GCPs. In any event, we have calculated the expected number counts for all seven broad band filters of NIRCam, and under the two assumptions for the mass budget factor. As an example, Figure 2 shows the counts for the 2μm (F200W) passband.

In this figure, the magenta line bordering the yellow band corresponds to all GCs forming at $z = 10$ and the blue line to all of them forming at $z = 3$. The red line in the middle of the band corresponds to GCs forming at constant rate (in time, not redshift) between $z = 10$ and 3. Clearly the expected number counts are fairly insensitive to the specific distribution of formation times. provided they are restricted to the first few Gyr of cosmic time. As extensively illustrated in our paper, this insensitivity to formation redshift is a direct consequence of the steep, near power-law shape of the rest-frame UV spectrum of young SSPs, with $F(\lambda) \propto \lambda^{-\beta}$ with $\beta \sim 2.5 - 3$, ensuring a strong negative K-correction effect, such that in all bands the peak luminosity of young GCP is fairly insensitive to formation redshift. Number counts are instead extremely sensitive to the adopted mass budget factor, with Figure 2 showing that the predicted counts drop by over two orders of magnitude for the MBF changing from 10 to 1. Actual counts will therefore set interesting constraints on the initial mass of GCs.

The first opportunity for astronomers to check these numbers will be offered by the JWST/NIRCam "Early release Science" (ERS) observations which will cover $\sim$100 arcmin^2 down to mag $\sim$29. As illustrated in Figure 2, $\sim$3,700 GCPs should be detected for MBF=10. This number falls down to $\sim$18 if MDF=1.

The next major deep sky coverage with NIRCam will be part of the guaranteed time observations (GTO), with the instrument team planning to reach mag $\sim$29.8 (at 10σ for point sources) over 46 arcmin^2 and mag $\sim$28.8 over additional 190 arcmin^2. Thus, all together ERS+GTO observations should detect (e.g., in F200W) from a minimum of $\sim$90 up to $\sim 13,500$ GCPs for a MBF between 1 and 10, as indicated in Figure 2. These figures should be cut by a factor of $\sim$2 if the no reddening assumption is valid only for the metal poor GCPs.

3. Future ELT Follow-up of NIRCam Candidate GCPs

Among all these NIRCam candidates, how will we distinguish true GCPs from dwarfs galaxies or clusters of GCPs? At these redshifts, the spatial resolution of NIRCAM is $\sim$200 pc, largely insufficient to make such distinction. Lensing already can! as demonstrated by Vanzella *et al.* (2019) and Bouwens *et al.* (2018), but lensing will not provide the required mass production of GCPs, anytime soon.

The best opportunity will instead be offered by MICADO, the first-light instrument at the ELT (Davies *et al.* 2016). Assisted by multi-conjugate adaptive optics (MCAO), MICADO will be diffraction limited, and thanks to a telescope aperture of 39m will reach a resolution of $\sim$30 pc. MICADO should then do a great job in following-up NIRCam candidates, separating massive and compact objects, qualifying as GCPs, from more fluffy dwarfs that may host them, such as the complex object shown in Figure 1.

MICADO has a field of view $53'' \times 53''$ ($\sim$0.8 arcmin^2) and on average should include $\sim$80 mag< 29.8 GCPs for MBF=10, or just $\sim$1 for MBF=1. But ERS+GTO observations will also give $\sim$30 $z > 3$ progenitors to massive clusters of galaxies, with their M87-like BCG progenitors, each with its swarm of $\sim 10,000$ GCPs. From Table 1 we expect, on average, one such system per NIRCam frame, so we will have a non negligible chance that one out of the 30 BCG progenitors will be caught near its firework peak of GC formation. With MICADO imaging all of them with $\sim$30 pc resolution!

MICADO will also be equipped with with a narrow slit (16 mas, or $\sim$80 pc) spectrograph, allowing us to get redshits of candidate GCPs from lines such as [OII] (up to $z \sim 4.4$), CIV (up to $z \sim 12$) and Lyα (for $z > 5.5$).

The other ELT first-light instrument is HARMONI (Thatte *et al.* 2016), the diffraction limited ($\sim$30 pc resolution) integral field spectrograph, which will allow the spectroscopic separation of GCPs from their dwarf hosts. Like MICADO, also HARMONI will be especially fast on point-like sources, as GCPs would be.

Next instrument on the line (and relevant in this context) will be MOSAIC Hammer *et al.* 2016, the multi-object spectrograph for the ELT. With its $0''.1$ arcsec spaxels ($\sim$500 pc), a multiplex of $\sim$100 and a patrol field of $\sim$40 arcmin^2, it will contain $\sim 4,000$ dwarf galaxies at $3 < z < 10$ brighter than mag $\sim$28, and some 40 GCPs to the same limiting magnitude (for MBF=10).

With these perspectives ahead, how can we avoid being impatient?

References

Bouwens, R. J., Illingworth, G. D., Oesch, P. A., Atek, H., Lam, D., & Stefanon, M. 2017, *ApJ*, 843, 41
Bouwens, R. J., Illingworth, G. D., Oesch, P. A., Labbé, I. *et al.* 2018, arXiv:1711.02090
Boylan-Kolchin, M. 2018, *MNRAS*, 479, 332
Carlberg, R. G. 2002, *ApJ*, 573, 60
Davies, R. *et al.* 2016, SPIE.9908E.1Z
Elmegreen, B. G. 2017, *ApJ*, 836, 80
Hammer, F. *et al.* 2016, SPIE.9908E.24
Harris, W. E. *et al.* 2014, *ApJ*, 797, 128
Katz, H. & Ricotti, M. 2013, *MNRAS*, 432, 3250
Katz, H. & Ricotti, M. 2014, *MNRAS*, 444, 2377
Leaman, R., VandenBerg, D. A., & Mendel, J. T. 2013, *MNRAS*, 436, 122
Madau, P. & Dickinson, M. 2014, *ARA&A*, 52, 415
Pozzetti, L., Maraston, C., & Renzini, A. 2019, *MNRAS*, 485, 5861
Ricotti, M. 2002, *MNRAS*, 336, L33
Renzini, A. 2017, *MNRAS*, 469, L63
Schraerer, D. & Charbonnel, C. 2011, *MNRAS*, 413, 2297
Thatte, N. A. *et al.* 2016, SPIE.9908E.1X
Trenti, M., Padoan, P., & Jimenez, R. 2015, *ApJ*, 808, L35
Vanzella, E. *et al.* 2016, *ApJ*, 821, L.27
Vanzella, E. *et al.* 2017a, *MNRAS*, 467, 4304
Vanzella, E. *et al.* 2017b, *MNRAS*, 465, 3803
Vanzella, E. *et al.* 2019, *MNRAS*, 483, 3618
Zick, T. O., Weisz, D. R., & Boylan-Kolchin, M. 2018, *MNRAS*, 477, 480

Uncovering Early Galaxy Evolution in the ALMA and JWST Era
Proceedings IAU Symposium No. 352, 2019
E. da Cunha, J. Hodge, J. Afonso, L. Pentericci & D. Sobral, eds.

doi:10.1017/S1743921319006963

Dynamical properties of Molecular Cloud Complexes at the Epoch of Reionization

T. K. Daisy Leung[1,2], **Andrea Pallottini**[3,4], **Andrea Ferrara**[4,5] **and Mordecai-Mark Mac Low**[2,6]

[1]Department of Astronomy, Cornell University, NY, USA
email: tleung@astro.cornell.edu

[2]Center for Computational Astrophysics, Flatiron Institute, NY, USA

[3]Centro Fermi, Rome, Italy

[4]Scuola Normale Superiore, Pisa, Italy

[5]Kavli Institute for the Physics and Mathematics of the Universe (IPMU), University of Tokyo, Japan

[6]American Museum of Natural History, NY, USA

Abstract. The Atacama Large (Sub-)millimeter Array (ALMA) has provided glimpse of the interstellar medium (ISM) properties of galaxies at the Epoch of Reionization (EoR); however, detailed understanding of their internal structure is still lacking. We present properties of molecular cloud complexes (MCCs) in a prototypical galaxy at this epoch studied in cosmological zoom-in simulations (Leung *et al.* 2019c). Typical MCC mass and size are comparable to nearby spirals and starburst galaxies ($M_{\rm gas} \sim 10^{6.5}\, M_{\odot}$ and $R \simeq 45$–$100\,{\rm pc}$). MCCs are highly supersonic, with velocity dispersion of $\sigma_{\rm gas} \simeq 20 - 100\,{\rm km\,s^{-1}}$ and pressure of $P/k_B \simeq 10^{7.6}\ K\ cm^{-3}$, which are comparable to gas-rich starburst galaxies. In addition, we perform stability analysis to understand the origin and dynamical properties of MCCs. We find that MCCs are globally stable in the main disk of Althæa. Densest regions where star formation is expected to take place in clumps and cores on even smaller scales instead have lower virial parameter and Toomre-Q values. Detailed studies of the star-forming gas dynamics at the EoR thus require a spatial resolution of < 40 pc ($\simeq 0.01''$), which is within reach of ALMA, to complement studies of stellar populations at EoR using the James Webb Space Telescope (*JWST*).

Keywords. galaxies: high-redshift – galaxies: ISM, galaxies: evolution, ISM: structure, ISM: kinematics and dynamics ISM: clouds

---

1. Introduction

Early galaxies have higher molecular gas fractions, star formation rate, and smaller sizes than present-day galaxies (e.g., Bouwens *et al.* 2011; Decarli *et al.* 2016, 2017; Leung *et al.* 2019c). As such, they are expected to be significantly more ionized, with more intense and harder interstellar radiation fields. Their metallicity and dust content are also expected to be lower, which in turn affect the thermal and chemical state of the multi-phase ISM. Here, we pose the question: *what are the physical properties of molecular cloud structures in early galaxies, and how do they differ from those found in local galaxy populations?*

2. Cosmological Zoom-in Simulations: SERRA

The simulations used are briefly summarized here (see Pallottini *et al.* 2017a,b for details). SERRA is a suite of cosmological zoom-in simulations performed using Eulerian hydrodynamics and adaptive mesh refinement (AMR) techniques, covering a comoving box of 20 Mpc h^{-1} in size and zooms in on a target halo of mass $M_{\rm DM} \simeq 10^{11} M_\odot$. The Lagrangian region of the halo (2.1 Mpc h^{-1}) has a dark matter mass resolution of $\simeq 6 \times 10^4 M_\odot$, and is spatially refined with the finest cell size of $l_{\rm cell} \simeq 30$ pc (at $z = 6$), i.e., comparable to the size of local giant molecular clouds. Our model includes a non-equilibrium chemical network (Grassi *et al.* 2014; Bovino *et al.* 2016), where abundances are calculated using an on-the-fly non-equilibrium formation of molecular hydrogen scheme (Pallottini *et al.* 2017b). The main zoom galaxy (Althæa) is a Lyman-break galaxy at $z \simeq 6$, with a stellar mass of $M_\star \simeq 3 \times 10^{10} M_\odot$, a metallicity of $Z \simeq 0.5 Z_\odot$, a molecular gas mass of $M_{\rm H2} \simeq 5 \times 10^7 M_\odot$, and a SFR of $30 - 80 M_\odot$ yr^{-1}.

3. Physical Properties and Stability of $z \sim 6$ Molecular Gas

The typical size and mass of MCCs are $R \simeq 50$ pc and $M_{\rm gas} \simeq 10^{6.5} M_\odot$, comparable to massive molecular structures observed in nearby star-forming and starburst galaxies (e.g., Leroy *et al.* 2015). MCCs are highly supersonic with an average Mach number of $\bar{\mathcal{M}} \simeq 6$. Their velocity dispersion and gas surface density are systematically higher than Milky Way clouds, but comparable to $z \sim 2$ starburst galaxies (e.g., Swinbank *et al.* 2011). High pressure ($\bar{P} \simeq 10^{7.6}$ K cm^{-3}) MCCs are found throughout the disk of Althæa and result from extra-planar flows and high velocity accretion/SN-driven outflows.

We also perform virial analysis, as motivated by observations, to assess the stability of MCCs. Virial parameter is lowest for MCCs in the densest regions, some of which are located in regions with Toomre $Q_{\rm eff} \lesssim 1$. These MCCs are unstable against collapse, where star formation is expected to take place within their gas *clumps* and *cores* on scales $\lesssim 40$ pc as energy quickly dissipates.

Contribution from the stellar component plays an important role in governing the stability of the MCCs against axisymmetric perturbations ($Q_{\rm eff}$), especially in the central part of Althæa. Similarly, stabilizing effect due to the thickness of its disk is also non-negligible. This illustrates the importance of accounting for both effects when examining the stability of molecular gas structures in relatively evolved and enriched systems at high redshift that are preferentially being observed now.

4. Summary and Outlook

We study the origin and dynamical properties of MCCs in prototypical galaxies at the EoR in numerical simulations to provide a framework within which upcoming observations can be compared against to aid in the interpretation. Details of our findings are reported in Leung *et al.* (2019c). Concerning the topic of this symposium, our results imply that spatially resolution better than $\simeq$40 pc are needed to examine the truly star-forming structures, and thus, star formation in the first galaxies. Such resolution is within reach of ALMA and will complement studies of stellar population in the first galaxies using *JWST*.

References

Bouwens, R. J., Illingworth, G. D., Labbe, I., *et al.* 2011, *Nature*, 469, 504
Bovino, S., Grassi, T., Capelo, P. R., Schleicher, D. R. G., & Banerjee, R. 2016, *A&A*, 590, A15
Decarli, R., Walter, F., Aravena, M., *et al.* 2016, *ApJ*, 833, 69
Decarli, R., Walter, F., Venemans, B. P., *et al.* 2017, *Nature*, 545, 457
Grassi, T., Bovino, S., Schleicher, D. R. G., *et al.* 2014, *MNRAS*, 439, 2386
Leroy, A. K., Bolatto, A. D., Ostriker, E. C., *et al.* 2015, *ApJ*, 801, 25
Leung, T. K. D., Riechers, D. A., Baker, A. J., *et al.* 2019a, *ApJ*, 871, 85
Leung, T. K. D., Pallottini, A., Ferrara, A., & Mac Low, M.-M. 2019c, *ApJ*, Submitted
Pallottini, A., Ferrara, A., Bovino, S., *et al.* 2017a, *MNRAS*, 471, 4128
Pallottini, A., Ferrara, A., Gallerani, S., *et al.* 2017b, *MNRAS*, 465, 2540
Swinbank, A. M., Papadopoulos, P. P., Cox, P., *et al.* 2011, *ApJ*, 742, 11

SESSION 2: Theoretical models and simulations

Uncovering Early Galaxy Evolution in the ALMA and JWST Era
Proceedings IAU Symposium No. 352, 2019
E. da Cunha, J. Hodge, J. Afonso, L. Pentericci & D. Sobral, eds.

doi:10.1017/S1743921320001106

Early galaxy formation and its large-scale effects

Pratika Dayal

Kapteyn Astronomical Institute, The Netherlands

Abstract. Galaxy formation in the first billion years mark a time of great upheaval in the history of the Universe: the first galaxies started both the 'metal age' as well as the era of cosmic reionization. I will start by reviewing the dust production mechanisms and dust masses for high-redshift galaxies which will be revolutionized in the ALMA era. I will then show how the JWST will be an invaluable experiment to shed light on the impact of reionization feedback on early galaxy formation. As we look forward towards the era of 21cm cosmology, I will highlight the crucial and urgent synergies required between 21cm facilities (such as the SKA) and galaxy experiments (JWST, E-ELT and Subaru to name a few) to understand the physics of the epoch of reionization that remains a crucial frontier in the field of astrophysics and physical cosmology. Time permitting, I will try to give a flavour of how the assembly of early galaxies, accessible with the forthcoming JWST, can provide a powerful testbed for Dark Matter models beyond 'Cold Dark Matter'.

Uncovering Early Galaxy Evolution in the ALMA and JWST Era
Proceedings IAU Symposium No. 352, 2019
E. da Cunha, J. Hodge, J. Afonso, L. Pentericci & D. Sobral, eds.

doi:10.1017/S1743921319009633

Modeling dust in a universe of galaxies

Desika Narayanan[1,2,3], Qi Li[1], Romeel Davé[4,5,6], Charlie Conroy[7], Benjamin D. Johnson[7] and Gergo Popping[8]

[1]Department of Astronomy, University of Florida, 211 Bryant Space Sciences Center, Gainesville, FL, 32611, USA
email: desika.narayanan@ufl.edu

[2]University of Florida Informatics Institute, 432 Newell Drive, CISE Bldg E251, Gainesville, FL, 32611, USA

[3]Cosmic Dawn Centre, Niels Bohr Institute, University of Copenhagen and DTU-Space, Technical University of Denmark

[4]Institute for Astronomy, Royal Observatory, University of Edinburgh, Edinburgh, EH9 3HJ, UK

[5]University of the Western Cape, Bellville, Cape Town, 7535, South Africa

[6]South African Astronomical Observatories, Observatory, Cape Town, 7925, South Africa

[7]Department of Astronomy, Harvard University, 60 Garden Street, Cambridge, MA, 02138, USA

[8]European Southern Observatory, Karl-Schartzchild-Strasse 2, 85748, Garching, Germany

Abstract. In this invited talk, we discuss the physics of the lifecycle of dust in the context of galaxy formation simulations. After outlining the basic physical processes, we apply algorithms for the formation, growth, and destruction of dust in the ISM to a state-of-the-art cosmological simulation to develop a model for the evolution of the dust to gas and dust to metals ratios in galaxies. We show that while modern simulations are able to match the observed dust mass function at redshift $z = 0$, most models underpredict the observed mass function at high-redshift ($z = 2$). We then show the power of these techniques by expanding our model to include a spectrum of dust sizes, and make initial predictions for extinction laws in local galaxies.

Keywords. ISM: dust, extinction, galaxies: ISM, galaxies: formation

---

1. Introduction

Dust plays a critical role in the physics of the interstellar medium (ISM) of galaxies, impacting both the chemistry of the ISM via serving as catalysts for molecular hydrogen (H_2) formation, as well serving as a sink/source of metals as grains are formed/destroyed (Draine (2011)). Similarly, dust impacts the thermal structure of the ISM by acting as an ISM coolant, as well as a shield from far ultraviolet (FUV) radiation (Goldsmith (2001); Draine (2003); Narayanan & Krumholz (2017)).

And yet, to date, most simulations of galaxy formation and evolution have not explicity included the physics of dust formation, growth, and destruction in the interstellar medium. Typically, dust is included in post-processing, via a simple constant dust mass to metal mass ratio, i.e. akin to $M_{\rm dust} = 0.4 \times M_{\rm metals}$ (e.g. Hayward *et al.* (2013); Narayanan *et al.* (2015, 2018); Abruzzo *et al.* (2018)), while some slightly more sophisticated parameterizations have included dust-to-metals or dust-to-gas ratios that are a function of the metallicity (e.g. Privon *et al.* (2018); Lagos *et al.* (2019)), that derive from observational scaling relations (Rémy-Ruyer *et al.* (2014); De Vis *et al.* (2019)).

The impact of including dust only in post-processing can be significant. First, by basing the dust content on linear scalings with metallicity that are derived from only a few galaxies (e.g. Dwek (1998); Watson (2011)), it is entirely possible that the dust content could be grossly miscalculated for galaxies that have drastically different physical properties (i.e. locations in metallicity-density-temperature field space) than the galaxies used to derive the observational scaling relations. Second, because dust serves as a sink for metals in the ISM (as well as a source in locations that the dust may be destroyed, e.g. in the warm circumgalactic medium [CGM]), post-processing the dust content may lead to miscalculations in the total metallicity of galaxies over cosmic time (indeed, this is shown explicitly in Davé *et al.* (2019)). Moreover, developing a predictive model for the evolution of the cosmic dust content, dust mass functions, or dust to gas ratios requires a self-consistent treatment of dust in galaxy evolution simulations (Li *et al.* (2019)).

In these proceedings, we describe a model for the inclusion of the formation, growth, and evolution of dust grains on-the-fly in cosmological hydrodynamic galaxy formation simulations. Much of the relevant work has been already discussed in Li *et al.* (2019), and these proceedings merely summarize these efforts, as well as ongoing efforts to develop a evermore sophisticated model.

2. Relevant Physics

In Figure 1, we summarize by schematic the relevant physical processes for dust formation, growth, and destruction. In what follows, we describe these processes in more detail. These algorithms follow work by McKinnon *et al.* (2016, 2018) and Popping *et al.* (2017), with some modificiations as described in Davé *et al.* (2019) and Li *et al.* (2019).

First, dust has to form. Dust is produced from the condensation of metals from both supernovae, as well as in AGB ejecta. In the formation equations, first developed by Dwek (1998), with updates by Popping *et al.* (2017), the mass of dust formed in each process is a fraction of the ejecta mass, modulated by a condensation efficiency, δ. This efficiency is a free parameter (with ranges constrained by theoretical ejecta models, e.g. Ferrarotti & Gail (2006); Bianchi & Schneider (2007)), and is dependent on the chemical composition of the ejecta (i.e. the C/O ratio).

After dust forms, it can either grow or erode. The growth processes are twofold: it can either accrete metals from surrounding gas, or collide with another dust particle (otherwise known as "coagulation") to grow in mass. The growth rate is of course dependent on the local metal mass, though is also inversely related the density, temperature and metallicity (all of which are subsumed into a characteristic accretion time scale for dust growth; see Li *et al.* (2019) for details). The zero-points of the density/temperature/metallicity are also free parameters. The coagulation rates are dependent on the masses of grain particles, their grain sizes, and their relative velocities.

Finally, dust can be destroyed via four processes in our model. First, electrons colliding with dust grains can erode away metals from the grain in a process known as "thermal sputtering". This is of course more efficient in ionized regions, where the time scales are controlled by the gas density and temperature. Second, dust can be destroyed in supernovae shocks. While SNe blast waves are rarely resolved in galaxy formation simulations, these can be treated in a subresolution manner, with destruction rates tied to the local type II SNe rate and local gas mass. Third, in a process known as "shattering", dust grains can smash together, and in a process analgous to coagulation, break apart. And finally in an entirely numerical effect, dust can be consumed by star formation. The latter process simply corresponds to algorithms where the dust is locked to gas particles (in particle-based codes), and the entire gas particle turns into a star particle in a star formation event.

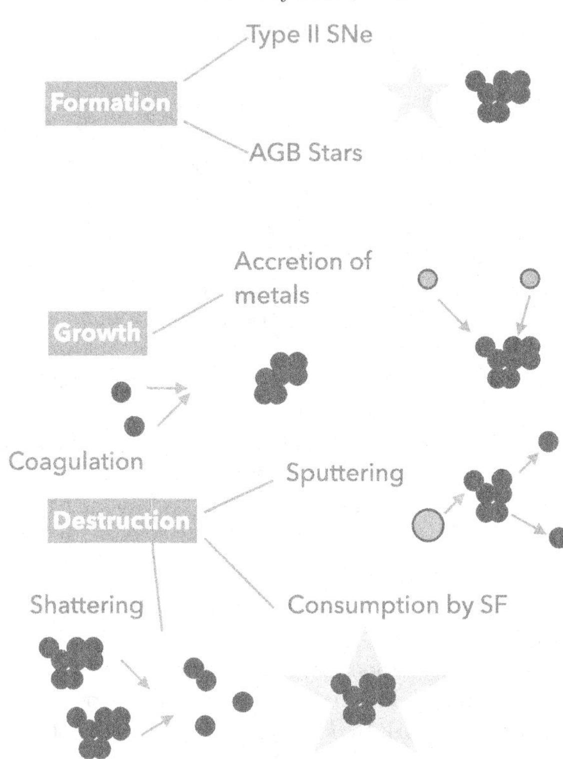

Figure 1. Schematic summarizing main physical processes of dust grains in cosmological hydrodynamic galaxy formation simulations. Dust grains form in supernovae and AGB stars (top section), and grow via the accretion of metals in the ISM, as well as coagulation (the process of sticking together). Grains are destroyed via thermal sputtering (when a hot electron collides with a grain), shattering (when grains collide with one another) or (due to numerical resolution) when they are consumed by a star-forming particle. Note, the latter process is entirely artificial, and simply stands as a proxy for other aforementioned processes in regions near forming stars.

3. Implementation of Self-Consistent Dust Physics into Cosmological Simulations

We have implemented the physics described in §2 and Figure 1 into the SIMBA cosmological galaxy formation model (Davé *et al.* (2019). In our initial implementation, we have put in a scaled-down version of the aforementioned physics for simplicity. Here, we have included just growth by accretion, and destruction by sputtering, supernovae, and star formation. In other words, in our initial implementation, we neglect coagulation and shattering. Why do we do this? The simplest model is one in which we attach the dust grains to the gas particles, and passively advect the dust alongside the gas. Here, we ideally want to minimize the number of grain sizes that we model since each bin in size adds additional memory requirements for every gas particle. So, in this simplest incarnation of the dust model, we do not model a spectrum of grain sizes, which prevents us from calculating the shattering and coagulation rates (that depend on the grain sizes). As we will discuss downstream, we are in the process of updating the algorithms to model a spectrum of grain sizes.

The primary SIMBA cosmological hydrodynamic simulation simulates a cube of $100/h$ Mpc side length with 1024^3 gas elements, and is run from $z = 249$ down to redshift $z = 0$. The main simulation details are described in Davé *et al.* (2019), and we summarize only the salient points here. The gas particle mass resolution is $1.82 \times 10^7 M_\odot$, while the

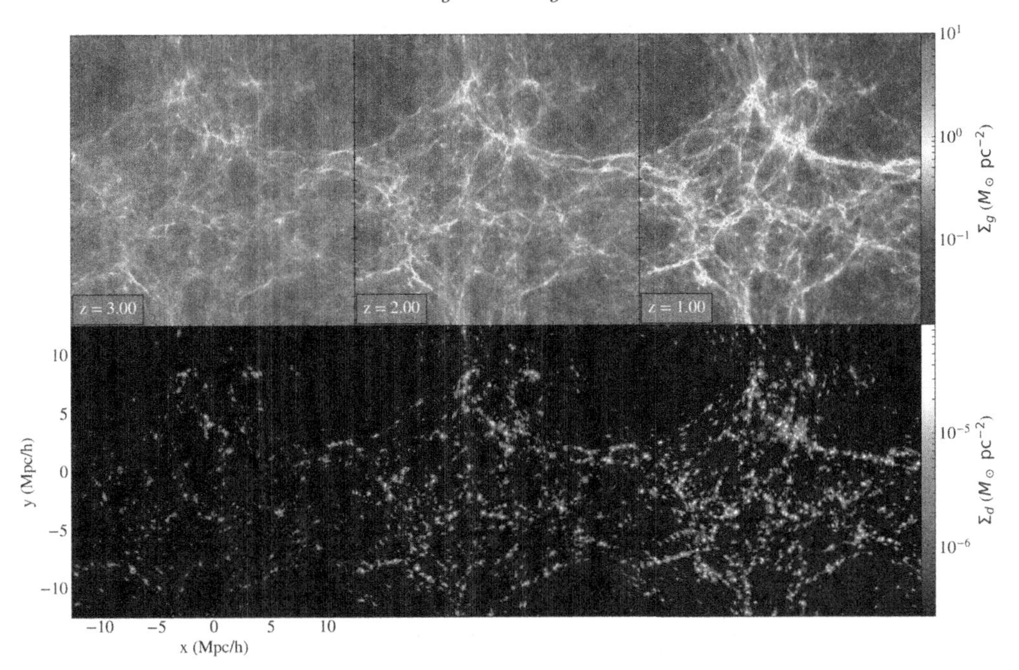

Figure 2. Gas and dust surface density projection maps of a pilot cosmological simulation at redshifts $z = 1, 2, 3$. Dust broadly traces the overdensities in which galaxies form, though at early times is relatively low abundance due to the lack of metals.

dark matter particle resolution is $9.6 \times 10^7 M_\odot$. The hydrodynamics are evolved using a forked version of the GIZMO hydrodynamics code (Hopkins (2014)) in meshless finite mass (MFM) mode.

Radiative cooling and heating from photoionization are calculated with the GRACKLE-3.1 libraries (Smith *et al.* (2017)), in which we include metal line cooling. The neutral gas is broken into an HI and H_2 component following the subresolution model of Krumholz *et al.* (2009), which balances H_2 formation rates against photodissocation by Lyman-Werner band photons. Stars form in this molecular gas with a rate SFR $= \epsilon_* \rho_{H2}/t_{\rm dyn}$, where $\epsilon_* = 0.02$. These stars feed back into the interstellar medium with kinetic energies, ejection probabilities, and mass loading factors designed to mimic the global galaxy wind scaling relations derived from the very high-resolution FIRE galaxy zoom simulation series (Hopkins *et al.* (2018)). Finally, SIMBA includes a model for black hole accretion and feedback. These included physics help shape the physical properties of the galaxies to produce a realistic population of galaxies.

4. Initial Results: The Dust to Gas and Dust to Metals Ratio in Galaxies

In Figure 2, we show three $25/h$Mpc projection plots in time at redshifts $z = 1, 2, 3$ showing the gas and dust surface density for a pilot simulation. These projections were made with the YT software (Turk *et al.* (2011)). Beyond serving as eye-candy (which, admittedly, is the primary concern of most galaxy formation theorists), Figure 2 shows how the large scale structure of dust in an evolving cosmological simulation grows with time.

As a first test of our model, we compare our simulated dust mass functions against observational constraints by Dunne *et al.* (2003); Eales *et al.* (2009); Dunne *et al.* (2011); Clemens *et al.* (2013) and Beeston *et al.* (2018). In Figure 3, we show the simulated and

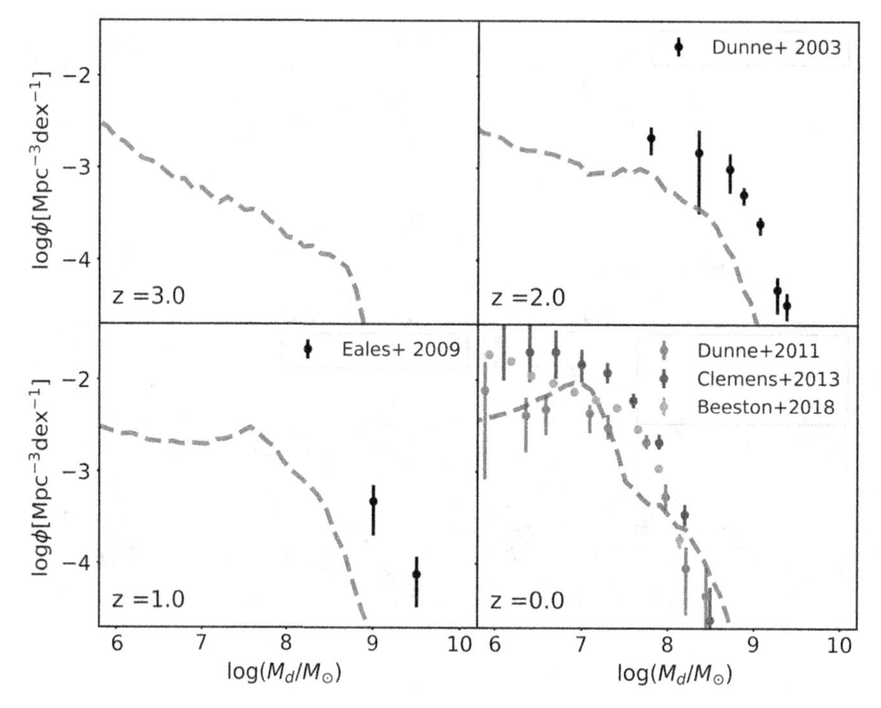

Figure 3. Model dust mass functions from the SIMBA $(100/h^3)$ Mpc volume compared to observations by Dunne *et al.* (2003); Eales *et al.* (2009); Dunne *et al.* (2011); Clemens *et al.* (2013) and Beeston *et al.* (2018). Our model does a reasonable job at matching the locus of observed points at low redshift, though slightly underpredicts the observed dust masses at redshifts $z = 1, 2$. This figure first published by Li *et al.* (2019).

observed dust mass functions at redshifts $z = 0 - 3$, as reproduced from Li *et al.* (2019). The models do a reasonable job matching the observed dust mass function at redshift $z = 0$, though slightly under-predict the observed constraints at redshifts $z = 1 - 2$. The underproduction of dust at high-redshift somewhat generic one amongst galaxy formation models that include dust (e.g. McKinnon *et al.* (2016); Popping *et al.* (2017). This may point to a number of issues, including incomplete physics in our model description of dust formation, growth and destruction, evolving parameters that we otherwise hold fixed (such as the condensation efficiencies), or issues with our understanding of stellar yields at high-redshift. Of course, as the last refuge of the scoundrel theorist, we also blame the observations: we note that the observational data points at high-redshift are scant few, and potential future observations may be in agreement with the growing number of simulations that underpredict the dust number counts at high-redshift.

We now turn from verification to prediction. What we would like to know is: "can we predict the quantity of dust a galaxy has, given a set of observed physical properties?". As a first point of investigation, we plot the dust to gas mass ratio (DGR) as a function of the galaxy gas-phase metallicity (normalized to solar) in Figure 4. Here, we color-code our simulated data based on the galaxy specific star formation rate (sSFR$\equiv$ SFR$/M_*$), and when available, show the observational constraints.

Generally, the dust to gas mass ratio decreases with metallicity, and follows a nearly constant relationship at all redshifts. The primary discriminant between high and low redshift snapshots is (a) the number of galaxies, and (b) the appearance of a red clump at late times, representing quenched galaxies with low sSFR. By and large, the models match the few observational constraints that exist at $z \approx 0$ (Rémy-Ruyer *et al.* (2014);

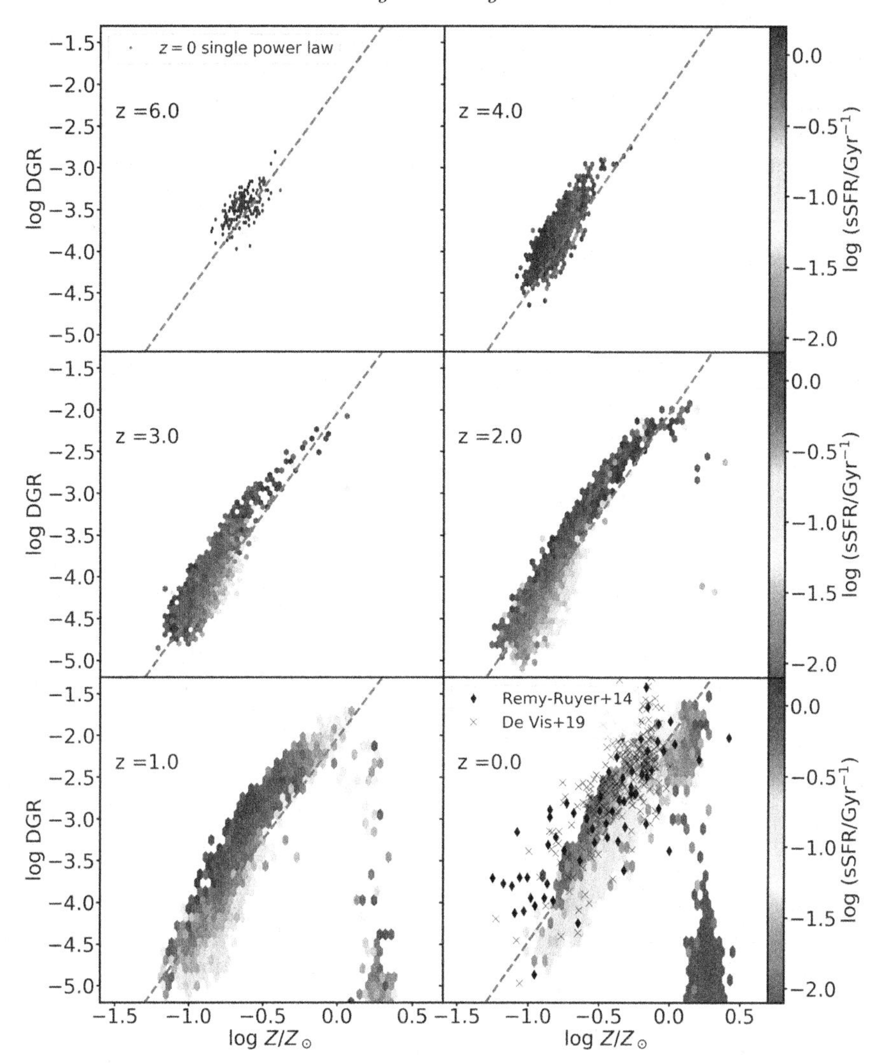

Figure 4. Dust to gas mass ratio of our model galaxies vs their gas phase metallicity. Colors show the sSFR of the galaxies, while black points show observational constraints when available (from Rémy-Ruyer *et al.* (2014); De Vis *et al.* (2019)). The dust to gas ratio for main sequence star-forming galaxies follows roughly a power-law relation with the metallicity, though as we discuss in the main text, there are secondary parameters that are non-negligible. Figure first published by Li *et al.* (2019).

De Vis *et al.* (2019)). This said, as we will discuss shortly, while a single powerlaw reasonably represents the relationship between the dust to gas ratio and metallicity, there are important secondary parameters that, when considered, can increase the accuracy of derived dust masses. We predict similar relationships for the dust to metals ratio, and show these in Li *et al.* (2019).

Figure 4 suggests that if one knows the gas-phase metallicity of a galaxy, within some dispersion, they can back out the galaxy dust to gas ratio. This said, the metallicity alone is not the best predictor for the dust to gas ratio. To demonstrate this, we employ a machine learning approach known as Extremely Randomized Trees (ERT) to develop a map between the dust to gas ratio and galaxy physical parameters. We search for a mapping between DGR (and the dust to metals ratio) and the metallicity, gas depletion time scaleback ($t_{\mathrm{dep}} = M_g/\mathrm{SFR}$), stellar mass, half-baryonic mass radius, gas mass fraction ($f_g = M_g/(M_g + M_*)$), and gas surface density (Σ_g). While the details are published in Li *et al.* (2019), we find that alongside the metallicity (which, in a normalized sense, has an importance level of ~ 0.5 in predicting the dust to gas ratio), the gas fraction and depletion time scale combine to a nearly comparable importance. At https://bitbucket.org/lq3552/dust_galaxy_analyzer, we provide a public code base to map from physical properties of galaxies to the dust to gas and dust to metals ratios. These python codes can be used by observers and theorists to derive the dust content of their observed/modeled galaxies.

5. Attenuation Laws and Extinction Laws

A natural application of modeling dust in galaxies is to understand the role of attenuation and extinction in galaxies. The two concepts are different. Extinction refers to the wavelength-dependent removal of photons from the line of sight, whereas attenuation folds in the complexities of the large scale geometry of a galaxy, and therefore accounts for scattering of light back into the line of sight, as well as the contribution by unobscured stars to the UV/optical flux. The extinction curve is determined by the combination of the dust grain size distribution with the optical properties of the grains themselves (including the fractional contribution to the grain population by silicates and graphites). Attenuation curves build on this by convolving the grain properties with the complexities of galaxy geometry. For unresolved or large scale observations of galaxies, the attenuation curve is the relevant concept in understanding how much UV/optical energy was removed.

Even for a constant set of grain properties, the attenuation curve in galaxies can vary dramatically. As an example, Figure 5, we take a $(25/h)^3$ Mpc cosmological simulation run with the MUFASA physics suite†, and assume an *fixed extinction curve* for the grains in the galaxy‡. We infer the attenuation curves by running the dust radiative transfer package POWDERDAY (Narayanan *et al.* (2015), which is publicly available (https://powderday.readthedocs.io/en/latest/). We show these for snapshots at $z = 0, 2, 4, 6$. The green squares are a hexbin denoting the number of galaxies with a given curve, while the pink show the median and 1σ standard deviation.

As is evident, even with a constant extinction curve, there is dramatic variation in the inferred attenuation curve of a galaxy due entirely to the complexities of the star-dust geometry. As we discuss in Narayanan *et al.* (2018), variations in the attenuation curve arise due to unobscured stars. For normalized curves, unobscured OB stars flatten (grey) attenuation curves, whereas unobscured old stars steepen normalized curves. The 2175Å bump in attenuation can be filled in due to an excess of UV flux from unobscured OB stars.

In principle, one could use this sort of study to predict the attenuation curve an observer would ideally assume for their observations. As reviewed by Walcher *et al.* (2011) and Conroy (2013), the dust attenuation curve is one of the more critical aspects to deriving correct physical parameters in fitting the SEDs of galaxies. One could imagine combining the attenuation curves derived from dust radiative transfer modeling with the synthetic SEDs and a machine learning algorithm to inform observers, based solely on their UV/OIR photometry, what the underlying attenuation curve of their observation is. Unfortunately, while the study of Narayanan *et al.* (2018) unpacked the origin of

† MUFASA is the parent simulation suite of SIMBA Davé *et al.* (2016).

‡ These simulations were run prior to our development of the dust physics suite in the SIMBA simulations, and as such use the now antiquated method of including dust via a constant dust to metals mass ratio across all galaxies.

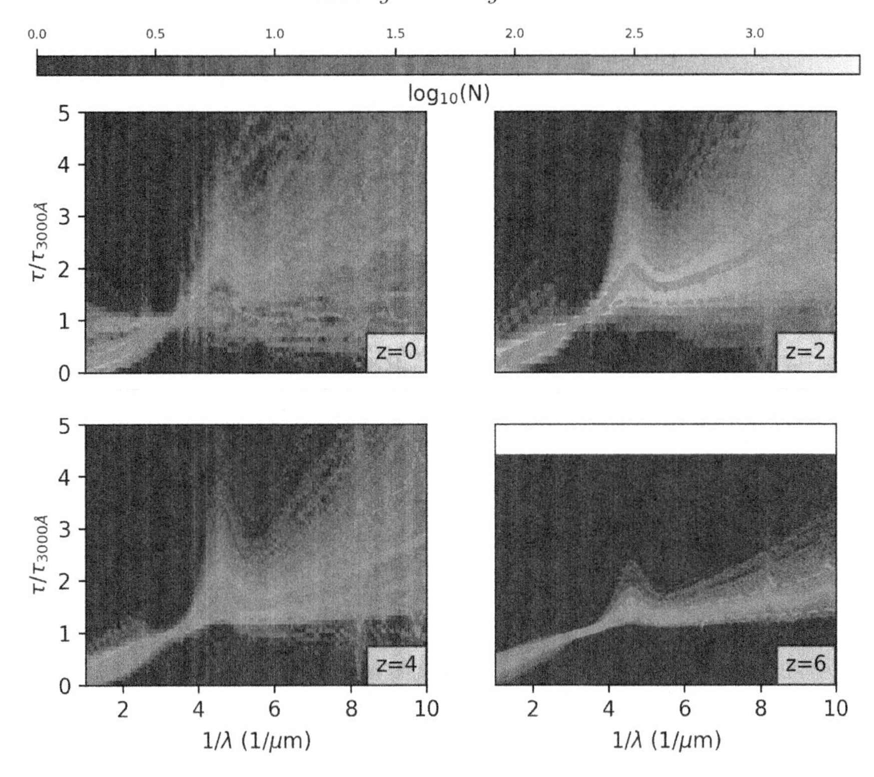

Figure 5. Attenuation curves from a $(25/h)^3$ Mpc MUFASA volume with a fixed underlying extinction curve (taken originally from Narayanan *et al.* (2018)). Shown are panels for redshifts $z = 0, 2, 4, 6$.

variations in an attenuation curve, it is not yet predictive for what sort of attenuation curve an observer can expect for a particular galaxy. This is because Narayanan *et al.* (2018) employed a *fixed* extinction curve. The next step in deriving realistic attenuation curves for galaxies is to develop a physical model for the underlying extinction curves of galaxies. This is where dust modeling comes in.

We have recently adapted our model to include a size distribution of dust grains. To do this, we have included a new particle type into the GIZMO simulation code that represents dust, and is therefore decoupled in its dynamics from the gas (Q. Li, Narayanan & Davé *et al.* in prep.). We do this following the algorithms outlined by McKinnon *et al.* (2018), though for the first time include these in bona fide cosmological simulations. With knowledge of the grain sizes, we include the physics of coagulation for grain growth, and shattering for destruction. The grains are treated as spherical objects with mass $m\,(a) = 4/3\pi\rho_{\rm grain}a^3$, where $\rho_{\rm grain}$ is the dust grain density, and a is the size. We assume a fixed grain density of $\rho_{\rm grain} = 2.4$ g cm^{-3} as reviewed by Draine (2003), and therefore infer the size, a from the grain mass. The box sizes for the active dust are necessarily smaller than the original SIMBA volumes due to the need for extremely high resolution to faithfully model coagulation and shattering processes. The initialized size distributions are assumed to follow the theoretical calculations of Asano *et al.* (2013) for the AGB ejecta-produced grains, and Nozawa *et al.* (2007) for the SNe produced grains. In Figure 6, we show the modeled extinction curves at $z = 0$ of ~ 50 galaxies of a pilot $(6/h)^3$ Mpc box with full SIMBA physics, and the aforementioned updated active dust model. We overplot the curve expected for a Mathis *et al.* (1977) distribution for the Galaxy in red. As is evident, a diverse range of extinction curves exist for local galaxies in this small

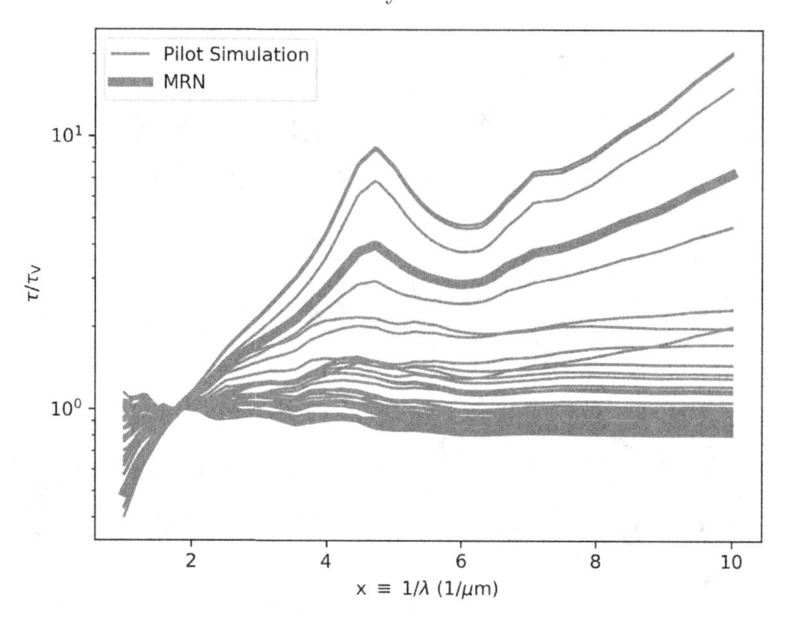

Figure 6. Model extinction curves for a pilot small box ($(6/h)^3$ Mpc) SIMBA volume, with a model for active dust grains with a spectrum of sizes. The model output is the final dust size distribution for each galaxy at $z = 0$ which, when convolved with an assumed set of optical properties for the galaxy, results in a model extinction curve.

box, though at least a handful have properties similar to the Milky Way's. We note that these results are a pilot study, and meant to be taken more as a proof of concept, rather than conclusive.

6. Relationship to other Models

A number of groups world-wide have published or are developing complementary models for including the physics of dust formation/growth/destruction in galaxy formation models, highlighting the importance of this physics in both understanding galaxy evolution, but also comparing models to observations across the electromagnetic spectrum.

McKinnon *et al.* (2016); Hou *et al.* (2017); Popping *et al.* (2017); Aoyama *et al.* (2018); Hou *et al.* (2019); Vijayan *et al.* (2019) and Vogelsberger *et al.* (2019) have all included similar dust physics in cosmological hydrodynamic simulations, or semi-analytic models. As previously mentioned, no model is able to match the $z = 2$ dust mass function to date, though our presented model appears to be the closest match, underpredicting by a factor ~ 3. These works model a single grain size, and therefore generally focus on global scaling relations, as in our work (Li *et al.* (2019).

More recently, a number of groups have turned their attention to simulating a spectrum of grain sizes. Hou *et al.* (2017, 2019) employ a 2-size approximation, binning grains as "small" or "large", with $a = 0.03\mu$m as the dividing line (Hirashita (2015)). While an approximation to a more finely binned spectrum of sizes, Hirashita (2015) have shown that in some test cases the two-size approximation can reproduce the results of more finely binned simulations. McKinnon *et al.* (2018) has developed a model for a full spectrum of grain sizes (that we adopt here in generating Figure 6) that has been applied to idealized disk galaxies with the AREPO hydrodynamic code.

7. Looking Forward

The relatively new field of including the physics of dust formation, growth, and destruction on the fly in galaxy formation simulations is a promising one. There are several important steps forward to take with these simulations:

• Galaxy formation models should begin to include the physics of dust cooling when modeling the low temperature cooling in simulations.

• A new generation of ultra high-resolution cosmological simulations, or zoom simulations will elucidate our understanding of the origin of variations in extinction laws, and the origin of the 2175Å bump in galaxies.

• In the long run, combining the results of these sorts of simulations with SED fitting software will enable a modern approach to accurately deriving the physical properties of galaxies from observed photometry.

• Not yet mentioned in this proceedings, though prevalent in both the community as well as this wonderful conference is the topic of the origin of high-redshift dusty galaxies. Simulations that include self-consistent models for dust physics will begin to address, for the first time, the origin of the highest-redshift dusty galaxies (e.g. Casey *et al.* (2014); Marrone *et al.* (2018)). The initial comparisons to redshift $z = 2$ dust mass functions of modern simulations may indicate a different formulation of the age-old tension between submillimeter-selected galaxies at high-redshift, and observations (e.g. Narayanan *et al.* (2010)).

Acknowledgements

The authors thank the organizers of this meeting for a lively and educational conference in a phenomenal location, Viana do Castelo, Portugal. This research made use of the YT code base, and the authors are grateful to the community developers of this software. D. N. would like to extend gratitude to Cristina Fernandes and Gabriel Menezes for taking him surfing on the first day of the meeting, and appreciates the organizers patience while he skipped the panel he was otherwise slated to participate in, to go surf.

References

Abruzzo, M. W., Narayanan, D., Davé, R., *et al.* 2018, arXiv e-prints, arXiv:1803.02374
Asano, R. S., Takeuchi, T. T., Hirashita, H., *et al.* 2013, *MNRAS*, 432, 637
Aoyama, S., Hou, K.-C., Hirashita, H., *et al.* 2018, *MNRAS*, 478, 4905
Beeston, R. A., Wright, A. H., Maddox, S., *et al.* 2018, *MNRAS*, 479, 1077
Bianchi, S. & Schneider, R. 2007, *MNRAS*, 378, 973
Casey, C. M., Narayanan, D., & Cooray, A. 2014, *PhR*, 541, 45
Clemens, M. S., Negrello, M., De Zotti, G., *et al.* 2013, *MNRAS*, 433, 695
Conroy, C. 2013, *ARA&A*, 51, 393
Davé, R., Thompson, R., & Hopkins, P. F. 2016, *MNRAS*, 462, 3265
Davé, R., Anglés-Alcázar, D., Narayanan, D., *et al.* 2019, *MNRAS*, 486, 2827
De Vis, P., Jones, A., Viaene, S., *et al.* 2019, *A&A*, 623, A5
Draine, B. T. 2003, *ApJ*, 598, 1017
Draine, B. T. 2011, Physics of the Interstellar and Intergalactic Medium by Bruce T. Draine. Princeton University Press
Dunne, L., Eales, S. A., & Edmunds, M. G. 2003, *MNRAS*, 341, 589
Dunne, L., Gomez, H. L., da Cunha, E., *et al.* 2011, *MNRAS*, 417, 1510
Dwek, E. 1998, *ApJ*, 501, 643
Eales, S., Chapin, E. L., Devlin, M. J., *et al.* 2009, *ApJ*, 707, 1779
Ferrarotti, A. S. & Gail, H.-P. 2006, *A&A*, 447, 553
Goldsmith, P. F. 2001, *ApJ*, 557, 736
Hirashita, H. 2015, *MNRAS*, 447, 2937
Hou, K.-C., Hirashita, H., Nagamine, K., *et al.* 2017, *MNRAS*, 469, 870

Hou, K.-C., Aoyama, S., Hirashita, H., *et al.* 2019, *MNRAS*, 485, 1727
Hopkins, P. F., Wetzel, A., Kereš, D., *et al.* 2018, *MNRAS*, 480, 800
Krumholz, M. R., McKee, C. F., & Tumlinson, J. 2009, *ApJ*, 699, 850
Hayward, C. C., Narayanan, D., Kereš, D., *et al.* 2013, *MNRAS*, 428, 2529
Hopkins, P. F. 2014, GIZMO: Multi-method magneto-hydrodynamics+gravity code, ascl:1410.003
Lagos, C. del P., Robotham, A. S. G., Trayford, J. W., *et al.* 2019, arXiv e-prints, arXiv:1908.03423
Li, Q., Narayanan, D., & Davé, R. 2019, arXiv e-prints, arXiv:1906.09277
Marrone, D. P., Spilker, J. S., Hayward, C. C., *et al.* 2018, *Nature*, 553, 51
Mathis, J. S., Rumpl, W., & Nordsieck, K. H. 1977, *ApJ*, 217, 425
McKinnon, R., Torrey, P., & Vogelsberger, M. 2016, *MNRAS*, 457, 3775
McKinnon, R., Vogelsberger, M., Torrey, P., *et al.* 2018, *MNRAS*, 478, 2851
Narayanan, D., Hayward, C. C., Cox, T. J., *et al.* 2010, *MNRAS*, 401, 1613
Narayanan, D., Turk, M., Feldmann, R., *et al.* 2015, *Nature*, 525, 496
Narayanan, D. & Krumholz, M. R. 2017, *MNRAS*, 467, 50
Narayanan, D., Conroy, C., Davé, R., *et al.* 2018, *ApJ*, 869, 70
Nozawa, T., Kozasa, T., Habe, A., *et al.* 2007, *ApJ*, 666, 955
Popping, G., Somerville, R. S., & Galametz, M. 2017, *MNRAS*, 471, 3152
Privon, G. C., Narayanan, D., & Davé, R. 2018, *ApJ*, 867, 102
Rémy-Ruyer, A., Madden, S. C., Galliano, F., *et al.* 2014, *A&A*, 563, A31
Smith, B. D., Bryan, G. L., Glover, S. C. O., *et al.* 2017, *MNRAS*, 466, 2217
Turk, M. J., Smith, B. D., Oishi, J. S., *et al.* 2011, *ApJS*, 192, 9
Vijayan, A. P., Clay, S. J., Thomas, P. A., *et al.* 2019, arXiv:1904.02196
Vogelsberger, M., McKinnon, R., O'Neil, S., *et al.* 2019, *MNRAS*, 487, 4870
Watson, D. 2011, *A&A*, 533, A16
Walcher, J., Groves, B., Budavári, T., *et al.* 2011, *Ap&SS*, 331, 1

Uncovering Early Galaxy Evolution in the ALMA and JWST Era
Proceedings IAU Symposium No. 352, 2019
E. da Cunha, J. Hodge, J. Afonso, L. Pentericci & D. Sobral, eds.

doi:10.1017/S1743921319009554

Galaxy evolution and radiative properties in the early universe: Multi-wavelength analysis in cosmological simulations

Shohei Arata[1], Hidenobu Yajima[2], Kentaro Nagamine[1,3,4], Yuexing Li[5,6] and Sadegh Khochfar[7]

[1]Theoretical Astrophysics, Department of Earth and Space Science, Graduate School of Science, Osaka University, Toyonaka, Osaka 560-0043, Japan
email: arata@astro-osaka.jp

[2]Center of Computational Sciences University of Tsukuba, Ibaraki 305-8577, Japan

[3]Department of Physics & Astronomy, University of Nevada, Las Vegas, 4505 S. Maryland Pkwy, Las Vegas, NV 89154-4002, USA

[4]Kavli IPMU (WPI), The University of Tokyo, 5-1-5 Kashiwanoha, Kashiwa, Chiba, 277-8583, Japan

[5]Department of Astronomy & Astrophysics, The Pennsylvania State University, 525 Davey Lab, University Park, PA 16802, USA

[6]Institute for Gravitation and the Cosmos, The Pennsylvania State University, University Park, PA 16802, USA

[7]SUPA, Institute for Astronomy, University of Edinburgh, Royal Observatory, Edinburgh, EH9 3HJ, UK

Abstract. Recent observations have successfully detected UV or infrared flux from galaxies at the epoch of reionization. However, the origin of their radiative properties has not been fully understood yet. Combining cosmological hydrodynamic simulations and radiative transfer calculations, we present theoretical predictions of multi-wavelength radiative properties of the first galaxies at $z = 6 - 15$. We find that most of the gas and dust are ejected from star-forming regions due to supernova (SN) feedback, which allows UV photons to escape. We show that the peak of SED rapidly shifts between UV and infrared wavelengths on a timescale of 100 Myr due to intermittent star formation and feedback. When dusty gas covers the star-forming regions, the galaxies become bright in the observed-frame sub-millimeter wavelengths. In addition, we find that the escape fraction of ionizing photons also changes between $1 - 40\,\%$ at $z > 10$. The mass fraction of H II region changes with star formation history, resulting in fluctuations of metal lines and Lyman-α line luminosities. In the starbursting phase of galaxies with a halo mass $\sim 10^{11}\,\mathrm{M}_\odot$ ($10^{12}\,\mathrm{M}_\odot$), the simulated galaxy has $L_{[\mathrm{O\,III}]} \sim 10^{42}$ (10^{43}) erg s^{-1}, which is consistent with the observed star-forming galaxies at $z > 7$. Our simulations suggest that deep [C II] observation with ALMA can trace the distribution of neutral gas extending over ~ 20 physical kpc. We also find that the luminosity ratio $L_{[\mathrm{O\,III}]}/L_{[\mathrm{C\,II}]}$ decreases with bolometric luminosity due to metal enrichment. Our simulations show that the combination of multi-wavelength observations by ALMA and JWST will be able to reveal the multi-phase ISM structure and the transition from starbursting to outflowing phases of high-z galaxies.

Keywords. hydrodynamics, radiative transfer, galaxies: evolution, galaxies: high-redshift

1. Introduction

Recently, galaxies in the reionization epoch ($z \gtrsim 6$) have been observed in the sub-mm wavelength; for example, dust continuum (Riechers *et al.* 2013; Watson *et al.* 2015;

Table 1. Parameters of our zoom-in cosmological hydrodynamic simulations: (1) $M_{\rm h}$ is the halo mass at $z = 6$. (2) $m_{\rm DM}$ is the mass of a dark matter particle in the corresponding simulation run. (3) $m_{\rm gas}$ is the initial mass of a gas particle. (4) $\epsilon_{\rm min}$ is the gravitational softening length in comoving units.

Halo ID	$M_{\rm h}$ [$h^{-1}{\rm M}_\odot$]	m_{DM} [h^{-1} ${\rm M}_\odot$]	$m_{\rm gas}$ [h^{-1} ${\rm M}_\odot$]	$\epsilon_{\rm min}$ [h^{-1} pc]
Halo-11	1.6×10^{11}	6.6×10^{4}	1.2×10^{4}	200
Halo-12	7.5×10^{11}	1.1×10^{6}	1.8×10^{5}	200

Laporte *et al.* 2017; Marrone *et al.* 2018; Hashimoto *et al.* 2019), metal emission lines of [O III] 88 μm (Inoue *et al.* 2016; Carniani *et al.* 2017; Hashimoto *et al.* 2018, 2019; Tamura *et al.* 2019) and [C II] 158 μm Carniani *et al.* (2017); Marrone *et al.* (2018); Hashimoto *et al.* (2019). These galaxies were originally high-z candidates identified by the UV photometric observations (e.g. Bouwens *et al.* 2015). In particular, recent ALMA observations have allowed us to detect many galaxies at $z > 6$ and showed wide variety of radiative properties of the *first galaxies*. However, the origin of the variety has not been understood yet. Therefore, we here investigate the relation between galaxy evolution and radiative properties by combining cosmological simulations and radiative transfer calculations.

Recent state-of-the-art simulations showed that the star formation in high-z galaxies occurred intermittently due to SN feedback and gas accretion (e.g. Kimm & Cen 2014; Yajima *et al.* 2017). In this work, we focus on how the intermittent star formation affects the radiative properties. Dust is also important for the SED because it absorbs UV photons and re-emits IR radiation. The dust absorption efficiency changes with the size distribution, composition, and spatial distribution of dusty clouds. We show that changes of dust distribution driven by SNe induce rapid transitions of radiative properties.

2. Simulations

We perform cosmological smoothed particle hydrodynamics (SPH) simulations using the GADGET-3 code (modified version of Springel 2005) with the sub-grid models developed in the OWLS project (Schaye *et al.* 2010) and FiBY project (Johnson *et al.* 2013). First, we conduct dark-matter only N-body simulations with $(20\,h^{-1}\,{\rm cMpc})^3$ and $(100\,h^{-1}\,{\rm cMpc})^3$ boxes, and identify the most massive haloes at $z = 6$ with the friends-of-friends method. Hereafter we call the haloes 'Halo-11' and 'Halo-12', respectively. Next we calculate these galaxies with zoom-in initial conditions and hydrodynamics. The details of sub-grid models are described in Yajima *et al.* (2017). Table 1 shows the properties of Halo-11 and Halo-12 runs.

For multi-wavelength radiative transfer (RT), we use the ART^2 code which calculates the propagation of photon packets considering hydrogen ionization, UV continuum, dust emission and Lyman-α line (Li *et al.* 2008; Yajima *et al.* 2012). This code utilizes the adaptive refinement grid following the simulated gas distribution, which enables the RT computation with the minimum physical scale of $\sim 2.7\,h^{-1}$ pc for Halo-11 at $z \sim 6$. The dust mass in each cell is proportional to the gas metallicity. We focus on how the RT results depend on the dust distribution. More details are described in Arata *et al.* (2019).

We also analyze the metal line emissions of [O III] 88 μm and [C II] 158 μm using ionization structure of hydrogen estimated by the RT calculation. We first calculate O^{2+} and C^{+} abundances in each cell assuming ionization equilibrium under the stellar radiation field. Next we calculate the rate equations of energy levels, and obtain the metal line luminosities. More details of the model will be described in Arata *et al.* (in prep.).

Figure 1 presents the images of Halo-11 at $z = 6.0$. The rest-UV surface brightness traces the stellar distribution, and the rest-FIR traces dust distribution. The galaxy has a clumpy structure, and each clump changes its color with intermittent star formation

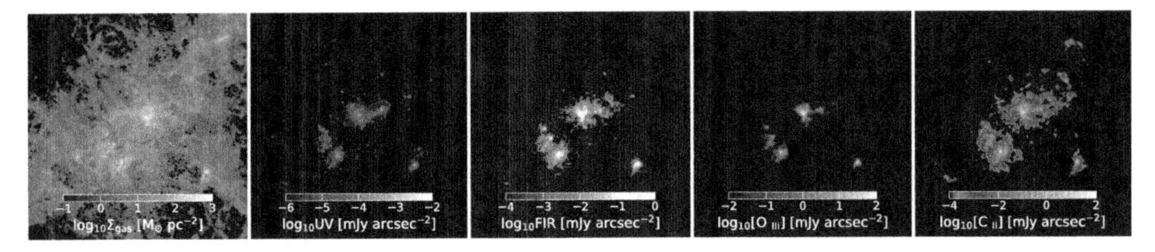

Figure 1. Maps of Halo-11 at $z = 6.0$. From left to right panels: gas surface density, surface brightness of UV, FIR, [O III] 88 μm, and [C II] 158 μm in the rest-frame, respectively. The field of view is ~ 51 physical kpc. The pixel size is ~ 0.07 arcsec.

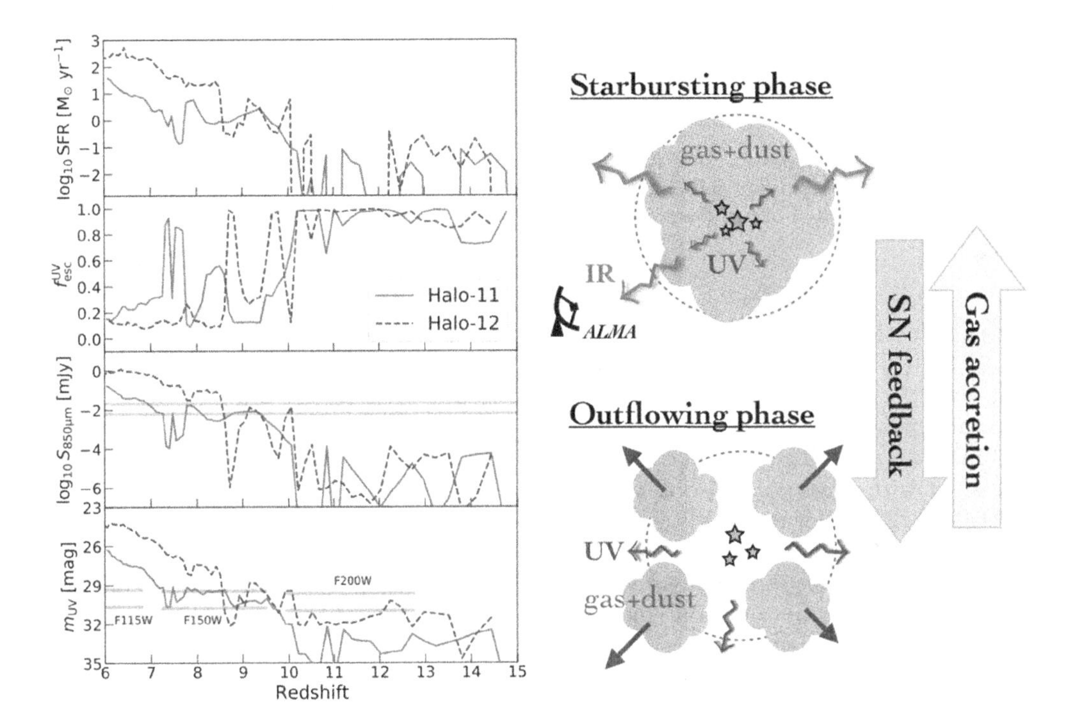

Figure 2. **Left**: Redshift evolution of SFR, escape fraction of UV photons, sub-mm flux and apparent UV magnitude in Halo-11 and Halo-12 runs. The gray horizontal lines in the third and forth panels show 3σ and 10σ detection thresholds for fully operated ALMA and JWST with 10-hour integration. **Right**: Schematic picture of relation between galaxy evolution and radiative properties. First galaxies rapidly make transitions between UV and IR bright phases due to intermittent star formation.

(see next section), which results in a large spatial offset of ~ 1.4 arcsec between the brightest pixels of UV and FIR wavelength. The [O III] and [C II] maps trace the gas distributions of ionized and neutral phases, respectively. Future deep [C II] observations ($\gtrsim 10^{-4}$ mJy arcsec^{-2}) may reveal an extended neutral gas over ~ 20 kpc (Fujimoto *et al.* 2019).

3. Rapid transition between UV and IR bright phases

Figure 2 shows the redshift evolution of SFR, escape fraction of UV photons, sub-mm flux and apparent UV magnitude for Halo-11 and Halo-12 at $z = 6 - 15$. As described in Yajima *et al.* (2017), star formation in the first galaxies occurs intermittently because of gas accretion and SN feedback processes. In the star-bursting phase, central dusty

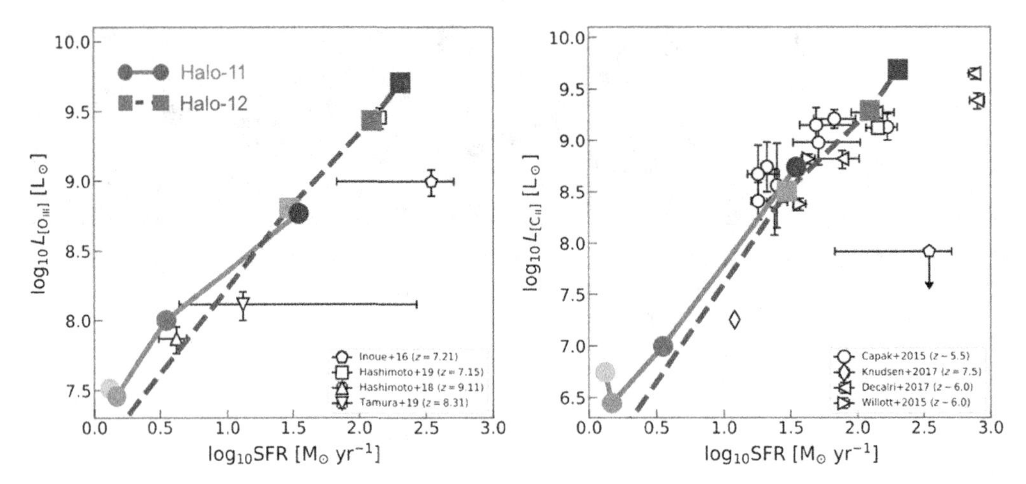

Figure 3. Relation between SFR and metal line luminosities of [O III] 88 μm (left) and [C II] 158 μm (right). Red solid line and circles represent evolution of Halo-11 at $z = 6, 7, 8, 9$ (darker to lighter). Blue dashed line and squares are for Halo-12. Open symbols represent observations.

gas efficiently absorbs UV photons, while most of the UV photons can escape from the halo in the outflowing phase, which results in the fluctuation of escape fraction ($f_{\rm esc} = 0.2 - 0.8$ at $z < 10$) as presented in the second panel (see also right picture). The time-scale of fluctuation is ~ 100 Myr, which corresponds to the free-fall time of the halo. The third panel shows that the sub-mm flux of Halo-11 (Halo-12) becomes $\sim 10^{-2}$ mJy (~ 1 mJy) at $z \sim 7$. The dust mass is $\sim 5 \times 10^6\,{\rm M}_\odot$ ($\sim 8 \times 10^7\,{\rm M}_\odot$). Our simulations are consistent with recent ALMA observations: Watson *et al.* (2015) observed dust emission of 0.61 ± 0.12 mJy for a galaxy at $z \approx 7.5$, and the estimated dust mass was $4 \times 10^7\,{\rm M}_\odot$, which are intermediate values of Halo-11 and Halo-12.

To investigate the sub-mm observability of high-z galaxies statistically, we also analyze the radiative properties of all satellite galaxies in Halo-11 and Halo-12 zoom-in regions. We find that the observability of $M_{\rm h} \gtrsim 10^{11}\,{\rm M}_\odot$ galaxies at $z < 7$ exceeds 50 % for a detection limit of 0.1 mJy. In addition, we predict that the number density of sub-mm sources observed by future deep survey ($> 10^{-2}$ mJy) will be $10^{-2}\,{\rm cMpc}^{-3}$, which agrees with present UV observations (see Arata *et al.* 2019 details).

4. Star-bursting [O III] emitters and quiescent [C II] emitters

The [O III] line is emitted only in star-bursting phase because O^{2+} ions exist in H II regions formed by massive stars. Thus intermittent star formation results in luminosity fluctuation in the range of $\sim 10^{39} - 10^{42}\,{\rm erg\ s^{-1}}$ at $z < 10$ for Halo-11. Meanwhile, the [C II] line is continuously emitted from neutral regions even in the outflowing phase.

Figure 3 presents the relation between SFR and metal line luminosities of Halo-11 and Halo-12. Our simulations well reproduce the observations of both [O III] and [C II] lines. Note that, however, the observational points may shift to higher SFR because actual dust temperature might be higher than the assumed one (Arata *et al.* 2019; Ma *et al.* 2019). We will examine the physical properties of very [C II]-faint galaxies (Inoue *et al.* 2016; Knudsen *et al.* 2017) in our future paper.

Finally, we focus on the luminosity ratio of $L_{\rm [O\,III]}/L_{\rm [C\,II]}$. Hashimoto *et al.* (2019) suggested a negative correlation between $L_{\rm [O\,III]}/L_{\rm [C\,II]}$ and the bolometric luminosity measured from UV and IR fluxes. Figure 4 shows the evolution of the ratio for Halo-11 (red circles) and Halo-12 (blue squares) at $z = 6 - 9$. We find that the ratio decreases

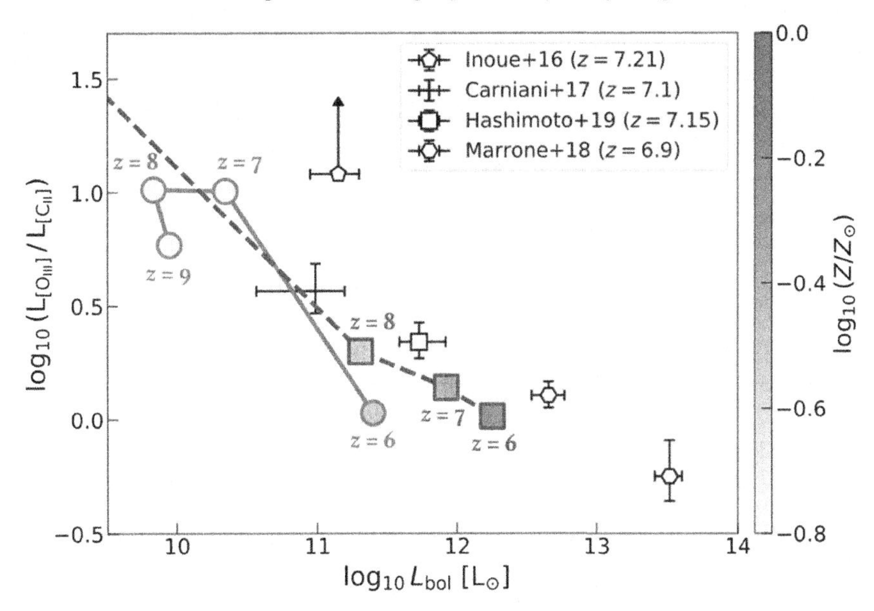

Figure 4. Relation between luminosity ratio $L_{[\mathrm{O\,III}]}/L_{[\mathrm{C\,II}]}$ and the bolometric luminosity ($L_{\mathrm{bol}} = L_{\mathrm{UV}} + L_{\mathrm{IR}}$). Red solid line and circles represent the evolution of Halo-11, and blue dashed line and squares are for Halo-12. The color indicates gas metallicity as shown in the color bar. Open symbols represent observed galaxies at $z \sim 7$.

about one order of magnitude as gas metallicity increases from sub-solar to solar metallicity. The log (O/C) abundance ratio is initially dominated by oxygen enrichment of Type-II SNe, and decreases from ~ 0.9 to ~ 0.5 due to carbon-rich winds from AGB stars. Thus, $L_{[\mathrm{O\,III}]}/L_{[\mathrm{C\,II}]}$ decreases with galaxy evolution and metal enrichment. Future deep [O III] and [C II] observations will reveal the connection between multi-phase ISM structure and the radiative properties for high-z galaxies.

References

Arata, S., Yajima, H., Nagamine, K., *et al.* 2019, *MNRAS*, 488, 2629

Bouwens, R. J., Illingworth, G. D., Oesch, P. A., *et al.* 2015, *ApJ*, 803, 34

Carniani, S., Maiolino, R., Pallottini, A., *et al.* 2017, *A&A*, 605, A42

Fujimoto, S., Ouchi, M., Ferrara, A., *et al.* 2019, arXiv e-prints, arXiv:1902.06760

Hashimoto, T., Laporte, N., Mawatari, K., *et al.* 2018, *Nature*, 557, 392

Hashimoto, T., Inoue, A., Mawatari, K., *et al.* 2019, *PASJ*, 71, 71

Inoue, A. K., Tamura, Y., Matsuo, H., *et al.* 2016, *Science*, 352, 1559

Johnson, J. L., Dalla Vecchia, C., & Khochfar, S. 2013, *MNRAS*, 428, 1857

Laporte N., Ellis R. S., Boone F., *et al.* 2017, *ApJ*, 837, L21

Li, Y., Hopkins, P. F., Hernquist, L., *et al.* 2008, *ApJ*, 678, 41

Ma, X., Hayward, C. C., Casey, C. M., *et al.* 2019, *MNRAS*, 487, 1844

Knudsen, K. K., Watson, D., Frayer, D., *et al.* 2017, *MNRAS*, 466, 138

Marrone, D. P., Spilker, J. S., Hayward, C. C., *et al.* 2018, *Nature*, 553, 51

Riechers, D. A., Bradford, C. M., Clements, D. L., *et al.* 2013, *Nature*, 496, 329

Schaye, J., Dalla Vecchia, C., Booth, C. M., *et al.* 2010, *MNRAS*, 402, 1536

Tamura, Y., Mawatari, K., Hashimoto, T., *et al.* 2019, *ApJ*, 874, 27

Watson, D., Christensen, L., Knudsen, K. K., *et al.* 2015, *Nature*, 519, 327

Yajima, H., Li, Y., Zhu, Q., & Abel, T., 2012, *MNRAS*, 424, 884

Yajima, H., Nagamine, K., Zhu, Q., *et al.* 2017, *ApJ*, 846, 30

Uncovering Early Galaxy Evolution in the ALMA and JWST Era
Proceedings IAU Symposium No. 352, 2019
E. da Cunha, J. Hodge, J. Afonso, L. Pentericci & D. Sobral, eds.

doi:10.1017/S1743921319009086

FirstLight: Cosmological simulations of first galaxies at cosmic dawn

Daniel Ceverino[1,2,3]

[1]Cosmic Dawn Center (DAWN)
[2]Niels Bohr Institute, University of Copenhagen, Lyngbyvej 2, 2100 Copenhagen Ø, Denmark
[3]Universitat Heidelberg, Zentrum für Astronomie, Institut fur Theoretische Astrophysik, Albert-Ueberle-Str. 2, 69120 Heidelberg, Germany

Abstract. Cosmological hydrodynamical simulations have become an important theoretical tool for understanding the formation and evolution of the first galaxies during cosmic dawn, between redshifts 5 and 15. I will introduce the FirstLight database of about 300 zoom-in simulations with a resolution of 10 parsecs. This database agrees well with observed UV luminosity functions and stellar mass functions. I will discuss the origin and evolution of the star-forming main sequence of galaxies and the main drivers of the star formation histories at these early epochs. I will show simulated SEDs from UV to IR, including stellar and nebular emission. The rest-frame UV spectra show steep slopes and a high production efficiency of Lyman continuum photons. These properties are consistent with young stellar populations with low metallicities. Simulated recombination lines allow us to link the physical conditions of the gas around these stellar populations with observables, like equivalent widths in OIII or Hα or BPT diagrams at high-z. These simulations are making predictions that will be tested for the first time in future deep fields with the James Webb Space Telescope (JWST). I will finally discuss preliminary results involving JWST mock fields and predictions for ALMA observations by post-processing FirstLight snapshots with Powderday radiative transfer code.

Keywords. galaxies: evolution – galaxies: formation – galaxies: high-redshift.

1. Introduction

The period of cosmic dawn, the first billion years in the history of the Universe, is the final frontier for galaxy formation. Theory roughly predicts that the first galaxies form in the first gravitationally bound structures at high redshifts, $z \geqslant 12$. They quickly start to ionize their surroundings, driving the reionization of the Universe, which ends around $z \simeq 6$. However, little is known about these primeval galaxies. One of their most important properties is their spectral energy distribution (SED). In this talk, I will discuss SEDs coming from the FirstLight database of cosmological simulations of first galaxies at cosmic dawn, $z = 5 - 15$ (Ceverino, Glover & Klessen 2017). These cosmological simulations provide complex SFHs that are coupled with the non-uniform gas accretion history of galactic halos (Section 2 and Ceverino *et al.* 2018). The simulated SEDs are therefore consistent with the cosmological growth of structures at early times (Section 3). The final section discusses some results that can be found in Ceverino *et al.* (2018), such as BPT diagrams.

2. Dwarfs at Cosmic Dawn ($z = 6 - 15$)

Galaxies at cosmic dawn look very compact (Fig. 1). The stellar half-mass radius of this example is only 0.5 kpc. This is a factor 10 smaller than local galaxies with a similar mass. They are gas rich and they have typical SFR $\simeq 20\, M_{\odot}\, \mathrm{yr}^{-1}$, a factor 10 higher

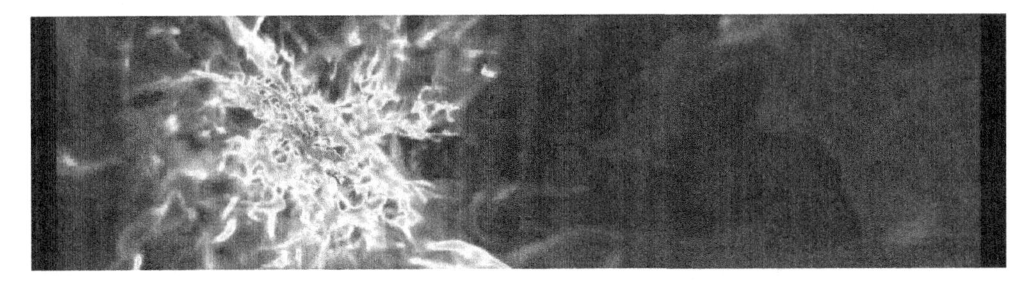

Figure 1. Gas distribution of a dwarf galaxy at $z=6$. The vertical size of the image is 3 kpc.

than their counterparts at $z=0$. This drives a clumpy and turbulent gas morphology, more similar to local starbursts. We have used the FirstLight database to study the Star Formation (SF) histories of ∼300 galaxies with a stellar mass between $M_* = 10^6$ and $3 \times 10^9 M_\odot$ during cosmic dawn ($z = 5-15$).

The evolution of the SFR in each galaxy is complex and diverse, characterized by bursts of SF. Overall, first galaxies spend about 70% of their time undergoing SF bursts at $z>5$. This diversity sets the mean and scatter of the SFMS at $z \simeq 5-13$. High gas fractions and short gas depletion times are common during the SF bursts. The typical bursts at $z \simeq 6$ have a sSFR maximum of $5-15\,\mathrm{Gyr}^{-1}$ with a duration of ∼100 Myr, one tenth of the age of the Universe. A quarter of the bursts populate a tail with very high sSFR maxima of $20-30\,\mathrm{Gyr}^{-1}$ and significantly shorter time-scales of $\sim 40-80$ Myr at all masses. The mean period of time between consecutive bursts is ∼200 Myr with a small mass dependence at $z \simeq 6$. The mean sSFR increases with redshift approximately as $sSFR \propto (1+z)^{5/2}$ at all masses, as predicted by ΛCDM models. This is consistent with existing observations at $z \leqslant 8$. The typical sSFR height of a SF burst also increases with redshift, but it is always a factor 2 ± 0.5 higher than the mean sSFR at that redshift. This implies typical sSFR maxima of $sSFR_{\mathrm{max}} = 20-30\,\mathrm{Gyr}^{-1}$ at $z=9-10$. The tail of the distribution reaches $sSFR_{\mathrm{max}} \simeq 60\,\mathrm{Gyr}^{-1}$ at these high redshifts. This evolution is driven by shorter time-scales at higher redshifts, proportional to the age of the Universe. These galaxies are the most efficient star formers in the history of the Universe.

3. Example SEDs

In this talk we focus on the intrinsic SEDs coming from the stellar populations and their surrounding, unresolved HII regions. They provide templates that could be used for a better understanding of the underlying stellar and gas properties in galaxies at cosmic dawn. Therefore, we ignore dust attenuation or any other radiative transfer effect from intervening gas. These effects will be considered in future works. We compute the SEDs of the simulated galaxies using publicly available tables from the Binary Population and Spectral Synthesis (BPASS) model (Eldrige *et al.* 2017) including nebular emission (Xiao *et al.* 2018).

Figure 2 shows two examples which represent the same galaxy with a stellar mass of $M_* \simeq 10^8 M_\odot$ at $z \simeq 6$, at two different phases of evolution. The left panel depicts the galaxy at the peak of a burst of star formation. The starburst brings the galaxy a factor 6 times above the star forming main sequence at that redshift (SSFR=30 Gyr^{-1}). Its luminosity at 1500 Å is correspondingly high, $M_{1500} \simeq -19$. The nebular emission dominates the SED at all wavelengths. Lines, such as Lyα, Hα, OIII or SiIII, are particularly prominent. For example, the equivalent width (EW) of Hα+NII exceeds 1000 Å. OIII+Hβ is the next prominent feature with EW$\simeq$ 700Å. These lines contaminate the continuum in the B and R bands. The nebular continuum is also particularly important in the U band and it is able to completely remove the mass -sensitive Balmer break at 4000 Å

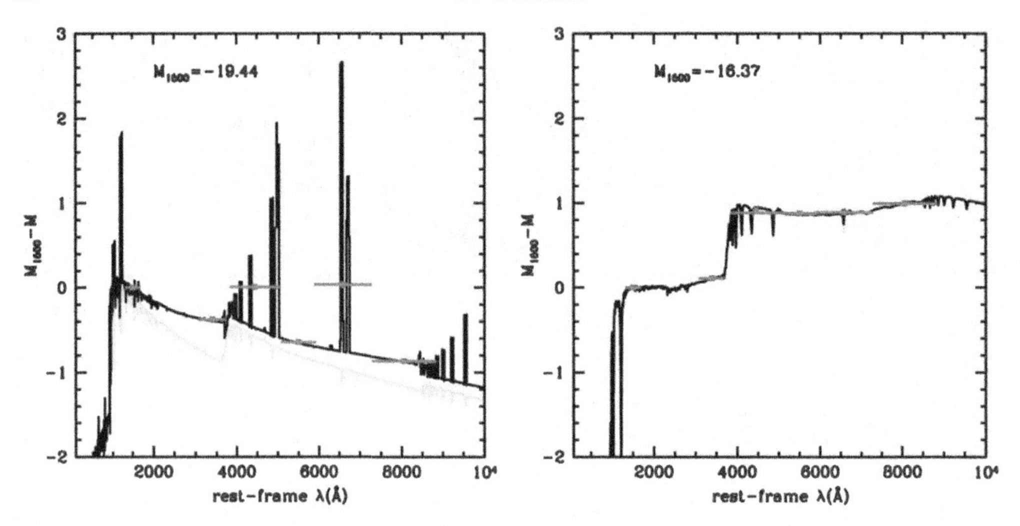

Figure 2. rest-frame SEDs of the same galaxy at the peak of a star formation burst at $z \simeq 6$ (left) and during the subsequent quiescent phase 200 Myr later (right). Cyan lines represent stellar light and black lines include nebular emission (lines+continuum). Red circles with bars represent six photometric bands. The monochromatic luminosity at each wavelength in erg/s/Hz is normalized to the absolute magnitude at 1500 $\AA$ (top label).

The stellar continuum is also very steep, consistent with a young stellar population with extremely blue colors, $V-I=-0.2$ and $U-V=-0.4$. These are the typical features of an extreme emission-line galaxy at cosmic dawn. The SED of the same galaxy 200 Myr after the starburst looks very different (right panel of Figure 2). The starburst and the subsequent feedback have quenched star formation significantly (SSFR$=0.02$ Gyr^{-1}), placing the galaxy well below the star-forming main sequence. The SED confirms this quiescent nature. The luminosity at 1500 Åis very low, $M_{1500} \simeq -16$, for that stellar mass. There is no significant nebular emission. Instead, there is a strong Balmer break with significantly red rest-frame colors, $V-I=0.2$ and $U-V=0.8$, typical of a mature population.

4. BPT Diagrams

The nebular emission lines give some clues about the properties of the HII regions in galaxies using line luminosity ratios like OIII(5007 $\AA$)/Hβ and NII(6584 $\AA$)/Hα. In Figure 3 we make predictions of BPT diagrams. The FirstLight galaxies form a clear sequence at $z=6$. The most massive galaxies, $M_* \simeq 1-2\times10^9$ M$_\odot$ occupy the tip of the sequence with OIII/H$\beta \simeq 0.5$ dex and NII/H$\alpha \simeq -1.5$ dex. These values are consistent with the line ratios of typical HII regions in low-metallicity dwarf galaxies at $z=0$ (van Zee & Haynes 2006). These regions have relatively high luminosities in OIII with respect to Hβ. Therefore, strong OIII-emitters at low-z could be good analogs of galaxies responsible for reionization (Fletcher *et al.* 2018), although none of our galaxies reaches the highest values of OIII/H$\beta \simeq$ 1 dex found in some of these analogs. This is mostly due to the low-metallicity of the primeval galaxies.

Low-mass galaxies have low luminosity ratios, reaching extremely low values for $M_* \simeq 10^6$ M$_\odot$: OIII/H$\beta \simeq -0.5$ dex and NII/H$\alpha \simeq -2.75$ dex. This is mostly driven by the low nebular metallicities of these low-mass galaxies (right panel of Figure 3), which reach $\log(Z_g/Z_\odot)=-2$ for the smallest galaxies in the sample. More massive galaxies, $M_* \simeq 10^8$ M$_\odot$, still have lower ratios than z=0 galaxies due to their low (10% solar) metallicities.

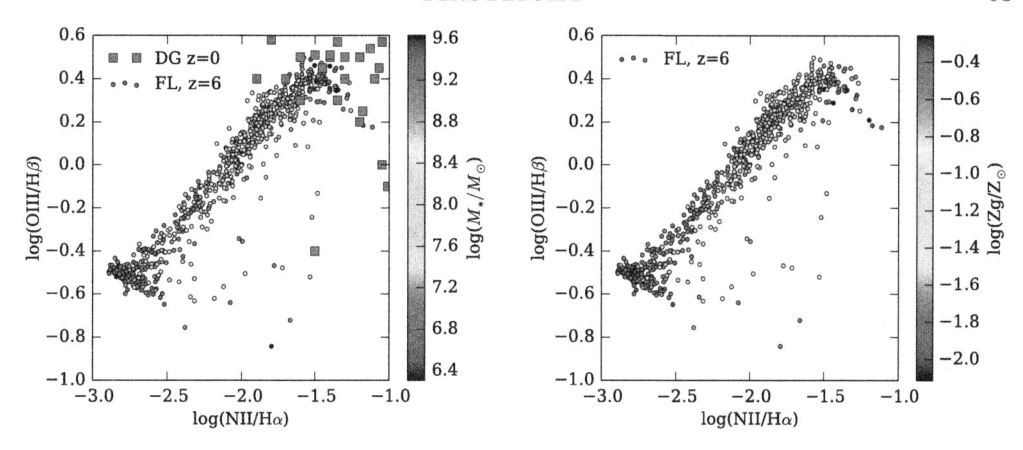

Figure 3. BPT diagrams at $z = 6$ coloured by stellar mass (left), and mean gas metallicity (right). Massive galaxies are consistent with local dwarfs galaxies of similar mass (van Zee & Haynes 2006). Low nebular metallicities drive the sequence towards very low metal-line luminosities.

References

Ceverino, D., Glover, S. C. O., & Klessen, R. S. 2017, *MNRAS*, 470, 2791

Ceverino, D., Klessen, R. S., Glover, S. C. O., *et al.* 2018, *MNRAS*, 480, 4842

Ceverino, D., Klessen, R., & Glover, S. 2019, *MNRAS*, 484, 1366

Eldridge, J. J., Stanway, E. R., Xiao, L., McClelland, L. A. S., Taylor, G., Ng, M., Greis, S. M. L., & Bray, J. C., 2017, *Publ. Astron. Soc. Australia*, 34, e058

Fletcher, T. J., Robertson, B. E., Nakajima, K., Ellis, R. S., Stark, D. P., & Inoue, A., 2018, preprint, p. arXiv:1806.01741

Xiao, L., Stanway, E. R., & Eldridge, J. J. 2018, *MNRAS*, 477, 904

van Zee, L. & Haynes, M. P. 2006, *ApJ*, 636, 214

Uncovering Early Galaxy Evolution in the ALMA and JWST Era
Proceedings IAU Symposium No. 352, 2019
E. da Cunha, J. Hodge, J. Afonso, L. Pentericci & D. Sobral, eds.

doi:10.1017/S1743921319008986

Understanding galaxy formation in the reionization era using the FIRE simulations

Xiangcheng Ma

Department of Astronomy and Theoretical Astrophysics Center,
University of California Berkeley, Berkeley, CA 94720, USA
email: xchma@berkeley.edu

Abstract. We present a suite of high-resolution cosmological zoom-in simulations of galaxies at $z \geqslant 5$ using the state-of-the-art models for the multi-phase ISM, star formation, and stellar feedback from the FIRE project. We present a series of key results from these simulations, including the stellar mass–halo mass relation, the ultraviolet luminosity functions, dust attenuation and dust temperatures, the ubiquitous formation of bound star clusters, morphology and clumpiness, and the escape fractions of ionizing photons from high-redshift galaxies. We discuss how different simulations in the literature agree and disagree and what observations are most useful for testing the models in the era of ALMA and JWST.

Keywords. binaries: general, stars: formation, dust, extinction, ISM: structure, galaxies: evolution, galaxies: formation, galaxies: high-redshift, galaxies: star clusters, cosmology: theory

---

1. Introduction

Understanding the formation of galaxies in the first billion years of the Universe is one of the most important questions in extragalactic astronomy and cosmology. It is believed that these high-redshift galaxies are the dominant sources for cosmic reionization (e.g., Robertson *et al.* 2013). Moreover, they open a new window for testing our understanding of galaxy formation physics. Limited by the capability of current observational facilities, our knowledge of high-redshift galaxies to date is far less extensive than those at low and intermediate redshifts. Thanks to the *Atacama Large Millimeter Array* (ALMA) and the upcoming *James Webb Space Telescope* (JWST), we expect to collect enormous data of galaxies at $z \gtrsim 5$ with unprecedented quality over the next few years. It is thus critical and timely, from a theoretical point of view, to make more detailed, realistic predictions for high-redshift galaxies that can be confronted with future observations.

We will present a sample of high-resolution, cosmological hydrodynamic zoom-in simulations of high-redshift galaxies we have been building for the past $\sim$2–3 years. These simulations adopt the Feedback in Realistic Environments (FIRE) models for the multi-phase interstellar medium (ISM), star formation, and stellar feedback (Hopkins *et al.* 2018). Our simulations complement the semi-analytic models (e.g., Yung *et al.* 2019) and large-volume cosmological simulations of high-redshift galaxies (e.g., Feng *et al.* 2016) by resolving small-scale structures and physical processes inside the galaxies. We note that the FIRE project also has a large simulation library run to $z=0$ and $z=2$, which other zoom-in simulations of high-redshift galaxies may not have (e.g., Pallottini *et al.* 2017). We will briefly describe the simulation sample and present the key results we have published. We will also discuss in what aspects simulations and models from different groups agree and disagree and what future observations can be used to test these predictions.

2. The simulations

These simulations use the well-established multi-scale cosmological zoom-in technique. In short, a dark-matter-only cosmological volume is simulated first. Halos of desired mass at desired redshift are selected. The region around the target halos is then re-simulated with baryons at much higher resolution, while the rest of the box is simulated at lower resolution to keep the large-scale tidal force. Our high-redshift simulation sample consists of 35 zoom-in regions run to $z = 5$–9 selected from a $(120\,\mathrm{Mpc}/h)^3$ box. The baryonic mass and spatial resolution adopted in these simulations are ~ 100–$7000\,M_\odot$ and ~ 0.1–1 pc. The entire sample contains over ~ 2000 halos from $M_{\mathrm{halo}} \sim 10^8$–$10^{12}\,M_\odot$ in all the zoom-in regions. We refer to Ma *et al.* (2019a) for details of the initial conditions.

These simulations use the FIRE-2 model for the multi-phase ISM, star formation, and feedback, which we briefly summarize here. Gas cooling is followed from 10–10^{10} K, including free-free, bound-free, atomic, and metallicity-dependent metal-line, fine-structure, and molecular cooling. Star formation is restricted to molecular, self-gravitating gas with density above $10^3\,\mathrm{cm}^{-3}$. The model also includes explicit treatments for photoionization, photo-heating, radiation pressure, stellar winds, and supernovae feedback. We consider metal enrichment from Type-II, Type-Ia SNe, and AGB winds and sub-resolution metal mixing. We refer to Hopkins *et al.* (2018) for details of the FIRE-2 model.

3. Results

We highlight some of the key results based on the high-redshift galaxy simulation suite from the FIRE project, but refer to the papers we have published in recent years for more details (Ma *et al.* 2015, 2016, 2018a,b, 2019a,b).

The stellar mass–halo mass (SMHM) relation at $z \gtrsim 5$. We find the SMHM relation for our simulated galaxies can be best described by $\log M_* = 1.58\,(\log M_{\mathrm{halo}} - 10) + 7.1$ for halo mass $M_{\mathrm{halo}} \sim 10^8$–$10^{12}\,M_\odot$ with *little* evolution in $z \sim 5$–12. We stress that the redshift evolution of the SMHM relation is still an open question. Very different trends have been reported among cosmological simulations, semi-analytic models, and empirical models (e.g., Behroozi & Silk 2015; Ceverino *et al.* 2017; Wilkins *et al.* 2017; Tacchella *et al.* 2018). The discrepancy can be resolved by better stellar mass measurements with improved data in the rest-frame optical bands from JWST.

Stellar mass functions (SMFs) and luminosity functions (LFs). In Ma *et al.* (2018a), we derive the SMFs and multi-band LFs from our simulations. The left panel of Figure 1 shows the rest-frame UV (1500 Å) LF at $z = 6$, where we compare our results (dashed lines) with most up-to-date observational constraints (color points). The bright-end UV LFs ($\mathrm{M_{UV}} < -17$) up to $z \sim 8$ are the most well constrained properties of high-redshift galaxies. Our simulations agree with observations very well, especially when dust attenuation is taken into account. We refer to Ma *et al.* (2018a) for more results.

Dust attenuation, dust emission, and dust temperature. We have conducted full three-dimensional Monte Carlo dust radiative transfer calculations on our simulations in Ma *et al.* (2019a), using the public SKIRT code (Baes *et al.* 2011). We reiterate dust attenuation is required for shaping the bright-end UVLFs even at $z \sim 8$. The right panel of Figure 1 shows the relation between peak wavelength of dust emission, λ_{peak}, and the total dust emission, L_{IR}. The color points represent individual galaxies at $z \geqslant 5$ from our simulated sample, while the red triangles show the $z = 0$ FIRE simulations for reference. The grey points, orange lines, and blue squares show the observed data at lower redshifts compiled by Casey *et al.* (2018). We find λ_{peak} decreases with redshift at fixed L_{IR}, which implies higher dust temperatures at higher redshifts. This can be understood by higher specific star formation rates and higher star formation surface densities at higher

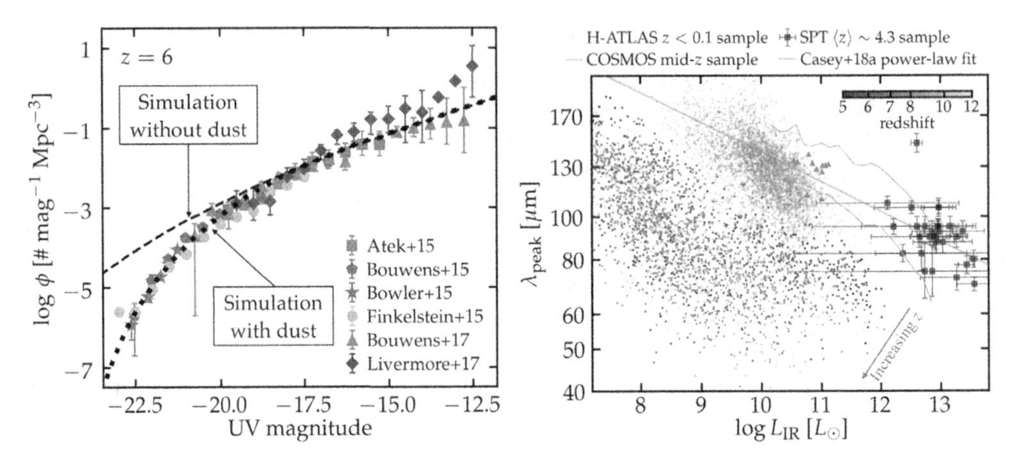

Figure 1. *Left*: The UVLF at $z = 6$. The dashed lines show predictions from our simulations, with (thick) and without (thin) dust attenuation. The color points show observational constraints from the literature. *Right*: The relation between peak wavelength of dust emission, $\lambda_{\rm peak}$, and the total dust emission, $L_{\rm IR}$.

redshifts. We refer to Ma *et al.* (2019a) for more results (e.g., the IRX–β relation), but emphasize the importance for distinguishing different definitions of dust temperature.

Ubiquitous star cluster formation. We identify a large number of bound star clusters in almost all of our simulations of high-redshift galaxies. In Ma *et al.* (2019b), we investigate how these star clusters form in these simulations by re-simulating a starburst in three galaxies spanning over three orders of magnitude in stellar mass with much finer output frequency. Figure 2 shows the formation sites of two clusters in a galaxy of $M_* \sim 10^9\, M_\odot$ (halo mass $M_{\rm halo} \sim 10^{11}\, M_\odot$) at $z \sim 5$. In general, high-redshift galaxies are highly gas-rich and turbulent. High-pressure clouds form efficiently in such environments via cloud-cloud collision and compression by feedback-driven outflows. Stars form in these clouds at near unity efficiency within a cloud free-fall time, so they tend to stay gravitationally bound at formation. We find that newly formed clusters broadly follow a power-law mass function of ${\rm d}N/{\rm d}M \propto M^{-2}$ (see Ma *et al.* 2019b for more details).

Clumpy morphologies. As high-redshift galaxies tend to form a large number of star clusters, one implication is that their rest-frame UV morphologies are very clumpy. The left panel of Figure 3 shows one example galaxy from our simulated sample, which has stellar mass $M_* \sim 2 \times 10^8\, M_\odot$ and UV magnitude $\rm M_{UV} \sim -16.7$. The rest-frame UV image is dominated by three star-forming clumps that do not contribute a large stellar mass. For observations at finite surface brightness depth, one may miss most of the UV light but only see the brightest clump. The right panel of Figure 3 illustrates this effect. We caution that the faint galaxies detected in the Hubble Frontier Fields are likely biased by individual star-forming clumps. This makes it more difficult to constrain the faint-end UVLFs robustly at high redshifts.

Escape fractions of ionizing photons from high-redshift galaxies. In Ma *et al.* (2015, 2016), we have conducted Monte Carlo radiative transfer calculations of Lyman-continuum photons on a small sample of high-redshift galaxy simulations run with the previous version of the FIRE model to study the escape fraction of hydrogen ionizing photons, $f_{\rm esc}$, from $z \geqslant 5$ galaxies. We find one key physics that determines $f_{\rm esc}$ is the competition of time-scales between birth cloud disruption and stellar evolution: most ionizing photons cannot escape before feedback from young stars disrupt and clear their birth clouds, but if this happens longer than a few Myrs, there will be fewer ionizing photons available as massive stars have left the main sequence. We find if binary stars

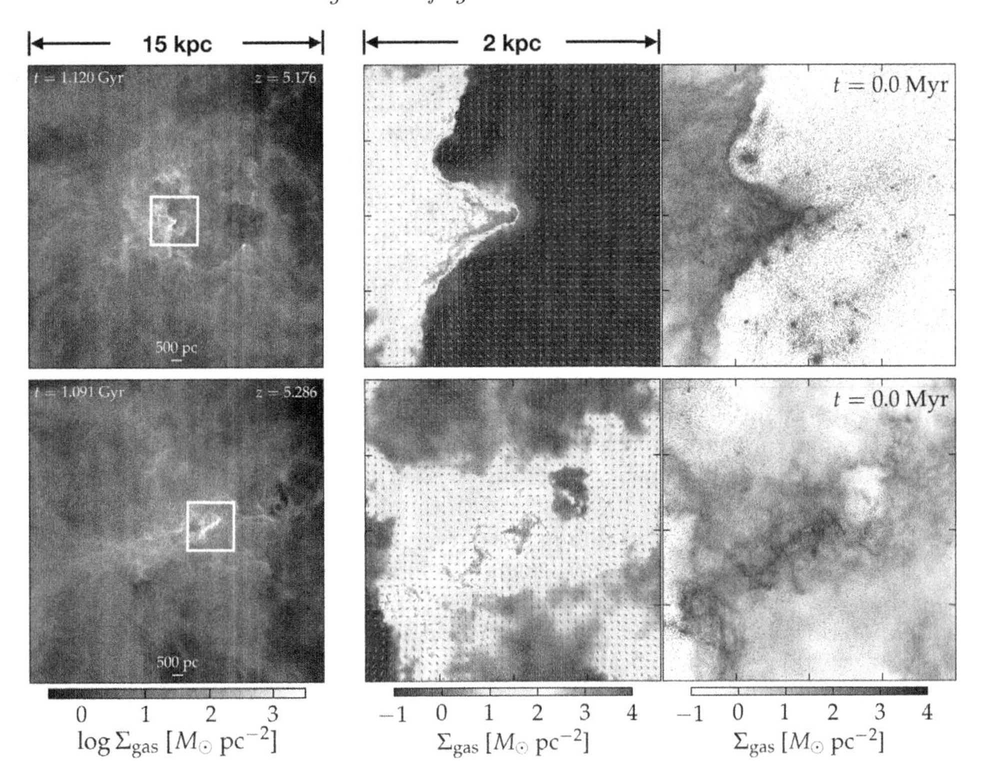

Figure 2. Bound star clusters formed in a simulated galaxy with stellar mass $M_* \sim 10^9\, M_\odot$ (halo mass $M_{\rm halo} \sim 10^{11}\, M_\odot$) at $z \sim 5$. These clusters form in high-pressure clouds with gas surface density $\Sigma_{\rm gas} \gg 10^3\, M_\odot\, {\rm pc}^{-2}$, which are formed in highly gas-rich, turbulent ISM via cloud-cloud collision and compression by feedback-driven winds.

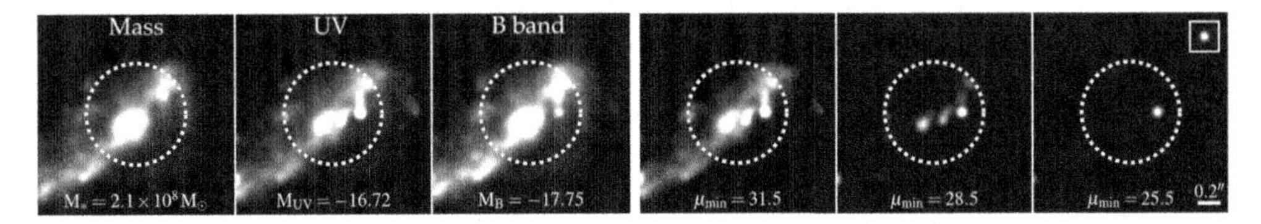

Figure 3. *Left*: Stellar mass, rest-frame UV, and optical maps of a simulated galaxy in our sample at $z \sim 6$. The UV image is dominated by three compact star-forming clumps that do not contribute much stellar mass. *Right*: Rest-frame UV images when the galaxy is observed at finite surface brightness depths.

are taken into account, $f_{\rm esc}$ will increase by a large factor, as binaries produce more ionizing photons on 10 Myr time-scales, which can escape from the galaxy much more easily.

4. Discussion

In the era of ALMA and JWST, there will be tremendous amount of new data coming in the next few years probing various properties of high-redshift galaxies that we cannot do in the past. We argue that such data make it possible to use high-redshift galaxies to test our understanding of galaxy formation physics.

First of all, we iterate the discrepancy on the SMHM relation and its redshift evolution between different theoretical predictions in the literature. The SMFs at $z \geqslant 5$ have not been well constrained so far (the discrepancy can be as large as an order of magnitude; e.g., Stefanon *et al.* 2017), partly due to the lack of high-quality data in the rest-frame optical bands. Better measurements on the stellar mass of high-redshift galaxies can be useful for testing the underlying assumptions (e.g., star formation efficiencies, feedback, the scatter of the SMHM relation, etc.) adopted in these models.

We find in our simulations, the galaxies are highly gas-rich and turbulence supported, even in halos as massive as $M_{\rm halo} \sim 10^{12}\,M_{\odot}$ (stellar mass up to $M_* \sim 2\times10^{10}\,M_{\odot}$). In such systems, star formation is bursty, star clusters form efficiently, and feedback is sufficiently strong to blow out nearly all the gas from the galaxy. There is no rotationally supported gas disk formed in these galaxies. Although such features are broadly in line with some work in the literature, we notice other simulations may produce very different galaxy properties (e.g., rotating disks form at very high redshift even in relatively lower-mass halos; see Pallottini *et al.* 2017; Katz *et al.* 2019). The discrepancy is largely due to the star formation and feedback models adopted in these simulations. We propose the following observational diagnostics that can be used to test the simulations in the era of ALMA and JWST: Hα- and UV-derived SFRs, Balmer break (as indicators of bursty star formation), morphology and clumpiness of high-redshift galaxies (to test star cluster formation), and gas kinematics measurements (cold gas from [C II] and ionized gas from Balmer lines, as diagnostics of rotational support). It would be extremely interesting to quantify burstiness and disk fraction as a function of stellar mass and redshift to form a coherent picture of galaxy formation across cosmic time. JWST and ALMA can definitely fill in the relatively massive at $z \gtrsim 6$.

References

Baes, M., Verstappen, J., De Looze, I., *et al.* 2011, *ApJS*, 196, 22
Behroozi, P. S. & Silk, J., 2015, *ApJ*, 799, 32
Casey, C. M., Zavala, J. A., Spilker, J., *et al.* 2018, *ApJ*, 862, 77
Ceverino, D., Glover, S. C. O., & Klessen, R. S., 2017, *MNRAS*, 470, 2791
Feng, Y., Di-Matteo, T., Croft, R. A., *et al.* 2016, *MNRAS*, 455, 2778
Hopkins, P. F., Wetzel, A., Kereš, D., *et al.* 2018, *MNRAS*, 480, 800
Katz, H., Kimm, T., Haehnelt, M. G., *et al.* 2019, *MNRAS*, 483, 2019
Ma, X., Kasen, D., Hopkins, P. F., *et al.* 2015, *MNRAS*, 453, 960
Ma, X., Hopkins, P. F., Kasen, D., *et al.* 2016, *MNRAS*, 459, 3614
Ma, X., Hopkins, P. F., Garrison-Kimmel, S., *et al.* 2018a, *MNRAS*, 478, 1694
Ma, X., Hopkins, P. F., Boylan-Kolchin, M., *et al.* 2018b, *MNRAS*, 477, 219
Ma, X., Hayward, C. C., Casey, C. M., *et al.* 2019a, *MNRAS*, 487, 1844
Ma, X., Grudić, M. Y., Quataert, E., *et al.* 2019b, arXiv e-print: 1906.11261
Pallottini, A., Ferrara, A., Gallerani, S., *et al.* 2017, *MNRAS*, 465, 2540
Robertson, B. E., Furlanetto, S. R., Schneider, E., *et al.* 2013, *ApJ*, 768, 71
Stefanon, M., Bouwens, R. J., Labbé, I., *et al.* 2017, *ApJ*, 843, 36
Tacchella, S., Bose, S., Conroy, C., *et al.* 2018, *ApJ*, 868, 92
Wilkins, S. M., Feng, Y., Di Matteo, T., *et al.* 2017, *MNRAS*, 469, 2517
Yung, L. Y. A., Somerville, R. S., Finkelstein, S. L., *et al.* 2019, *MNRAS*, 483, 2983

Uncovering Early Galaxy Evolution in the ALMA and JWST Era
Proceedings IAU Symposium No. 352, 2019
E. da Cunha, J. Hodge, J. Afonso, L. Pentericci & D. Sobral, eds.

doi:10.1017/S1743921320001118

Shedding light on high-redshift galaxies with the 21cm signal

Anne Hutter

Kapteyn Astronomical Institute, The Netherlands

Abstract. Reionization represents an important epoch in the history in the Universe, when the first stars and galaxies gradually ionize the neutral hydrogen in the intergalactic medium (IGM). Understanding the nature of the ionizing sources, the associated ionization of the IGM, and its impact on subsequent structure formation and galaxy evolution by means of radiative feedback effects, represent key outstanding questions in current astrophysics. High-redshift galaxy observations and simulations have significantly extended our knowledge on the nature of high-redshift galaxies. However, essential properties such as the escape fraction of ionizing photons from galaxies into the IGM and their dependency on galactic properties remain essentially unknown, but determine significantly the distribution and time evolution of the ionized regions during reionization. Analyzing this ionization topology by means of the neutral hydrogen sensitive 21cm signal with radio interferometers such as SKA offers a complementary and unique opportunity to determine the nature of these first galaxies. I will show results from a self-consistent semi-numerical model of galaxy evolution and reionization, and discuss the potential of inferring galactic properties with the 21cm signal as well as the impact of reionization on the high-redshift galaxy population and its evolution.

Uncovering Early Galaxy Evolution in the
ALMA and JWST Era
Proceedings IAU Symposium No. 352, 2019
E. da Cunha, J. Hodge, J. Afonso, L. Pentericci &
D. Sobral, eds.

doi:10.1017/S174392132000112X

New constraints on reionization from a redshift-independent efficiency model

Rohan Potham Naidu

Harvard University, USA

Abstract. We present an empirical model built on a high-resolution N-body dark matter simulation. We assume a redshift-independent star-formation efficiency for each halo to convert the accretion rate into a star-formation rate. Our model is calibrated using the $z = 4$ UV luminosity function (UVLF) and successfully predicts the observed UVLF at $z = 5 - 10$. We present predictions at $z = 5 - 10$ for UV luminosity and stellar mass functions, JWST number counts, the stellar-to-halo mass relation and star-formation histories. We combine this model with bleeding-edge reionization constraints (from $z > 7$ quasars, $z \sim 7$ Lyα line-profiles, the updated Planck τ) to find new perspectives on the Epoch of Reionization (EoR). We find $M_{\rm UV} < -13.5$ galaxies need an average $f_{\rm esc} = 0.22 \pm 0.05$ to drive reionization and a highly compressed timeline: the IGM neutral fraction is $[0.9, 0.5, 0.1]$ at $z = [8.4 \pm 0.2, 7.0 \pm 0.2, 6.3 \pm 0.2]$. Inspired by the newly assembled sample of Lyman Continuum leakers that unanimously displays higher-than-average star-formation surface density (sigma), we fit a model tying $f_{\rm esc}$ to sigma. Since sigma grows by > 2.5dex over $z = 0 - 8$, our model explains the humble values of $f_{\rm esc}$ at low-z. We find, strikingly, that <5% of galaxies with $M_{\rm UV} < -18$ account for $> 80\%$ of the reionization budget. We predict leakers like COLA1 ($z = 6.6$, $M_{\rm UV} = -21.5$) become common towards the EoR and that the protagonists of reionization are not hiding across the faint-end of the luminosity function but are already known to us.

Uncovering Early Galaxy Evolution in the ALMA and JWST Era
Proceedings IAU Symposium No. 352, 2019
E. da Cunha, J. Hodge, J. Afonso, L. Pentericci & D. Sobral, eds.

doi:10.1017/S174392131900841X

Theoretically modelling photoionised regions with fractal geometry in three-dimension

Yifei Jin, Lisa Kewley and Ralph Sutherland

Research School of Astronomy & Astrophysics, The Australian National University, Cotter Road, Weston Creek, 2611, ACT, Australia
email: Yifei.Jin@anu.edu.au

Abstract. Accurate predictions of the physics of interstellar medium (ISM) are vital for understanding galaxy formation and evolution. Modelling photoionized regions with complex geometry produces realistic ionization structures within the nebulae, providing the necessary physical predictions to interpret observational data. 3D photoionization codes built with Monte Carlo techniques provide powerful tools to produce the ionizing radiation field with fractal geometry. We present a high-resolution Monte Carlo modelling of a nebula with fractal geometry, and will further show how nebular geometry influences the emission-line behaviours. Our research has important implications for studies of emission-line ratios in high redshift galaxies.

Keywords. ISM: HII regions, galaxies: ISM, galaxies: starburst, galaxies: high-redshift.

1. Nebular Geometry in Photoionisation Models

The accurate interpretations of emission-line behaviours in galaxies largely depend on photoionisation models. Over the past three decades, the CLOUDY (Ferland *et al.* 2017) and the MAPPINGS codes (Sutherland *et al.* 2017) have been two widely-recognized self-consistent radiative transfer codes. The MAPPINGS code was also developed to resolve the physical structure and emission-line spectrum of both fast and slow shocks. However, both the CLOUDY code and the MAPPINGS code are built in plane parallel or spherical simplifications of nebular geometry, which limit their usage in modelling the realistic photoionised regions.

The Monte Carlo technique is a promising solution to 3D radiative transfer calculations. Several Monte Carlo radiative transfer codes have been developed and used in the nebular geometric studies, like MOCASSIN3D (Ercolano *et al.* 2005) and DIRTY (Gordon *et al.* 2001). So far, these nebular geometric studies focus on modelling the density bounded photoionised regions with simple density distributions, like bipolar or porous spherical density distributions. In this work, we model a photoionised region with complex three-dimensional geometry, in order to investigate the geometric effect on the behaviour of nebular lines.

2. Modelled Radiation Bounded Photoionised Region with Complex Geometry

We adopt the MOCASSIN3D code to model a radiation bounded photoionised region in a three-dimensional ISM cube. The ISM has a lognormal density distribution in order to describe the turbulence in ISM. We place an O-star at the centre of the ISM cube. The ionising spectrum is selected from the CMFGEN stellar library.

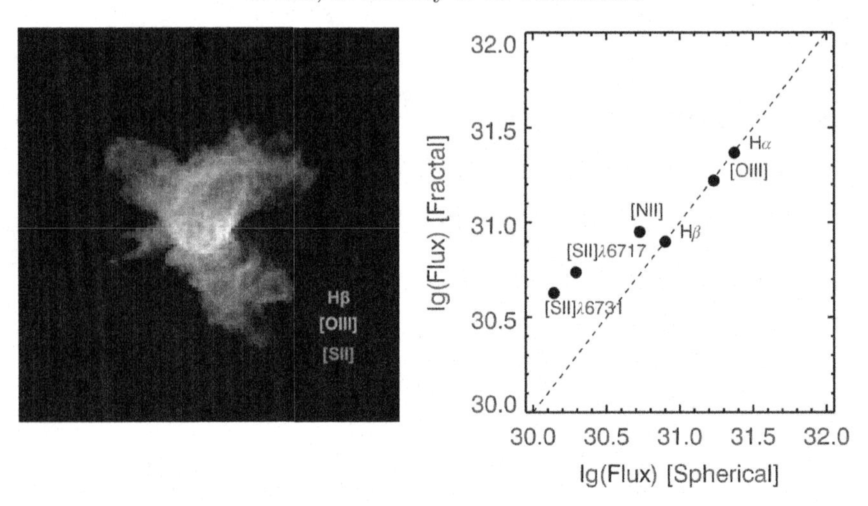

Figure 1. *Left:* The projected distributions of Hβ, [O III] and [S II] emissions. *Right:* The comparison of integrated emission-line fluxes between the fractal model and the spherical simplified model.

Results. Figure 1 presents the projected distributions of Hβ, [O III] and [S II] emissions of our modelled photoionised region. The photoionised region has a twisted boundary, which is shaped by the fractal density distribution of ISM. Similar to the predication of 1D models, Hβ and [O III] emissions spread across the entire photoionised region, while [S II] and [N II] preferentially locate at the boundary of the photoionised region. Compared with the equal-mass spherically simplified model, our geometric model shows higher [S II] and [N II] fluxes but similar Hα, Hβ and [O III] fluxes. The increasing complexity of nebular geometry enhances the integrated flux of [S II] and [N II] but has little influence on Hβ and [O III] fluxes.

Implications and future. The 3D Monte Carlo photoionisation model with complex geometry provides a better predication of the emission-line behaviours in realistic galaxies, especially for high-redshift galaxies where the ISM is highly-turbulent. In order to fully understand the geometric effect on emission-line behaviours, systematic studies considering various nebular geometries are required in future.

References

Ercolano, B., Barlow, M. J., & Storey, P. J. 2005, *MNRAS*, 362, 1038

Ferland, G. J., Chatzikos, M., Guzmán, F., Lykins, M. L., van Hoof, P. A. M., Williams, R. J. R., Abel, N. P., Badnell, N. R., Keenan, F. P., Porter, R. L., & Stancil, P. C. 2017, *Rev. Mexicana AyA*, 53, 385

Gordon, K. D., Misselt, K. A., Witt, A. N., & Clayton, G. C. 2001, *ApJ*, 551, 269

Sutherland, R. S. & Dopita, M. A., *et al.* 2017, *ApJS*, 229, 34

Uncovering Early Galaxy Evolution in the ALMA and JWST Era
Proceedings IAU Symposium No. 352, 2019
E. da Cunha, J. Hodge, J. Afonso, L. Pentericci & D. Sobral, eds.

doi:10.1017/S1743921320000654

Lyα generation in intermediate velocity shock waves

Andrew Lehmann[1], Benjamin Godard[1,2], Guillaume Pineau des Forêts[2] and Edith Falgarone[1]

[1]ENS, Université PSL, CNRS, Sorbonne Université, Paris, France
[2]LERMA, Observatoire de Paris, France
email: andrew.lehmann@ens.fr

Abstract. We update the Paris-Durham shock model, a state-of-the-art magnetohydrodynamic (MHD) shock code developed with a focus on molecular chemistry, in order to account for the self-generated UV field produced in shocks at velocities in the range 25-50 km/s. In these shocks there is significant excitation of atomic Hydrogen, with a large flux of Lyα photons escaping ahead of the shock to heat, ionize and drive molecular chemistry in a large slab of preshock gas.

Keywords. Lyα, shock waves, UV

Introduction

Recent ALMA observations of CH^+ emission from a sample of high redshift starburst galaxies (Falgarone *et al.* 2017) have uncovered molecular gas with extreme velocity dispersions ($\sim$1000 km/s). The presence of molecules at such high velocities is difficult to explain, as shocks at these speeds produce an X-ray emitting hot plasma devoid of H_2. However, a picture has emerged of large-scale outflows triggering a turbulent cascade in a multiphase medium with significant mechanical energy dissipating in molecular shocks at much lower velocities (Lesaffre *et al.* 2013).

Paris-Durham Shock Code

The Paris-Durham public† shock code (Flower & Pineau des Forêts 2015) solves the steady-state plane-parallel MHD equations coupled with chemical equations using an extensive reaction network appropriate to the insterstellar medium. The latest version is described in Godard *et al.* (2019). In molecular shocks with velocities $V_s \gtrsim 25$ km/s, collisional dissociation destroys H_2 and the region where atomic H starts to dominate is where the majority of the UV production takes place. To overcome computational challenges caused by extreme optical depths, we have implemented an accelerated Λ-iteration scheme to solve for the radiative transfer of a 3-level hydrogen atom to compute the Lyα (1216 Å), Lyβ (1026 Å) and 2-photon continuum emission. Fig. 1 shows temperature profiles of shocks with velocities 25-50 km/s propagating into molecular gas at $n = 10^4$ cm^{-3}. The emergent UV radiation is strong enough in these shocks to produce a warm slab ($\sim$100 K) of molecular gas over $\sim 10^{17}$ cm. Shocks with velocities 30 km/s and under produced negligible UV.

Lyman-α and other chemical probes

In fig. 2 (left) we show Lyα spectra at three positions in a representative 40 km/s shock: emerging from the shock front (black), deep in the postshock (blue), and in the preshock

† available on the ISM platform https://ism.obspm.fr

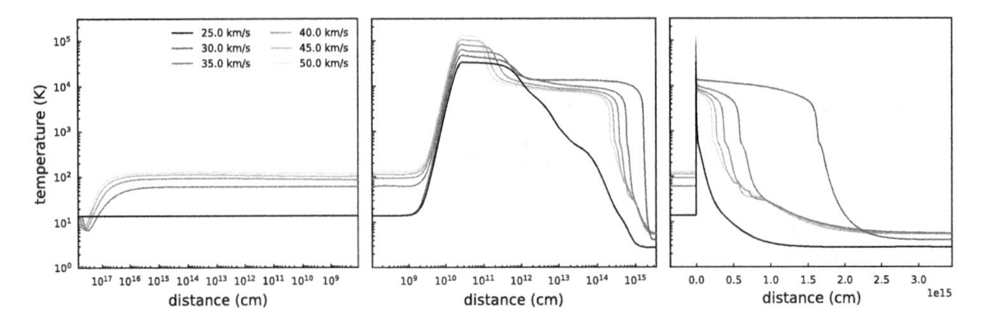

Figure 1. Temperature profiles of shocks from 25-50 km/s propagating into molecular gas at $n = 10^4$ cm^{-3}. (Left) preshock, with distance measured from the shock front in log scale, (middle) postshock in log scale (right) postshock in linear scale.

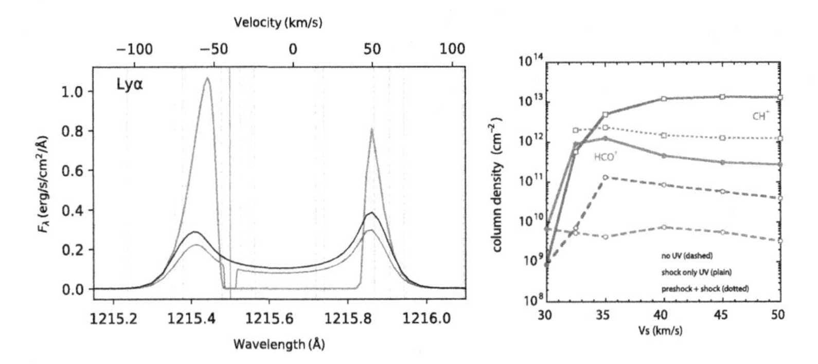

Figure 2. (Left) Lyα profiles at three positions in a 40 km/s shock. (Right) Column densities of CH^+ and HCO^+.

(red). The solid grey vertical line shows the Lyα central wavelength doppler shifted by the shock velocity, and the fainter vertical lines show line centres of H_2 Lyman and Werner band rovibrational lines. The cold hydrogen in the preshock cannot significantly attenuate the broad shock emission, and so the size of the preshock is determined by dust absorption. The photoelectric effect due to these Lyα absorptions causes the preshock heating to $\sim$100 K seen in fig. 1. Photodissociation by the UV radiation can heavily affect the abundances of several species both inside and ahead of the shock. In fig. 2 (right) we show how column densities of CH^+ and HCO^+ vary with shock velocity including (solid) or not including (dashed) the self-generated UV. These column densities increase by orders of magnitude at these velocities, showing the importance of treating Lyα in these shocks. Recent ALMA observations of CH^+ in high-z starburst galaxies (Falgarone *et al.* 2017) may require these kind of shocks to be explained.

References

Falgarone, E., Zwaan, M. A., Godard, B., *et al.* 2017, *Nature*, 548, 430
Flower, D. R. & Pineau des Forêts, G. 2015, *A&A*, 578, A63
Godard, B., Pineau des Forêts, G., Lesaffre, P., *et al.* 2019, *A&A*, 622, A100
Lesaffre, P., Pineau des Forêts, G., Godard, B., *et al.* 2013, *A&A*, 550, 106

SESSION 3: Spectral energy distribution models

Uncovering Early Galaxy Evolution in the ALMA and JWST Era
Proceedings IAU Symposium No. 352, 2019
E. da Cunha, J. Hodge, J. Afonso, L. Pentericci & D. Sobral, eds.

doi:10.1017/S1743921320001131

A review of spectral energy distribution modeling at high-redshift

Stéphane Charlot

Institut d'Astrophysique de Paris, France

Abstract. I will review recent developments in the modeling of high-redshift galaxy spectra, focusing in particular on the rest-frame ultraviolet and optical emission from young stellar populations and the interstellar medium.

Uncovering Early Galaxy Evolution in the ALMA and JWST Era
Proceedings IAU Symposium No. 352, 2019
E. da Cunha, J. Hodge, J. Afonso, L. Pentericci & D. Sobral, eds.

doi:10.1017/S1743921320001143

A VLT/MUSE analysis of HeIIλ1640 emitters at $z = 2 - 4$

Themiya Nanayakkara

Leiden University, The Netherlands

Abstract. In the quest to study early star-formation physics in the universe, one of the most sought after tracers is HeIIλ1640, with its presence in the lack of other metal emission/absorption lines generally being interpreted as evidence for metal-poor stellar populations. HeII ionizing photons are produced via sources of hard ionizing radiation and requires photons with energies $\geqslant 54.4$eV, however, traditional stellar population models lack sufficient ionising photons to match with current observations. Our analysis of $z = 2 - 4$ HeIIλ1640 emitters from deep 10-30h pointings from MUSE has shown that ISM properties inferred from multiple rest-UV diagnostics are not compatible with requirements necessary to reproduce HeIIλ1640 equivalent-widths. Thus, we have used latest generation of single, rotational, and binary stellar population models with realistic dust physics to explore rest-UV emission line diagnostics and link with H and He^+ ionisation photon production efficiencies ($\xi_{\rm ion}$(H,He^+)) in a variety of stellar/gas metallicities and star-formation histories. I will discus our latest results and show that including 'exotic' stellar phenomena such as extreme low-metallicity binary stars, X-ray binaries, and dust dissociation physics may be necessary to lessen the tension between models and observations.

Uncovering Early Galaxy Evolution in the ALMA and JWST Era
Proceedings IAU Symposium No. 352, 2019
E. da Cunha, J. Hodge, J. Afonso, L. Pentericci & D. Sobral, eds.

doi:10.1017/S1743921319009645

New insight on the far-UV SED and He II emission from low metallicity galaxies

Daniel Schaerer[1,2], **Yuri Izotov**[3] **and Tassos Fragos**[1]

[1]Observatoire de Genève, Université de Genève, 51 Ch. des Maillettes, 1290 Versoix, Switzerland
email: daniel.schaerer@unige.ch

[2]CNRS, IRAP, 14 Avenue E. Belin, 31400 Toulouse, France

[3]Bogolyubov Institute for Theoretical Physics, National Academy of Sciences of Ukraine, 14-b Metrolohichna str., Kyiv, 03143, Ukraine

Abstract. Understanding the ionizing spectrum of low-metallicity galaxies is of great importance for modeling and interpreting emission line observations of early/distant galaxies.

Although a wide suite of stellar evolution, atmosphere, population synthesis, and photoionization models, taking many physical processes into account now exist, all models face a common problem: the inability to explain the presence of nebular He II emission, which is observed in many low metallicity galaxies, both in UV and optical spectra. Several possible explanations have been proposed in the literature, including Wolf-Rayet (WR) stars, binaries, very massive stars, X-ray sources, or shocks. However, none has so far been able to explain the major observations.

We briefly discuss the He II problem, available empirical data, and observed trends combining X-ray, optical and other studies. We present a simple and consistent physical model showing that X-ray binaries could explain the long-standing nebular He II problem. Our model, described in Schaerer *et al.* (2019), successfully explains the observed trends and strength of nebular He II emission in large samples of low metallicity galaxies and in individual galaxies, which have been studied in detail and with multi-wavelength observations. Our results have in particular important implications for the interpretation of galaxy spectra in the early Universe, which will be obtained with upcoming and future facilities.

Keywords. galaxies: high-redshift, ultraviolet: galaxies, (ISM:) HII regions, X-rays: binaries

1. The nebular He II problem

Nebular He II λ4686 emission in optical spectra of nearby sources has been discovered in the late 1980ies (see e.g. Pakull & Angebault (1986); Garnett *et al.* (1991)). Except for planetary nebulae (PN), where such emission is quite common, such objects are very rare in the local group, and the sources known then cover a diversity of objects, including a Wolf–Rayet star, a O type, a massive binary system, and one X-ray source (Garnett *et al.* (1991)). Since their discovery, the origin of this He II emission has been puzzling. Only stars with very high effective temperatures $T_{\rm eff} \gtrsim 80 - 100$ kK emit non-negligible amounts of He^+ ionizing photons above 54 eV (e.g. Schaerer (2002)), and because such temperatures are only reached in very peculiar evolutionary phases (e.g., in the WR or PN phase).

Zooming out of the local group, large numbers of star-forming galaxies showing nebular He II emission have been found, and these high-ionization lines are also present in AGN. E.g. using the Sloan survey, Shirazi & Brinchmann (2012) have found 2865 galaxies with nebular He II. Excluding AGN, they have ∼200 star-forming galaxies, some of

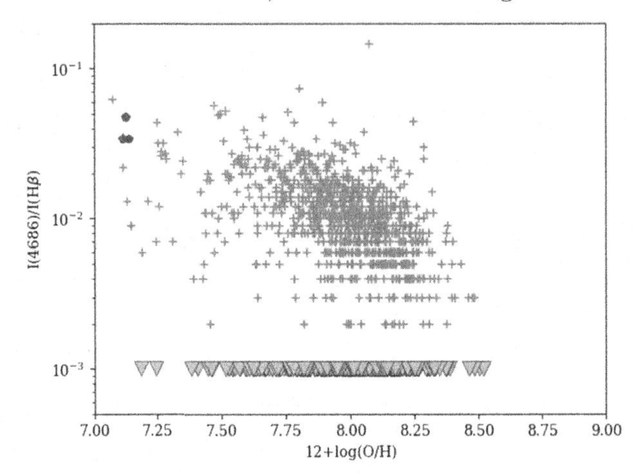

Figure 1. Observed $I(4686)/I(\mathrm{H}\beta)$ relative nebular line intensities as a function of metallicity. Observations of low-metallicity star-forming galaxies from the SDSS DR12-14 and observations of Izotov and collaborators are shown as red plusses. Measurements of the NW region of I Zw 18 are plotted with blue diamonds. The triangles show the non-detections ($\sim$1/3 of the sources).

them showing also the presence of broad emission lines, indicative of WR stars, in these galaxies. Since "normal" stars should be the most numerous ones in star-forming regions/galaxies, and their UV/ionizing spectra are expected to be dominated by massive stars, very hot stars (cf. above) are at best very rare in these objects. Therefore no or very weak nebular He II is expected in general. In Figure 1 we show a compilation of more than 1400 star-forming galaxies or regions thereof showing nebular He II emission from Izotov *et al.* (2016). This sample reprensents a significant jump in statistics. They are selected based on the quality of the spectra, allowing direct metallicity determinations via the auroral line method. Active galaxies are excluded based on BPT diagrams.

Nebular He II emission traced by the UV He II λ1640 line has also been observed in significant numbers of galaxies, some at low redshift and more at high-z (e.g. Cassata *et al.* (2013), Berg *et al.* (2018), Nanayakkara *et al.* (2019)). The same trend of increasing frequency and strength of nebular He II λ1640 with decreasing metallicity seems to be observed (Senchyna *et al.* (2017)), as for the optical line. This is expected since the He II lines are recombination lines from a simple hydrogenic atom, whose relative strengths between the UV and optical lines are to first order determined by atomic properties.

Since the discovery of He II emitters, other sources and processes that could emit more energetic photons than stellar sources have been suggested. These include X-ray binaries (XRBs), photoionization by X-rays, and strong shocks (see e.g. Pakull & Angebault (1986); Garnett *et al.* (1991); Thuan & Izotov (2005); Kehrig *et al.* (2015)). Other studies have explored if rotation and binarity could significantly alter the evolution of massive stars and create sufficiently hot stars and hence a harder ionizing spectrum (e.g. Szécsi *et al.* (2015), Götberg *et al.* (2018)). None of them has so far been able to quantitatively explain the observed intensity of the He II emission in low-metallicity galaxies. For example, the latest BPASS binary population and synthesis models underpredict the observed He II $\lambda4686/\mathrm{H}\beta$ intensities by $\sim$one order of magnitude, as shown in Fig. 2 (right). Shocks and X-rays seem to be able to explain specific cases, but appear insufficient in others (see above references). However, no predictive model that would allow linking shock models to other galaxy properties exists, and especially it is unclear how shocks would reproduce

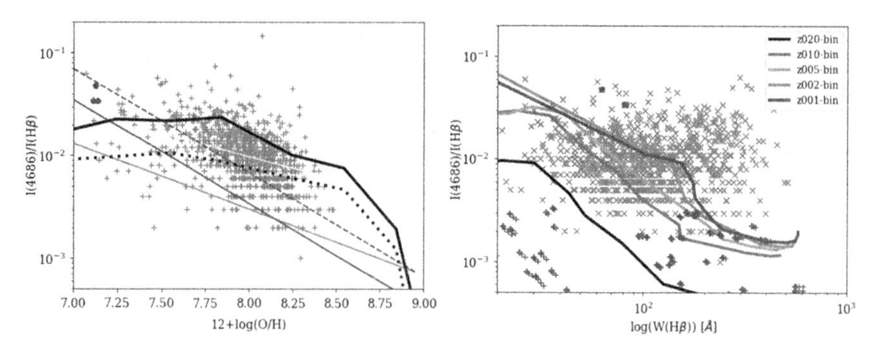

Figure 2. Observed and predicted $I(4686)/I(\mathrm{H}\beta)$ relative nebular line intensities as a function of metallicity (left) and the Hβ equivalent width (right) as an age indicator. Observations as in Fig. 1 *Left:* The observations are fitted by the yellow line (linear regression to the data points, detections only). Assuming $q = 2 \times 10^{10}$ photon/erg, the empirical L_X/SFR–O/H relations of Douna *et al.* (2015) and Brorby *et al.* (2016) translate into He II intensities shown by the blue and magenta solid lines, respectively, assuming constant SFR. The black lines show the predicted He II intensity adopting L_X/SFR predicted from the XRB synthesis models for a constant SFR over 10 Myr (dotted) and 0.1 Gyr (solid), and the same value of q. The blue dashed line differs from the solid line by assuming a value of q that is a factor of two higher. *Right:* Instantaneous burst models for different metallicities, showing the predicted age dependence. Blue crosses show the BPASS models from Xiao *et al.* (2018). Figs. from Schaerer *et al.* (2019).

the trends of He II intensity with metallicity (Fig. 1). In short, the He II problem remains overall an unsolved problem calling for new/different approaches.

2. A fresh look at the He II problem – empirical results

The literature on X-ray emission in and from star-forming galaxies (SFGs) provides interesting insights, which could help understand the nature of He II emission. Indeed, the observations show overall the following picture (e.g. Mineo *et al.* (2012), Douna *et al.* (2015), Brorby *et al.* (2016)):
(*a*) First, the X-ray emission of SFGs is dominated by point-like sources, whose luminosity follows a power-law, including ultra-luminous X-ray sources (ULX) at the bright end ($L_X > 10^{40}$ erg s^{-1}), high-mass X-ray binaries (HMXBs) and others.
(*b*) The total (spatially integrated) X-ray luminosity correlates to *first order* with the total star formation rate (SFR).
(*c*) There is an excess of X-ray emission per unit SFR, L_X/SFR, at low metallicity.
(*d*) The observed scatter in L_X/SFR is due to stochasticity of the ULX+HMXB at the high luminosity end of the X-ray luminosity function, which dominates the integrated X-ray emission.

These empirical findings suggest that X-ray bright sources (ULX and/or HMXB) could naturally explain the observed trends of He II emission, in particular the increase of He II/Hβ with decreasing metallicity, for the following reasons and in the following way: First, the Hβ (or any other H recombination line) luminosity scales linearly with the SFR, i.e. $L(\mathrm{H}\beta)/SFR \approx$ const. Second, assuming that bright X-ray sources emit a constant fraction of ionizing photons per X-ray luminosity above 54 eV, i.e. $q = Q(\mathrm{He}^+)/L_X$, the number of He$^+$ ionizing photons per L_X, is constant. Then the He II line luminosity is proportional to L_X, and the observed scaling of L_X with SFR (point b) translates immediately to a relation between the He II/Hβ line luminosities, hence also their relative intensity, which is the main observable. Finally, the observed increase of L_X/SFR towards low metallicity (point c) implies that $I(4686)/I(\mathrm{H}\beta)$ increases with decreasing O/H, as observed (Fig. 1). Furthermore, the observed scatter in L_X/SFR (point d) would also lead to scatter in $I(4686)/I(\mathrm{H}\beta)$. This is in essence what could explain nebular He II at low metallicity, the observed metallicity trend, and possibly also the scatter, as we proposed in Schaerer *et al.* (2019).

To relate the observed trends of nebular He II with X-ray emission, we basically have to make only a single assumption, namely that He II emission originates from or is related to the bright sources which dominate in X-rays, i.e. ULX and/or HMBX. In fact, this assumption appears very reasonable and is supported by some empirical evidence. Indeed He II emission has been observed in several ULX and bright HMXB nebulae (e.g. Kaaret & Corbel (2009), Pakull & Mirioni (2012)). Furthermore such sources are present in SFGs, and they significantly modifiy the ionizing spectrum (and hence affect certain emission lines), as shown e.g. in the very detailed study of the metal-poor He II emitting galaxy I Zw 18 by Lebouteiller *et al.* (2017). Indeed in this galaxy, Chandra observations have shown the presence of a ULX and demonstrated that it is located in the main HII region (NW region), where the bulk of He II emission originates from. This strongly supports a physical connection between bright X-ray sources and nebular He II. This finding is also confirmed by a new study of Heap *et al.* (2019), who also show that a well-motivated ULX spectral model is able to quantitatively explain the He II intensity in I Zw 18.

In Schaerer *et al.* (2019) we have used the observed X-ray and He II $\lambda4686$ line luminosities of I Zw 18 to empirically derive $q = Q(\mathrm{He}^+)/L_X = (1.0 - 3.4) \times 10^{10}$ photon/erg and adopt a "typical" value of $q = 2 \times 10^{10}$ photon/erg. Predicted $I(4686)/I(\mathrm{H}\beta)$ relative intensities obtained using this simple approach are plotted in Fig. 2. They show that the observations can be reproduced approximately with these simple assumptions.

3. Synthetic models including He II emission from X-ray binaries

To go beyond the simple empirical attempt, we have combined X-ray binaries (XRB) population synthesis models with models describing spectral evolution of "normal" stellar populations. The main results, taken from Schaerer *et al.* (2019), are shown in Fig. 2.

Concretely, we have used models developed by Fragos *et al.* (2013ab) to study the cosmological evolution of XRB populations that were recently recalibrated to updated measurements of the cosmic star-formation history and metallicity evolution (Madau & Fragos (2017). An important prediction of these models is the strong dependence of the XRB population on metallicity, both in terms of the formation efficiency of XRBs and the integrated X-ray luminosity of the whole population. This means that at low O/H more XRBs are predicted and their X-ray luminosity is higher. Therefore these XRB models predict a higher L_X/SFR at low O/H, as shown by the observations discussed above (point c). Additionnaly, these models predict a strong dependence of L_X on the age of the stellar population (see Schaerer *et al.* (2019)).

To bracket a range of star formation histories, we have examined instantaneous bursts and constant SFR. The predicted $I(4686)/I(\mathrm{H}\beta)$ intensities are shown in Fig. 2. They broadly cover the range of the observations, showing that our models can fairly well reproduce the observed $I(4686)/I(\mathrm{H}\beta)$, possibly with a superposition of populations.

As already mentioned, binary populations synthesis models such as BPASS do not predict sufficiently hard ionizing spectra to solve the He II problem (cf. Fig. 2). This shows that even hot/rejuvenated stars, which are created through binary interactions, are not sufficient, and that other sources (e.g. HMBX and ULX) need to be included. The earlier synthesis models of Cervinõ *et al.* (2002) including X-rays have also failed in this respect, since they primarily include soft X-ray emission from stellar winds, supernovae, and their remnants.

4. Conclusions and outlook

Based on empirical X-ray data for star-forming galaxies showing a very similar behaviour of increasing L_X/SFR as nebular He II emission with metallicity (Fig. 1), on the observational finding of a ULX in the strong He II $\lambda 4686$ emitting galaxy I Zw 18, and on the detailed modeling of this galaxy by Lebouteiller *et al.* (2017) and Heap *et al.* (2019), we have proposed that ULXs/HMXBs are the prime source of nebular He II emission low metallicity SFGs. A simple quantitative model using both empirical data and recent X-ray binary population synthesis models has recently been proposed by Schaerer *et al.* (2019) to solve the long-standing problem of nebular He II emission.

Obviously further applications, and tests are welcome and required, and this model does not exclude the contribution from other processes to the observed He II emission. For example, establishing observationally more firmly the correlation between X-ray and He II emission, both on spatially resolved scales and in integrated populations, should be a useful test. It also needs to be worked out if/how much the high-energy emission affects other emission lines, also in relation with the apparent hardening of the SEDs at high-z inferred by several studies (see e.g. Stark (2016)). In any case observations show that X-rays are ubiquitous in star-forming galaxies (even if weaker than in AGN), as discussed amply in the literature (cf. Sect. 1). Therefore, their effects on the SEDs need to be examined, a task which should be more important for high-z and low-metallicity galaxies, since HMXBs, ULXs, and their X-ray emission becomes stronger at low O/H.

References

Berg, D. A., Erb, D. K., Auger, M. W., Pettini, M., & Brammer, G. B. 2018, *ApJ*, 859, 164
Brorby, M., Kaaret, P., Prestwich, A., & Mirabel, I. F. 2016, *MNRAS*, 457, 4081
Cassata, P., Le Fèvre, O., Charlot, S., *et al.* 2013, *A&A*, 556, A68
Cervinõ, M., Mas-Hesse, M., & Kunth, D. 2002, *A&A*, 392, 19
Douna, V. M., Pellizza, L. J., Mirabel, I. F., & Pedrosa, S. E. 2015, *A&A*, 579, A44
Fragos, T., Lehmer, B., Tremmel, M., *et al.* 2013a, *ApJ*, 764, 41
Fragos, T., Lehmer, B. D., Naoz, S., Zezas, A., & Basu-Zych, A. 2013b, *ApJL*, 776, L31
Garnett, D. R., Kennicutt, Jr., R. C., Chu, Y.-H., & Skillman, E. D., 1991, *ApJ*, 373, 458
Götberg, Y., de Mink, S. E., Groh, J. H., *et al.* 2018, *A&A*, 615, A78
Heap, S. R., Hubeny, I., Bouret, J.-C., & Lanz, T. 2019, in Radiative signatures from the cosmos, ed. K. Werner & T. Rauch, ASP Conference Series, in press
Izotov, Y. I., Guseva, N. G., Fricke, K. J., & Henkel, C. 2016, *MNRAS*, 462, 4427
Kaaret, P. & Corbel, S. 2019, *ApJ*, 697, 950
Kehrig, C., Vílchez, J. M., Pérez-Montero, E., *et al.* 2015, *ApJL*, 801, L28
Lebouteiller, V., Péquignot, D., Cormier, D., *et al.* 2017, *A&A*, 602, A45
Madau, P. & Fragos, T. 2017, *ApJ*, 840, 39
Mineo, S. *et al.* 2012, *MNRAS*, 419, 2095
Nanayakkara, T. *et al.* 2019, *A&A*, 624, A89
Pakull, M. W. & Angebault, L. P. 1986, *Nature*, 322, 511 EP
Pakull, M. & Mirioni, L. 2012, arXiv:0202488
Schaerer, D. 2002, *A&A*, 382, 28
Schaerer, D., Fragos, T., & Izotov, Y. I. 2019, *A&A*, 622, L10
Shirazi, M. & Brinchmann, J. 2012, *MNRAS*, 421, 1043
Senchyna, P. *et al.* 2017, *MNRAS*, 472, 2608
Stanway, E. R. & Eldridge, J. J. 2018, arXiv e-prints, 1811.03856
Stark, D. P. 2016, *ARAA*, 54, 761
Szécsi, D. *et al.* 2015, *A&A*, 581, A15
Thuan, T. X. & Izotov, Y. I. 2005, *ApJS*, 161, 240
Xiao, L., Stanway, E. R., & Eldridge, J. J. 2018, *MNRAS*, 477, 904

Uncovering Early Galaxy Evolution in the ALMA and JWST Era
Proceedings IAU Symposium No. 352, 2019
E. da Cunha, J. Hodge, J. Afonso, L. Pentericci & D. Sobral, eds.

doi:10.1017/S1743921319009050

Interpreting galaxy properties with improved modelling

E. R. Stanway[1] and **J. J. Eldridge**[2]

[1]Dept. of Physics, University of Warwick, Gibbet Hill Road, Coventry, CV7 4AL, UK
email: e.r.stanway@warwick.ac.uk

[2]Dept. of Physics, University of Auckland, Private Bag 92019, Auckland, New Zealand
email: j.eldridge@auckland.ac.nz

Abstract. Observations of star-forming galaxies in the distant Universe have confirmed the importance of massive stars in shaping galaxy emission and evolution. Distant stellar populations are unresolved, and the limited data available must be interpreted in the context of stellar population models. Understanding these populations, and their evolution with age and heavy element content is key to interpreting processes such as supernovae, cosmic reionization and the chemical enrichment of the Universe. With the upcoming launch of JWST and observations of galaxies within a billion years of the Big Bang, the uncertainties in modelling massive stars - particularly their interactions with binary companions - are becoming increasingly important to our interpretation of the high redshift Universe. In turn, observations of distant stellar populations provide ever stronger tests against which to gauge the success of, and flaws in, current massive star models. Here we briefly review the current status binary stellar population synthesis.

Keywords. binaries: general, methods: numerical, galaxies: high-redshift, X-rays: binaries

1. Background

The galaxies observed in the distant Universe ($z > 2$) differ in significant respects from those seen more locally. It has been known for many years that the typical star formation rates of massive galaxies were much higher in the past than at the current time (e.g. Madau *et al.* 1996), and the typical stellar populations rather younger. Since the Universe is enriched by supernovae and winds from old stars, it also follows that the average metallicity of galaxies decreases as one moves towards higher redshifts.

The star-forming nature of typical galaxies in the distant Universe has practical advantages. While the optical emission from these sources is redshifted beyond the capabilities of most ground-based instrumentation, star-forming galaxies are ultraviolet luminous. This has allowed rest-frame ultraviolet (rest-UV) observations to identify large numbers of photometrically selected galaxy candidates at high redshift, and also enabled spectroscopic confirmation or even characterisation of a significant fraction of these based on ultraviolet emission and absorption features.

The rest-UV emission of star-forming galaxies at high redshift, and indeed at any redshift, is entirely dominated by the most massive, hottest and hence most UV-luminous and shortest-lived stars in a stellar population. These emit the stellar continuum shortwards of the Balmer break, and are also responsible for powering nebular emission from photoionized H II regions. These are impossible to resolve in the very distant Universe, but their presence can be inferred from the strength of narrow (low velocity) emission features in both rest-UV and rest-optical integrated light spectra of unresolved sources.

In addition to the indirect evidence from the continuum and nebular emission, certain key features have been identified as diagnostic of the presence of massive stars at high redshift. Prominent amongst these is the He II 1640Å emission line which occurs both in stacked spectra of typical ultraviolet-selected galaxies, and in individual sources where sufficiently bright. While a subset of these appear to show narrow lines (suggesting an emission source embedded in highly ionized nebular gas, e.g. Erb *et al.* 2010), many detected He II lines appear to be velocity broadened (e.g. Shapley *et al.* 2003) - a feature associated with the strong winds of massive, stripped-atmosphere Wolf-Rayet stars in the local Universe.

Such massive star dominated, young stellar populations are rare in the local Universe. High mass Galactic star forming regions typically lie in the Galactic plane and are often heavily obscured by dust, making observations (particularly in the rest-UV) challenging. They tend to be limited in size, forming stars at a very different rate to those seen in the distant population. Perhaps more importantly, very little local star formation occcurs in regions less metal rich than the Sun. The typical metallicities expected in distant galaxies ($Z < 0.2\,Z_\odot$) are not represented in the resolved, Galactic star forming population.

Amongst systems sufficiently close to resolve and analyse the contribution of individual stars to the integrated light, perhaps the closest analog to a distant galaxy can be seen in the 30 Doradus region of the large Magellanic Cloud. This super-starbust has been subject to intense study in recent years and has revealed abundant populations of massive stars (e.g. Schneider *et al.* 2018a,b), including 'Very Massive Stars' (with initial masses extending up to almost $300\,M_\odot$), and a near-ubiquitous population of stellar binaries (e.g. Sana *et al.* 2012). Indeed, the average number of companions for massive stars ($M_{\rm init} > 30\,M_\odot$) actually exceeds 1, suggesting significant numbers of triple and higher order multiple systems. It has been estimated that 70% of such massive stars will interact with a companion during their lifetime (Sana *et al.* 2012). Such interactions can include mass transfer through Roche lobe overflow, stripping of atmospheres, spin-orbit angular momentum transfer and, in extreme cases, merger. All of these have the potential to modify the spectrum of affected stars, and so will impact the interpretation of light from an integrated stellar population.

Such considerations have resulted in a growing recognition that the stellar population synthesis models widely used to interpret galaxy properties in the local Universe require modification to confront the very different physical conditions more commonly encountered at high redshift. It has also highlighted the considerable uncertainties that remain in massive star evolution, particularly at low metallicities where local exemplars are rare or non-existant, and when binary interaction effects are included. As ALMA opens new windows on the infrared properties of distant galaxies, and JWST promises unprecedented sensitivity to both the rest-UV and rest-optical emission of distant sources, the need to explore these uncertainties and improve our modelling of stellar populations is becoming acute.

2. Population and Spectral Synthesis

When comparing observations to galaxy models, care must be taken to account for the contributions of stellar types which may not have an obvious impact on the observed integrated light, but which nonetheless contribute to the physical interpretation. A widely-used approach to doing this is the technique of stellar population synthesis.

The key elements of this technique is illustrated schematically in the central region of figure 1 (within the dashed line). The evolution of individual stars is assumed to be a well understood function of initial stellar mass alone. Stellar evolution models (occasionally combined with empirical approximations) are used to establish the temperature, luminosity and perhaps surface gravity of a star as a function of its age. Alternately isochrones, which give the position of stars in luminosity-temperature space at fixed age as a function of initial mass, can be substituted. These stellar models are then combined,

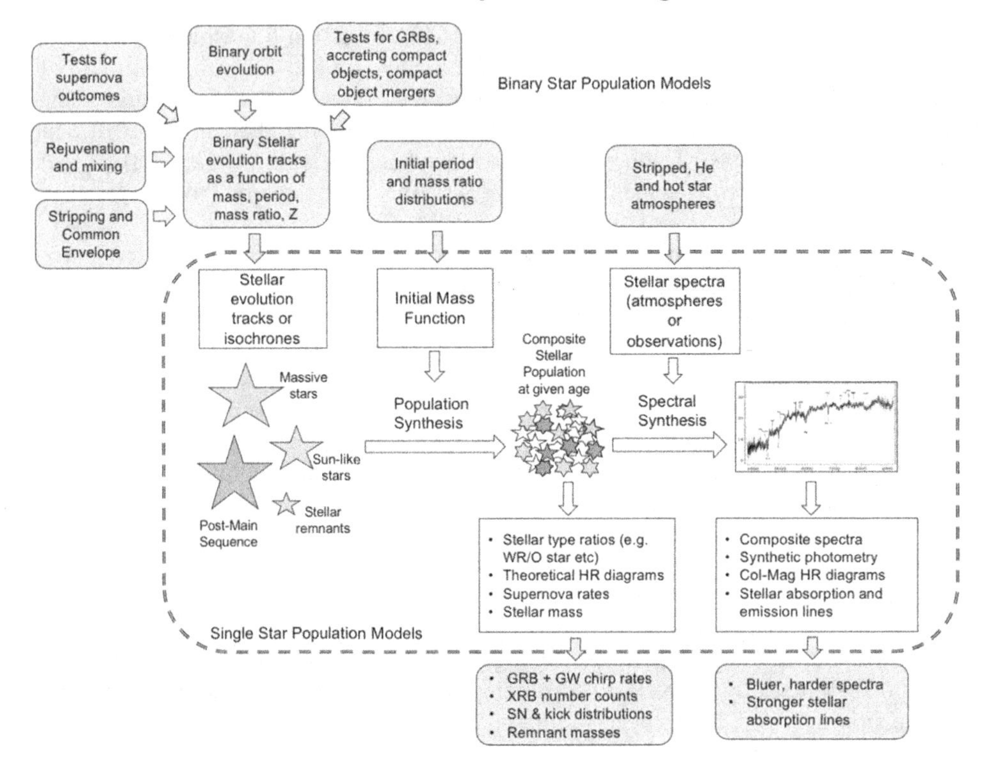

Figure 1. A schematic diagram indicating the key elements, inputs and outputs of stellar population synthesis models, and how these are modified when binary stellar evolution pathways are considered.

and the relative contibution of stars with a given initial mass determined by an initial mass function (IMF). This accounts for the presence of many low mass stars, as well as the relatively few luminous and massive stars.

Such a population synthesis models can yield information on the expected ratio of different stellar types as a function of age, as well as on the number of massive stars reaching the end of their lifetimes (i.e. type II core-collapse supernova rates), and luminosity-temperature Hertzsprung-Russell diagrams. Often comparison of models with data requires a further step. Spectral synthesis models assign a theoretical atmosphere or empirically derived spectrum to each modelled star on the basis of its physical properties. These are summed with the appropriate weightings determined by the population synthesis, to produce the integrated light spectum of an entire stellar population as a function of age, as well as synthetic photometry in desired wavebands.

The basic output of such models is known as a simple stellar population or SSP (e.g. Bruzual & Charlot 2003; Maraston 2005; Leitherer *et al.* 1999, 2014). It estimates the properties of a group of stars sufficiently large to fully sample the IMF, which formed simultaneously and instantaneously at a known time before the point of observation. These SSPs can be combined with a star formation history to create a complex or composite stellar population (CSP) and further processed using radiative transfer calculations to account for modifications to the stellar spectrum by dust or gas, either in the circumstellar medium or along the line of sight (e.g. Ferland *et al.* 1998; da Cunha *et al.* 2008). To reproduce the characteristics of unresolved stellar populations, star formation history, stellar properties, gas properties and dust extinction or emission characteristics may need to be fit to the data simultaneously.

3. Binary Population Models

Binary population synthesis models have been developed by the stellar physics community over a considerable timespan (e.g. van Bever & Vanbeveren 1998; Van Bever & Vanbeveren 2000; Hurley *et al.* 2002; De Donder & Vanbeveren 2004; Willems & Kolb 2004; Belczynski *et al.* 2008; Toonen & Nelemans 2013. The majority of these have focussed on the effects of binary interactions on massive stars, since these are both more likely to be found in a binary, and also show the most dramatic effects of binary interactions. However their use has been restricted to a narrow range of stellar evolution studies until recently, due both to the added computational burden they introduce and the additional assumptions and uncertainties they highlight.

Incorporating binary evolution pathways in a synthesis model is challenging. In the regions of Figure 1 outside the dashed lines, we illustrated key elements that must be considered above and beyond those in a standard single-star population synthesis. Binary stellar evolution tracks are no longer a simple function of mass, but also depend on the mass ratio between primary and secondary and the initial binary separation. Evolution in the binary orbit and Roche lobes must be tracked as the stars age and interact. In addition to solving the equations of stellar structure and hydrostatic equilibrium for each layer of the star at each time step, mass transfer either to a companion or out of the system entirely must be calculated. This can lead to substantial changes in the surface composition of the stars (e.g. stripping), to rejuvenation of the star through rotational spin-up and subsequent mixing, and in extreme cases to phases of common envelope evolution. At the end of each stellar core-burning lifetime, the system must be tested for the occurrence of supernovae which may dissociate the binary.

Given the wide range of initial masses, mass-dependent binary fractions, periods and mass ratios represented in a stellar population, the number of individual models contributing to an output spectrum increases from perhaps hundreds to high thousands, and each model incorporates more steps with more physical parameters and effects to be calculated and considered. A range of approaches exist to address these difficulties.

The strategy adopted by some models, known as rapid-pop-synth models, is to take a semi-analytic approach. These sample the parameter spaces occupied by binaries with detailed models, and fit analytic approximations to the behaviour of key parameters at intermediate masses and separations (Hurley *et al.* 2002; Pols *et al.* 1995). A strong advantage of such approaches is that a population can be rapidly constructed by integrating the analytic formula with appropriate weighting factors rather than by combining individual models. A disadvantage is that such approaches risk neglecting the impact of stellar populations which, while having a significant impact on the integrated light at a given age or metallicity, may be small in number or occupy a narrow region of parameter space. Given that many of the responses to binary interactions are very sensitive to initial conditions, this is a concern when interpreting extreme or rare objects.

An alternate strategy is to work with a fixed grid of detailed (usually 1D) stellar evolution models which is sufficiently finely sampled that interpolation between grid elements is not required (e.g. Han *et al.* 2007; Zhang *et al.* 2005; Eldridge *et al.* 2017). Since these models contain information on the structure, mass transfer rates, composition and other physical properties of individual stars, more potentially observable properties of a population can be calculated (e.g. compact remnant mass distributions and accreting binary number counts, mass ranges for stripped stars etc) but at a computational cost. Such detailed models are relatively slow and resource-intensive to calculate.

When a binary population has been synthesised (incorporating assumptions for the binary star occurance fraction, the initial mass ratio and separation distributions) the individual stellar models can be wedded to atmosphere models in a spectral synthesis step as before. This often requires a wider range of atmospheres (incorporating, for example, stripped star and extremely hot massive stars) than in the single star population synthesis case. Binary models yield a wider range of output products which can be compared to observational constraints. Like any model, they will also only be as good as the assumptions made when they are calculated.

4. Uncertainties in Population Synthesis

A wide range of assumptions must be made in calculating spectral synthesis models of stellar populations. Most of these are discussed extensively in the literature (see e.g. Conroy 2013; Eldridge *et al.* 2017) but we summarise key points below.

4.1. *Uncertainties affecting all population synthesis*

Initial mass function

The presumption of an initial mass function presents both practical and philisophical problems as discussed in detail by Hopkins (2018). There is increasing evidence that a single, universal stellar mass function that describes the fragmentation pattern of all molecular gas clouds cannot explain the observed stellar populations in different galaxies. Reasons for variation may include the composition of the molecular clouds and their temperature profile. Since the initial mass function determines the ratio of the stars dominating the emission (i.e. UV luminous, massive stars in the high redshift case) to those which are difficult to observe (e.g. low mass stars which may contribute significantly to the rest-infrared), this has consequences for determining parameters such as the total stellar mass of a population, its absorption line indices or its population of compact remnants at late times.

Post-main sequence evolution

Often overlooked inputs to the stellar models on which all population synthesis codes are built is the prescription adopted for the opacity of the stellar atmospheres and for stellar winds. These becomes particularly significant at high metallicities and during post-main sequence evolution. ALMA observations have provided direct evidence for the superwinds driven by asymptotic giant branch stars (e.g. Nhung *et al.* 2019), but the mass loss rates of individual stars show a wide variation and it remains unclear which prescription for this best describes the average behaviour of stars, and how that depends on mass and other physical properties (e.g. Rosenfield *et al.* 2014). While the AGB stars which dominate old stellar populations are unlikely to be important in the distant Universe, massive stars evolve very rapidly and their behaviour on the giant branch, while short-lived, may include superwind phases which substantially alter their evolution.

Short-lived instability effects

The calculation of stellar evolution models requires an inevitable compromise between resolution in both stellar evolution timescale and mass or radius, and the computational time required. A model which follows the evolution of a star though a Hubble time cannot probe the stellar properties of that star on timescales of hours, weeks, months or even a few years without being prohibitively expensive. While most codes are able to adjust their time steps adaptively, spanning a very large dynamic range in time tends to cause numerical instability. As a result, most evolution tracks fail to correctly follow very rapid, large magnitude changes in luminosity or mass. Examples of this include thermal pulsations of stars occurring in the instability strip, or rapid changes in internal structure such as occur at the onset of core burning for Helium, Carbon etc. Analytic or algorithmic prescriptions are used to bridge these phases of rapid evolution, but their impact on the integrated light of a population may be under or overestimated (see e.g. Maraston 2005).

Extrapolation to low metallicity

As discussed in the Introduction, most of our empirical evidence for the evolution of individual stars is limited to the very local Universe, where massive stars forming at low metallicities are rare. The Extremely Metal Poor stars found in the Milky Way are very old, and therefore low mass, remnants of early star formation, and can provide little insight into processes which only occur in high mass stars. While theoretical stellar evolution models are largely based on well-understood physical principles valid at all metallicities, they are routinely cross checked and key assumptions calibrated based on such local observations. Extrapolations to significantly sub-Solar (and particularly any extremely-low Population III-like) metallicity are therefore often unsupported by observations.

Rotation

Very few stellar evolution tracks exist for rotating stars. There is fair evidence that most massive stars are born rotating at an appreciable fraction of their break-up speed (see e.g. Dufton *et al.* 2019, and references therein). At high metallicities, stars rapidly spin down due to stellar wind-driven mass and angular momentum loss. However it is possible that some systems continue to rotate for an appreciable fraction of their evolutionary lifetime. This induces additional mixing of the stellar interior, feeding additional light elements to the core, and mixing heavy elements into the stellar atmosphere. As a result the star burns both brighter and hotter. In the extreme case, a star can evolve as a chemically homogenous system (i.e. with uniform composition) rather than through the standard shell burning prescription. The result is a hotter, bluer spectrum than is obtained for non-rotating stars (See e.g. Topping & Shull 2015; Leitherer *et al.* 2014).

Extrapolation to the far ultraviolet

The ionizing emission spectrum of any source is extremely difficult, if not impossible, to determine observationally. Neutral hydrogen in the interstellar and integalactic medium efficiently scatters hard ultraviolet and X-ray photons, while any trace of dust reprocesses the emission into the thermal infrared. The hardest radiation (extreme X-ray/gamma-ray) has such a long mean-free path that while it escapes absorption by hydrogen, it also has little impact on its surroundings. In the local Universe, it is possible to probe directly to a little shortwards of the hydrogen ionization edge at 912Å, but the far-ultraviolet remains out of reach even here. As a result, the ionizing emission spectrum of stars of a given temperature, composition and structure is usually calculated from theory (e.g. Gräfener *et al.* 2012; Götberg *et al.* 2018) and may be calibrated against the spectrum of ultraviolet and optical emission lines that results from reprocessing by nebular gas. Such calibration is dependent on the physical conditions of the gas, and again extrapolation to low metallicity is challenging.

Uncertainties in post-processing

Each stage of post-processing required before comparison of simple stellar populations to observations (e.g. star formation history, nebular gas, dust etc) requires assumptions regarding the physical conditions in a given galaxy and introduces unavoidable degeneracies in the interpretation of any integrated light source. In some cases, these may be broken by the combination of detailed data across a wide wavelength range.

4.2. *Uncertainties specific to binary models*

Binary parameters

Key inputs to a binary population synthesis are the fraction of stars in binaries, their initial period distribution as a function of initial mass, and the distribution of initial mass ratios as a function of both mass and period. These distributions have only recently been empirically determined for stellar populations in the local Universe (Moe & Di Stefano 2017). Inevitably these are derived from stellar populations which are at near-Solar metallicity, and also not at zero age. This is an area in its observational infancy, where much higher sensitivities and angular resolutions will be required to make substantial progress. There is some evidence that the binary fraction may be a sensitive function of metallicity, as well as mass (Moe *et al.* 2019), but there is insufficient data to characterise this function. As a result, initial binary parameter functions derived in the local Universe are typically applied at all metallicities.

Mass transfer efficiency

While Roche lobe overflow can be calculated from a pair of stellar models (using a spherical approximation in the 1D model case), the subsequent evolution of the stars will also depend on how efficiently that mass is transfered to the binary companion as opposed to being ejected from the system. The efficiency is likely stellar mass dependent, and may also be sensitive to other parameters such as metallicity and magnetic fields (which no current population models consider). Since mass transfer is also associated with angular momentum transfer between stars, the prescription used here will also interact with treatment of rotational mixing and also tidal interactions between stars (again often neglected in current models, although see de Mink *et al.* 2013, Chrimes *et al.*, in prep).

Common Envelope Evolution

The common envelope evolution phase has a dramatic effect on the evolution of a binary, stripping stellar envelopes and hardening the binaries into very close systems with strong gravitational and tidal interactions. However this phase falls firmly into the category of rapid processes with an extreme dynamic range of scales. As already discussed, these present a challenge to all stellar models. Common envelope phases are usually approximated in evolution models using analytic prescriptions informed by observational constraints and by the few detailed smooth particle hydrodynamics models that attempt to follow this process (see Ivanova *et al.* 2013, for a review).

Accretion-related emission

In addition to processing binary models through nebular gas and dust, as in the single star case, a potentially important additional emission component arises directly from the stellar population but is not captured by the stellar spectral synthesis. Stellar populations naturally give rise to binaries in which a compact remnant left after the death of the primary undergoes episodic accretion from the secondary. This can contribute a very high luminosity, very hot spectral component, which is short-lived in each individual binary but can nonetheless contribute significant hard UV and X-ray flux when averaged across a population. Again, this component has not been considered to date by any detailed binary population and spectral synthesis model.

Important:
Before leaving the issue of population synthesis uncertainties, there is an important point to be made: modelling stellar populations with single star evolution tracks **does not remove** the uncertainties associated with binary evolution. Instead it makes an implicit assumption **that we know to be wrong**: that the binary fraction is zero at all masses. Populations constructed from binary tracks **have different luminosities and spectral features** than single star populations at the same mass and age. Thus while any given binary model may be making one or more faulty assumptions, **all** single star models are *known* be making faulty assumptions which will lead to **incorrect interpretation** of observational data. The offset between the single star model and the 'true' model is likely a function of age, stellar mass, stellar metallicity, gas and dust properties. It is small in the high metallicity, old stellar population regime common in the local Universe, but cannot be neglected at high redshift or in extreme galaxies nearby.

5. Binary Population and Spectral Synthesis (BPASS)

The Binary Population and Spectral Synthesis (BPASS) project† is, as the name suggests, a grid of synthesis models which generates both population and spectral data products for simple stellar populations incorporating binary evolution pathways. The inputs are a custom grid of detailed 1D stellar evolution models generated using the BPASS stellar evolution code (a binary variant of the older STARS code, Eggleton 1971). These are combined with the binary stellar population parameters of Moe & Di Stefano (2017), a supernova kick prescription (we currently use that of Hobbs *et al.* 2005), and a compact-binary gravitational wave evolution algorithm. The models are combined with a high resolution grid of stellar atmospheres for normal stars, together with the Potsdam Wolf Rayet models for stripped stars (PoWR, Hamann & Gräfener 2003; Hamann *et al.* 2006; Gräfener *et al.* 2012) and a custom set of hot O-star models, interpolating the atmosphere models in luminosity, surface gravity and metallicity where required. Populations are generated at 13 metallicities, with nine initial mass functions in the current distribution.

The key physical inputs into the v2.1 BPASS stellar evolution and synthesis models are described in full in Eldridge *et al.* (2017). Refinements to the binary prescription and treatment of low mass stars resulted in the current v2.2 distribution, which is described in Stanway & Eldridge (2018).

Initialy developed as stellar models for supernova progenitors (e.g. Eldridge *et al.* 2008, 2013) and as a population synthesis tool for interpreting distant galaxies (Eldridge & Stanway 2009, 2012), BPASS has also been applied to nearby massive stellar populations (e.g. Wofford *et al.* 2016), old stellar populations (e.g. Stanway & Eldridge 2018) and cosmic reionization (e.g. Stanway *et al.* 2016; Ma *et al.* 2015). Recent development work has considered the predicted rate and properties of gravitational wave transients (e.g. Eldridge *et al.* 2019), explored gamma-ray burst progenitors (Chrimes et al, in prep), and also returned to the initial focus on low metallicity, young star forming galaxies in both the distant Universe and the rare local extreme galaxy population (see next section).

6. The Mystery of the Hard Ionizing Spectra

Amongst the most significant effects of incorporating binary pathways in a stellar population is the production of longer lived, hotter stars than are seen in single star

† See bpass.auckland.ac.uk or warwick.ac.uk/bpass for full data release and details.

populations. This leads to a harder and stronger rest-frame far ultraviolet radiation field being inferred from the (observable) near ultraviolet. This prediction of far-UV flux is important for calculations of the reionization of the Universe, in which the ionizing photon production efficiency, $\xi_{\rm ion}$, is a key parameter (see e.g. Stanway *et al.* 2016). It is also key to interpreting the nebular emission spectra of distant galaxies - an area which is currently challenging even binary population synthesis models.

6.1. *Observational Evidence*

Analysis of the ionizing spectra of distant galaxies is undertaken primarily through study of those ionizing photons reprocessed by nebular gas in the interstellar medium. Detection of multiple, high signal-to-noise absorption and emission features is now routinely possible in galaxies at $z \sim 2-4$ (e.g. Du *et al.* 2018; Steidel *et al.* 2016; Maseda *et al.* 2017), and even higher redshifts in rare cases, a situation which has been markedly improved by the advent of near-infrared sensitive multi-object spectrographs on some of the world's largest telescopes. As a result it is now possible to study spectral line ratios and construct photoionization models of the physical conditions in these star-forming systems. While there is always degeneracy between assumed gas conditions and the spectral energy distribution of the ionizing source, a large number of distant systems are now showing evidence for the existence of a very hard ionizing radiation field. This is observed both in rest-optical line ratios such as [O III λ5007Å]/[O II λ3727Å] or [O III λ5007Å]/H β λ4681Å, and directly in the strength of nebular emission lines including C III] λ 1909Å (which requires a 48 eV ionization potential) and He II λ 1640Å (at 54.4 eV) in the ultraviolet (Du *et al.* 2018; Erb *et al.* 2010; Berg *et al.* 2018).

Such work is not limited to the very distant Universe. A large and somewhat disparate class of low redshift ultraviolet luminous and extreme emission line galaxies has now been identified as having similar physical conditions to galaxies in the distant Universe (see e.g. Heckman *et al.* 2005; Greis *et al.* 2016; Amorín *et al.* 2015). Of these a subset are 'Lyman-continuum leakers', in which a measurable fraction of their flux just shortwards of 912 Å can be measured directly (e.g. Izotov et al. 2018). Many of these sources also show evidence for exceptionally hard ionizing spectra, inferred from their nebular emission.

One measure of this spectral hardness can be provided by the ratio of the recombination lines of hydrogen (typically the H I Balmer lines at 4861 or 6563Å) and helium (He II at 1640 or 4686Å). These are only weakly sensitive to the temperature and electron density of the nebular gas and so can be converted to a ratio of ionizing photon flux: Q(H I)/Q(He II) (see Stanway & Eldridge 2019, for details). Figure 2 indicates the inferred photon flux ratios for an indicative sample of extreme emission line sources observed at both high and low redshift: SL2S J021737-051329 at $z = 1.8$ (Berg *et al.* 2018), Q2343-BX418 ($z = 2.3$, Erb *et al.* 2010), SGAS J105039.6+001730 ($z = 3.6$, Bayliss *et al.* 2014), SBS 0335-052E ($D = 54$ Mpc, Kehrig *et al.* 2018), HS1442+4250, J0940+2935, J119+5130 and UM133 ($D = 11, 8, 22$ and 29 Mpc respectively, Senchyna *et al.* 2019) and the mean properties of stacked $z \sim 2.3$ Lyman break galaxies (Steidel *et al.* 2016). While the 'typical' distant galaxy is less extreme than some individual examples, it is clear that very low Q(H I)/Q(He II) ratios must be possible in some cases, implying exceptionally hard spectra ionizing the nebular gas.

6.2. *IMF variation*

IMF variations can substantially affect the ionizing flux. An IMF which increases the fraction of massive stars in a population will boost the ionizing flux below both the $\lambda = 912$ Å hydrogen and 227.9 Å helium ionization edges, with the ratio between them sensive to the slope of the IMF in the massive and very massive star regime. The He II

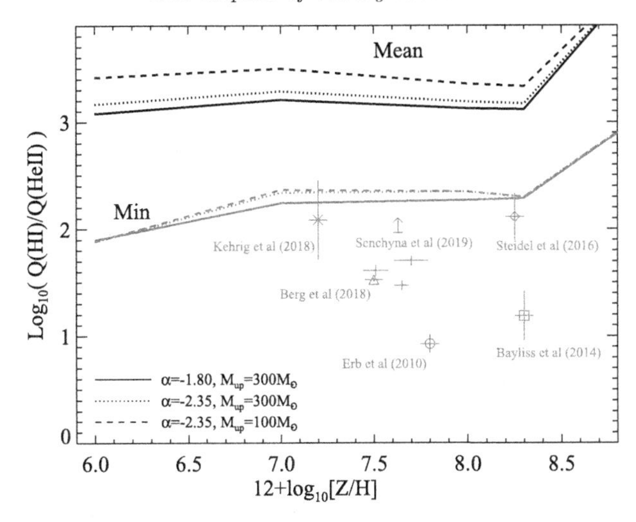

Figure 2. The hardness of the ionizing spectrum originating from a young starburst galaxy, as measured by the ratio of photons capable of ionizing hydrogen and helium. Lines indicate the minimum and mean photon flux ratios obtained in binary stellar population models at ages up to 10^8 years, with three different presumed initial mass functions. Points are calculated from observational data for extreme ionizing galaxies in the references labelled. Adapted from Stanway & Eldridge (2019).

line in particular is also sensitive to the upper mass cut-off in the stellar population, particularly at low metallicity and log(age/years) = 6.3–6.5. At these ages very massive stars are approaching the end of their life, swelling to become very luminous giants and (when stripped by winds or binary interactions) Wolf-Rayet stars. A single such star can dominate the emission of a stellar population. The presence or absence of these stars is strongly dependent on the initial mass distribution and how well the massive star IMF is sampled.

Stanway & Eldridge (2019) explored a range of plausible initial mass functions in the BPASS stellar population model formalism, varying both the IMF slope (α) and its upper mass cut-off (M_{up}). The ionizing photon flux ratio of three representative models are shown overplotted on figure 2. A slight subtlety arises in the direct comparison to the observations, as the latter are calibrated in terms of oxygen abundance while the stellar models are driven by the iron abundance. The models have been offset accordingly to account for a 0.3 dex abundance enhancement in oxygen (e.g. Steidel *et al.* 2016). As the figure demonstrates, while binary models can reproduce the high He II line strengths of some galaxies, including that typical of star-forming galaxies at $z \sim 2-3$, the exceptionally low photon flux ratios inferred from some extreme galaxies lie an order of magnitude below those inferred from models, regardless of IMF. Indeed, even a synthetic population comprised entirely of very massive 300 $M_{\odot}$ stars would not produce the observed line ratios. As discussed in Stanway & Eldridge (2019) and Eldridge *et al.* (2017), reproducing the hardest ionizing spectra will require exploring ionizing sources besides stellar photospheres.

6.3. *X-Ray Binaries*

Accreting binaries are a natural product of binary stellar population synthesis. The potentially important role of accreting compact objects and their associated accretion disks in ionizing photon production has been recognised for a long time (e.g. Van Bever & Vanbeveren 2000), and has been discussed in a high redshift context at conferences in

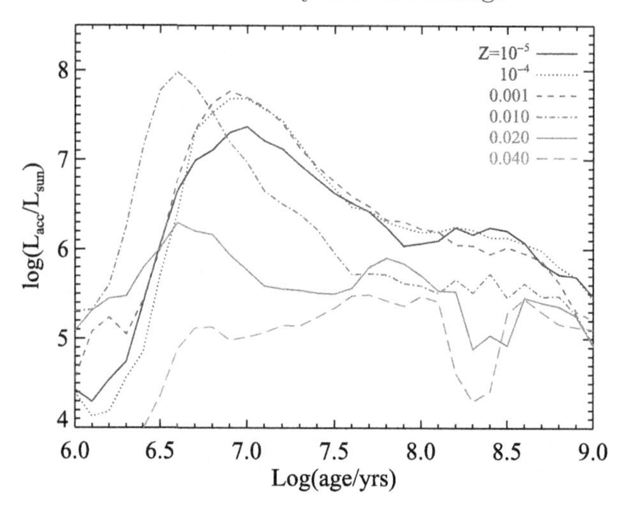

Figure 3. The evolution of bolometric accretion-powered luminosity from X-ray binaries in BPASS v2.2.1 models, illustrating the different behaviour as a function metallicity. Calibrating these preliminary estimates and the spectral shape of XRB emission is a work in progress (Eldridge *et al.*, in prep; Stanway *et al.*, in prep).

recent years†, motivated in part by the observational evidence that local Lyman break analogue galaxies tend to show an excess of resolvable X-ray binaries (Basu-Zych *et al.* 2013). Recently Schaerer *et al.* (2019) demonstrated that incorporating an ad hoc prescription for the X-ray emission from accreting binaries, informed by observations of extreme local sources, alongside the stellar emission generated from population synthesis models may help to explain the strength of observed He II emission in some low metallicity galaxies.

Limited information on the X-ray binary population generated by BPASS models has been available as an unverified, 'on request' data product since the BPASS v2.1 release in 2017 (Eldridge *et al.* 2017). In Figure 3 we illustrate the time evolution of the accretion-powered luminosity component calculated by BPASS v2.2.1 (Stanway & Eldridge 2018) as a function of metallicity. Tracks assume a 10% efficiency in conversion of gravitational potential energy to luminosity, and give the average population bolometric luminosity in each time bin for a simple stellar population with initial mass of $10^6\,M_{\odot}$. Models are smoothed across three time bins to account for stochasticity introduced by the underlying grid of stellar models. The figure illustrates the key motivation for exploring X-ray binaries as a source of ionizing photons at high redshift: low metallicity stars have weaker winds, they retain more of their initial mass at the end of their lifetime, form more massive compact remnants, and so are more likely to form highly luminous high mass X-Ray Binaries. Rather than the hundreds of millions of years it takes for higher metallicity, lower mass X-ray binaries to form, low metallicity systems establish a substantial accretion luminosity within a few Myr of the onset of star formation, and remains luminous for almost a Gyr.

Analysis to model the accretion disk dominated emission spectrum and calibrate the efficiency of these systems against observational evidence is ongoing. We expect the next BPASS data release to include calibrated accretion-powered ionizing spectra and data on the X-ray binary population, derived consistently with the underlying stellar population and integrated stellar spectra (Eldridge *et al.*, in prep; Stanway *et al.* in prep.)

† see e.g. http://www.iastro.pt/research/conferences/lyman2018/pdfs/ElizabethStanway_Lyman2018.pdf

7. Final Thoughts

The evolution of population and spectral synthesis models has been rapid and dramatic in recent years. While in part driven by advances in computational capacity, perhaps more important has been the development and implementation of prescriptions for ever more complex aspects of stellar evolution, and the observational instrument developments which have informed these. The recognition that distant galaxies provide a test bed for modelling the behaviours of high mass, low metallicity stellar populations has formed a virtuous circle with the growing realisation that such populations cannot be neglected when interpreting the distant Universe. Local exemplars of extreme galaxies providing opportunities for more detailed studies, an opportunity which is being widely exploited by the extragalactic community, and which has begun to be realised by the stellar evolution community in turn.

The contribution of ALMA has not been discussed in detail in this work. This should not be taken to indicate that it does not play a role. Long baseline imaging, particularly of lensed arcs, has begun to explore the physical conditions in distant star forming galaxies in exquisite detail and to redshifts inconceivable just a few years ago (e.g. Smit *et al.* 2018; Hashimoto *et al.* 2018, 2019). Intriguingly, one result of such observations has been to identify higher than expected ratios between the [O III] 88μm and [C II] 158μm emission lines in such sources, and potentially also elevated dust temperatures. While interpreting these features is challenging given the observational constraints involved, both hint at the same hard ionizing spectrum and excess of hot photons implied by the He II observations in the rest-frame ultraviolet and optical. Just as importantly, ALMA observations may be key to resolving some of the uncertainties in massive star evolution, particularly in terms of the mass loss rates and outflows from giant stars observed locally.

The launch of the James Webb Space Telescope presents both an opportunity and a challenge for the population synthesis and binary modelling communities. The large primary mirror of JWST will enable resolved stellar population studies to be extended to far greater distances than currently possible with Hubble or ground-based imaging. This will bring more low metallicity and young, massive star forming regions within reach, and enable better characterisation of close binary populations and their interactions. On the other hand, JWST's infrared sensitivity range is not optimal for exploring the hottest stars or their ionization regions nearby, for which a future ultraviolet mission will be needed. Where JWST really does promise to revolutionize our understanding is in the rest-ultraviolet and rest-optical spectral characteristics of distant galaxies. Spectroscopic observations currently carried out with great difficulty and expense in telescope time from the ground, have the potential to become routine with the massively multiplexed, near-to-mid infrared spectrographs mounted on JWST (NIRSPEC and MIRI), and to be pushed to far higher redshift. At the same time, improved infrared (rest-optical) photometry will allow the low mass stellar populations of these galaxies to be better characterised by spectral energy distribution fitting than is currently possible.

On the other hand, the expected improvement in signal-to-noise of observations of these extreme systems is likely to bring into sharp relief the limitations and uncertainties in existing stellar and spectral synthesis models. It is likely that the modelling effort will evolve hand-in-hand with the insights yielded by observational data, as some model parameters are ruled out, and others adjusted. Crucially, this effort will require an understanding of the role and impact of binary star evolution. As Figure 4 illustrates, we are now firmly within regimes in which binary evolution pathways can significantly affect the interpretation of stellar populations. Neglecting them would be a collective act of voluntary myopia on the part of the community akin to wearing dark and clouded glasses to an exhibition of fine art - while the broadest strokes might be visible, we would be

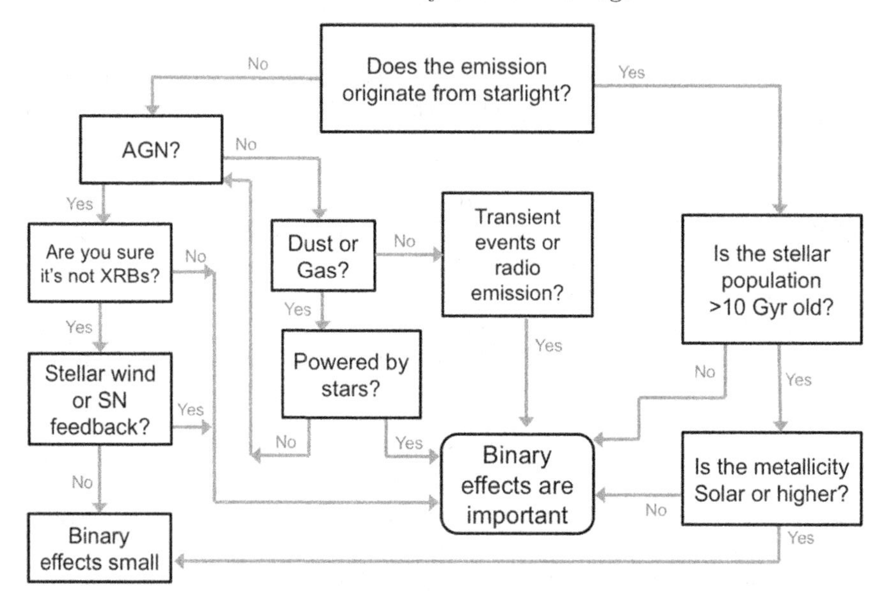

Figure 4. A non-exhaustive decision tree indicating regions of observational parameter space in which considering the effects of stellar binaries may have a significant effect on the interpretation of results.

ignorant of the exquisite detail, intellectual, and indeed emotional, stimulation which the complex Universe we inhabit has to offer.

Acknowledgements

ERS acknowledges support from the University of Warwick, and under UK STFC consolidated grant ST/P000495/1. JJE acknowledges support fom the University of Auckland and also the Royal Society Te Apārangi of New Zealand under Marsden Fund grant UOA1818.

References

Amorín, R., Pérez-Montero, E., Contini, T., *et al.* 2015, *A&A*, 578, A105
Basu-Zych, A. R., Lehmer, B. D., Hornschemeier, A. E., *et al.* 2013, *ApJ*, 774, 152
Bayliss, M. B., Rigby, J. R., Sharon, K., *et al.* 2014, *ApJ*, 790, 144
Belczynski, K., Kalogera, V., Rasio, F. A., *et al.* 2008, *ApJS*, 174, 223
Berg, D. A., Erb, D. K., Auger, M. W., *et al.* 2018, *ApJ*, 859, 164
Bruzual, G. & Charlot, S. 2003, *MNRAS*, 344, 1000
Conroy, C. 2013, *ARA&A*, 51, 393
da Cunha, E., Charlot, S., & Elbaz, D. 2008, *MNRAS*, 388, 1595
De Donder, E. & Vanbeveren, D. 2004, New Astronomy Review, 48, 861
de Mink, S. E., Langer, N., Izzard, R. G., *et al.* 2013, *ApJ*, 764, 166
Du, X., Shapley, A. E., Reddy, N. A., *et al.* 2018, *ApJ*, 860, 75
Dufton, P. L., Evans, C. J., Hunter, I., *et al.* 2019, *A&A*, 626, A50
Eggleton, P. P. 1971, *MNRAS*, 151, 351
Eldridge, J. J., Izzard, R. G., & Tout, C. A. 2008, *MNRAS*, 384, 1109
Eldridge, J. J. & Stanway, E. R. 2009, *MNRAS*, 400, 1019
Eldridge, J. J. & Stanway, E. R. 2012, *MNRAS*, 419, 479
Eldridge, J. J., Fraser, M., Smartt, S. J., *et al.* 2013, *MNRAS*, 436, 774
Eldridge, J. J., Stanway, E. R., Xiao, L., *et al.* 2017, *PASA*, 34, e058
Eldridge, J. J., Stanway, E. R., & Tang, P. N. 2019, *MNRAS*, 482, 870

Erb, D. K., Pettini, M., Shapley, A. E., *et al.* 2010, *ApJ*, 719, 1168
Ferland, G. J., Korista, K. T., Verner, D. A., *et al.* 1998, *PASP*, 110, 761
Götberg, Y., de Mink, S. E., Groh, J. H., *et al.* 2018, *A&A*, 615, A78
Gräfener, G., Owocki, S. P., & Vink, J. S. 2012, *A&A*, 538, A40
Greis, S. M. L., Stanway, E. R., Davies, L. J. M., *et al.* 2016, *MNRAS*, 459, 2591
Hamann, W.-R. & Gräfener, G. 2003, *A&A*, 410, 993
Hamann, W.-R., Gräfener, G., & Liermann, A. 2006, *A&A*, 457, 1015
Han, Z., Podsiadlowski, P., & Lynas-Gray, A. E. 2007, *MNRAS*, 380, 1098
Hashimoto, T., Laporte, N., Mawatari, K., *et al.* 2018, *Nature*, 557, 392
Hashimoto, T., Inoue, A. K., Mawatari, K., *et al.* 2019, *PASJ*, 70
Heckman, T. M., Hoopes, C. G., Seibert, M., *et al.* 2005, *ApJ*, 619, L35
Hobbs, G., Lorimer, D. R., Lyne, A. G., *et al.* 2005, *MNRAS*, 360, 974
Hopkins, A. M. 2018, *PASA*, 35, 39
Hurley, J. R., Tout, C. A., & Pols, O. R. 2002, *MNRAS*, 329, 897
Ivanova, N., Justham, S., Chen, X., *et al.* 2013, *A&A Rev.*, 21, 59
Izotov, Y. I., Worseck, G., Schaerer, D., *et al.* 2018, *MNRAS*, 478, 4851
Kehrig, C., Vílchez, J. M., Guerrero, M. A., *et al.* 2018, *MNRAS*, 480, 1081
Leitherer, C., Schaerer, D., Goldader, J. D., *et al.* 1999, *ApJS*, 123, 3
Leitherer, C., Ekström, S., Meynet, G., *et al.* 2014, *ApJS*, 212, 14
Ma, X., Kasen, D., Hopkins, P. F., *et al.* 2015, *MNRAS*, 453, 960
Madau, P., Ferguson, H. C., Dickinson, M. E., *et al.* 1996, *MNRAS*, 283, 1388
Maraston, C. 2005, *MNRAS*, 362, 799
Maseda, M. V., Brinchmann, J., Franx, M., *et al.* 2017, *A&A*, 608, A4
Moe, M. & Di Stefano, R. 2017, *ApJS*, 230, 15
Moe, M., Kratter, K. M., & Badenes, C. 2019, *ApJ*, 875, 61
Nhung, P. T., Hoai, D. T., Tuan-Anh, P., *et al.* 2019, arXiv e-prints, arXiv:1908.03311
Pols, O. R., Tout, C. A., Eggleton, P. P., *et al.* 1995, *MNRAS*, 274, 964
Rosenfield, P., Marigo, P., Girardi, L., *et al.* 2014, *ApJ*, 790, 22
Sana, H., de Mink, S. E., de Koter, A., *et al.* 2012, Science, 337, 444
Schaerer, D., Fragos, T., & Izotov, Y. I. 2019, *A&A*, 622, L10
Schneider, F. R. N., Ramírez-Agudelo, O. H., Tramper, F., *et al.* 2018, *A&A*, 618, A73
Schneider, F. R. N., Sana, H., Evans, C. J., *et al.* 2018, Science, 359, 69
Senchyna, P., Stark, D. P., Chevallard, J., *et al.* 2019, *MNRAS*, 488, 3492
Shapley, A. E., Steidel, C. C., Pettini, M., *et al.* 2003, *ApJ*, 588, 65
Smit, R., Bouwens, R. J., Carniani, S., *et al.* 2018, *Nature*, 553, 178
Stanway, E. R., Eldridge, J. J., & Becker, G. D. 2016, *MNRAS*, 456, 485
Stanway, E. R. & Eldridge, J. J. 2018, *MNRAS*, 479, 75
Stanway, E. R. & Eldridge, J. J. 2019, *A&A*, 621, A105
Steidel, C. C., Strom, A. L., Pettini, M., *et al.* 2016, *ApJ*, 826, 159
Toonen, S. & Nelemans, G. 2013, *A&A*, 557, A87
Topping, M. W. & Shull, J. M. 2015, *ApJ*, 800, 97
van Bever, J. & Vanbeveren, D. 1998, *A&A*, 334, 21
Van Bever, J. & Vanbeveren, D. 2000, *A&A*, 358, 462
Willems, B. & Kolb, U. 2004, *A&A*, 419, 1057
Wofford, A., Charlot, S., Bruzual, G., *et al.* 2016, *MNRAS*, 457, 4296
Zhang, F., Li, L., & Han, Z. 2005, *MNRAS*, 364, 503

Uncovering Early Galaxy Evolution in the ALMA and JWST Era
Proceedings IAU Symposium No. 352, 2019
E. da Cunha, J. Hodge, J. Afonso, L. Pentericci & D. Sobral, eds.

doi:10.1017/S1743921320001155

Measuring the stellar initial mass function

Andrew Hopkins

AAO/Macquarie University, Australia

Abstract. The birth of stars and the formation of galaxies are cornerstones of modern astrophysics. While much is known about how galaxies globally and their stars individually form and evolve, one fundamental property that affects both remains elusive. This is problematic because this key property, the stellar initial mass function (IMF), is a key tracer of the physics of star formation that underpins almost all of the unknowns in galaxy and stellar evolution. It is perhaps the greatest source of systematic uncertainty in star and galaxy evolution. The past decade has seen a growing number and variety of methods for measuring or inferring the shape of the IMF, along with progressively more detailed simulations, paralleled by refinements in the way the concept of the IMF is applied or conceptualised on different physical scales. This range of approaches and evolving definitions of the quantity being measured has in turn led to conflicting conclusions regarding whether or not the IMF is universal. Here I summarise the growing wealth of approaches to our understanding of this fundamental property that defines so much of astrophysics, and highlight the importance of considering potential IMF variations, reinforcing the need for measurements to quantify their scope and uncertainties carefully. I present a new framework to aid the discussion of the IMF and promote clarity in the further development of this fundamental field.

Uncovering Early Galaxy Evolution in the ALMA and JWST Era
Proceedings IAU Symposium No. 352, 2019
E. da Cunha, J. Hodge, J. Afonso, L. Pentericci & D. Sobral, eds.

doi:10.1017/S1743921319009025

An older, more quiescent universe from panchromatic SED fitting of the 3D-HST survey

Joel Leja[1], Benjamin D. Johnson[1], Charlie Conroy[1], Pieter van Dokkum[2], Joshua S. Speagle[1] and the 3D-HST Team

[1]Harvard-Smithsonian Center for Astrophysics, 60 Garden St. Cambridge, MA 02138, USA

[2]Department of Astronomy, Yale University, New Haven, CT 06511, USA

Abstract. Galaxies are complicated physical systems which obey complex scaling relationships; as a result, properties measured from broadband photometry are often highly correlated, degenerate, or both. Therefore, the accuracy of basic properties like stellar masses and star formation rates (SFRs) depend on the accuracy of many second-order galaxy properties, including star formation histories (SFHs), stellar metallicities, dust properties, and many others. Here, we reassess measurements of galaxy stellar masses and SFRs using a 14-parameter physical model built in the Prospector Bayesian inference framework. We find that galaxies are $\sim$0.2 dex more massive and have $\sim$0.2 dex lower star formation rates than classic measurements. These measurements lower the observed cosmic star formation rate density and increase the observed buildup of stellar mass, finally bringing these two metrics into agreement at the factor-of-two level at $0.5 < z < 2.5$.

Keywords. galaxies: evolution, galaxies: formation, galaxies: fundamental parameters

---

1. Summary

This IAU Proceeding is a summary of work published in the Astrophysical Journal under Leja *et al.* (2019b). Here we apply `Prospector`, a new code for interpreting galaxy spectral energy distributions (SED) described in Leja *et al.* (2017), to the photometry from the 3D-HST survey Skelton *et al.* (2014). The 3D-HST photometric catalog provides observed-frame UV to MIR photometry for $\sim 2 \times 10^5$ galaxies over five extragalactic fields totaling $\sim$ 900 arcmin^2. Here we use `Prospector` to fit objects in the redshift range $0.5 < z < 2.5$ above the observed mass completeness limit for a total of 58,461 objects. We fix the redshift to the photometric redshifts derived with the EAZY code Brammer *et al.* (2008). We compare the inferred stellar masses and star formation rates from `Prospector` to classic inference methods for stellar mass and star formation rates, namely the FAST SED-fitting code Kriek *et al.* (2009) and UV+IR star formation rates Whitaker *et al.* (2014).

The comparison with the FAST stellar masses can be found in Figure 1. `Prospector` infers larger stellar masses in low mass galaxies, with the largest systematic offsets around 0.3 dex. This offset decreases with increasing redshifts. A small part of this offset, $\sim$0.05 − 0.1 dex, is due to differing stellar populations codes Leja *et al.* (2019b). The bulk of the offset is because `Prospector` infers much older mass-weighted ages than FAST. This is a natural consequence of the nonparametric star formation histories used in `Prospector`, which remove the bias towards younger ages often found in parametric

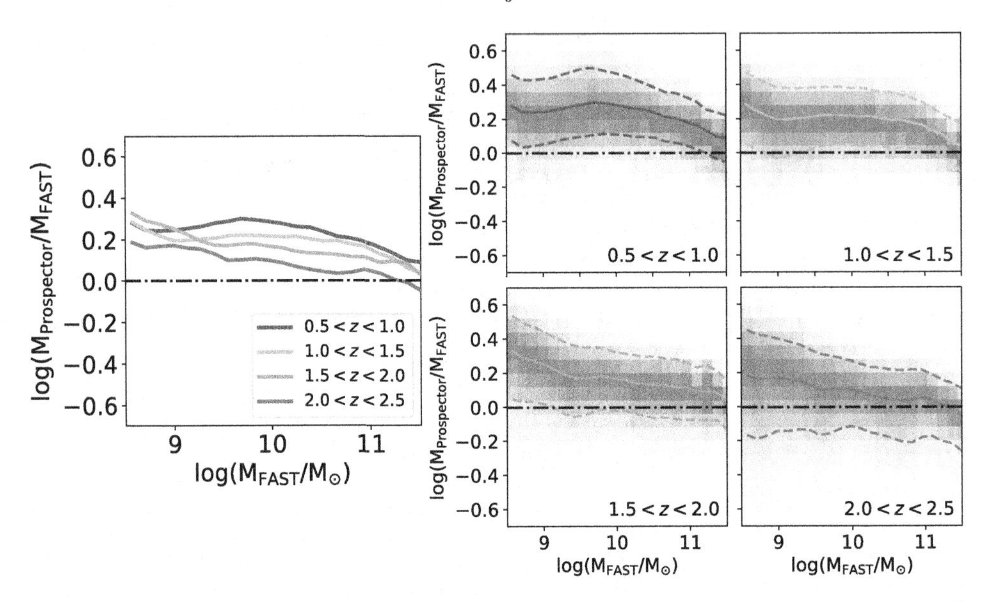

Figure 1. The relative difference between stellar mass derived with `Prospector` and stellar mass derived with FAST is shown as a function of redshift and stellar mass.

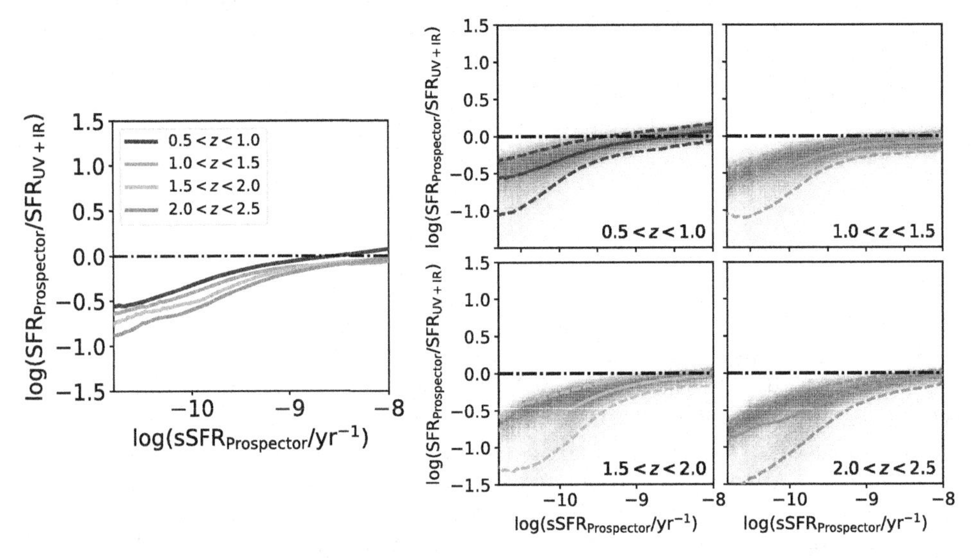

Figure 2. The relative difference between star formation rates derived with `Prospector` and star formation rates derived with L_{UV} and L_{IR} luminosities is shown as a function of redshift and specific star formation rates.

formulae Carnall *et al.* (2019), Leja *et al.* (2019a). These older ages create a higher mass-to-light ratio and therefore higher masses.

The comparison with SFRs inferred from UV and IR luminosities can be found in Figure 2. `Prospector` infers similar star formation rates to UV- and IR-based estimates at high specific star formation rates (sSFRs), but shows an increasing offset towards lower star formation rates at low sSFRs. The median offset is larger at higher redshifts and ranges up to an order-of-magnitude at low sSFRs. Several factors contribute to this

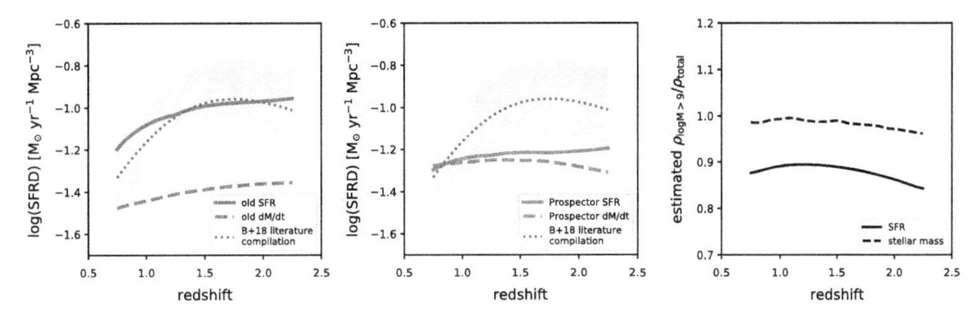

Figure 3. The estimated change in the star formation rate density as inferred independently from the instantaneous star formation rate and from the rate of change of the stellar mass density. The left panel shows the previous estimates while the center panel shows the estimates from `Prospector`. The right panel shows the estimated incompleteness from the Universe Machine.

effect at a moderate level, including `Prospector`'s accounting for both effects of variable metallicity and active galactic nucleus (AGN) emission in the rest-frame mid-infrared. However, the bulk of the offset comes from the fact that `Prospector` also models dust heating and UV emission from old stars (i.e. $t > 100$ Myr) on an object-by-object basis, while classic estimates based on UV and IR luminosities either neglect or fix this contribution. This explains the increase in the effect at higher redshifts: in the early Universe, the 'old' stars are relatively younger and more luminous. It also explains the increase in the size of the offset at low specific star formation rates, as objects with low sSFRs have relatively more old stars.

We couple these measured systematic offsets with the stellar mass function from Tomczak *et al.* (2014) in order to estimate the change in the star formation rate density (SFRD). This study estimates the mass function with the FAST SED-fitting code, allowing an apples-to-apples comparison. The SFRD is estimated from two independent sources: the redshift evolution of the mass function and the instantaneous star formation rate of galaxies in some cosmological volume.

For the mass function estimate, we use the smoothed parameterization of the redshift evolution of the stellar mass function from Leja *et al.* (2015) and the measured $\mathrm{M_{Prospector}/M_{FAST}}$ as a function of $\mathrm{M_{FAST}}$. We differentiate the mass function with respect to time and integrate over mass in order to measure $\mathrm{d}\rho_*/\mathrm{dt}$. Finally we correct for the mass loss rate R in order to turn this into an estimate of ρ_{SFRD}, the instantaneous star formation rate per unit volume. For the estimate directly from the instantaneous star formation rate, we measure $\mathrm{SFR_{Prospector}(M_{FAST})}$ and $\mathrm{SFR_{UV+IR}(M_{FAST})}$, multiply by the stellar mass function, and integrate in order to estimate ρ_{SFRD}. The incompleteness in both of these quantities due to the lower stellar mass completeness limit of the survey is estimated with the Universe Machine, a semi-empirical model of galaxy formation. This incompleteness is shown in the right-hand panel of Figure 3 and the total SFRDs are corrected for this effect.

The previous inferences of the SFRD are compared with the new inferences from `Prospector` in Figure 3. The classic inferences are offset from one another at the $\sim$0.3 dex level, such that there is more observed star formation than stellar mass growth. The `Prospector` measurements lower the instantaneous star formation rate density by $\sim 0.1 - 0.2$ dex while increasing the growth of the stellar mass density by ~ 0.2 dex. The combination of these effects brings these two independent measurements of the growth of stellar mass into agreement. This suggests that the `Prospector` values are more reflective of the underlying parameters than the classic estimates, though not conclusively: the behavior below the low-mass limit must be explored to settle this issue, and the effect of other cancelling systematic effects such as variation in the high-mass initial mass function may confound the results of this study.

References

Behroozi, P., Wechsler, R. H., Hearin, A. P., *et al.* 2019, *MNRAS*, 488, 3143
Brammer, G. B., van Dokkum, P. G., & Coppi, P. 2008, *ApJ*, 686, 1503
Carnall, A. C., Leja, J., Johnson, B. D., *et al.* 2019, *ApJ*, 873, 44
Kriek, M., van Dokkum, P. G., Labbé, I., *et al.* 2009, *ApJ*, 700, 221
Leja, J., Carnall, A. C., Johnson, B. D., *et al.* 2019, *ApJ*, 876, 3
Leja, J., Johnson, B. D., Conroy, C., van Dokkum, P., Speagle, J. S., Brammer, G., Momcheva, I., Skelton, R, Whitaker, & K. E., Franx, M. 2019, *ApJ*, 877, 140L
Leja, J., Johnson, B. D., Conroy, C., *et al.* 2017, *ApJ*, 837, 170
Leja, J., van Dokkum, P. G., Franx, M., *et al.* 2015, *ApJ*, 798, 115
Skelton, R. E., Whitaker, K. E., Momcheva, I. G., *et al.* 2014, *ApJ Supplement*, 214, 24
Tomczak, A. R., Quadri, R. F., Tran, K.-V. H., *et al.* 2014, *ApJ*, 783, 85
Whitaker, K. E., Franx, M., Leja, J., *et al.* 2014, *ApJ*, 795, 104

Uncovering Early Galaxy Evolution in the ALMA and JWST Era
Proceedings IAU Symposium No. 352, 2019
E. da Cunha, J. Hodge, J. Afonso, L. Pentericci & D. Sobral, eds.

doi:10.1017/S1743921320000708

Synthetic nebular emission lines of simulated galaxies over cosmic time

Michaela Hirschmann

DARK, Niels Bohr Institute, University of Copenhagen, Lynbyvej 2,
2100 Copenhagen, Denmark
email: michaela.hirschmann@nbi.ku.dk

Abstract. This article presents an up-dated analysis of synthetic optical and UV emission lines of simulated galaxies over cosmic time. The strong emission lines are derived from self-consistently coupling novel spectral models accounting for nebular emission from young stars, AGN and Post-AGB stars to cosmological zoom-in as well as large-scale simulations. Investigating the evolution of optical line-ratios in the BPT diagrams, the simulations can successfully reproduce the observed trend of [OIII]/Hβ ratio increasing from low to high redshifts, due to evolving star formation rate and gas metallicity. Standard selection criteria in the BPT diagrams can appropriately distinguish the main ionising source(s) of galaxies at low redshifts, but they are less reliable for metal-poor galaxies, dominating the early Universe. To robustly classify the ionising radiation of such metal-poor galaxies, diagnostic diagrams based on luminosity ratios of UV lines are discussed. The novel interface between simulations and observations is potentially important for the interpretation of high-quality spectra of very distant galaxies to be gathered by next-generation telescopes, such as the James Webb Space Telescope.

Keywords. methods: numerical, galaxies: evolution, galaxies: formation, galaxies: high-redshift, galaxies: ISM, (galaxies:) quasars: emission lines

---

1. Introduction

The emission from ionized interstellar gas contains valuable information about the nature of the ionizing radiation and the physical conditions in the interstellar medium (ISM) in a galaxy. In fact, prominent optical emission lines are routinely used to estimate the density, chemical abundances and dust content of the ISM and whether ionization is dominated by young massive stars (tracing the star formation rate hereafter SFR), an active galactic nucleus (hereafter AGN) or evolved, post-asymptotic giant branch (hereafter post-AGB) stars (e.g. Izotov *et al.* (1999); Kauffmann *et al.* (2003); Kewley & Ellison (2008); Morisset *et al.* (2016)). One of the most widely used line-ratio diagnostic diagrams, originally defined by Baldwin, Phillips & Terlevich (1981, hereafter BPT), relate the [OIII]λ5007/Hβ ratio to the [NII]λ6584/Hα ratio, allowing for an identification of the nature of the ionizing radiation in large samples of galaxies in the local Universe (e.g. Kewley *et al.* (2001); Kauffmann *et al.* (2003)).

Interestingly, several recent observational studies indicate that star-forming (SF) galaxies at $z > 1$ have systematically larger [O III]/Hβ ratios, at fixed [N II]/Hα ratio, than their present-day counterparts from the Sloan Digital Sky Survey (SDSS; see e.g. Yabe *et al.* (2014); Steidel *et al.* (2014); Shapley *et al.* (2015); Strom *et al.* (2017)). The physical origin of this intriguing observational feature is being heavily debated and several explanations have been proposed (higher ionisation parameter, higher electron density, contribution from AGN, harder ionising radiation from stars).

In addition, at the very low metallicities expected in the youngest galaxies at high redshifts (e.g. Maiolino *et al.* (2008)), emission-line ratios for SF- and AGN-dominated models tend to occupy similar regions of the [OIII]/Hβ–[NII]/Hα diagram (Feltre *et al.* (2016)) so that their usefulness to constrain the nature of ionizing sources in the early universe is uncertain. In recent years, interest has grown in ultraviolet (UV) nebular emission lines, such as CIII]λ1908, CIVλ1550, and HeIIλ1640 lines, which tend to be particularly prominent in metal-poor, actively SF dwarf galaxies at all redshifts (e.g. Stark *et al.* (2014); Berg *et al.* (2016)).

To reach more specific conclusions on the (i) origin of the evolution of optical line-ratios, and (ii) how to best identify the main ionising sources in metal poor galaxies requires the self-consistent modelling of nebular emission from different gas components ionized by different sources in simulated galaxies. In this analysis, the methodology introduced by Hirschmann *et al.* (2017) is considered to model in a self-consistent way the nebular emission from different regions in the ISM ionized by various sources in simulated galaxies. This is achieved by coupling photoionization models for AGN (Feltre *et al.* (2016)), young stars (Gutkin *et al.* (2016)) and post-AGB stars (Hirschmann *et al.* (2017)) with cosmological hydrodynamic simulations – both zoom-in simulations of massive galaxies (Choi *et al.* (2017); Hirschmann *et al.* (2017)) and large-scale IllustrisTNG simulations (Pillepich *et al.* (2018); Nelson *et al.* (2018)). The latter provides – for the first time – nebular emission line catalogues of full galaxy populations, allowing for a *first statistical* analysis of the outlined questions.

2. Theoretical framework

2.1. *Cosmological simulations*

Part of the analysis is based on a set of 20 high-resolution, cosmological zoom-in simulations of massive galaxies with present-day halo masses between $7 \times 10^{11} M_{\odot}/h$ and $2.7 \times 10^{13} M_{\odot}/h$, adopting a WMAP3 cosmology (Choi *et al.* (2017); Hirschmann *et al.* (2017)). These simulations were performed with a modified version of the highly parallel, smoothed particle hydrodynamics (SPH) code GADGET3 (Springel, Di Matteo & Hernquist 2005) and SPHGal (Hu *et al.* 2014; Choi *et al.* 2017; Nunez *et al.* 2017).

In addition, the publicly available large-scale cosmological "IllustrisTNG" simulation is employed, providing a box length of 100 Mpc, and adopting a Planck cosmology (Pillepich *et al.* (2018), Nelson *et al.* (2018)). This simulation was performed with the magneto-hydrodynamic code Arepo (Springel *et al.* 2010). Note only galaxies with a stellar mass above $3 \times 10^{9} M_{\odot}$ are considered (due to resolution limits).

For more details on the cosmological simulations, the readers are referred to the original works as cited above.

2.2. *Coupling to Photo-ionisation models*

The re-simulations of 20 galaxies as well as the IllustrisTNG-100 simulation presented in Section 2.1 are post-processed to include nebular emission. Recent prescriptions of Gutkin *et al.* (2015), Feltre *et al.* (2016) and Hirschmann *et al.* (2017) are adopted to compute the nebular emission arising from young massive stars, narrow-line regions of AGN and the environment of post-AGB stars. All emission-line models presented here were computed using version c13.03 of the photoionization co de CLOUDY (Ferland *et al.* 2013).

This extensive grid of nebular-emission models is coupled with the cosmological simulations, by selecting the SF/AGN/PAGB models appropriate for each galaxy by self-consistently matching all model parameters possibly available from the simulations (e.g. different BH, star and gas properties). Those parameters that cannot be retrieved from the simulation are set to standard values (i.e. ξ_d and n_H).

For more details on the grids of photo-ionisation models and on the coupling methodology, the reader is referred to Hirschmann *et al.* (2017).

3. The [OIII]/Hβ–[NII]/Hα diagram in the low redshift Universe

In this section, optical line-ratios of low-redshift simulated galaxies are explored. Specifically, the top row of Fig. 2 of Hirschmann *et al.* (2017) and the top row of Fig. 1 show the locations of galaxies in the [O III]/Hβ and [N II]/Hα diagnostic diagram for different galaxy types. Thereby, different galaxy types are theoretically distinguished on the basis of the predicted ratio of BH accretion rate (BHAR) to SFR and the Hβ-line luminosity (see Hirschmann *et al.* (2017) for exact definitions). For reference, the location of local (z $\sim$ 0.1) SDSS galaxies is indicated in these line-ratio diagrams (black contours), together with standard observational criteria to distinguish SF galaxies (below the dashed line) from composites (between the dashed and dotted lines), AGN (above the dotted line) and LI(N)ER (in the bottom-right quadrant defined by dot-dashed lines), according to Kewley *et al.* (2001); Kauffmann *et al.* (2003).

As already discussed in Hirschmann *et al.* (2017/2019) for the set of zoom-in simulations, the top row of Fig. 1 now *statistically* confirms that, in the low-redshift Universe, simulated galaxies occupy the same areas as SDSS galaxies in the [O III]/Hβ–[N II]/Hα plane. In addition, simulated galaxies of SF, AGN, and PAGB types appear to be located in the same regions of the diagram as the observationally defined SF, AGN, and LI(N)ER classes. Only composite galaxies appear to be distributed more widely than the observations, extending to higher than observed [OIII]/Hβ ratios at the highest AGN luminosities. None the less, the overall agreement between models and observations in Fig. 1 is striking given that, in our approach, different galaxy types are linked to physical parameters, such as the fraction of total Hβ luminosity and the BHAR/SFR ratio.

4. Evolution of the [OIII]/Hβ–[NII]/Hα diagram

Turning towards higher redshifts, the bottom row of Fig. 1 shows the analogue of the first row, but for z=3. As expected from the evolution of the 20 re-simulated galaxies Fig. 1 in Hirschmann *et al.* (2019), the bottom row of Fig. 1 confirms for a *statistical* population of simulated galaxies that [OIII]/Hβ tends to increase and [NII]/Hα to decrease from low to high redshift – qualitatively consistent with observations by Yabe *et al.* (2014) and Steidel *et al.* (2014). The physical origin of the cosmic evolution [OIII]/Hβ and [NII]/Hα is a complex mix of different evolving ISM and ionizing-radiation properties governing the nebular emission from young stars, AGN and post-AGB stars (see Hirschmann *et al.* (2017) for a more in-depth discussion).

In addition, the drop in [NII]/Hα from low to high redshifts makes the different galaxy types less distinguishable towards high redshift, implying that the traditional optical selection criteria become less reliable. The main reason for composite/AGN galaxies to move towards the SF region in the [OIII]/Hβ–[NII]/Hα plane is the lower typical metallicity of high- redshift compared to low-redshift galaxies. But note that the validity of standard optical selection criteria can be preserved by focusing on pre-selected subsamples of metal-rich galaxies at any redshift (see Hirschmann *et al.* (2017)).

5. UV diagnostic diagrams to identify the main ionising sources of metal-poor galaxies

This section is focused on diagnostic diagrams based on rest-frame UV emission lines to discriminate between ionizing sources in metal-poor galaxies at high redshifts

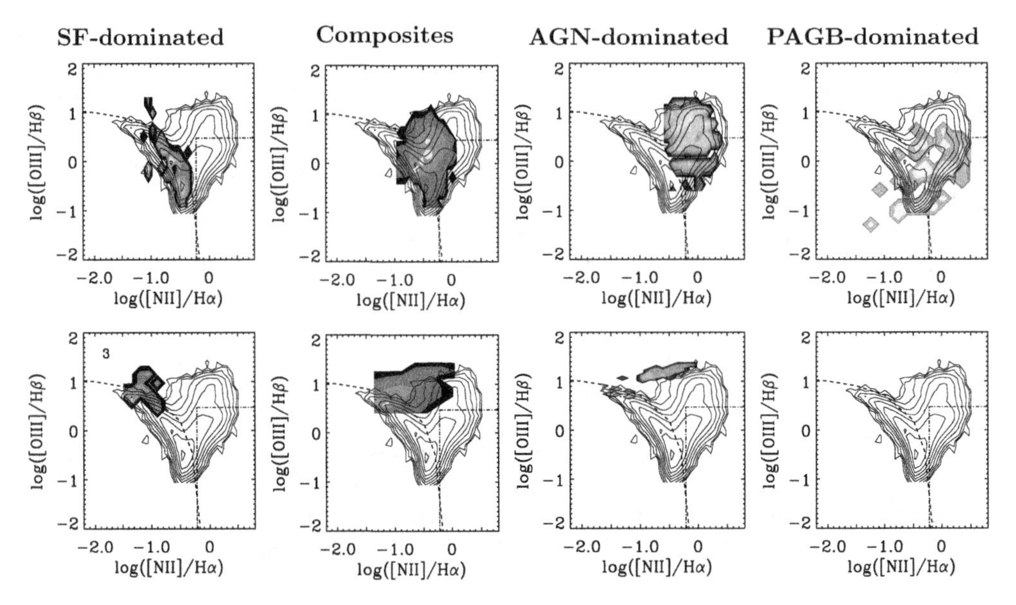

Figure 1. Location of optical line-ratios of IllustrisTNG-100 galaxies in the the [OIII]/Hβ–[NII]/Hα diagram at z = 0 (top row) and z = 3 (bottom row), distinguishing between different galaxy types (blue: SF-dominated, red: composites, green: AGN, yellow: PAGB-dominated).

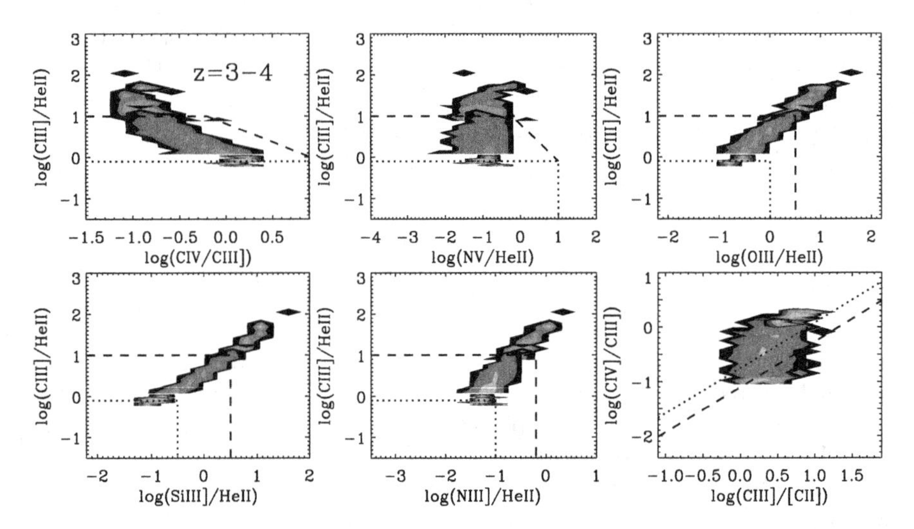

Figure 2. Six UV diagnostic diagrams for metal-poor galaxies of the IllustrisTNG-100 simulation at z = 3-4, distinguishing between SF-dominated (blue), composite (red) and AGN-dominated galaxies (green).

(where optical diagnostics become less reliable). Specifically, the usefulness of different combinations of UV line ratios, proposed by Hirschmann *et al.* (2019), are studied in a *statistical context*. Fig. 2 shows six (out of the 14 proposed) different UV diagnostic diagrams of metal-poor galaxies at z = 3-4, distinguishing the nature of the main ionising sources according to the BHAC/SFR ratio (as in section 3): (i) CIII]λ1908/HeIIλ1640 versus CIII]λ1908/CIVλ1550; (ii) CIII]λ1908/HeIIλ1640 versus NVλ1240/HeIIλ1640; (iii) CIII]λ1908/HeII λ1640 versus OIII]λ1663/HeIIλ1640; (iv) CIII]λ1908/HeIIλ1640 versus SiIII]λ1888/HeIIλ1640; (v) CIII]λ1908/HeIIλ1640 versus NIII]λ1750/HeIIλ1640; (vi) CIVλ1550/CIII]λ1908 versus CIII]λ1908/C II]λ2326. The selection criteria, derived in Hirschmann *et al.* (2019), are indicated by the black dashed (to separate SF from composite galaxies) and by the black dotted lines (to distinguish between composite and AGN-dominated galaxies). Except for the bottom right UV diagram, the separability of different types of metal-poor galaxies in the other UV diagnostic diagrams is fairly clear for the statistical metal-poor galaxy populations.

6. Conclusions

This article is dedicated to the exploration of optical and UV emission-line diagnostic diagrams over cosmic time. As described in Hirschmann *et al.* (2017/2019), synthetic optical and UV emission lines of galaxies have been computed in a cosmological framework by coupling in post-processing newly developed spectral evolution models, based on photoionization calculations, with a set of 20 high-resolution cosmological zoom- in simulations of massive galaxies and, for the first time, with the large-scale cosmological IllustrisTNG simulation. The latter allows for a *statistical* investigation of optical and UV line-ratios of *full* galaxy populations. The following results can be summarised:

- The synthetic [OIII]/Hβ and [NII]/Hα emission-line ratios predicted by the simulations are in excellent agreement with observations of both star-forming and active SDSS galaxies in the local universe.
- Towards higher redshifts [OIII]/Hβ is predicted to increase and [NII]/Hβ to decrease qualitatively consistent with recent observations. The latter makes the standard optical selection criteria in the [OIII]/Hβ–[NII]/Hα diagram less reliable, at least for metal-poor galaxies.
- To robustly classify the ionizing radiation of metal-poor galaxies, which dominate in the early Universe, we can statistically confirm some of the UV diagnostic diagrams and corresponding selection criteria introduced by Hirschmann *et al.* (2019).

Despite these promising results, it is important to keep in mind that a self-consistent relative contribution by radiative shocks to UV- and optical-line ratios is neglected so far, which I intend to add in future work. In addition, I plan to explore the contribution by different ionizing sources to nebular emission in different regions of a galaxy, i.e. producing spatially resolved emission-line maps to improve the interpretation of modern integral-field spectroscopic observations in terms of galaxy physical parameters.

References

Berg, *et al.* 2016, *ApJ*, 827, 126
Choi, *et al.* 2017, *ApJ*, 472, 2468
Feltre, *et al.* 2016, *MNRAS*, 456, 3354
Gutkin, *et al.* 2016, *MNRAS*, 462, 1757
Hirschmann, *et al.* 2017, *MNRAS*, 472, 2468
Izotov, *et al.* 1999, *ApJ*, 511, 639
Kauffmann, *et al.* 2003, *MNRAS*, 346, 1055
Kewley, *et al.* 2001, *ApJ*, 556, 121
Kewley & Ellison 2008, *ApJ*, 681, 1183
Maiolino, *et al.* 2008, *A & A*, 488, 463
Morisett, *et al.* 2016, *A & A*, 594, A37
Nelson, *et al.* 2018, *MNRAS*, 475,624
Pillepich, *et al.* 2018, *MNRAS*, 473,4077
Shapley, *et al.* 2015, *ApJ*, 801, 88
Stark, *et al.* 2014, *MNRAS*, 445, 3200
Steidel, *et al.* 2014, *ApJ*, 795, 165
Strom, *et al.* 2017, *ApJ*, 836, 164
Yabe, *et al.* 2014, *MNRAS*, 437, 3647

Uncovering Early Galaxy Evolution in the ALMA and JWST Era
Proceedings IAU Symposium No. 352, 2019
E. da Cunha, J. Hodge, J. Afonso, L. Pentericci & D. Sobral, eds.

doi:10.1017/S1743921319009402

FADO: A novel self-consistency spectral population synthesis tool for the exploration of galaxy evolution at high redshift

Jean Michel Gomes

Instituto de Astrofísica e Ciências do Espaço, CAUP, Rua das Estrelas, PT4150-762 Porto, Portugal
email: jean@astro.up.pt

Abstract. Despite significant progress over the past decades, all state-of-the-art population synthesis (PS) codes suffer from deficiencies limiting their potential of gaining sharp insights into the star formation history (SFH) and Chemical Enrichment History (CEH) of star-forming galaxies, i.e. the neglect of nebular continuum and, the lack of a mechanism to ensure consistency between the best-fitting SFH and the observed nebular characteristics (ONC; Balmer-lines, Balmer/Paschen jumps). These introduce biases in their recovered physical properties (stellar mass $M_\star$ and sSFR). FADO is a novel self-consistent PS code employing genetic optimization, publicly available (http://www.spectralsynthesis.org), capable of identifying the SFH & CEH that reproduce the ONC of a galaxy, alleviating degeneracies in the spectral fits. The current version of FADO (v1.b) uses standard BPT emission-line ratios for the classification of low redshift (z) galaxies. Whereas this permits a reliable distinction between star-forming, Composite, Seyfert and LINERs, it is inapplicable to many intermediate-z galaxies. We present an adaptation of FADO (version v1.c) to classify higher z galaxies employing the "Blue Diagram" (e.g., Lamareille 2010) for which the most prominent blue emission-lines (< [OIII]5007Å) are observable while the Hα and [NII] are inaccessible. FADO v1.c was applied to synthetic spectra simulating the evolution of galaxies formed at higher-z with different SFHs. FADO can recover the physical and evolutionary properties of galaxies, such as $M_\star$ and mean age / metallicity, with an accuracy significantly better than purely-stellar codes. An outline of FADO v1.c and applications to local and intermediate-z galaxies will be presented.

Keywords. Spectral Synthesis, Stellar Populations, Nebular Emission, Galaxy Evolution, Galaxy Formation

---

1. Introduction

Spectral synthesis is one of the most fundamental and powerful tools available in modern extragalactic astronomy. It has been extensively used and improved over the past 50 years and has as its main goal to recover the mass-assembly history of galaxies from observed galaxy spectra. Two inverse yet complementary techniques have been developed so far: evolutionary and population synthesis, which are not seldom combined in one hybrid approach that involves, e.g., a search for the best-fitting model among a grid of synthetic spectral energy distributions (SEDs). The first one aims to shed light into galaxy evolution through comparison of observed galaxy SEDs with synthetic ones computed with simple assumptions on the star-formation and chemical enrichment history (SFH & CEH, respectively) of galaxies. The second approach, on the contrary, attempts to infer its elementary stellar building blocks (e.g., star clusters formed instantaneously with a given age and metallicity).

Crucial and much-debated issues in spectral synthesis are the uniqueness of best-fitting SFHs & CEHs solutions in the presence of the notorious age-metallicity-attenuation degeneracy, the requirements in signal-to-noise (S/N) for an accurate retrieval of the SFH, and the possible modeling biases due to an incomplete description of physical ingredients (e.g., nebular emission) of the galaxy SED.

2. Overview of the new version of FADO v1.c

FADO stands for **F**itting **A**nalysis using **D**ifferential evolution **O**ptimization (Gomes & Papaderos 2017, 2018). This population synthesis code has two innovative concepts related to its mathematical and physical foundations. It is built using genetic differential evolution optimization and unsupervised clustering machine learning techniques to find the best-fit linear combination of stellar populations from galaxy spectra. Additionally, it incorporates both stellar & nebular emission (continuum plus Balmer line-emission) in a self-consistent manner, i.e. FADO finds the best-match SFH that reproduces the nebular characteristics of a given galaxy, which alleviates known degeneracies that plague stellar population modelling (Cardoso, Gomes & Papaderos 2019). So, FADO is well suited for the study of both young & old stellar populations in galaxies.

Here we present the new upgraded version v1.c of FADO. There are several improvements related to the spectral fitting (Gomes *et al.*, in prep.). We will briefly explain two main improvements related to intermediate and high-redshift galaxy spectral fitting.

FADO v1.b uses the standard BPT (Baldwin, Phillips & Terlevich 1981) diagram for the spectral pre-classification of emission-line galaxies at low redshift in the optical. This has proven to be extremelly reliable to distinguish between star-forming, Composite, Seyfert 2 and LINERs. FADO v1.c implements a new classification scheme that uses the "Blue Diagram", i.e. for spectra redshifted out of the wavelength range of the red optical emission-lines ($H\alpha$ & [NII] lines), but still containing the most prominent blue emission-lines in galaxies ([OIII] 5007Å, $H\beta$, [OII]3727,3729Å). This calibration comes from Lamareille (2010) and minimizes the contamination from LINERs & Seyfert 2 galaxies. It is defined as: $\log([\mathrm{OIII}]5007/H\beta) = 0.11/\{\log([\mathrm{OII}]3727, 3729/H\beta) - 0.92\} + 0.85$

Therefore, from this work it was defined the galaxies that have a contamination from an active galactic nucleus: $\log([\mathrm{OIII}]/H\beta) > 0.3$. The second important improvement in FADO is related to the pre-fitting of the underlying continuum in order to reliably estimate the emission-lines since the beginning of the fitting procedure. The new continuum determination uses a sliding box 'crap fitting'. This permits to have an idea of the non-systematic errors of the spectrum and also to measure accuratelly emission-lines and equivalent widths, which in turn can be used with the self-consistent fitting in FADO from the very first run, helping towards convergence.

3. Mock-galaxies: SED simulations test

We have computed and subsequently modelled synthetic SEDs with the FADO population spectral synthesis code. The synthetic optical spectra were computed using distinct SFHs. For this, we adopted the Simple Stellar Population (SSP) models of the evolutionary synthesis code GALAXEV by Bruzual & Charlot (2003), with Chabrier (2003) stellar initial mass function (IMF) between 0.1 and 100 $M_\odot$ and Padova evolutionary tracks (Alongi *et al.* 1993; Bressan *et al.* 1993; Fagotto *et al.* 1994a,b,c; Girardi *et al.* 1996).

The synthetic SEDs were then fitted from 2175 to 6000Å simulating observations of the VLT/MOONS spectral range in the low-resolution mode for galaxies at redshifts ~ 2. We have constructed the FADO base out of SSPs from the GALAXEV evolutionary synthesis code by Bruzual & Charlot (2003). It contains 25 ages from 1 Myr to ~13.7 Gyr and 4 metallicities ($Z = 0.2$, 0.4, 1 & 2.5 $Z_\odot$). Note that these base elements have the

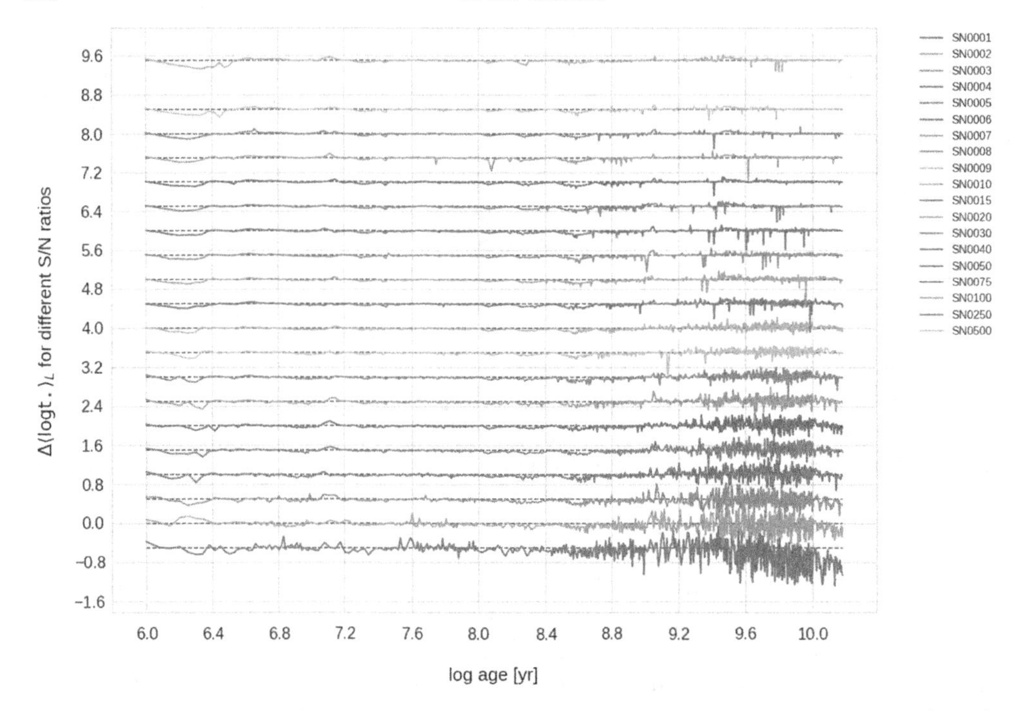

Figure 1. Difference between the output & input luminosity-weighted mean stellar age $\langle \log t_{\star} \rangle_L$ as a function of the evolutionary age for an instantaneous burst model. Several S/N ratios at around 5020Å are shown increasing from bottom to top (see labels). The curves are shifted in the y-direction for better appreciation of the results. The spectral range in the FADO fits goes from 2175 to 6000 Å.

same ingredients as the generated mock-galaxy spectra. This is to try to map the intrinsic degeneracies of the fits. We show in figure 1 the difference (output−input) in the mean stellar age light-weighted as a function of the age for an instantaneous burst model (values similar to other SFHs tested, e.g., constant, exponentially declining) & several S/N ratios spanning values from 1 to 500 at around 5020Å. FADO v1.c recovers the parameters within 0.2 dex on average, however, it may reach ∼0.6 dex for old ages if the S/N is low (∼ 1). A thorough investigation will be presented in a future article (Gomes *et al.*, in prep.).

4. FADO applications to LEGA-C: stellar populations

The Large Early Galaxy Astrophysics Census (van der Wel *et al.* 2016; LEGA-C) is an ESO Public Spectroscopic Survey available for the astronomical community that has obtained deep continuum galaxy spectra of several thousand galaxies from redshifts between 0.6 to 1, which corresponds to ages of the universe 5.86 and 7.92 Gyr, respectively. The adopted cosmology uses: $\Omega_{\rm M} = 0.3$, $\Omega_\Lambda = 0.7$, and $H_0 = 72$ km s^{-1} Mpc^{-1} (Spergel *et al.* 2007).

The galaxy spectra were downloaded from the LEGA-C DR2 survey (Straatman *et al.* 2018). These spectra were extracted using a pipeline to correct for Galactic attenuation assuming a uniform dust screen model and the Cardelli, Clayton & Mathis (1989) reddening curve. The dust IR emission map from our galaxy was used in the estimation of the attenuation. This map was taken from Schlegel, Finkbeiner & Davis (1998) with the new recalibration from Schlafly & Finkbeiner (2011). All spectra were corrected to rest-frame wavelengths and fluxes prior to spectral modelling. The luminosity distances were

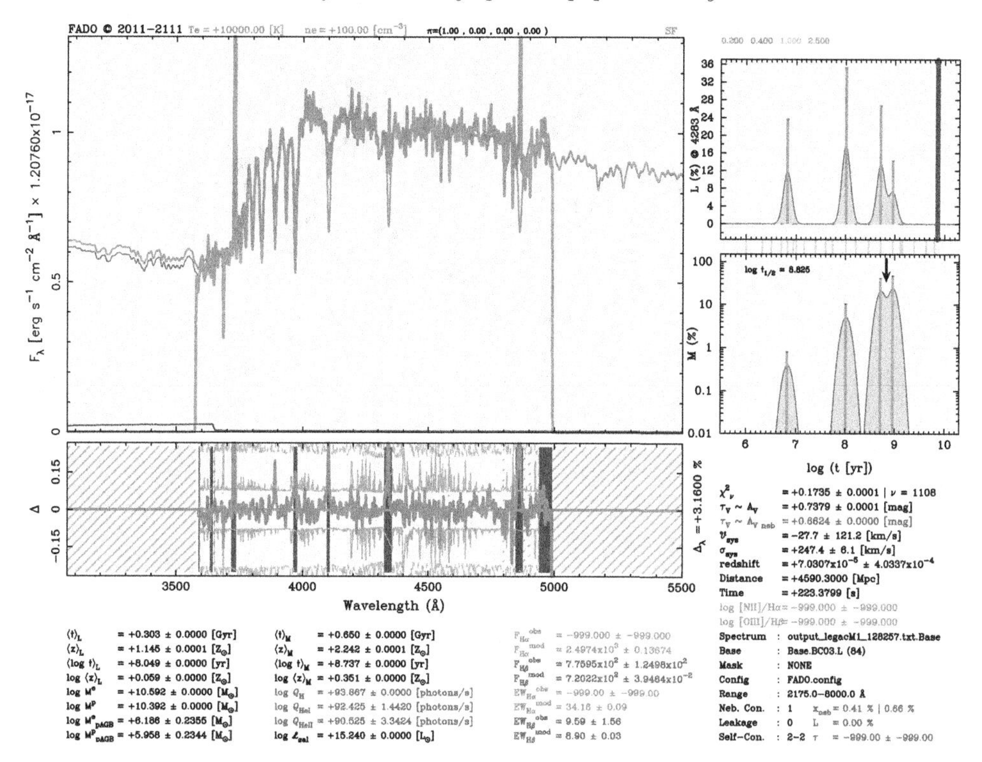

Figure 2. Example of a spectral fit with FADO of the star-forming galaxy M1 128257 from the LEGA-C survey. Top left panel: The LEGA-C spectrum of the source, corrected for Galactic attenuation (orange curve), reveals nebular emission in the blue region of the spectrum with a Hβ equivalent width of about 9.4 Å. Bottom left panel: residuals between fit and observed spectrum, with the shaded area and the dashed curve delineating, respectively, the $\pm 1\sigma$ and $\pm 3\sigma$ error spectrum. Right panels are the luminosity fraction (top) at the normalization wavelength (4283 Å) and the stellar mass fraction (bottom) of the SSPs composing the best-fitting population vector as a function of their age. The color-coding depicts the metallicity and the vertical bars $\pm 1\sigma$ uncertainties. The thin-gray vertical lines connecting both diagrams correspond to the ages of the SSPs in the used library. The light-blue shaded area in both panels shows smoothed version of the SFH using the Akima (1970) interpolation. The dark red line in the top right panel showing the light-fraction star formation history diagram depicts the maximum SSP age at the redshift 0.7269 of the source, i.e. the age of the universe at that redshift is ~7.16 Gyr.

computed taking into account the infall onto Virgo supercluster, the Great Attractor, and the Shapley concentration using the methodology described in Mould *et al.* (2000). This was done to obtain absolute quantities from the spectral synthesis (e.g., stellar masses and luminosities).

To retrieve the star formation histories of our sample galaxies we used the population spectral synthesis code FADO v1.c. The spectra were not rebinned prior to the fitting since FADO has an automatic routine that flux-conserves the spectra. The FADO fits were performed using the full wavelength range of the LEGA-C spectra. The SSPs were the same as in Sect. 3. The upper bound on the age of the SSPs was chosen to be consistent with the maximum age of the universe, however, FADO automatically selects only stellar populations that are consistent with the redshift of the galaxy, i.e. the age of the universe at the redshift of observed galaxy. In our fits we have adopted the cosmology (Spergel *et al.* 2007). The SSPs encompass four metallicities: 0.2, 0.4, 1 & 2.5 Z$_\odot$.

In Figure 2, we show an example of the spectral pixel-by-pixel modelling for galaxy M1 128257 of the LEGA-C survey, while in Figure 3 we show the mean stellar age versus

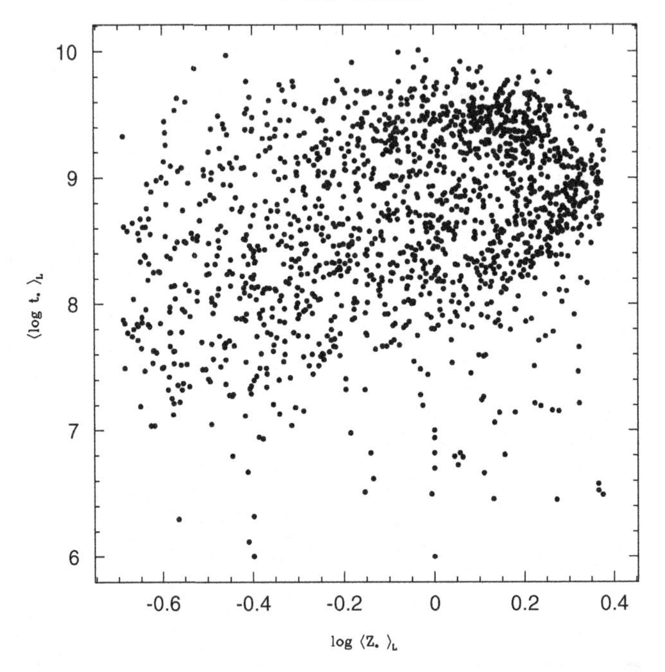

Figure 3. Mean stellar age versus mean stellar metallicity both luminosity-weighted for all LEGA-C galaxies where FADO has converged to a solution.

the mean stellar metallicity mass-weighted for all LEGA-C galaxies where FADO has converged. However, some of the galaxies are too noisy to be fitted.

Acknowledgements

J.M.G. acknowledges support by Fundação para a Ciência e a Tecnologia (FCT) through national funds (UID/FIS/04434/2013), FCT/MCTES through national funds (PIDDAC) by this grant UID/FIS/04434/2019 and by FEDER through COMPETE2020 (POCI-01-0145-FEDER007672). J.M.G. is supported by the DL 57/2016/CP1364/CT0003 contract and acknowledges the previous support by the fellowships CIAAUP-04/2016-BPD in the context of the FCT project UID/FIS/04434/2013 & POCI-01-0145-FEDER-007672, and SFRH/BPD/66958/2009 funded by FCT and POPH/FSE (EC).

References

Akima, H. 1970, *Journal of the ACM (JACM)*, 17, Issue 4, 589–602
Alongi, M., Bertelli, G., Bressan, A., *et al.* 1993, *A&AS*, 97, 851
Bressan, A., Fagotto, F., Bertelli, G., & Chiosi, C. 1993, *A&AS*, 100, 647
van der Wel, A. *et al.* 2016, *ApJS*, 223, 29V
Baldwin, J. A., Phillips, M. M., & Terlevich, R. 1981, *PASP*, 93, 5
Bruzual, G. & Charlot, S. 2003, *MNRAS*, 344, 1000
Cardelli, J. A., Clayton, G. C., & Mathis, J. S. 1989, *ApJ*, 345, 245
Chabrier, G. 2003, *PASP*, 115, 763
Cardoso, L. S. M., Gomes, J. M., & Papaderos P. 2019, *A&A*, 622, A56
Fagotto, F., Bressan A., Bertelli, G., & Chiosi, C. 1994a, *A&AS*, 104, 365
Fagotto, F., Bressan, A., Bertelli, G., & Chiosi C. 1994b, *A&AS*, 105, 29
Fagotto, F., Bressan, A., Bertelli, G., & Chiosi, C. 1994c, *A&AS*, 105, 39
Girardi, L., Bressan, A., Chiosi, C., Bertelli, G., & Nasi, E. 1996, *A&AS*, 117, 113
Gomes, J. M. & Papaderos P. 2017, *A&A*, 603, A63
Gomes, J. M. & Papaderos, P. 2018, *A&A*, 618, C3

Lamareille, F. 2010, *A&A*, 509, A53
Mould, J. R. *et al.* 2000, *ApJ*, 545, 547
Schlafly, E. F. & Finkbeiner, D. P. 2011, *ApJ*, 737, 103
Schlegel, D. J., Finkbeiner, D. P., & Davis, M. 1998, *ApJ*, 500, 525
Spergel, D. N. *et al.* 2007, *ApJS*, 170, 377
Straatman, C. M. S. *et al.* 2018, *ApJS* Series, 239, 27

Uncovering Early Galaxy Evolution in the ALMA and JWST Era
Proceedings IAU Symposium No. 352, 2019
E. da Cunha, J. Hodge, J. Afonso, L. Pentericci & D. Sobral, eds.

doi:10.1017/S1743921320001167

Modelling the mass-SFR relation at high redshifts; predicted constraints from JWST

Emma Curtis-Lake

KICC, University of Cambridge, United Kingdom

Abstract. The mass-SFR relation of galaxies encodes information of present and historical star formation in the galaxy population. We expect the intrinsic scatter in the relation to increase to low mass where SFR becomes more stochastic. Measurements at $z \gtrsim 4$ from the Hubble Frontier fields have hinted at this (Santini *et al.*, 2017), however, with the added uncertainty of lensing magnification we await JWST to provide robust measurements. Even with datasets provided by JWST, uncertainties on mass and SFR estimates are often large, potentially covariant and dependent on assumptions used. I will present our method of Bayesian hierarchical modelling of the mass-SFR relation that self-consistently propagates uncertainties on mass and SFR estimates to uncertainties on the mass-SFR relation parameters. I will expose the biases imposed by standard SED-modelling practices, and address to what significance we can measure an increase in intrinsic scatter to low masses with JWST.

Uncovering Early Galaxy Evolution in the ALMA and JWST Era
Proceedings IAU Symposium No. 352, 2019
E. da Cunha, J. Hodge, J. Afonso, L. Pentericci & D. Sobral, eds.

doi:10.1017/S1743921319009256

Star-formation efficiency at 600Myr of cosmic time

Mauro Stefanon[1], Ivo Labbé[2], Rychard Bouwens[1] and Pascal Oesch[3]

[1]Leiden Observatory, University of Leiden, Niels Bohrweg, 2 - 2333CA Leiden - The Netherlands
email: stefanon@strw.leidenuniv.nl

[2]Centre for Astrophysics and SuperComputing, Swinburne, University of Technology, Hawthorn, Victoria, 3122, Australia

[3]Observatoire de Genève, 51 Ch. des Maillettes, 1290 Versoix, Switzerland

Abstract. Current observations suggest an accelerated evolution of the cosmic star formation rate density for $8 < z < 10$, indicating that galaxy assembly experienced an extremely intense phase during the first ~ 600 Myr years of cosmic time. We performed a systematic search of ultrabright star-forming galaxies at $z \sim 8$ over the COSMOS/UltraVISTA survey, identifying 16 candidate Lyman-break galaxies. The still large uncertainties on the associated volume density do not yet allow us to ascertain whether a different star-formation efficiency (SFE) existed at early cosmic epochs. Leveraging the deepest Spitzer/IRAC data available from the GREATS program over the CANDELS/GOODS fields, we also constructed stacked SEDs of sub-L^* LBGs at $z \sim 8$. We find extreme nebular line emission ($\mathrm{EW_0(H\alpha)} \sim 1000$ Å), high specific star-formation rates (~ 10/Gyr) and indication of an inverse Balmer break. These results point toward very young ages (< 100 Myr), and, combined with measurements at lower redshifts, that the SFE evolved only marginally during the first ~ 1.5 Gyr of cosmic history.

1. Introduction

In the last decade, the synergy between the major Observatories combined with efficient photometric selection techniques have allowed us to probe galaxy formation as far as the early re-ionization epoch, identifying ~ 10k galaxies at $z > 4$, with $\gtrsim 500$ candidates at $8 \lesssim z \lesssim 11$ (Ellis *et al.* 2013; Oesch *et al.* 2013, 2014, 2018; Schenker *et al.* 2013; Schmidt *et al.* 2014; Bouwens *et al.* 2015, 2016; Finkelstein *et al.* 2015b; McLeod *et al.* 2016; Ishigaki *et al.* 2018). These remarkable observational efforts have allowed us to gain insights on the evolution of the star-formation rate density (SFRD), stellar mass density and specific star-formation rate (sSFR) up to $z \sim 8 - 10$ (Labbé *et al.* 2013; Stark *et al.* 2013; González *et al.* 2014; Smit *et al.* 2014; Salmon *et al.* 2015; Faisst *et al.* 2016; Duncan *et al.* 2014,; Song *et al.* 2016; Davidzon *et al.* 2018).

The redshift range $z \sim 8 - 11$ is of particular interest. Controversial results on whether the efficiency of star formation evolved or not at early cosmic epochs still exist both among observations (e.g., Behroozi *et al.*, 2013; Finkelstein *et al.* 2015a; Harikane *et al.* 2016, 2018; Stefanon *et al.* 2017a) and among models (Behroozi *et al.* 2018; Tacchella *et al.* 2018). Further constraints to the star-formation efficiency come from the study of the bright end of the UV luminosity function (e.g., Finlator *et al.* 2011; Mason *et al.* 2015; Mashian *et al.* 2016). Indeed, the recent spectroscopic confirmation of one ultra-bright $M_{\mathrm{UV}} = -22.1$ mag galaxy at $z = 11.1$ (GN-z11 - Oesch *et al.* 2016) suggests that galaxy assembly could have been well under way at just ~ 400 Myr of cosmic time, challenging current models of galaxy evolution (e.g., Mutch *et al.*, 2016).

To further our understanding of galaxy assembly at early cosmic epochs, we need to successfully overcome challenges at different luminosity regimes. For $L \lesssim L^*$, one factor that is currently limiting the estimates of the physical parameters of $z \gtrsim 8$ galaxies is the lack of deep observations at rest-frame optical wavelengths, currently only probed by Spitzer/IRAC. Observational progress at the very bright end, instead, has been relatively slow due to the combined effects of the small field of view of the HST/WFC3 camera and of the low surface densities of bright galaxies at $z \gtrsim 8$, with remarkable exceptions from the BoRG/HIPPIES programs (Trenti *et al.* 2011; Yan *et al.* 2011) and from the Hubble Frontier Fields, CLASH and RELICS initiatives.

2. Bright galaxies at $z \sim 8$

An alternative approach in the search for bright LBGs at $z \gtrsim 8$ consists in analyzing the ground-based wide-field surveys such as COSMOS/UltraVISTA and UKIDSS/UDS, with deep (~ 26 mag) broad wavelength coverage ($0.3 - 5\mu$m - e.g., Bowler *et al.* 2012, 2014, 2015, 2017; Stefanon *et al.* 2017b).

Our group has performed a systematic search for the brightest, most luminous, star-forming galaxies at $z \sim 8$ over the ~ 0.7 square degrees of COSMOS/UltraVISTA field, selecting them as Y-dropouts. Our search identified 16 bona-fide galaxies with redshifts in the range $7.4 \lesssim z \lesssim 8.7$ and $-22.5 \lesssim M_{\rm UV} \lesssim -21.4$ mag. These luminosities make them among the most luminous galaxies identified at these redshifts, the brightest identifications based on such a significant range of multi-wavelength observations. Our SED analysis shows that 10 of these galaxies have a confidence of $> 95\%$ of being robust $z \sim 8$ star-forming galaxies. One of the currently open questions is whether the bright end of the UV LF follows a Schechter or a double power-law form. A double power-law form would imply a more efficient star-formation compared to the faster decline of the Schechter exponential form. However, our analysis combining our results with the volume densities of Bouwens *et al.* (2015) does not yet allow us to distinguish between these two forms (Stefanon *et al.*, 2019a, in press).

3. Probing the specific star-formation rate at $z \sim 8$

The recently completed GOODS Re-ionization Era wide-Area Treasury from Spitzer (GREATS) program (PI: Labbé - Stefanon *et al.* 2019b, in preparation) provides full-depth coverage in the 3.6μm and 4.5μm bands with depth of ~ 150 hours over ~ 150arcmin^2, corresponding to nominal ultradeep limits of ~ 27.0 and 26.7 mag (AB, 5σ), enabling simultaneous detection in the 3.6μm and 4.5μm bands for more than half of the $z \sim 8$ LBGs identified in the GOODS fields. Using these data we built stacked SEDs from 119 robust LBG candidates at $z \sim 8$ segregated into four different luminosity bins to study their main physical parameters (SFR, age and stellar mass). The stacked SEDs are characterized by two main features:

Red [3.6] – [4.5] color. The first notable feature is a red $[3.6] - [4.5] > 0$ mag color. Previous works have interpreted a red color at similar redshifts as the effect of nebular Hβ+[O III] emission entering the 4.5μm band at $z \sim 7.3$ (e.g., Schaerer & de Barros 2009; Labbé *et al.* 2013). Under this hypothesis, we estimate extreme rest-frame equivalent width EW(Hβ+[OIII]) ~ 1000Å (assuming the line ratios of Anders & Fritze-v. Alvensleben 2003 for sub-solar metallicity ($Z = 0.2 Z_\odot$), and assuming a flat f_ν continuum), consistent with other estimates at similar redshifts (Labbé *et al.* 2013; Smit *et al.* 2014, 2015), and suggesting either young ages, low metallicities high rates of star formation or a combination thereof.

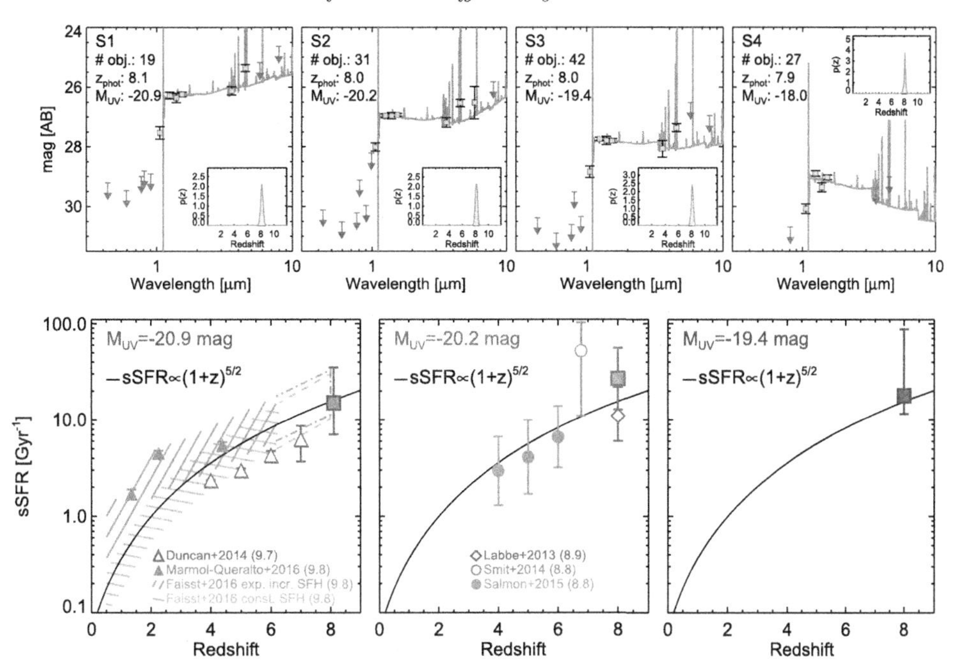

Figure 1. **Top row:** Stacked SEDs resulting from our analysis of $L < L^*$ LBGs at $z \sim 8$. Each panel refers to a stack at a different luminosity, as indicated by the label at the top-left corner of each panel. In each panel, the filled red squares with errorbars mark the stacked photometry, while the red arrows represent 2σ upperlimits. All SEDs present red $[3.6] - [4.5] > 0$ mag colors; the SEDs of the three less luminous stacks present blue $H_{160} - [3.6] < 0$ mag color, suggesting young stellar populations. **Bottom row:** Evolution of the sSFR with redshift. The measurements from the stacking analysis are presented in the panels according to their $M_{\rm UV}$, indicated at the top. In each panel we also plot recent estimates of the sSFR at high redshifts from the literature, as indicated by the legends at the bottom, segregated according to their stellar mass, quoted in parenthesis in log scale. The black curve corresponds to the evolution of the sSFR from the toy model of Dekel *et al.* (2013), of the form sSFR $\propto (1+z)^{5/2}$, as expected from cold gas inflow that follows the hierarchical merging of the dark matter halos. The overall good match of the curve to the observations suggests a scenario where the star-formation efficiency did not strongly evolve with cosmic time.

Blue $H - [3.6]$ color. Perhaps, the most intriguing feature of the stacked SEDs is the blue $H_{160} - [3.6] < 0$ color. At redshifts $z \sim 8$ the 3.6μm band covers the rest-frame wavelengths around the Balmer/4000Å break; the measured blue color suggests very young ages of the stellar populations. More accurate estimates of the stellar continuum red-wards of the Balmer Break would benefit from 8.0μm-band measurements, poorly constrained in our sample.

An SED analysis assuming a constant SFH results in ages of < 100 Myr and correspond to sSFR of $10 - 35\,\mathrm{Gyr}^{-1}$, consistent with those observed at $z \sim 7-8$ LBGs from broad band photometry (Labbé *et al.* 2013; Smit *et al.* 2014; Castellano *et al.* 2017). When considered together with estimates at lower redshifts, our new measurements indicate an increasing sSFR with redshift out to $z \sim 8$. Remarkably, the similarity between the evolution of the sSFR and that of the specific accretion rate of the dark matter halos at $3 < z < 8$, that we show in Figure 1, suggests that at high redshifts galaxy formation might be dominated by the assembly of cold gas, driven by the hierarchical formation of the dark matter halos and therefore that the evolution with redshift of the $M_\star/M_{\rm halo}$ ratio should be marginal.

Acknowledgements

We are delighted to acknowledge the contributions of our collaborators Ivo Labbé, Rychard Bouwens, Pascal Oesch, Garth Illingworth and the UltraVISTA team.

References

Anders, P. & Fritze-v. Alvensleben, U. 2003, *A&A*, 401, 1063
Behroozi, P. & Silk, J. 2018, *MNRAS*, 477, 5382
Bernard, S. R., Carrasco, D., Trenti, M., *et al.* 2016, *ApJ*, 827, 76
Bouwens, R. J., Illingworth, G. D., Oesch, P. A., *et al.* 2015, *ApJ*, 803, 34
Bouwens, R. J., Oesch, P. A., Labbé, I., *et al.* 2016, *ApJ*, 830, 67
Bowler, R. A. A., Dunlop, J. S., McLure, R. J., & McLeod, D. J. 2017, *MNRAS*, 466, 3612
Bowler, R. A. A., Dunlop, J. S., McLure, R. J., *et al.* 2012, *MNRAS*, 426, 2772
—. 2014, *MNRAS*, 440, 2810
—. 2015, *MNRAS*, 452, 1817
Calvi, V., Trenti, M., Stiavelli, M., *et al.* 2016, *ApJ*, 817, 120
Castellano, M., Pentericci, L., Fontana, A., *et al.* 2017, *ApJ*, 839, 73
Davidzon, I., Ilbert, O., Faisst, A. L., *et al.* 2018, *ApJ*, 852, 107
de Barros, S., Oesch, P. A., Labbé, I., *et al.* 2018 - submitted, *ApJ*
Dekel, A., Zolotov, A., Tweed, D., *et al.* 2013, *MNRAS*, 435, 999
Duncan, K., Conselice, C. J., Mortlock, A., *et al.* 2014, *MNRAS*, 444, 2960
Ellis, R. S., McLure, R. J., Dunlop, J. S., *et al.* 2013, *ApJL*, 763, L7
Faisst, A. L., Capak, P., Hsieh, B. C., *et al.* 2016, *ApJ*, 821, 122
Finkelstein, S. L., Ryan, R. E., Papovich, C., *et al.* 2015a, *ApJ*, 810, 71
Finkelstein, S. L., Song, M., Behroozi, P., *et al.* 2015b, *ApJ*, 814, 95
Finlator, K., Oppenheimer, B. D., & Davé, R. 2011, *MNRAS*, 410, 1703
González, V., Bouwens, R., Illingworth, G., *et al.* 2014, *ApJ*, 781, 34
Harikane, Y., Ouchi, M., Ono, Y., *et al.* 2016, *ApJ*, 821, 123
Harikane, Y., Ouchi, M., Ono, Y., *et al.* 2018, *PASJ*, 70, S11
Ishigaki, M., Kawamata, R., Ouchi, M., *et al.* 2018, *ApJ*, 854, 73
Labbé, I., Oesch, P. A., Bouwens, R. J., *et al.* 2013, *ApJL*, 777, L19
Livermore, R. C., Trenti, M., Bradley, L. D., *et al.* 2018, *ApJL*, 861, L17
Lotz, J. M., Koekemoer, A., Coe, D., *et al.* 2017, *ApJ*, 837, 97
Mashian, N., Oesch, P. A., & Loeb, A. 2016, *MNRAS*, 455, 2101
Mason, C. A., Trenti, M., & Treu, T. 2015, *ApJ*, 813, 21
McLeod, D. J., McLure, R. J., & Dunlop, J. S. 2016, *MNRAS*, 459, 3812
McLeod, D. J., McLure, R. J., Dunlop, J. S., *et al.* 2015, *MNRAS*, 450, 3032
Morishita, T., Trenti, M., Stiavelli, M., *et al.* 2018, *ApJ*, 867, 150
Oesch, P. A., Labbé, I., Bouwens, R. J., *et al.* 2013, *ApJ*, 772, 136
Oesch, P. A., Bouwens, R. J., Illingworth, G. D., *et al.* 2014, *ApJ*, 786, 108
Oesch, P. A., Brammer, G., van Dokkum, P. G., *et al.* 2016, *ApJ*, 819, 129
Oesch, P. A., Bouwens, R. J., Illingworth, G. D., *et al.* 2018, *ApJ*, 855, 105
Salmon, B., Papovich, C., Finkelstein, S. L., *et al.* 2015, *ApJ*, 799, 183
Schenker, M. A., Robertson, B. E., Ellis, R. S., *et al.* 2013, *ApJ*, 768, 196
Schmidt, K. B., Treu, T., Trenti, M., *et al.* 2014, *ApJ*, 786, 57
Smit, R., Bouwens, R. J., Labbé, I., *et al.* 2014, *ApJ*, 784, 58
Song, M., Finkelstein, S. L., Ashby, M. L. N., *et al.* 2016, *ApJ*, 825, 5
Stark, D. P., Schenker, M. A., Ellis, R., *et al.* 2013, *ApJ*, 763, 129
Stefanon, M., Labbé, I., Bouwens, R. J., *et al.* 2017b, *ApJ*, 851, 43
Stefanon, M., Labbé, I., Bouwens, R. J., *et al.* 2019a, arXiv e-prints, arXiv:1902.10713
Stefanon, M. *et al.* 2019b - in prep., *ApJ*
Tacchella, S., Bose, S., Conroy, C., *et al.* 2018, *ApJ*, 868, 92
Trenti, M., Bradley, L. D., Stiavelli, M., *et al.* 2011, *ApJL*, 727, L39
Waters, D., Wilkins, S. M., Di Matteo, T., *et al.* 2016, *MNRAS*, 461, L51
Yan, H., Yan, L., Zamojski, M. A., *et al.* 2011, *ApJL*, 728, L22

Uncovering Early Galaxy Evolution in the ALMA and JWST Era
Proceedings IAU Symposium No. 352, 2019
E. da Cunha, J. Hodge, J. Afonso, L. Pentericci & D. Sobral, eds.

doi:10.1017/S1743921319008470

Galaxy Evolution through spectral fitting tools: A comparative study between STECKMAP and FADO

Ciro Pappalardo, Polychronis Papaderos, Jean Michel Gomes, Leandro Cardoso, Ana Afonso, Israel Matute and Stergios Amarantidis

Instituto de Astrofísica e Ciências do Espaço, Portugal
email: ciro@oal.ul.pt

Abstract. Spectral analysis is nowadays a widely used tool to investigate the evolution of galaxies. Assessing the reliability of this approach is crucia, motivating a through analysis. In this poster, a comparative study between two widely tools, FADO and STECKMAP, is performed, focusing on the discrepancies between the different approaches. Both codes use different methods to extract the best fit, allowing the possibility to disentangle possible biases introduced in the analysis. Our analysis showed that where nebular emission is not negligible, the results obtained with methods taking into account such a component are more reliable, and this can be very important when moving at higher redshift, where stellar populations are younger. In particular, this is true for starburst systems, where a huge amount of stars are forming almost at the same epoch. This is an important aspect to take into account the future survey, as JWST for example, which will provide the community with medium resolution spectra of galaxies at redshift 3-4 and even higher.

Keywords. Spectral fitting tools, Galaxy Evolution, Nebular emission

---

A fundamental tool to investigate the evolution of galaxies is the analysis of the spectra obtained through the decomposition of the light received from these objects. Spectra contain crucial information about the stellar and gas emission, chemistry, and kinematics. However, to apply such an approach we must be absolutely confident about the reliability of the results. With the increase of technical capabilities, and with all a new generation of instruments, different spectral analysis tools have been proposed to the scientific community in order to extract such parameters, e.g. STECKMAP, FireFly, FADO, and VESPA Ocvirk *et al.* (2006); Wilkinson *et al.* (2017); Gomes & Papaderos (2017); Tojeiro *et al.* (2009). But still, many biases and inconsistencies in the results are present, with scientists finding different results using different tools.

In this poster, a comparative study for some of these spectral fitting tools is performed, focusing on the discrepancies between the different approaches. Using the evolutionary stellar population code REBETIKO, a set of different mock spectra, with different SNR, and star formation histories are produced (Fig. 1). These spectra are analyzed with two different tools: FADO and STECKMAP (Fig. 2). Both codes use different methods to extract the best fit, allowing the possibility to disentangle possible biases introduced in the analysis. STECKMAP reproduces quite well the spectra and the star formation history of the mock data, even if for some particular model a secondary peak appears. This double peak is due to a bad spectral coverage within the time bin of the secondary peak since the proposed spectral synthesis population basis does not have a sufficient number of spectra covering that specific time bin. Moreover, we note that the typical double

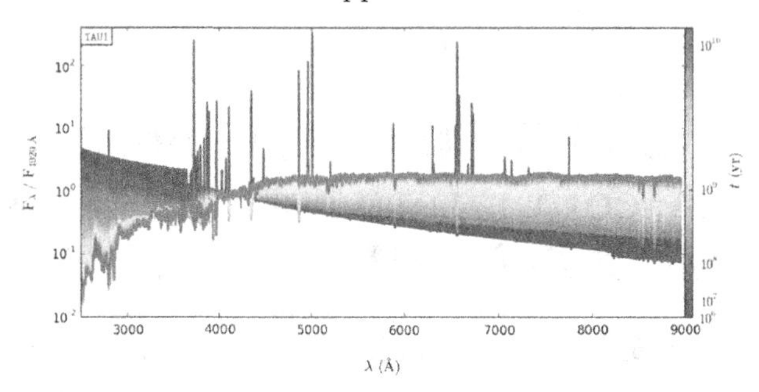

Figure 1. Mock spectra produced using `rebetiko` code, considering an exponentially decreasing star formation history (τ = 1 Gyr), solar metallicity, and no extinction, color coded according to the age (color bar on the right). Spectra are normalized at 4020 Å.

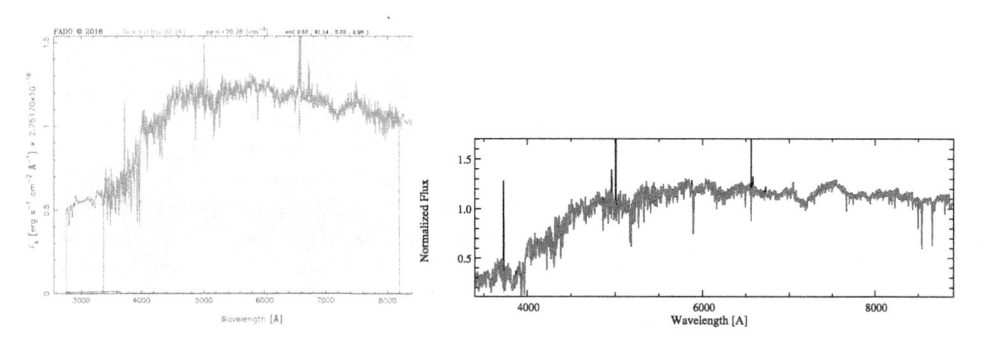

Figure 2. Examples of spectral fit for FADO (left panel) and STECKMAP (right panel).

peak is on average shifted towards the younger or the older time bin. This introduces an interesting feature because the oldest time bin corresponds to a stellar population with high weight in terms of total mass and low weight in terms of total light, while the young one has a low weight in terms of mass but high weight in terms of light. This trend can have different origins: for example, these secondary peaks are particularly evident when considering stellar populations at an epoch where the nebular emission is still relevant. In those cases, FADO reproduces better the trend for both nebular and stellar continuum emission, confirming that at these stages of star formation the nebular emission must be properly taken into account. Where nebular emission is not negligible, the results obtained with methods taking into account such a component are more reliable, and this can be very important when moving at higher redshift, where the stellar populations are young. In particular, this is true for starburst systems, where a huge amount of stars are forming almost at the same epoch. This is an important aspect to take into account the future survey, as JWST for example, which will provide the community with medium resolution spectra of galaxies at redshift 3-4 and even higher. For these objects, it is important to quantify these effects, as the emission due to the ionized gas can be very high compared to the emission of the stellar continuum.

References

Gomes, J. M. & Papaderos, P. 2017, *A&A*, 603, A63

Ocvirk, P., Pichon, C., Lançon, A., & Thiébaut, E. 2006, *MNRAS*, 365, 74

Wilkinson, D. M., Maraston, C., Goddard, D., Thomas, D., & Parikh, T. 2017, *MNRAS*, 472, 4297

Tojeiro, R., Wilkins, S., Heavens, A. F., Panter, B., & Jimenez, R. 2009, *ApJS*, 185, 1

Uncovering Early Galaxy Evolution in the ALMA and JWST Era
Proceedings IAU Symposium No. 352, 2019
E. da Cunha, J. Hodge, J. Afonso, L. Pentericci & D. Sobral, eds.

doi:10.1017/S1743921319009347

Constraints on the production and escape of ionizing radiation from the emission-lines of metal-poor star-forming galaxies

A. Plat[1], S. Charlot[1], G. Bruzual[2], A. Feltre[1,3,4], A. Vidal-García[1,5], C. Morisset[6] and J. Chevallard[1]

[1]Sorbonne Université, CNRS, UMR7095, Institut d'Astrophysique de Paris, F-75014, Paris, France

[2]Instituto de Radioastronomía y Astrofísica, UNAM, Campus Morelia, Michoacan, México, C.P. 58089, México

[3]SISSA, via Bonomea 265, I-34136 Trieste, Italy

[4]Univ. Lyon, Univ. Lyon1, ENS de Lyon, CNRS, Centre de Recherche Astrophysique de Lyon, UMR5574, 69230 Saint-Genis-Laval, France

[5]LERMA, Observatoire de Paris, Ecole Normale Supérieure, PSL Research University, CNRS, UMR8112, F-75014 Paris, France

[6]Instituto de Astronomía, UNAM, Apdo. postal 106, C.P. 22800 Ensenada, Baja California, México

Abstract. To understand how the nature of the ionizing sources and the leakage of ionizing photons in high-redshift galaxies can be constrained from their emission-line spectra, we compare emission-line models of star-forming galaxies including leakage of ionizing radiation, active galactic nuclei (AGN) and radiative shocks, with observations of galaxies at various redshifts with properties expected to approach those of primeval galaxies.

Keywords. galaxies: general, galaxies: high-redshift, galaxies: ISM

We investigate the properties of young galaxies by following the approach of Charlot & Longhetti (2001) and Gutkin *et al.* (2016) to compute the emission from star-forming galaxies, for both ionization-bounded and density-bounded models to account for the leakage of ionizing photons. We also include the emission from AGN narrow-line regions using an updated version of the Feltre *et al.* (2016) models and the emission from radiative shocks by appealing to the models of Alarie *et al.* (2019). All models are computed with the same physically-consistent description of element abundances and depletion onto dust grains. We assemble an observational sample of 68 metal-poor star-forming galaxies, 6 confirmed and 23 candidate LyC leakers and a few more quiescent star-forming galaxies and AGN, at redshifts up to 7.1. These observations are compared with the models in various diagnostic diagrams involving ultraviolet and optical emission lines. While these data had been interpreted independently in previous studies, with only a few lines typically available at once, our approach allows the simultaneous investigation of multiple diagnostics with a wide collection of homogenous models. Fig. 1 shows an example of diagram comparing this observational sample with the models, the effects of different physical parameters on emission lines being illustrated by segments of different colours. Observations are plotted in grey. The black circle corresponds to a 'standard' model (with constant star formation of 3 Myr, upper IMF cutoff $m_{\rm up} = 300\,{\rm M}_\odot$, metallicity $Z = 0.002$, ionization parameter $\log\langle U\rangle = -2$, gas density $n_{\rm H} = 10^2\,{\rm cm}^{-3}$, dust-to-metal

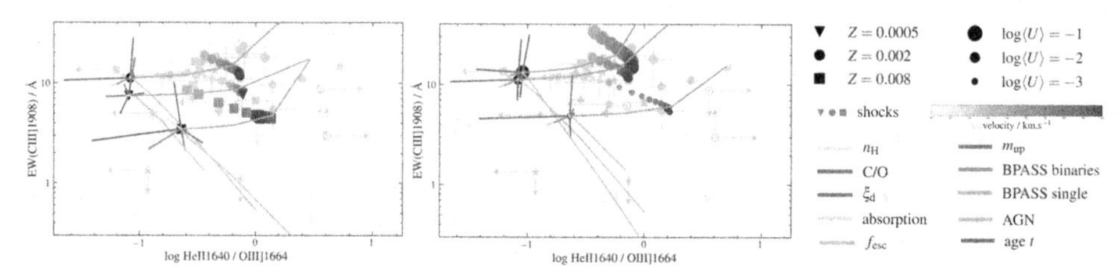

Figure 1. Segments of different colours show the effect of altering a single parameter at the time with respect to the standard model: rise in C/O ratio from 0.17 to $(\mathrm{C/O})_\odot = 0.44$ (blue); drop in dust-to-mass ratio from $\xi_\mathrm{d} = 0.3$ to 0.1 (dark green); rise in n_H from 10^2 to $10^3\,\mathrm{cm}^{-3}$ (yellow); inclusion of interstellar-line absorption in the H II interiors and outer H I envelopes of stellar birth clouds (light green); increase in stellar population age from 3 to 10 Myr (brown); rise in m_up from 100, to 300, to 600 $\mathrm{M}_\odot$ (dark purple); adopting the BPASS single- (light purple) and binary-star (magenta) models in place of the C&B model (BPASS models are not available for $Z = 0.0005$); drop in the optical depth to LyC photons at 570 Å from +1.0 to −1.0 (light blue); inclusion of an AGN component contributing from 0 to over 90% of the total He II λ1640 emission (orange); and inclusion of a radiative-shock component contributing 90% of the total He II λ1640 emission [red symbols, with shape corresponding to the metallicity of the associated benchmark model, and darkness to the shock velocity, from $10^2\,\mathrm{km\,s}^{-1}$ (light) to $10^3\,\mathrm{km\,s}^{-1}$ (dark)]. See Plat *et al.* (2019) for details.

mass ratio $\xi_\mathrm{d} = 0.3$, C/O ratio of 0.17), while the black upside-down triangle and black square are benchmark models with the same parameters, but with $Z = 0.0005$ and 0.008, respectively, and the different sizes correspond to different ionization parameters.

From this analysis, we confirm that models of star-forming galaxies powered by either single-star population (as in the latest version of the Bruzual & Charlot (2003) model) or binary-star population (as in the latest version of the Eldridge *et al.* (2017) models) overall reproduce the observed emission-line properties of young metal-poor star-forming galaxies, except in diagrams involving He II (λ1640 and λ4686) recombination lines. The models can be brought into agreement with the observations in all line-ratio diagrams by invoking an AGN or a radiative-shock component, which increases He II emission and the line equivalent width (see, e.g., Fig. 1). We also confirm that, while a rise in the escape fraction of Lyman continuum photons in density-bounded models triggers a drop in the equivalent widths of low-ionization potential lines, and in the ratios of low- to high-ionization potential lines, these signatures are degenerate with several other parameters, such as the nature of the ionizing source, the age of the stellar population, the metallicity and ionization parameter of interstellar gas, in such a way that most regions of ultraviolet and optical line-ratio diagrams sampled by observations can be covered by both ionization-bounded and density-bounded photoionization models. Hence, firm conclusions about the escape of LyC photons cannot be drawn without detailed comparisons of models with observations in several diagnostic diagrams at once.

References

Alarie, A. S., Morisset, C., & Binette, L. 2019, *Rev. Mex. Astron. Astrofis.*, submitted
Bruzual, G. & Charlot, S. 2003, *MNRAS*, 344, 1000
Charlot, S. & Longhetti, M. 2001, *MNRAS*, 323, 887
Eldridge, J. J., Stanway, E. R., Xiao, L., McClelland, L. A. S., Taylor, G., Ng, M., Greis, S. M. L., & Bray, J. C. 2017, *Publ. Astron. Soc. Australia*, 34, e058
Feltre, A., Charlot, S., & Gutkin J. 2016, *MNRAS*, 462, 1757
Gutkin, J., Charlot, S., & Bruzual, G. 2016, *MNRAS*, 462, 1757
Vidal-Garc'a, A., Charlot, S., Bruzual, G., & Hubeny, I. 2017, *MNRAS*, 470, 3532

SESSION 4: Massive galaxy assembly, and the effects of AGN and environment

Uncovering Early Galaxy Evolution in the ALMA and JWST Era
Proceedings IAU Symposium No. 352, 2019
E. da Cunha, J. Hodge, J. Afonso, L. Pentericci & D. Sobral, eds.

doi:10.1017/S1743921320001179

The most distant quasars and their environments

Eduardo Bañados

Max Planck Institute for Astronomy, Germany

Abstract. The number of quasars known within the first billion years of the universe ($z > 6$) has increased significantly over the last five years. Many of these recently discovered quasars are ideal targets for observatories in the southern hemisphere such as ALMA. I will review the current status of the highest-redshift quasars and their environments, highlighting main achievements and limitations. I will then discuss how synergistic JWST/ALMA observations will shed light onto the properties and formation of some of the most extreme environments in the early universe.

Uncovering Early Galaxy Evolution in the ALMA and JWST Era
Proceedings IAU Symposium No. 352, 2019
E. da Cunha, J. Hodge, J. Afonso, L. Pentericci & D. Sobral, eds.

doi:10.1017/S1743921320001180

A rapidly evolving quasar population at the epoch of reionzation

Xiaohui Fan

University of Arizona, USA

Abstract. I will present results from our on-going large area survey of high-redshift quasars, which has discovered more than 20 new quasars at $z > 6.5$, at the epoch of reionization, forming the first large statistical sample of EoR quasars. I will discuss the rapid evolution of quasar density at that epoch, which suggests that we are witnessing the emergence of the first supermassive black hole population. I will also present multiwavelength followup observation results, especially from ALMA and Chandra, which reveals a diverse environment of quasar activities and yields new insights into the supermassive black hole/massive galaxy co-evolution.

Uncovering Early Galaxy Evolution in the ALMA and JWST Era
Proceedings IAU Symposium No. 352, 2019
E. da Cunha, J. Hodge, J. Afonso, L. Pentericci & D. Sobral, eds.

doi:10.1017/S1743921319009037

Massive quasar host galaxies in the reionisation epoch

Bram P. Venemans[1], Fabian Walter[1], Marcel Neeleman[1], Mladen Novak[1] and Roberto Decarli[2]

[1]Max Planck Institute for Astronomy, Königstuhl 17, D-69117, Heidelberg, Germany
email: venemans@mpia.de

[2]Osservatorio di Astrofisica e Scienza dello Spazio di Bologna
via Gobetti 93/3, 40129 Bologna, Italy

Abstract. Luminous quasars are powered by accretion onto supermassive black holes. Such luminous quasars have been discovered up to the highest redshifts, $z > 7$. Here we discuss recent observations of the host galaxies of luminous quasars at $z \gtrsim 6$. We do not find a correlation between ongoing black hole growth and star-formation rate in the high redshift quasars, possibly indicating that black holes and their hosts do not co-evolve. We further show that even with high spatial resolution observations of the gas kinematics, dynamical mass estimates remain highly uncertain and should be used with caution.

Keywords. cosmology: observations, galaxies: high-redshift, galaxies: ISM, galaxies: active

---

1. Introduction

In the local universe, a fundamental observational result in extragalactic astronomy is the tight correlation between the mass of the central black hole and the mass of the bulge of its host galaxy (Kormendy & Ho 2013). At first glance, such a correlation suggests that bulges and black holes co-evolve, increasing their mass simultaneously. However, recent work has questioned this co-evolution, especially at low redshift, and suggests this correlation is the end result of concomitant growth of bulges and supermassive black holes at high redshift (Kormendy & Ho 2013). It is therefore important to study this correlation at the highest redshifts, $z > 6$, to explore if the tight correlation between the mass of the black hole and the mass of the bulge already has been established in the early universe. Unfortunately, the mass of high redshift quasar hosts can not be measured directly as the stellar light is outshone by the bright, central quasar. However, the host galaxy can be detected by the redshifted emission of gas and cold dust.

In the last decade, we have been targeting $z \gtrsim 6$ quasars with mm facilities such as the PdBI/NOEMA interferometer and ALMA to characterise the host galaxies. Recently, we used ALMA to survey the [C II] line and the underlying dust continuum emission in 27 quasar host galaxies at $z > 5.94$. Even with less than 10 minutes on-source integration time, a very large fraction of the targeted quasars was detected: >85% of the quasar hosts showed significant [C II] emission (with [C II] luminosities between 10^9 and $10^{10}\,L_\odot$; Decarli *et al.* 2018) and all were detected in the dust continuum (with continuum flux densities between 0.1 and 6.0 mJy; Venemans *et al.* 2018). This high detection rate allowed us for the first time to study quasar host galaxies at the end of the reionisation epoch as a population. In this proceedings we will focus on the star-formation rates and masses of the quasar host galaxies, and compare them with the mass and growth of the black hole.

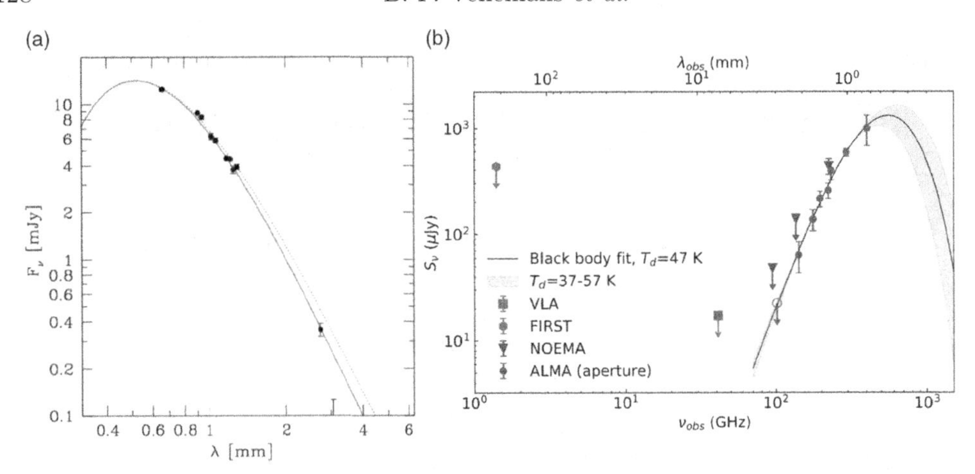

Figure 1. Dust spectral energy distribution of two quasar host galaxies at $z > 6$. **(a)** Dust SED of P183+05 at $z = 6.44$ (Decarli *et al.* 2018). The continuum has been detected at 4 different wavelengths ranging from 88 μm to 370 μm in the rest-frame. A modified black body fit gives a temperature of $T_d = 45 \pm 1$ K (Decarli *et al.* in prep). **(b)** Dust SED of J1342+0942 at $z = 7.54$ (Bañados *et al.* 2018; Venemans *et al.* 2017b). A modified black body with $T_d = 47$ K fits the data well (Novak *et al.* 2019).

2. Dust spectral energy distribution

First of all, to estimate the total infrared luminosity and to obtain a constraint on the star-formation rate in the host galaxy, we need to know the shape of the dust spectral energy distribution (SED). The dust SED is often described as a modified black body with a dust temperature T_d and emissivity index β. Based on submm and mm observations of very luminous and lensed quasars at $2 < z < 6$, Priddey & McMahon (2001) estimated $T_d = 41$ K and $\beta = 1.95$ while Beelen *et al.* 2006 measured an average $T_d = 47$ K and $\beta = 1.6$. These values have been subsequently used in the literature to derive FIR luminosities of $z > 6$ quasar host galaxies from single continuum band measurements (e.g. Wang *et al.* 2013; Willott *et al.* 2015; Venemans *et al.* 2016; Izumi *et al.* 2018). Recently, we tested this assumption by measuring the dust continuum of several $z > 6$ quasar host galaxies at multiple frequencies, which allowed us to constrain the shape of the dust SED and to verify whether the canonical values of $T_d = 47$ K and $\beta = 1.6$ are valid very high redshift quasar hosts. Two dust SEDs of $z > 6$ quasar hosts are shown in Figure 1. Both dust SEDs are consistent with the canonical value $T_d = 47$ K, assuming a fixed slope of $\beta = 1.6$. The best fit temperature of the dust in the luminous quasar P183+05 at $z = 6.44$ is $T_d = 45 \pm 1$ K (Figure 1a; Decarli et al. in prep), while for the fainter quasar host J1342+0942 at $z = 7.54$, a dust temperature $T_d = 47$ K fits the dust SED well (Figure 1b; Novak *et al.* 2019). These results justify the use of a single modified black body with $T_d = 47$ K and $\beta = 1.6$ to derive the FIR luminosity of quasar host galaxies at $z > 6$.

We can now investigate how the black hole and galaxy relate to each other by comparing the luminosity of the quasar $L_{\rm bol}$ derived from the absolute UV luminosity (a proxy for black hole growth), and the FIR luminosity (a proxy for star formation). In Figure 2 we plot the FIR luminosity derived from observed 1 mm continuum detections of $z > 5.7$ quasar host galaxy (assuming that the dust SED can be described by a modified black body with $T_d = 47$ K and $\beta = 1.6$, see discussion above) as a function of quasar brightness. We do not find a correlation between the brightness of the central source and that of the dust: for a given quasar brightness, the FIR luminosity of the host galaxy can differ more than an order of magnitude. First of all, this argues against a strong contribution of the

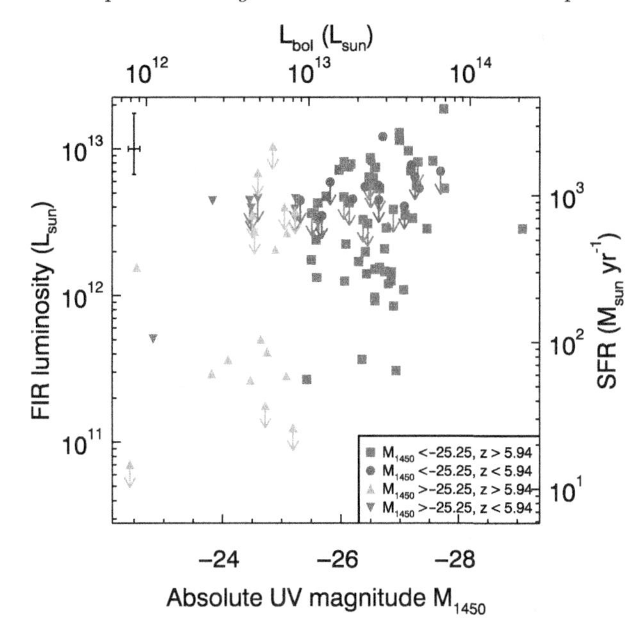

Figure 2. FIR luminosity of $z > 5.7$ quasars as a function of the absolute UV magnitude $M_{\rm UV}$. The green symbols are objects in a region of the parameter space that is highly complete. No correlation is apparent between the brightness of the quasar and the strength of the dust emission (adapted from Venemans *et al.* 2018).

quasar to the heating of the dust. The apparent lack of correlation between black hole growth and the stellar mass growth (as traced by the dust emission) indicates that at these high redshifts, the black hole and its host galaxy do not grow simultaneously. A possible explanation of our results is that the time scales of black hole accretion and star formation are vastly different. Furthermore, strong feedback from the central quasar could suppress the star formation in the host galaxy by removing the dusty interstellar medium (e.g. Hickox *et al.* 2014; Lapi *et al.* 2014; see Venemans *et al.* 2018 for a detailed discussion).

3. The mass of quasar host galaxies

While we do not find a clear correlation between the growth of the black hole and the star-formation rate, locally the correlation has been found between the **mass** of the black hole and that of the host galaxy. With current instrumentation, we cannot measure the stellar mass in the quasar host directly as the central quasar vastly outshines the light of the stars (e.g. Decarli *et al.* 2012). This will change in the near future, as observations with the *James Webb Space Telescope* will open up the potential to finally detect the stars in $z > 6$ quasar host galaxies. As an alternative, dynamical masses of the host galaxies (used as proxy for the stellar mass) have been computed in the literature from the extent and kinematics of the [C II] emission line (e.g. Wang *et al.* 2013; Willott *et al.* 2015; Venemans *et al.* 2016; Izumi *et al.* 2018). Based on these dynamical mass estimates, several groups concluded that the host galaxies of luminous quasars with black holes of $> 10^9\,M_\odot$ are less massive than expected based on the local relation (or the black holes are too massive, see e.g. Decarli *et al.* 2018). In contrast, other groups mostly targeting fainter quasars (powered by black holes $\ll 10^9\,M_\odot$) found that such $z \sim 6$ quasars are already located on the local relation (e.g. Izumi *et al.* 2018).

This lack of consensus between different groups might be a result of the large uncertainty in the dynamical mass measurements. Typically, dynamical mass estimates are derived from unresolved [C II] measurements under the assumption that the gas is located in a thin, rotating disk. In this case, the dynamical mass $M_{\rm dyn}$ can be computed using:

$$M_{\rm dyn} = R_{\rm [CII]}\,(0.75\,{\rm FWHM}/\sin i)^2\,/\,G, \tag{3.1}$$

with $R_{\rm [CII]}$ the radius of the [C II] emitting region and FWHM the full width at half maximum of the line (e.g. Walter *et al.* 2003; Wang *et al.* 2013; Willott *et al.* 2015; Venemans *et al.* 2016). Alternatively, if the gas is dispersion-dominated, the dynamical mass would be:

$$M_{\rm dyn} = 3/2\,R_{\rm [CII]}\,\sigma^2\,/\,G, \tag{3.2}$$

with σ the with of the [C II] line (e.g. Venemans *et al.* 2017a; Decarli *et al.* (2018)). The range in dynamical mass estimates from these two methods varies by a factor of 3, and the uncertainty on the inclination i often contributes another factor of a few uncertainty. High spatial resolution observations of the [C II] emission are required to reduce the large uncertainties in the dynamical mass estimates.

To resolve the structure of a distant host galaxy, explore its detailed kinematics and constrain the dynamical mass we obtained high spatial resolution (0.076″) ALMA observations of the host galaxy of quasar J0305–3150 at $z = 6.6$ (Venemans *et al.* 2019). The results are shown in Figure 3. Both the continuum and the [C II] emission are spatially resolved, and the emission is extended over ~5 kpc. The gas distribution and kinematics, as traced by the [C II] emission, are highly complex. There is a pronounced lack of [C II] emission toward the east of the quasar. This cavity in the [C II] emission is also seen in the dust continuum observations. In general, the continuum and [C II] emission trace similar structures, the main difference being the bright peak in the continuum.

From the mean velocity map of the [C II] emission (third panel in Figure 3) it is evident that the position of the accreting black hole coincides with the kinematic center of the [C II] emission. It is also located at the center of the global [C II] emission. Generally, the gas exhibits some ordered motion along the line of sight, with the gas having positive line–of–sight velocities toward the east and negative velocities toward the west. The gas with the highest velocities — that was already seen at positive velocities toward the northeast in the low-resolution data (Venemans *et al.* 2016) — is now clearly separated, both spatially and in frequency. It is therefore likely that this is a companion galaxy close to the quasar host. Furthermore, the new data with the improved sensitivity reveal the presence of two additional [C II] emitters within 40 kpc at the same redshift as the quasar (Venemans *et al.* 2019).

The unique distribution and kinematics of the [C II] emission (Figure 3) cannot be explained by a simple model. Plausible scenarios are that the gas is located in a truncated or warped disk, or the holes are created by interactions with nearby galaxies or due to energy injection into the gas (Venemans *et al.* 2019). In the latter case, the energy required to form the cavities must originate from the central active galactic nucleus, as the required energy far exceeds the energy output expected from supernovae. This energy input into the gas, however, does not inhibit the high rate of star formation of 1500 $M_\odot\,{\rm yr}^{-1}$ (Venemans *et al.* 2016). Both star formation and black hole activity could have been triggered by interactions with satellite galaxies that are discovered in our ALMA data via their [C II] emission.

To summarize, our high spatial resolution imaging of a quasar host galaxy at $z = 6.6$ shows that its formation is a complex and chaotic process. We find that the interstellar medium (ISM) in the quasar host has not yet settled in a simple disk. Based on this study, we conclude that deriving dynamical masses of quasar host galaxies from low-spatial resolution observations is highly uncertain and generally cannot be used to compare the mass of the black hole with that of its host galaxy.

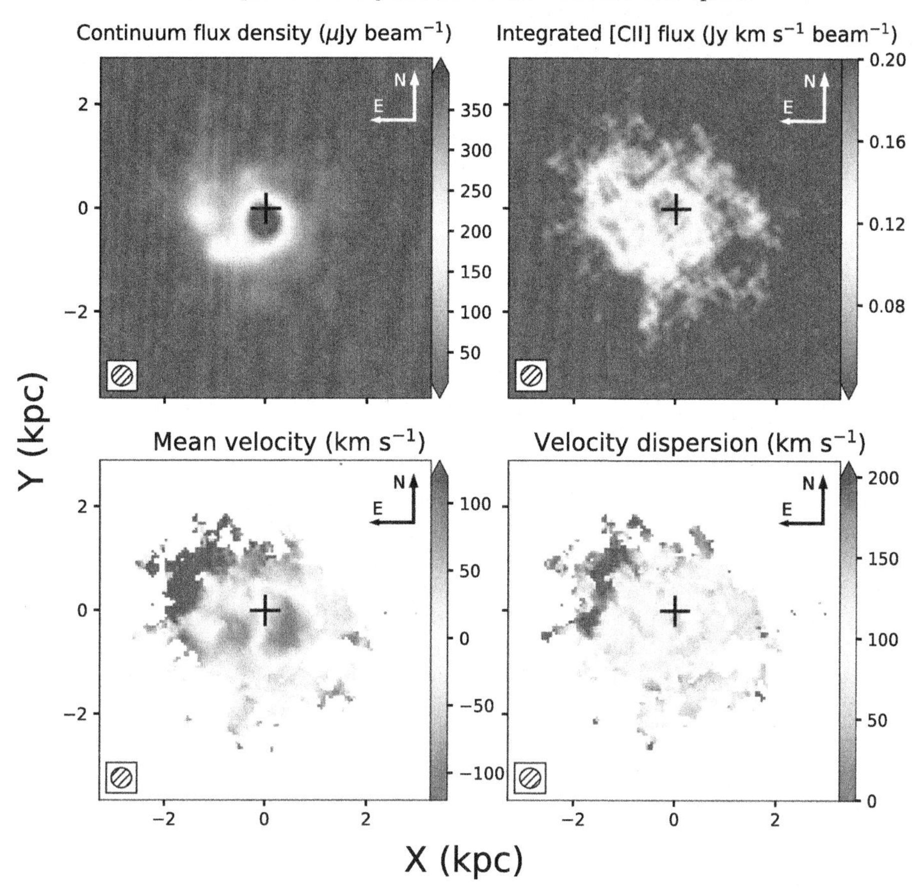

Figure 3. Dust and [C II] intensity maps (top rows) and [C II] velocity and dispersion maps (bottom row) of the host galaxy of the quasar J0305–3150 at $z = 6.6$. The cross indicates the location of the quasar. Both the gas distribution and the kinematics point to a complex system. Adapted from Venemans *et al.* (2019).

References

Bañados, E. *et al.* 2018, *Nature*, 553, 473
Beelen, A. *et al.* 2006, *ApJ*, 642, 694
Decarli, R. *et al.* 2012, *ApJ*, 756, 150
Decarli, R. *et al.* 2018, *ApJ*, 854, 97
Hickox, R. C. *et al.* 2014, *ApJ*, 782, 9
Izumi, T. *et al.* 2018, *PASJ*, 70, 36
Kormendy, J. & Ho, J. C. 2013, *ARAA*, 51, 511
Lapi, A. *et al.* 2014, *ApJ*, 782, 69
Novak, M. *et al.* 2019, *ApJ*, 881, 63
Priddey, R. S. & McMahon, R. G. 2001, *MNRAS* (Letters), 324, L17
Venemans, B. P. *et al.* 2016, *ApJ*, 816, 37
Venemans, B. P. *et al.* 2017a, *ApJ*, 837, 146
Venemans, B. P. *et al.* 2017b, *ApJ* (Letters), 851, L8
Venemans, B. P. *et al.* 2018, *ApJ*, 866, 159
Venemans, B. P. *et al.* 2019, *ApJ* (Letters), 874, L30
Walter, F. *et al.* 2003, *Nature*, 424, 406
Wang, R. *et al.* 2013, *ApJ*, 773, 44
Willott, C. J. *et al.* 2015, *ApJ*, 801, 123

Uncovering Early Galaxy Evolution in the ALMA and JWST Era
Proceedings IAU Symposium No. 352, 2019
E. da Cunha, J. Hodge, J. Afonso, L. Pentericci & D. Sobral, eds.

doi:10.1017/S1743921320000666

Connecting black holes and galaxies in faint radio populations at cosmic noon

Stacey Alberts[1], Wiphu Rujopakarn[2] and George H. Rieke[1]

[1]Steward Observatory, University of Arizona, 933 North Cherry Avenue, Tucson, AZ 85721, USA
email: salberts@email.arizona.edu

[2]Department of Physics, Faculty of Science, Chulalongkorn University
254 Phayathai Road, Pathumwam, Bangkok 10330, Thailand

Abstract. We leverage new ultra-deep, high resolution, multi-frequency radio imaging at 6 and 3 GHz with the unique datasets available in the GOODS-S/HUDF region in order to assess the AGN fraction in a faint radio-selected sample. For AGN identification, we adopt a multi-wavelength approach, combining X-ray and (mid-)infrared (IR) selections with radio identification such as X-ray to radio excess, flat radio spectral slopes, and the radio-IR correlation. We identify AGN in 43% of our radio sample, yielding an AGN source density of ~ 1 arcmin^{-2}. This AGN fraction is likely underestimated, as 1) our shallower 3 GHz data is biased against flat radio spectrum sources and 2) all of our selections may be biased against the most heavily obscured AGN. The *James Webb Space Telescope's* Mid-Infrared Instrument (MIRI) will address the latter issue and we briefly outline our Cycle 1 Guaranteed Time Observation (GTO) program to search for heavily obscured AGN.

Keywords. Galaxy: evolution, radio continuum: galaxies, galaxies: active

---

1. Introduction

Understanding the role of supermassive black hole growth in shaping galaxy populations across cosmic time remains an open and pressing issue. Locally, the tight relation between black hole mass and galaxy properties has long been established (e.g., Magorrian *et al.* 1998), a link which is supported at earlier epochs by the strikingly similar evolution of the volume-averaged cosmic star formation (SF) and black hole accretion histories (see Madau & Dickinson 2014, for a review). Though existing on very disparate physical scales, SF and Active Galactic Nuclei (AGN) activity may be connected through a common gas supply (Vito *et al.* 2014), with AGN playing a key role in the self-regulation of massive galaxy growth through feedback (e.g. Kormendy & Ho 2013). However, establishing the causal link between these phenomena remains challenging, in no small part hampered by difficulties in obtaining a complete census of AGN within a given statistical galaxy population. AGN selection methods are well known to be sensitive to distinct AGN subpopulations, with minimal overlap (Mendez *et al.* 2013; Delvecchio *et al.* 2017). This necessitates a multi-wavelength approach to avoid strong selection biases.

In this study, we leverage new multi-frequency, ultra-deep, high resolution radio imaging with the deep X-ray-mid-infrared (IR) imaging and spectroscopy available in the GOODS-S/HUDF region in order to obtain a census of AGN within faint radio galaxies and place constraints on the AGN population that may still be missing due to extreme obscuration (see Hickox & Alexander 2018, for a review). Radio imaging provides a unique complement to common X-ray, optical, and (mid-)IR AGN selections, supplying an extinction-free tracer of SF and/or AGN activity, with multiple AGN indicators:

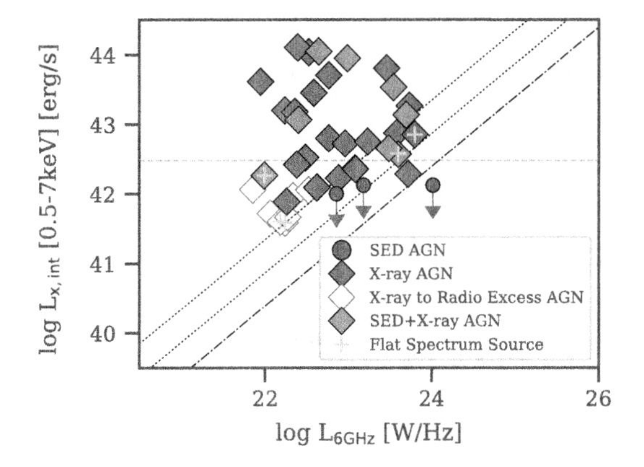

Figure 1. The intrinsic (absorption corrected, see Luo *et al.* 2017) 0.5-7 keV X-ray luminosity as a function of 6 GHz luminosity for radio sources classified as AGN. Multiple selection techniques are used: X-ray properties (diamonds), optical-mid-IR SED decomposition (orange diamonds and red circles), and flat radio spectrum sources (yellow crosses). The threshold for AGN selection via high intrinsic X-ray luminosity (dashed) as well as the expected $L_x - L_{6\mathrm{GHz}}$ relation for SFGs (dot-dashed) and 3 and 6σ outliers (dotted) from this relation are indicated.

radio jets, flat (radio) spectrum sources, and outliers from the radio-IR correlation (see Padovani 2016, for a review).

2. Data

Our sample is selected at 6 GHz using deep C-band (4-8 GHz) imaging from the Karl G. Jansky Very Large Array within the half-power radius of a single pointing centered on the Hubble Ultra Deep Field (Rujopakarn, in prep.). This imaging reaches a depth of 0.32 μJy beam^{-1} with an angular resolution of 0.61″x0.31″, $\sim$10 times deeper and $\sim$2 times higher resolution then comparable surveys (i.e. Delvecchio *et al.* (2017); Gim *et al.* (2019)). The data reduction and source extraction will be presented in Rujopakarn, in prep. Our 6 GHz sample is then matched to the Chandra X-ray Observatory 7 Ms GOODS-S catalog (Luo *et al.* 2017) and the UV-mid-IR 3D-HST (v4.1) photometric catalog (Skelton *et al.* 2014), providing X-ray properties, 0.3-24μm photometry, redshifts, and stellar masses. Our final sample is 100 6 GHz sources with secure counterparts at $z \sim 0.75 - 3$. Lastly, this sample is matched to new 3 GHz S-band (2-4 GHz) imaging (Rujopakarn, in prep.), centered on the same area, which reaches a depth of 0.7 μJy beam^{-1} and an angular resolution of 0.7″. 75/100 of the 6 GHz sources have a robust 3 GHz counterpart.

3. AGN Selection

X-ray Selection: X-ray based AGN classifications are adopted following the criteria outlined in Luo *et al.* (2017). Briefly, these are 1) a high intrinsic X-ray luminosity, 2) a hard X-ray spectrum indicating obscured AGN, 3) excess X-ray to optical or near-IR emission, and 4) excess X-ray to radio emission.

Out of our 100 6GHz sources, 31 are classified as AGN in Luo *et al.* (2017). Their intrinsic X-ray luminosity as a function of radio luminosity (assuming $\alpha = -0.7$; Condon 1992) can be seen in Figure 1. While most of the X-ray selected AGN satisfy the high intrinsic X-ray luminosity criteria, a threshold chosen to lie comfortably above the potential X-ray emission generated by SF, about a quarter are identified via X-ray excess relative to optical, near-IR, or radio emission. We expand on the latter criteria – X-ray to radio excess

– with our 6 GHz imaging, a factor of 9 deeper than the radio used in Luo *et al.* (2017). To do so, we look for sources with a significant excess (>5x) of X-ray emission over the expected level from pure SF given the $L_{\rm x} - L_{\rm radio}$ relation for star forming galaxies (SFGs) (i.e. Alexander *et al.* 2005; Lehmer *et al.* 2016). We find an additional 7 AGN candidates based on this criteria.

Optical-mid-IR Selection via SED Fitting: Mid-IR color-color diagnostics (i.e. Stern *et al.* 2005) have proven a useful tool in identifying AGN sub-populations which are often missed in X-ray surveys due to obscuration. These color diagnostics isolate regions of a galaxy spectrum where obscured AGN may dominate over the host galaxy emission. Optical-mid-IR SED fitting is an expansion on this technique, utilizing photometric data over a long baseline to identify AGN (i.e. Assef *et al.* 2010; Chung *et al.* 2014; Delvecchio *et al.* 2017.

In this work, we utilize the extensive optical-mid-IR coverage provided by 3D-HST (Skelton *et al.* 2014) and perform SED fitting and decomposition through an adapted version of the publicly available code from Assef *et al.* (2010). SED fits are performed as a non-negative linear combination of templates over the range 0.3-30μm, fitting galaxy templates first and then galaxy+AGN templates. An F-test is then applied to see if the AGN template significantly improves the fit (see Chung *et al.* (2014) for details). We identify 12 AGN via this method, 3 of which where not previously identified (orange diamond and red circles, Figure 1). Our X-ray identified AGN fraction and overlap between X-ray and optical-mid-IR selection is high relative to other studies (i.e. Delvecchio *et al.* 2017) due to the uniquely deep and complete X-ray dataset available in the GOODS-S/HUDF region.

Radio Selection via Jets: There are no large scale radio jets apparent in our radio imaging as determined via visual inspection.

Radio Selection via the Radio Spectral Index: The low frequency radio spectrum of SFGs is typically parameterized as a superposition of two power laws: the dominant steep synchrotron spectrum and a flatter free-free component contributing $\sim 10\%$ at 1.4 GHz (Condon 1992). For AGN with compact, optically thick radio cores, the radio spectrum – in SFGs a synchrotron spectrum, $S_\nu \propto \nu^\alpha$ with $\alpha \sim -0.7$ arising from the acceleration of cosmic rays in supernova remnants – experiences synchrotron self-absorption, producing a telltale flat spectrum with $\alpha \geq -0.5$ (Padovani 2016).

In Figure 2, the 3-6 GHz radio spectral slope, $\alpha^{\rm 3GHz}_{\rm 6GHz}$, is shown for our AGN. We identify 8 flat spectrum sources, including 2 which were not selected as AGN via other methods, constituting 18% of our AGN. This number is likely a lower limit on the fraction of flat spectrum sources as one quarter of our radio sample are not detected at 3 GHz. Additional 3 GHz imaging is being gathered by our team; in the meantime, we stack the 3 GHz undetected sources (Figure 2, green star), finding that their average slope is $\alpha^{\rm 3GHz}_{\rm 6GHz} \sim 0$, suggesting that there are indeed additional flat spectrum sources in our sample.

Radio Selection via Radio-Infrared Correlation: The radio-IR correlation (i.e. Yun *et al.* 2001) is a tight relation for SFGs arising due to the similar origins of radio and IR emission, with radio synchrotron generated via supernova remnants and IR from dust-reprocessed starlight from young stars. In the presence of an AGN, the radio may be elevated relative to the IR and such sources identified as outliers from the radio-IR correlation. Here we compare radio emission with the observed 24μm flux via the parameter $\mathrm{q}_{24,\rm obs} = \log\,(S_{24\mu{\rm m,obs}}/S_{\rm radio})$ (i.e. Bonzini *et al.* 2013). In order to define $\mathrm{q}_{24,\rm obs}$ for SFGs, we adopt the Rieke *et al.* (2009) empirical local templates with log $\mathrm{L_{IR}}$ = 10-13$\mathrm{L}_\odot$ with 0.2 dex scatter. Figure 3 shows $\mathrm{q}_{24,\rm obs}$ for our radio sample up to $z \sim 3$, beyond

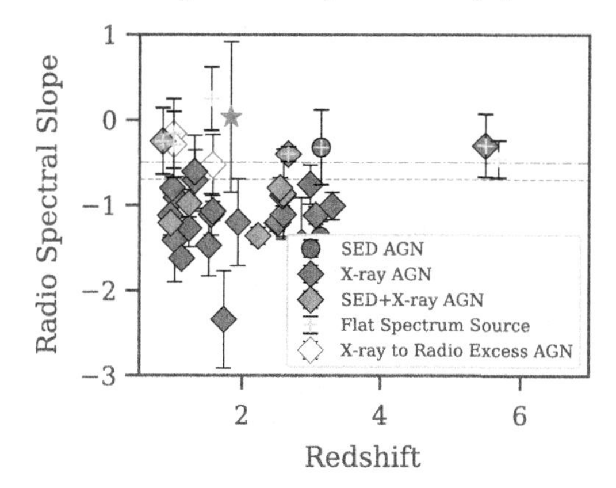

Figure 2. The radio spectral slope as a function of redshift for AGN. Symbols and colors are as in Figure 1. The green star is the stacked radio spectral slope of sources undetected at 3 GHz. The dashed line indicates the slope typically assumed for SFGs ($\alpha = -0.7$; Condon 1992), while the dot-dashed line indicates the threshold for flat spectrum AGN ($\alpha \geqslant -0.5$; Padovani 2016).

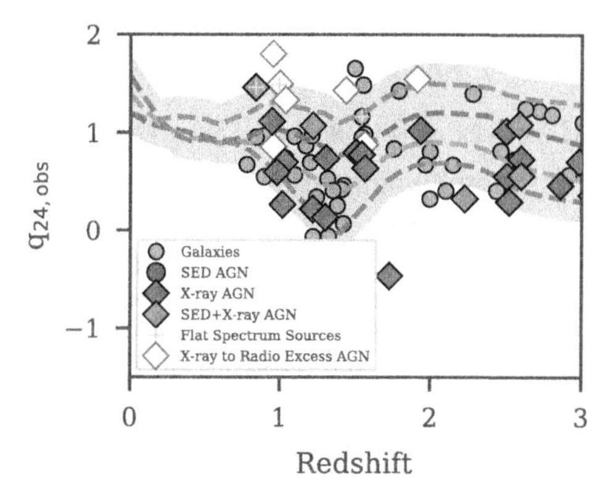

Figure 3. The radio-IR correlation, parameterized via $q_{24\mu m,obs}$ as a function of redshift for our radio sample, including SFGs (gray circles). Symbols and colors are as in Figure 1. The dashed lines represent $q_{24\mu m,obs}$ for the Rieke *et al.* (2009) local templates with log L_{IR} = 10, 11, 12, 13 $L_\odot$ with 0.2 dex scatter (shaded regions).

which 24μm no longer traces SF. We find only one source, previously identified as an X-ray AGN, is an outlier from the SFG distribution.

4. A Census of AGN

We identify AGN in 43/100 sources in our radio sample, an AGN source density of ~1 arcmin^{-2}. This fraction is a lower limit pending our approved program for deeper 3 GHz data, which may reveal additional flat spectrum sources.

What AGN might we be missing? The contribution of heavily obscured and Compton Thick (CT) AGN to the total AGN population is still disputed, with current estimates at $\sim 30 - 50\%$ (Hickox & Alexander 2018). Exploiting our deep X-ray imaging and SED fitting, we compare our sources to the intrinsic relation (Lansbury *et al.* 2015) between X-ray luminosity and the 6μm AGN luminosity – measured via SED decomposition – in Figure 4. We find that by this measure approximately a quarter of our AGN are heavily obscured or CT, slightly below the lower end of estimates of the CT fraction. For a complete analysis of these results, the reader is referred to Alberts *et al.*, in prep.

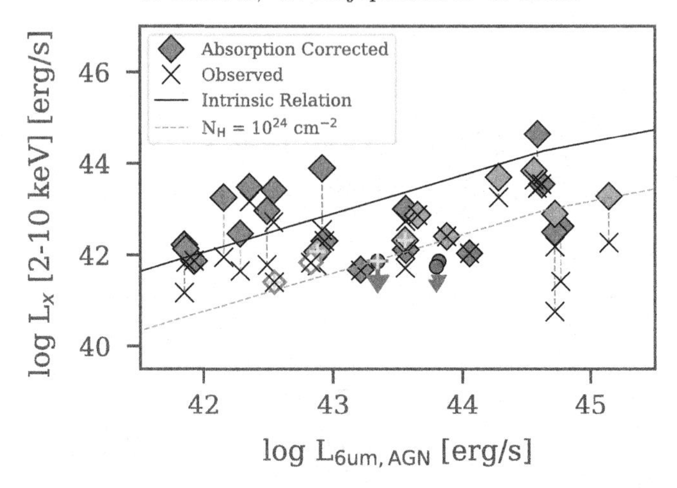

Figure 4. The X-ray luminosity (converted to 2-10 keV) as a function of the luminosity at 6μm, derived from SED decomposition for our AGN. Symbols represent the absorption corrected X-ray (Luo *et al.* 2017), as in Figure 1. X's mark the observed X-ray luminosity. The intrinsic $L_x - L_{6\mu m}$ relation for unobscured AGN (solid) and the relation given heavy obscuration ($N_H = 10^{24}$ cm^{-2}) (dashed; i.e. Lansbury *et al.* 2015) is indicated.

5. Finding Obscured AGN with Future Facilities

A complete accounting of the AGN population and their influence on their host galaxies remains elusive, in part because of the difficulty in identifying obscured and Compton Thick AGN in large, unbiased photometric samples. A solution to this problem will be available upon the launch of the *James Webb Space Telescope* (JWST). JWST's Mid-Infrared Instrument (MIRI) will provide 9 photometric bands covering 5.6-25.5μm with high sensitivity and resolution (Bouchet *et al.* 2015). This wavelength coverage is ideally suited to identify AGN through mid-IR diagnostics (i.e. Kirkpatrick *et al.* 2017). This capability is the focus of our Cycle 1 GTO HUDF imaging program (PI: G. Rieke), which samples the full MIRI photometric coverage in order to take advantage of the ubiquitous minimum in stellar emission found in non-AGN galaxies at rest 3-5μm (Sorba & Sawicki 2010). In the presence of an AGN, even heavily obscured, this minimum will be filled in by warm torus emission, providing a new and highly effective AGN selection technique that will enable a more complete census of AGN up to cosmic noon.

Acknowledgements

The authors thank Kristina Nyland and Preshanth Jagannathan for their work on the radio imaging described in this proceedings.

References

Alexander, D. M., Bauer, F. E., Chapman, S. C., *et al.* 2005, *ApJ*, 632, 736
Assef, R. J., Kochanek, C. S., Brodwin, M., *et al.* 2010, *ApJ*, 713, 970
Bonzini, M., Padovani, P., Mainieri, V., *et al.* 2013, *MNRAS*, 436, 3759
Bouchet, P., García-Marín, M., Lagage, P.-O., *et al.* 2015, *PASP*, 127, 612
Condon, J. J. 1992, *ARA&A*, 30, 575
Chung, S. M., Kochanek, C. S., Assef, R., *et al.* 2014, *ApJ*, 790, 54
Delvecchio, I., Smolčić, V., Zamorani, G., *et al.* 2017, *A&A*, 602, A3
Elvis, M., Wilkes, B. J., McDowell, J. C., *et al.* 1994, *ApJS*, 95, 1
Gim, H. B., Yun, M. S., Owen, F. N., *et al.* 2019, *ApJ*, 875, 80
Hickox, R. C. & Alexander, D. M. 2018, *ARA&A*, 56, 625
Kirkpatrick, A., Alberts, S., Pope, A., *et al.* 2017, *ApJ*, 849, 111

Kormendy, J. & Ho, L. C. 2013, *ARA&A*, 51, 511
Lansbury, G. B., Gandhi, P., Alexander, D. M., *et al.* 2015, *ApJ*, 809, 115
Lehmer, B. D., Basu-Zych, A. R., Mineo, S., *et al.* 2016, *ApJ*, 825, 7
Luo, B., Brandt, W. N., Xue, Y. Q., *et al.* 2017, *ApJS*, 228, 2
Madau, P. & Dickinson, M. 2014, *ARA&A*, 52, 415
Magorrian, J., Tremaine, S., Richstone, D., *et al.* 1998, *AJ*, 115, 2285
Mendez, A. J., Coil, A. L., Aird, J., *et al.* 2013, *ApJ*, 770, 40
Padovani, P. 2016, *A&A Rev.*, 24, 13
Rieke, G. H., Alonso-Herrero, A., Weiner, B. J., *et al.* 2009, *ApJ*, 692, 556
Skelton, R. E., Whitaker, K. E., Momcheva, I. G., *et al.* 2014, *ApJS*, 214, 24
Sorba, R. & Sawicki, M. 2010, *ApJ*, 721, 1056
Stern, D., Eisenhardt, P., Gorjian, V., *et al.* 2005, *ApJ*, 631, 163
Tisanić, K., Smolčić, V., Delhaize, J., *et al.* 2019, *A&A*, 621, A139
Vito, F., Maiolino, R., Santini, P., *et al.* 2014, *MNRAS*, 441, 1059
Yun, M. S., Reddy, N. A., & Condon, J. J. 2001, *ApJ*, 554, 803

Uncovering Early Galaxy Evolution in the ALMA and JWST Era
Proceedings IAU Symposium No. 352, 2019
E. da Cunha, J. Hodge, J. Afonso, L. Pentericci & D. Sobral, eds.

doi:10.1017/S1743921320001192

Uncovering QSO-driven outflows and galaxy assembly at cosmic Dawn with ALMA

Manuela Bischetti

INAF Roma, Italy

Abstract. I will present evidence of QSO-driven outflows in the early Universe, resulting from the stacking analysis of ALMA observations of the [CII] emission line for a sample of 50 QSOs at $z \sim 5-7$. The high sensitivity reached by our analysis allowed us to assess that very broad wings are on average present in the [CII] profile, and extend beyond velocities of 1000 km/s in systems with low and high SFR. Such wings are therefore tracing QSO-driven [CII] outflows, with associated mass outflow rates of $100-200 M_{\odot}/\mathrm{yr}$. I will discuss how these outflows relate to those observed in lower-z AGNs and give an estimate of their spatial extent. Furthermore, I will focus on the high-resolution ALMA observation of a hyper luminous QSO at $z \sim 4.5$, revealing an exceptional overdensity with multiple companions as close as 2 kpc. These crowded surroundings, and the QSO host galaxy itself, are forming stars at a very high rate (hundreds of $M_{\odot}/\mathrm{yr}$), suggesting that a significant fraction of the stellar mass assembly at early epochs might have taken place in the companions. I will discuss how the BH and host-galaxy masses are growing in this multi- source system, which likely represents the cradle of what would be a giant galaxy at $z = 0$.

Uncovering Early Galaxy Evolution in the ALMA and JWST Era
Proceedings IAU Symposium No. 352, 2019
E. da Cunha, J. Hodge, J. Afonso, L. Pentericci & D. Sobral, eds.

doi:10.1017/S1743921319009487

Rapid evolution and transformation into quiescence?: ALMA view on $z > 6$ low-luminosity quasars

Takuma Izumi[1,2], Masafusa Onoue[3], Yoshiki Matsuoka[4], Tohru Nagao[4], Michael A. Strauss[5], Masatoshi Imanishi[1,2], Nobunari Kashikawa[6], Seiji Fujimoto[7], Kotaro Kohno[8], Yoshiki Toba[9,4], Hideki Umehata[10,8], Tomotsugu Goto[11], Yoshihiro Ueda[9], Hikari Shirakata[12], John D. Silverman[13], Jenny E. Greene[14], Yuichi Harikane[7], Yasuhiro Hashimoto[15], Soh Ikarashi[16], Daisuke Iono[1,2], Kazushi Iwasawa[17], Chien-Hsiu Lee[1], Takeo Minezaki[8], Kouichiro Nakanishi[1,2], Yoichi Tamura[18], Ji-Jia Tang[19] and Akio Taniguchi[18]

[1]National Astronomical Observatory of Japan, Mitaka, Tokyo, 181-8588 Japan
email: takuma.izumi@nao.ac.jp

[2]Department of Astronomical Science, SOKENDAI, Mitaka, Tokyo, Japan

[3]Max Planck Institut für Astronomie, Königstuhl 17, D-69117 Heidelberg, Germany

[4]Research Center for Space and Cosmic Evolution, Ehime University, Ehime, Japan

[5]Princeton University Observatory, Peyton Hall, Princeton, NJ 08544, USA

[6]Department of Astronomy, School of Science, The University of Tokyo, Tokyo, Japan

[7]Institute for Cosmic Ray Research, The University of Tokyo, Chiba, Japan

[8]Institute of Astronomy, Graduate School of Science, The University of Tokyo, Tokyo, Japan

[9]Department of Astronomy, Kyoto University, Kyoto, Japan

[10]RIKEN Cluster for Pioneering Research, Saitama, Japan

[11]Institute of Astronomy and Department of Physics, National Tsing Hua University, Taiwan

[12]Department of Cosmosciences, Hokkaido University, Hokkaido, Japan

[13]Kavli Institute for the Physics and Mathematics of the Universe, Chiba, Japan

[14]Department of Astrophysics, Princeton University, Princeton, NJ, USA

[15]Department of Earth Sciences, National Taiwan Normal University, Taipei, Taiwan

[16]Kapteyn Astronomical Institute, University of Groningen, Groningen, Netherlands

[17]ICREA and Institut de Ciències del Cosmos, Universitat de Barcelona, Barcelona, Spain

[18]Division of Particle and Astrophysical Science, Nagoya University, Aichi, Japan

[19]Australian National University, Weston Creek, Australia

Abstract. We present ALMA [CII] line and far-infrared (FIR) continuum observations of seven $z > 6$ low-luminosity quasars ($M_{1450} > -25$ mag) discovered by our on-going Subaru Hyper Suprime-Cam survey. The [CII] line was detected in all targets with luminosities of $\sim (2-10) \times 10^8$ $L_\odot$, about one order of magnitude smaller than optically luminous quasars. Also found was a wide scatter of FIR continuum luminosity, ranging from $L_{\rm FIR} < 10^{11}$ $L_\odot$ to $\sim 2 \times 10^{12}$ $L_\odot$. With the [CII]-based dynamical mass, we suggest that a significant fraction of low-luminosity quasars are located on or even below the local Magorrian relation, particularly at the massive end of the galaxy mass distribution. This is a clear contrast to the previous finding that luminous quasars tend to have overmassive black holes relative to the relation. Our result is expected to show a less-biased nature of the early co-evolution of black holes and their host galaxies.

Keywords. quasars: general, galaxies: evolution, galaxies: high-redshift

1. Introduction

In the local universe, it is clear that the mass of a supermassive black hole (SMBH) is tightly correlated with that of a spheroidal component of its host galaxy (e.g., Kormendy & Ho 2013). This relation, as well as the remarkable similarity between global star formation and SMBH mass accretion histories (e.g., Madau & Dickinson 2014), suggest that black holes and their host galaxies have *co-evolved*. Postulated physical mechanisms of co-evolution include mergers of galaxies and feedback of active galactic nuclei (AGN) to terminate star formation in the host (e.g., Di Matteo *et al.* 2005; Hopkins *et al.* 2008).

Observations of physical properties of both SMBHs and their host galaxies over cosmic time are essential to test such models, as they specifically predict time evolution of the systems (e.g., Gallerani *et al.* 2017). High redshift quasars provide the unique sites for this purpose. Indeed, the last two decades have witnessed the discovery of > 200 quasars at $z > 5.7$ owing to wide-field optical and near-infrared (NIR) surveys like the Sloan Digital Sky Survey (SDSS; e.g., Fan *et al.* 2003; Jiang *et al.* 2016). These quasars are luminous at optical ($M_{1450} < -25$ mag), with SMBH masses of $M_{\rm BH} \gtrsim 10^9$ $M_\odot$ (e.g., De Rosa *et al.* 2014). Far-infrared (FIR) to submillimeter observations of cool interstellar medium have provided crucial information of their host galaxies as they are hard to detect at rest-frame ultraviolet-to-optical wavelength, due to the outshining quasar nuclei. These host galaxies typically possess copious amount of dust ($M_{\rm dust} \gtrsim 10^8$ $M_\odot$) and gas ($M_{\rm H_2} \gtrsim 10^{10}$ $M_\odot$), with FIR continuum-based star formation rate (SFR) of $\gtrsim 100 - 1000$ $M_\odot$ yr^{-1} (e.g., Wang *et al.* 2011a,b). The dusty starburst regions appear to be compact, with sizes of a few kpc or less (e.g., Venemans *et al.* 2016).

Recent high resolution submillimeter observations by the Atacama Large Millimeter/submillimeter Array (ALMA) have provided detailed properties of the high redshift quasar host galaxies, including their dynamical masses measured via [CII] emission line. They have found that $z \gtrsim 6$ luminous quasars have ratios of $M_{\rm BH}$ to host galaxy dynamical mass ($M_{\rm dyn}$) ~ 10 times larger than the $z \sim 0$ relation (e.g., Wang *et al.* 2013; Venemans *et al.* 2016) if we equate $M_{\rm dyn}$ to the bulge mass ($M_{\rm bulge}$). However, there is an observational bias, whereby more luminous quasars are powered by more massive SMBHs at high redshifts (Lauer *et al.* 2007). Indeed, early observations of low-luminosity ($M_{1450} \gtrsim -25$ mag) quasars showed that they are powered by less massive SMBHs ($\sim 10^8$ $M_\odot$), and showed $M_{\rm BH}/M_{\rm dyn}$ (surrogate of $M_{\rm BH}/M_{\rm bulge}$) ratios roughly consistent with local galaxies (e.g., Willott *et al.* 2015, 2017).

With this in mind, we have been conducting ALMA observations toward optically low-luminosity quasars, originally discovered by our on-going optical survey with the Subaru Hyper Suprime-Cam (HSC): we have thus far discovered > 80 low-luminosity quasar at $z \gtrsim 6$ down to $M_{1450} \sim -22$ mag (Matsuoka *et al.* 2016, 2018a,b). We then organized an intensive multi-wavelength follow-up consortium: *Subaru High-z Exploration of Low-Luminosity Quasars (SHELLQs)*. Here we present an overview of our ALMA programs.

2. ALMA Observations and Data Analysis

We have observed [CII] 158 μm line and the underlying rest-FIR continuum emission toward seven $z \gtrsim 6$ HSC quasars in ALMA Cycle 4 (4 objects) and Cycle 5 (3 objects), respectively (Izumi *et al.* 2018, 2019). The band 6 receiver was used to cover the redshifted [CII] line at ~ 230 GHz. During the observations, 40–47 antennas were available and the range of baseline lengths was between 15–784 m, which resulted in a good *uv* coverage

Table 1. Overview of the HSC quasars observed by ALMA.

Name	$z_{[CII]}$	M_{1450} (mag)	M_{BH} ($M_\odot$)	$L_{[CII]}$ (10^8 $L_\odot$)	L_{FIR} (10^{11} $L_\odot$)	SFR ($M_\odot$ yr^{-1})	M_{dyn} (10^{10} $M_\odot$)
J0859+0022	6.3903	−24.1	0.3	4.6 ± 0.5	3.4 ± 0.5	71 ± 10	5.6
J1152+0055	6.3637	−25.3	6.3	3.8 ± 0.8	4.1 ± 0.7	86 ± 14	1.4
J2216−0016	6.0962	−23.8	7.0	10.2 ± 0.8	2.8 ± 0.6	58 ± 11	8.2
J1202−0057	5.9289	−22.8	> 0.4	6.2 ± 0.4	4.8 ± 0.2	100 ± 5	4.4
J1208−0200	6.1165	−24.3	7.1	2.7 ± 0.5	1.6 ± 0.4	34 ± 8	1.3
J2228+0152	6.0805	−24.0	> 1.1	2.4 ± 0.6	< 0.9	< 20	2.0
J2239+0207	6.2497	−24.6	11	9.5 ± 0.9	22 ± 1	453 ± 10	29

Notes: M_{1450} and M_{BH} are quoted from Matsuoka *et al.* (2016, 2018a) and Onoue *et al.* 2019. For J1202−0057 and J2228+0152, we assumed the Eddington-limited accretion to compute the lower-limit of their M_{BH}.

and a typical synthesized beam size of $\sim 0''.5$. Reduction and calibration of the data were performed with the Common Astronomy Software Applications package (CASA, McMullin *et al.* 2007) in a standard manner. All images were reconstructed with the CASA task `clean`. The typical noise level is $\sim 0.10 - 0.24$ mJy beam^{-1} for the [CII] cubes, depending on the adopted velocity resolution (50 or 100 km s^{-1}). Their [CII] line luminosities ($L_{[CII]}$) were computed with the standard equation. To derive their FIR continuum luminosities (L_{FIR}), we assumed a modified black body spectrum with a dust temperature of 47 K and an emissivity index of 1.6. These returned $L_{[CII]} \simeq (3 - 10) \times 10^8$ $L_\odot$ and $L_{FIR} \simeq (1 - 20) \times 10^{11}$ $L_\odot$. The spatial extents of [CII] emitting regions were measured by applying a 2-dimensional Gaussian profile to the velocity-integrated intensity distributions, yielding $\sim 2 - 3$ kpc FWHM (major axis). Properties of the targets are listed in Table 1.

3. Results and Discussion

Dynamical mass (M_{dyn}). We measured M_{dyn} of our HSC quasar host galaxies based on their [CII] spatial extents and FWHM, and by following the standard procedure described in Wang *et al.* (2010). The resultant values are listed in Table 1, but formal errors are omitted as there are multiple *unconstrained* uncertainties including the inclination angles and the true geometry of the line emitting region. Despite those uncertainties, we use M_{dyn} as a surrogate for M_{bulge} (e.g., Wang *et al.* 2013; Venemans *et al.* 2016). As seen in other $z \gtrsim 6$ quasars, M_{dyn} of our low-luminosity HSC quasars exceed 10^{10} $M_\odot$, or even 10^{11} $M_\odot$, which lie at the massive end of the stellar mass distribution for $z \sim 6$ galaxies in general (e.g., Grazian *et al.* 2015). Therefore, the host galaxies of these HSC quasars are among the most evolved systems known to date at $z \sim 6$.

Comparison with the star-forming main sequence. We estimated star formation rates (SFR) of our HSC quasars based on their total infrared (TIR; 8–1000 μm) luminosities, which were computed by assuming the same modified black body profile as described in § 2. The calibration of Murphy *et al.* (2011), $SFR/M_\odot$ yr$^{-1} = 1.49 \times 10^{-10}$ $L_{TIR}/L_\odot$, was applied for this purpose. The resultant values span over a wide range, from < 20 $M_\odot$ yr^{-1} to ~ 450 $M_\odot$ yr^{-1}. The same calibration was applied to some other $z \gtrsim 6$ quasars known to date, which are compiled from the literature (see details in Izumi *et al.* 2018, 2019). By equating the [CII]-based M_{dyn} to stellar mass of the host galaxies, we can place these $z \gtrsim 6$ quasars on the M_{star}–SFR plane (Figure 1). Our particular interest is paid on the comparison of these quasars with the $z \sim 6$ star-forming main sequence (MS; Salmon *et al.* 2015). Figure 1 suggests that, while optically-luminous quasars show *starburst* class SFR on this plane (i.e., above the MS), low-luminosity quasars including our HSC quasars are located on or even below the MS. This would indicate that these low-luminosity quasars are transforming into a quiescent population.

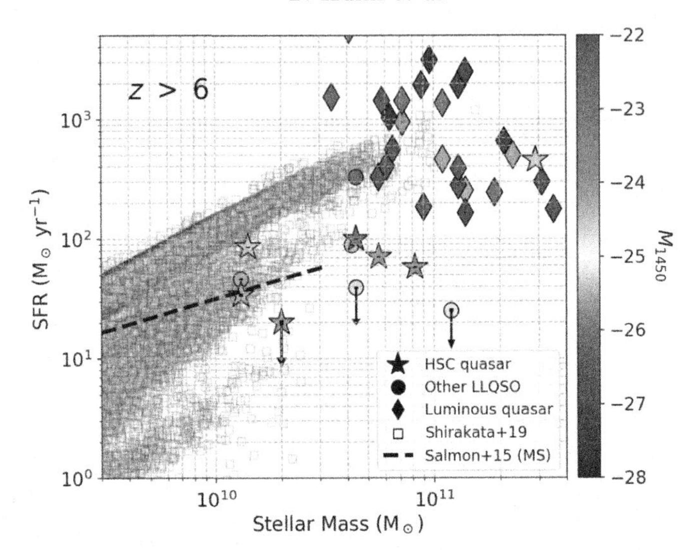

Figure 1. SFR plotted as a function of stellar mass (updated from Izumi *et al.* 2018). The dashed line indicates the star-forming main sequence (MS) for $z \sim 6$ galaxies (Salmon *et al.* 2015). Background gray squares show simulated $z \sim 6$ galaxies based on our semi-analytic model (ν^2GC model; Shirakata *et al.* 2019), which show two sequences: MS and starburst sequence. By equating $M_{\rm dyn}$ to stellar mass, we plot $z > 6$ quasars, which are color-coded by their M_{1450}. Many of the HSC quasars are located on or even below the MS, suggesting that they are transforming into a quiescent population.

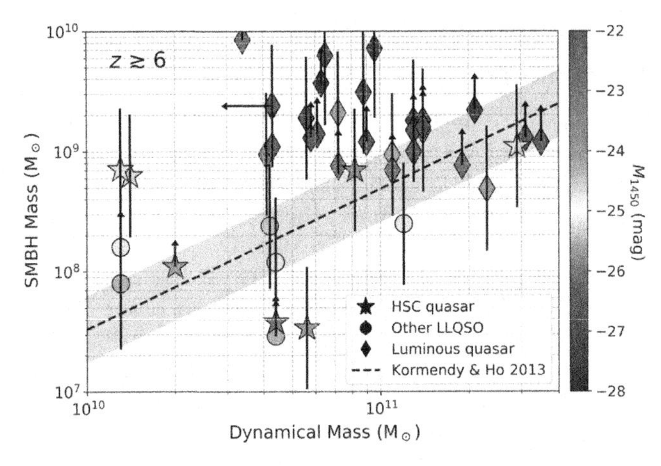

Figure 2. Black hole mass ($M_{\rm BH}$) vs host galaxy dynamical mass ($M_{\rm dyn}$) relationship for $z \gtrsim 6$ quasars, color-coded by their M_{1450} magnitude (Izumi *et al.* 2019). The diagonal dashed line and the shaded region indicate the local $M_{\rm BH} - M_{\rm bulge}$ relationship and its 1σ scatter, respectively (Kormendy & Ho 2013). It is clear that optically luminous quasars ($M_{1450} \lesssim -25$ mag) typically show overmassive $M_{\rm BH}$ relative to the local relation, whereas low-luminosity quasars lie close to, or even below, that relation.

Less-biased early co-evolution. We also compiled MgII-based (i.e., single-epoch method) $M_{\rm BH}$ data and/or M_{1450} data of $z \gtrsim 6$ quasars known to date from the literature (see details in Izumi *et al.* 2019). The latter data is used to compute the lower-limit of $M_{\rm BH}$ by assuming the Eddington-limited mass accretion. The total number of the quasars compiled here is 40. The $M_{\rm BH}$ measurements of our HSC quasars, which are also based on the single-epoch method with MgII line, are described in Onoue *et al.* (2019).

In Figure 2 we display the relation between $M_{\rm BH}$ and $M_{\rm dyn}$ for the above-mentioned quasars, overlaid with the local $M_{\rm BH} - M_{\rm bulge}$ relation after equating $M_{\rm dyn}$ to $M_{\rm bulge}$ (Kormendy & Ho 2013). Regarding the optically luminous quasars ($M_{1450} \lesssim -25$ mag), this figure supports conclusions in previous works (e.g., Wang *et al.* 2013; Venemans *et al.* 2016), i.e., the luminous quasars typically have overmassive SMBHs relative to the local relation, although the discrepancy becomes less evident at $M_{\rm dyn} \gtrsim 10^{11}\ M_{\odot}$. On the other hand, most of the low-luminosity quasars ($M_{1450} \gtrsim -25$ mag) show comparable ratios to, or even lower ratios than, the local relation. The existence of the undermassive SMBHs even implies an evolutionary path, in which galaxies grow earlier than SMBHs, such as expected in a standard merger-induced evolution model (Hopkins *et al.* 2008).

Particularly at a high-mass range ($M_{\rm dyn} \gtrsim 4 \times 10^{10}\ M_{\odot}$), our result demonstrates that previous works on luminous quasars have been largely biased toward the most massive SMBHs, easily resulting in objects lying above the local relation. Therefore, our results highlights the power of the sensitive Subaru HSC survey, as well as the importance of probing their host galaxy nature, in order to depict the less-biased, more genuine shape of early co-evolution of SMBHs and galaxies.

References

De Rosa, G. *et al.* 2014, *ApJ*, 790, 145

Di Matteo, T., Springel, V., & Hernquist, L. 2005, *Nature*, 433, 604

Fan, X. *et al.* 2003, *AJ*, 125, 1649

Gallerani, S., Fan, X., Maiolino, R., & Pacucci, F. 2017, *PASA*, 34, e022

Grazian, A. *et al.* 2015, *A&A*, 575, A96

Hopkins, P. F., Hernquist, L., Cox, T. J., & Kereš, D. 2008, *ApJS*, 175, 356

Izumi, T. *et al.* 2018, *PASJ*, 70, 36

Izumi, T. *et al.* 2019, *PASJ*, in press

Jiang, L. *et al.* 2016, *ApJ*, 833, 222

Kormendy, J. & Ho, L. C. 2013, *ARAA*, 51, 511

Lauer, T. R., Tremaine, S., Richstone, D., & Faber, S. M. 2007, *ApJ*, 670, 249

Madau, P. & Dickinson, M. 2014, *ARAA*, 52, 415

Matsuoka, Y. *et al.* 2016, *ApJ*, 828, 26

Matsuoka, Y. *et al.* 2018a, *PASJ*, 70, S35

Matsuoka, Y. *et al.* 2018b, *PASJ*, 237, 5

McMullin, J. P., Waters, B., Schiebel, D., Young, W., & Golap, K. 2007, in *ASP Conf. Ser.* 376, Astronomical Data Analysis Software and Systems XVI, ed. R. A. Shaw, F. Hill, & D. J. Bell (San Francisco, CA: ASP), 127

Murphy, E. J. *et al.* 2011, *ApJ*, 737, 67

Onoue, M. *et al.* 2019, *ApJ*, 880, 77

Salmon, B. *et al.* 2015, *ApJ*, 799, 183

Shirakata, H. *et al.* 2019, *MNRAS*, 482, 4846

Venemans, B. P., Walter, F., Zschaechner, L., Decarli, R., De Rosa, G., Findlay, J. R., McMahon, R. G., & Sutherland, W. J. 2016, *ApJ*, 816, 37

Wang, R. *et al.* 2011a, *AJ*, 142, 101

Wang, R. *et al.* 2011b, *ApJ*, 739, 34

Wang, R. *et al.* 2013, *ApJ*, 773, 44

Willott, C. J., Bergeron, J., & Omont, A. 2015, *ApJ*, 801, 123

Willott, C. J., Bergeron, J., & Omont, A. 2017, *ApJ*, 850, 108

Uncovering Early Galaxy Evolution in the ALMA and JWST Era
Proceedings IAU Symposium No. 352, 2019
E. da Cunha, J. Hodge, J. Afonso, L. Pentericci & D. Sobral, eds.

doi:10.1017/S1743921319009517

The AGN-galaxy connection: Low-redshift benchmark & lessons learnt

Stéphanie Juneau

National Optical Astronomy Observatory, Tucson AZ 85719, USA
email: `sjuneau@noao.edu`

Abstract. Several scenarios have been proposed to describe the physical connection between galaxies and their central active galactic nuclei (AGN). This connection could act on a range of spatial scales and vary across cosmic time. In these proceedings, we consider black hole and galaxy growth and whether that growth is affected by AGN feedback both based on statistical approaches – which reveal general population trends – and based on an individual case study – which gives us a more detailed insight on the physical processes at play. For the statistical approach, we showcase a low-redshift ($0.04 < z < 0.2$) SDSS sample with AGN classification based on a combination of emission-line diagnostic diagrams, and for which we account for sample selection by using a V/V_{max} approach. The trends on the star formation rate - stellar mass ($SFR - M_{\star}$) plane suggest that the most likely connection is a common gas reservoir for star formation and AGN, and that they both decline as the gas reservoir is consumed. The trends established at low-redshift could act as a local benchmark against which to compare higher redshift studies. As a complementary approach, we use a detailed case study of a nearby AGN host with integral field spectroscopy from the VLT/MUSE instrument in order to spatially resolve the interplay between AGN feedback and the host galaxy. We find that the galaxy substructure likely plays a role by collimating and/or obscuring the outflows and radiation from the central engine. Ongoing and future work with 3D spectroscopy will enable us to learn more about galaxy and black hole coevolution. Lastly, we briefly discuss lessons learnt from both approaches.

Keywords. galaxies: active, galaxies: fundamental parameters (classification), galaxies: Seyfert, galaxies: evolution, galaxies: individual (NGC 7582)

---

1. Introduction: AGN fueling & feedback

The growth of galaxies and that of their supermassive black holes (SMBHs) may be physically connected given the observed scaling relations between them (e.g., Magorrian *et al.* 1998; Ferrarese & Merritt 2000, but also see Jahnke & Macćio 2011). Therefore one can ask whether the mechanism responsible for the BH growth itself (AGN fueling) is related to host galaxies properties, and vice versa. Alternatively, it could be that the culprit is a mechanism regulating or stopping the growth of BHs and galaxies, such as AGN feedback. Cosmological simulations need a source of energy to regulate and suppress star formation in galaxies in order to reproduce the observed galaxy distribution functions (e.g., Bower *et al.* 2006; Gabor & Davé 2012; Hirschmann *et al.* 2014), which is often attributed to AGN feedback (e.g., Croton *et al.* 2006; Schaye *et al.* 2015). The flavors of feedback include a maintenance (or radio) mode according to which AGN inject mechanical energy into their surrounding medium, keep it too hot to form stars (e.g., Ciotti *et al.* 2010; Karouzos *et al.* 2014), and a radiative (or quasar) mode that is postulated to expel material in the form of outflows which likely originate as accretion disk winds (e.g., Beckmann *et al.* 2017; Harrison *et al.* 2018; also see reviews by Fabian 2012 and Harrison 2017).

In order to learn about the processes driving AGN fueling and AGN feedback, one can consider AGN activity as a function of host galaxy properties. So far, different results have been obtained for various galaxy and/or AGN selection methods, which emphasizes the need to understand sample selection. For example, some studies found that AGN were predominantly hosted in star forming galaxies (e.g., Silverman *et al.* 2009; Rosario *et al.* 2013), while others found that their host galaxies had on average lower star formation rates (SFRs) compared to normally star-forming galaxies (e.g., Mullaney *et al.* (2015)) and that they tend to occupy the transition region between star-forming and passive galaxies in terms of their colors and/or SFRs (e.g., Schawinski *et al.* 2007). Yet other studies reported that galaxies' stellar mass may be a better determinant of AGN fueling than SFRs (e.g., Bongiorno *et al.* 2012; Tanaka 2012). Furthermore, there are disagreements about the role of galaxy mergers in affecting (or not) the presence and feedback from AGN (e.g., Kartaltepe *et al.* 2010; Treister *et al.* 2012; Grogin *et al.* 2005; Cisternas *et al.* 2011; Steinborn *et al.* 2018). Different AGN selection methods based on various wavebands (optical, infrared, radio, X-rays, etc.) can also lead to different results (e.g., Hickox *et al.* 2009; Juneau *et al.* 2013; Ellison *et al.* 2016).

In these proceedings, we revisit the situation at low redshift with a well-defined selection function in order to establish a reference benchmark to compare with other work. We first consider a statistical approach by looking at the galaxy evolution context as quantified by the location of galaxies on the star formation rate (SFR) - stellar mass ($M_\star$) plane (Section 2). We then move on to the detailed case study where we investigate the gas ionization properties and the kinematics of the stars and gas in order to paint a picture of AGN feedback in relation to the host galaxy substructure (Section 3). We conclude each section with lessons learnt and a few open questions that motivate future effort.

2. AGN fueling & feedback: Low-Redshift Benchmark

The first part of these proceedings consists of a statistical approach in order to identify global trends of AGN fueling and feedback. Given that AGN variability can occur on much shorter timescales relative to galaxies' dynamical time, averaging large ensembles can give us an overview of the overall trends (e.g., Chen *et al.* 2013).

2.1. *Sample Selection*

Motivated by the need to revisit a low-redshift benchmark to understand the AGN-galaxy connection, we selected a sample based on the Sloan Digital Sky Survey survey (SDSS) DR7 (Abazajian *et al.* 2009). We opted to use SDSS-II rather than SDSS-III because the main galaxy sample is more uniformly selected with magnitude limits of $14.5 < r < 17.77$, which helps to compute volume correction factors to account for selection effects. Because we are interested in understanding the host galaxy properties, the bulk of the analysis concerns Type 2 AGNs, for which the nuclear light is optically obscured, therefore giving us access to spectral and photometric diagnostics of the host galaxy stellar content. A complete account of AGN including Type 1 is estimated by applying a correction factor for the Type 1/Type 2 ratio, which is observed to depend on AGN luminosity (e.g., Simpson 2005; Lusso *et al.* 2013), and assuming that this ratio does not otherwise vary strongly as a function of SFR and $M_\star$.

The sample selection was done similarly to the study by Juneau *et al.* (2014, hereafter J14) by requiring the signal-to-noise ratio of the emission line ratios [O III]/Hβ and [N II]/Hα to be above $3/\sqrt{2}$ ($\approx$2.12), after applying the corrections to the reported Hβ fluxes for Balmer absorption correction (Groves *et al.* 2012), and after applying the emission line ratio uncertainty scalings that J14 determined from comparing duplicate observations (their Appendix A). These emission line ratios are used to classify the ionization source of the galaxies as described in the *Method* below. We further applied a redshift cut of $0.04 < z < 0.2$, and required an available value of stellar mass ($\log(M_\star/M_\odot) > 6$). We used the MPA-JHU Value-Added Catalogs (VACs)† generated following the methods described by Kauffmann *et al.* (2003a), Brinchmann *et al.* (2004), and Tremonti *et al.* (2004).

2.2. *Method*

Starting from the observed and derived properties from the SDSS dataset and associated VACs as described above, we need to identify the presence of AGN, derive additional properties such as the AGN bolometric luminosity (L_{bol}), and to compute correction factors to account for selection effects in order to obtain an overall census of the galaxy populations. We describe these steps below.

AGN Classification. We identify the presence of AGN in galaxies according to emission line diagnostic diagrams. Namely, we use their locus on the BPT (Baldwin *et al.* 1981) and VO87 (Veilleux & Osterbrock 1987) diagrams. These diagnostic diagrams consist of [O III]/Hβ as a function of, respectively, [N II]/Hα, and [S II]/Hα. On the BPT, we use the demarcation line determined by Kauffmann *et al.* (2003b) in order to distinguish between star-forming galaxies (lower left) and the rest. The higher demarcation from Kewley *et al.* (2001) further distinguishes between AGN (above and to the right), and composite galaxies which are located between the two demarcations. The VO87 diagram is more sensitive to shock and/or post-AGB ionization, and therefore can more readily separate Seyfert-like ionization from LINER-like ionization. We used the demarcations from Kewley *et al.* (2006) in this case. Below, we consider the galaxy population separated according to AGN bolometric luminosities.

AGN Bolometric Luminosity. Low-redshift, massive galaxies tend to be metal-rich (e.g., Tremonti *et al.* 2004). As a result, their [O III] emission lines are dominated by AGN activity when present (~93% AGN contribution; Kauffmann *et al.* 2003b). We use the [O III] luminosity as a proxy, and apply a bolometric correction factor to estimate the AGN bolometric luminosities. We correct the emission line for foreground dust attenuation following Bassani *et al.* (1999) and assuming an intrinsic Balmer Decrement Hα/Hβ value of 3.0 appropriate for NLR gas (Osterbrock & Ferland 2006). We then employ the Lamastra *et al.* (2009) bolometric correction factor

Type 1 AGN Correction. Type 1 AGN were not explicitly included in our sample due to additional uncertainties in accessing the host stellar properties in the presence of a point source. The brightest type 1 AGNs (quasars) were also targeted separately from the selection function of the SDSS *main* galaxy sample. However, we estimate their contribution to the total number (and fraction) of AGN based on the assumption that they share similar host galaxies as Type 2 AGNs, and that their contribution can be derived from the obscured (Type 2) fraction. Following the work by Lusso *et al.* (2013), we adopt the functional form of a modified receding torus model proposed by Simpson 2005, which predicts the following relationship between the obscured fraction f_{obsc} and AGN bolometric luminosity:

$$f_{obsc} = [1 + 3(L_{bol}/L_0)^{1-2\xi}]^{-0.5}, \tag{2.1}$$

where we use the values found by Lusso *et al.* (2013) for their main sample (optically thin case) of $L_0 = 10^{46.16}$ erg s^{-1}and $\xi = 0.37$. We convert into statistical weights $w_{obsc} = 1/f_{obsc}$ to use when computing the number and fraction of AGN based on the obscured (Type 2) number and fraction. We note that the optically-thick case predicts

† http://www.mpa-garching.mpg.de/SDSS/DR7/

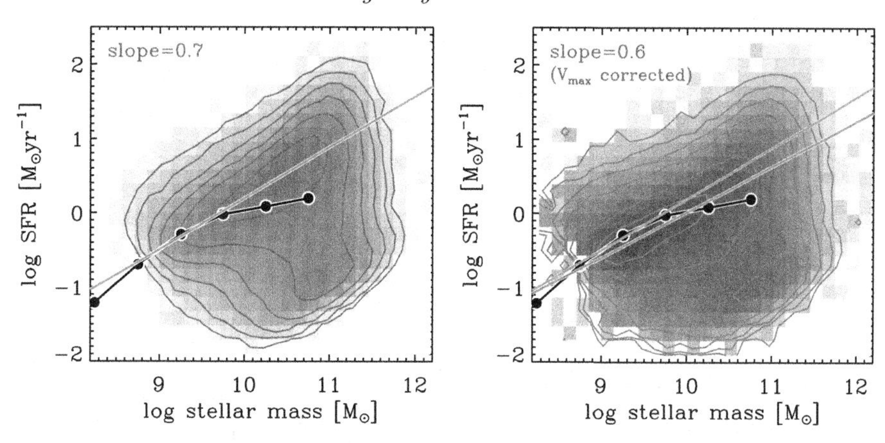

Figure 1. SFR-$M_\star$ plane for the SDSS $0.04 < z < 0.2$ galaxy sample. *Left.* Bivariate distribution of our sample before applying the V/V_{max} correction factors. The contours are logarithmically spaced by 0.5 dex. The best-fit linear relationship calculated for the non-AGN galaxies with $9.2 < \log(M_\star/M_\odot) < 10.7$ is shown with the orange line. The black circles show the low-mass sample of Gavazzi *et al.* (2015). *Right.* Same as the left-hand panel, except after applying the V/V_{max} correction factors. In addition to the current best-fit linear relation to the mode of the non-AGN galaxies (solid blue line), we overlay the best-fit relationship from the left-hand panel (solid orange line) for comparison. We find a slope that is flatter by 0.1 after correcting for the detection limits using the V/V_{max} technique.

lower obscured fractions, and consequently higher total number and fraction of AGN relative to what is reported here. However, our implementation of the statistical weights mostly affects the normalization of the bivariate distributions on the SFR-$M_\star$ plane, and not much the locus or shape of the contours (Section 2.3).

V/V_{max} *Correction.* In order to account for the selection functions of both the SDSS *main* galaxy sample, which is magnitude limited, and of the emission-line criterion, we define separately the maximum volume within which a given object would be detected (V_{max}) for both selection steps, and keep the minimum value as the most stringent limit. The photometric selection is $14.5 < r < 17.77$. For each galaxy in our sample, we compute the k-correction using the dereddened *ugriz* photometry converted to AB magnitudes to predict the observed r band magnitude over a range of redshifts in order to determine z_{min} – the minimum redshift allowed before the object is too bright ($r < 14.5$) – and z_{max} – the maximum redshift before the object is too faint ($r > 17.77$). We store these values to compare with the maximum redshift at which the [O III]/Hβ and [N II]/Hα emission line ratios are still detected with $S/N > 3/\sqrt{2}$, which we define as z_{max}^{lines}, and we keep $z_{max} = min(z_{max}, z_{max}^{lines})$.

Lastly, we compute the maximum detectable co-moving volume element within – and normalized by – the co-moving volume from the redshift range of interest ($0.04 < z < 0.2$). The resulting V/V_{max} values are used as statistical weights when generating bivariate distributions such as the SFR-$M_\star$ plane, which will be the main parameter space to set our analysis within the galaxy evolution context. In Figure 1, we show the bivariate distribution of the full sample before and after applying the V/V_{max} correction factors. In both panels, there is a fairly tight sequence of star-forming galaxies with rising SFRs as a function of $M_\star$, and a population of passive (and/or quenched) galaxies with comparatively lower SFRs. Applying the V/V_{max} correction changes the relative proportions of galaxies within the parameter space. As expected, the correction is much more pronounced at the low-mass end (until the original sample runs out of galaxies altogether, which explains the sharp cutoff at the low SFR-$M_\star$ portion of the right-hand side panel). Also unsurprisingly, at a given stellar mass, the correction is more important

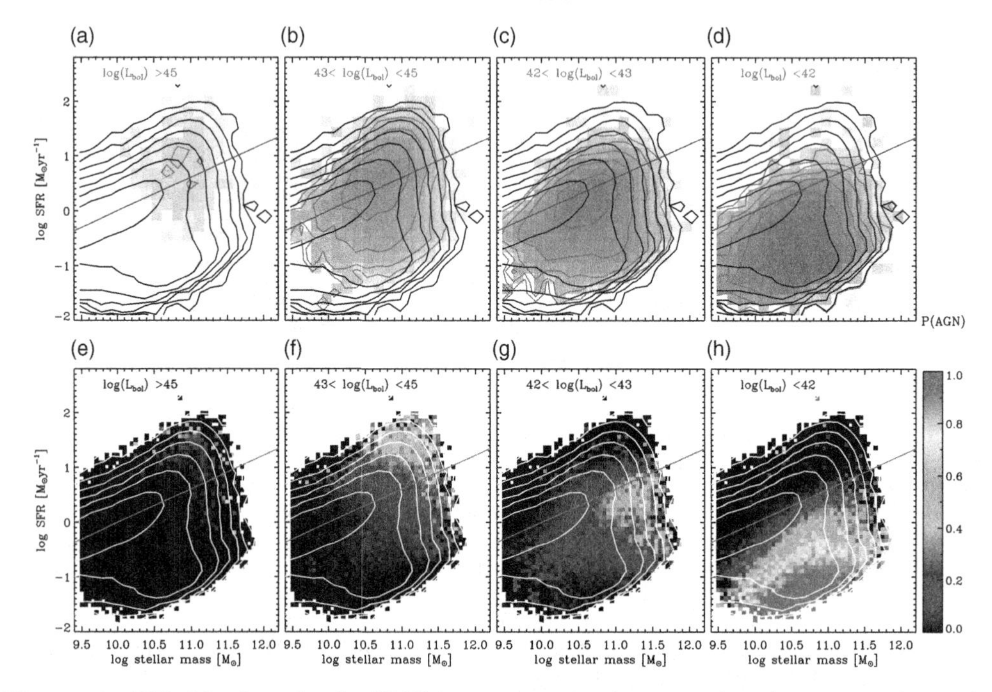

Figure 2. SFR-$M_\star$ plane for the SDSS $0.04 < z < 0.2$ galaxy sample. The top row shows the bivariate distribution of AGN in bins of bolometric luminosity as labeled (grey shaded and red contours), overlaid on the bivariate distribution of the full galaxy population (black contours), which is identical on all panels. The bottom row displays the same galaxy population except that in each panel, the SFR-$M_\star$ bins are color-coded by the fraction of galaxies hosting an AGN within the labeled AGN luminosity range (color bar). This AGN fraction could correspond to a duty cycle assuming that all galaxies in a given bin of SFR and $M_\star$ go through *on* and *off* phases.

for objects with a lower SFR. As a result, the sequence is slightly lower after correcting for V/V_{max}. A linear fit to the mode of the log(SFR)-log($M_\star$) distribution for BPT-SF galaxies is shown before (orange) and after (blue) applying the correction. The final slope is slightly flatter at 0.6 relative to the value of 0.7 found before. We note that previous work by, e.g., Salim *et al.* (2007) also investigated the effect of statistical weighting on the sequence of star-forming galaxies. Their resulting best-fit slope for their UV-based SFRs is 0.65, between the two values reported here.

2.3. *AGN on the SFR-Stellar Mass Plane*

In this section, we examine the location of galaxies with AGN on the SFR-$M_\star$ plane relative to the full galaxy population. We split them in bins of AGN bolometric luminosity in order to assess whether their is a trend that we could interpret to learn about AGN fueling and/or feedback. We use both the V/V_{max} correction factors, and the obscured AGN fraction weights described previously in order to estimate the total AGN population. We do not distinguish between AGN types (Seyferts, Composites, LINERs; but see, e.g., Salim *et al.* 2007; Leslie *et al.* 2015). Figure 2 shows the resulting bivariate distributions for AGN separated in bins of bolometric luminosity. As we can see on the top row, the more luminous AGN ($L_{bol} > 45$) are preferentially located above or on the star forming sequence, while the lower luminosity bins occupy regions with decreasing SFR. At the low AGN luminosity end ($L_{bol} < 42$), AGN hosts are located below the star forming sequence, and overlapping with the population of passive galaxies.

However, we note that the lowest AGN luminosity bin includes an important number of LINERs, some of which may be so-called *retired* galaxies which can occupy the same regions of the BPT and VO87 as true LINERs even though their ionization emission is dominated by old stellar populations and/or shocks rather than AGN (e.g., Cid Fernandes *et al.* 2011; Yan & Blanton 2012). This would be consistent with LINER galaxies being predominantly colocated with passive galaxies. On the other hand, there are indications that a significant fraction of LINERs do host an AGN when following them up in different wave bands and/or stacking their X-ray emission (e.g., Ho 2008; Trouille *et al.* 2011; Trump *et al.* 2011, also see Section 3.2 by Juneau *et al.* 2011). Therefore, we consider the result of the lowest AGN luminosity to be ambiguous. An additional source of uncertainty at the low luminosity end comes from the difficulty of identifying weak AGN among strongly star-forming galaxies due to their lower fractional contribution to emission line fluxes, leading to a dilution of the AGN spectral signatures, and decreasing the completeness of AGN identification at low luminosities (Trump *et al.* 2015).

In contrast, the results at the high-luminosity end are much less subject to dilution of AGN emission line signatures. So we will focus our interpretation on these more reliable results. The fact that we find the most active AGN to reside in star-forming hosts supports a picture where the black hole fueling and star formation activity use a common gas reservoir and are both facilitated by the presence of a high gas content likely together with mechanisms that reduce the angular momentum of the gas. Conversely, while one would expect that the most luminous AGN would be the most capable of generating feedback that reduces or halts star formation in their hosts, their location on the SFR-$M_\star$ plane suggests that either they are not generating strong feedback, or that the effects from the feedback would be substantially delayed. It could also be the case that only in exceptional circumstances, such as major galaxy mergers, is strong AGN feedback playing an important role. This is supported by the scenario of two modes of quenching with a fast quenching in the case of major mergers aided with AGN feedback, while the bulk of galaxies would go through a slow quenching as they slowly use up their gas reservoir or as the latter gets hotter and no longer sustains star formation (e.g., Schawinski *et al.* 2014). However, these results are also consistent with cosmological simulations for which AGN feedback is a key ingredient to control both galaxy growh and black hole growth (e.g., Hirschmann *et al.* 2014)

Next, we divide the number of AGN hosts by the total number of galaxies in our sample over the SFR-$M_\star$ plane in order to calculate the AGN fraction (bottom row of Figure 2). From the left-hand side to right-hand side panels, we see an overall trend where the fraction of galaxies with luminous AGN is higher in hosts with elevated SFRs, which is qualitatively similar to the top row showing the bivariate distributions. However, an interesting distinction is that the AGN fraction is higher above and below the star-forming sequence than directly over it. This means that there is a larger spread of SFR values for AGN hosts at a given $M_\star$ compared to the bulk of the galaxy population. A physical origin of this difference would imply a process other than a strictly negative AGN feedback as the AGN hosts can be both over and under the sequence, or potentially it indicates the presence of more than one process at play.

If all galaxies in a given bin of SFR & $M_\star$ host a similarly massive black hole with episodes of accretion interspersed with episodes of inactivity, the fraction of AGN can be interpreted as a duty cycle (Novak *et al.* 2011; Gabor & Bournaud 2013; Schawinski *et al.* 2015). In the bottom row of Figure 2, the duty cycle would be defined as the fraction of the time that a given BH is accreting within the L_{bol} range of interest. As mentioned earlier, the two caveats at the low-luminosity end (nature of LINER-like emission, and incompleteness of AGN identification in star-forming hosts) prevent us from making a robust interpretation for that regime. The most interesting and also reliable trend is seen in the $43 < log(L_{bol}) < 45$ bin (panel f), where we can see that galaxies with elevated SFR relative to the star-forming sequence are more likely to experience frequent AGN episodes. This result could be interpreted as a the presence of a gas reservoir facilitating both star formation and BH accretion.

2.4. *Implications & Lessons Learnt*

We found that the locus of AGN on the $SFR - M_{\star}$ plane varies as a function of AGN luminosity, itself a proxy of the black hole accretion rate. This means that there would not be a single answer to a question such as: *where are AGN located with respect to the star forming sequence?* Similarly, previous work reported a different level of SFR at a given $M_{\star}$ for AGN selected in different wavebands (e.g., radio, optical, IR; Ellison *et al.* 2016), and/or with different optical classification (LINER, composite, Seyferts; Leslie *et al.* 2015).

The division in terms of AGN luminosity can be used to infer information about AGN fueling and feedback. One would expect that higher luminosities imply both a more efficient fueling and the potential for more effective feedback (at least for radiation-driven feedback such as the *quasar* mode). Regarding the former, we found that AGN in the high luminosity bins live in hosts with elevated SFR, therefore supporting the idea of a common gas reservoir, and that the presence of this gas reservoir helps to sustain more frequent AGN episodes (from the fraction of galaxies with an AGN). In terms of feedback, the observed elevated SFRs in galaxies hosting luminous AGN instead suggest that the feedback is either unimportant or delayed. A limitation of this interpretation is that we do not know the past star formation history, and whether the SFRs were yet higher before the AGN was present or before it reached its current luminosity. For that reason, it is helpful to compare with cosmological simulations that include black hole growth and AGN feedback, such as the MAGNETICUM simulations with a revised AGN implementation (Hirschmann *et al.* 2014), as well as emission-line signature predictions (Hirschmann *et al.* 2017).

In addition, the relationship between AGN fueling, AGN feedback, and host galaxies properties may vary across cosmic time. Delvecchio *et al.* (2015) explored this topic by computing the average AGN accretion rate across the $SFR - M_{\star}$ plane between redshift $0 < z < 2.5$ for a sample of *Herschel* infrared-selected galaxies. The main conclusion from that work is that the accretion of the BH and the host growth track each other across cosmic time, and possibly more closely at $z > 0.8$ compared to low-redshift galaxies. While the sample size was limited (in particular in the lowest redshift bin), this work is a useful example to guide future statistical studies, though broadening to multiple selection techniques would be useful to understand variations with respect to other work. A complete picture requires an understanding of the AGN and host galaxy selection, and possibly splitting the samples into physically-motivated categories. Based on previous effort, the work presented here, and on discussion at the IAU Symposium, we list a few lessons:

- Correcting for the selection function arising from photometry and/or spectroscopy selection allows us to obtain more accurate representation of the relative galaxy (and AGN) subpopulations, for example using the V/V_{max} method;
- Performing multi-wavelength AGN analyses will remain key for sample completeness albeit it will require considering the various selection functions together;
- Relatedly, we might need to re-write the big questions to be more physically-specific rather than observationally-specific as we attempt to paint the global picture of AGN and galaxy co-evolution.

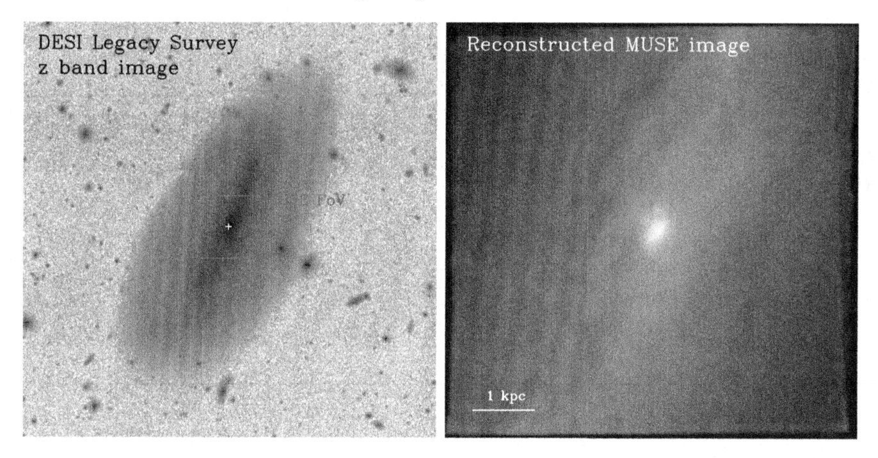

Figure 3. Images of galaxy NGC 7582. *(Left)* z-band image from the Dark Energy Spectroscopic Instrument (DESI) imaging Legacy Survey (Dey *et al.* 2019) to show the full extent of the galaxy. This image is 6.66' on a side, and we overlay the MUSE field-of-view with a $1' \times 1'$ square centered on the target. *(Right)* Color image constructed from the MUSE datacube by selecting 3 spectral windows for the red, green and blue channels that are free of emission lines as described by J19. We can see large-scale dust lanes as well as the highly reddened central region, which corresponds to a starburst surrounding the central AGN. North is up. East is to the left.

3. AGN fueling & feedback: A case study of NGC 7582

After considering a statistical study of AGN in nearby galaxies, we focus on a case study to investigate the relationship between AGN feedback and the host galaxy in a spatially resolved manner. Numerical simulations of isolated host galaxies with resolved ISM (reaching scales of a few pc) showed that the galaxy substructure can influence not only AGN fueling but also the impact of AGN feedback onto its surroundings (outflows and radiation; see, e.g., Gabor & Bournaud 2013 and Roos *et al.* 2015). Observational studies have also suggested a closer relationship between AGN fueling and the resolved, central SFR than with the total, galaxy-wide SFR (e.g., Diamond-Stanic & Rieke 2012).

We employed integral field spectroscopy to spatially resolve and map the ionized gas and stellar content of nearby AGN host NGC 7582, which is a barred spiral galaxy known to host both a central starburst and a heavily obscured AGN. Previous work with narrow-band imaging and integral field spectroscopy revealed the presence of extended narrow line regions (e.g., Morris *et al.* 1985; Riffel *et al.* 2009; Davies *et al.* 2016), which could in turn suggest a possible widespread influence of the AGN. The additional observations presented below augment previous effort with a combination of high spatial resolution (~ 1") over a 1 arcmin field-of-view spanning a scale of ~ 8 kpc at the distance to the target galaxy (Figure 3).

3.1. *Observations & Method*

NGC 7582 was observed with the Multi-Unit Spectroscopic Explorer (MUSE) instrument (Bacon *et al.* (2010)) on the Very Large Telescope (see Figure 3 for the MUSE field-of-view overlaid on the target galaxy). The observations, data reduction, spectral measurements, and inferred physical properties are described in detail elsewhere (Juneau *et al.*, in prep.; hereafter J19). In summary, the integration time on source was 40 minutes, with two short (1 minute) sky exposures offset from the target. The data were processed with the Reflex MUSE pipeline (Freudling *et al.* 2013, v1.0.5). The resulting datacube covers approximately a field-of-view of one square arcminute, with spaxels of

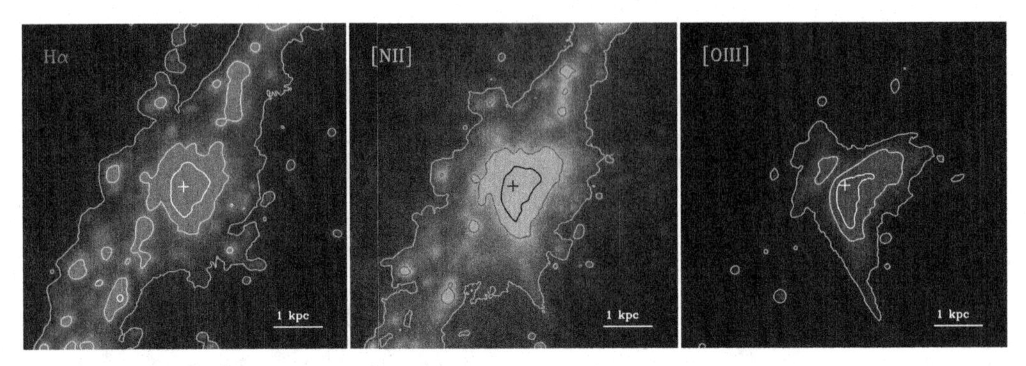

Figure 4. Intensity map for strong emission lines Hα (red), [N II] (green), and [O III] (blue). For each emission-line, we show the total line flux across the MUSE field-of-view. North is up; East is left. The scale bar corresponds to 1 kpc at the distance to NGC 7582. The plus symbol marks the center of the galaxy. Figure adapted from J19.

0.2" × 0.2" each. The spatial resolution is limited by the $\sim$1" seeing, and the spectral elements are 1.25 Å, which corresponds to $\sim 55-75$ $km\ s^{-1}$ over the wavelength range of interest.

We fit the reduced datacube with the LZIFU package (Ho *et al.* 2016) in order to account for both the stellar continuum with spectral synthesis models (with pPXF; Cappellari & Emsellem 2004), and for the emission lines with Gaussian profiles. The fitting procedure was run in separate instances to fit the emission lines with one, and with two Gaussian components. We compared the results and assigned in post-processing the spaxels for which two components are needed to describe the observed [O III] λ5007 line profiles, which are the most likely tracers of the secondary ionized gas component associated with an outflow rather than with the systemic velocity. We identified such cases by applying a condition for the second component to be detected with a signal-to-noise ratio $S/N > 3$ in [O III].

3.2. *Maps of Ionized Gas*

The MUSE datacube includes several emission lines. In this section, we consider some of the commonly used strong lines, namely Hα, [N II] λ6584, and [O III] λ5007 (Figure 4). Hα traces well the total ionized gas, including gas ionized from young stars in H II regions, and other sources of ionization such as shocks and AGN photoionization. The Hα emission is distributed along the disk and bar of NGC 7582 spanning the entire MUSE field-of-view, as well as showing an important component from the central starburst and extending toward the AGN narrow line region. The clumpy or patchy appearance indicates the presence of star forming regions along the disk (and/or bar), and possible patchy obscuration although the Hα-bright regions also tend to exhibit elevated dust obscuration according to the Hα/Hβ Balmer Decrement. As described by J19, we converted the Balmer Decrement to gas attenuation by assuming an intrinsic ratio $(\mathrm{H}\alpha/\mathrm{H}\beta)_0 = 2.86$ ($n_e = 100$ cm^{-3}; $T_e = 10^4$ K; Osterbrock & Ferland 2006), and the Calzetti extinction law (Calzetti *et al.* 2000).

On the middle panel of Figure 4, we can see that the [N II] emission is largely co-located with Hα except that it is comparatively less prominent along the disk and bar, and more prominent on the central starburst and AGN ionization cone region. For H II regions, the [N II]/Hα ratio is commonly used as an indicator of the gas-phase metallicity. For AGN narrow-line regions, the ratio can be further enhanced due to a combination of high metallicity and harder ionizing spectrum (e.g., Kewley *et al.* (2013)). The [O III] emission (right-hand panel of Figure 4) is largely concentrated on the AGN ionization cones. In this

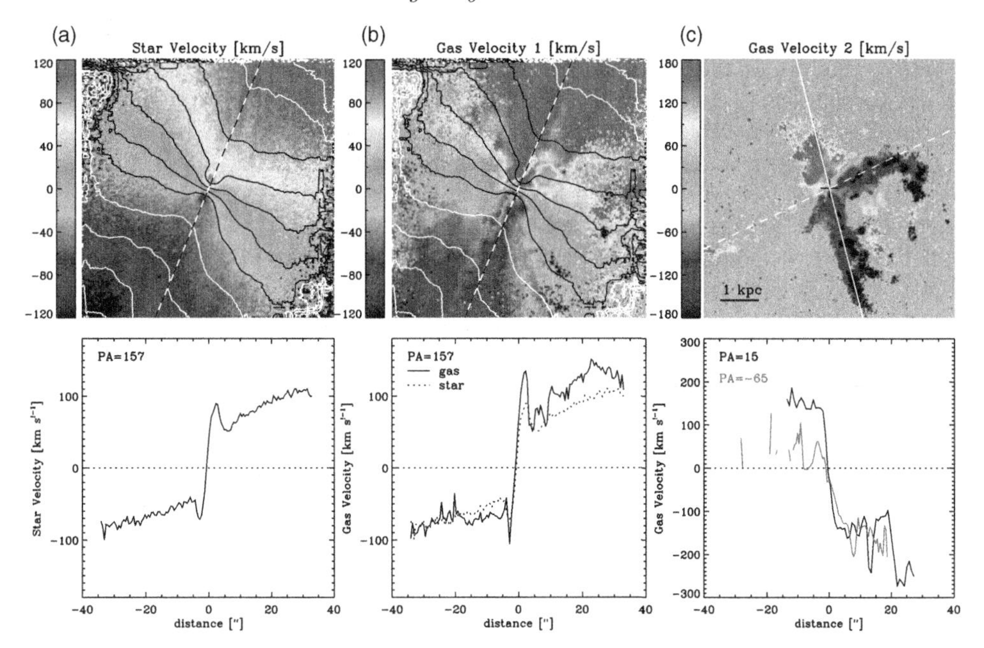

Figure 5. Top row: velocity maps with iso-contours of the stellar velocity field. The PA of 157 is shown with the dashed line. (a) Stellar velocity map (in km s^{-1}); (b) Gas velocity map (in km s^{-1}) for Component 1; (c) Gas velocity map (in km s^{-1}) for Component 2. North is up; East is left. The scale bar corresponds to 1 kpc at the distance of the target. Bottom row: velocity profile (in km s^{-1}) along PA $= 157$ for the stellar and gas components. (d) Stellar velocity profile, which shows a clear velocity excess at the KDC while the main curve is slowly rising in amplitude with distance away from the center. (e) Nebular gas velocity profile for Component 1 (solid line) compared to the stellar profile from panel d (dotted line). (f) Nebular gas velocity profile for Component 2 along axes defined to follow the edges of the cones with PA $= 15$ (black line; solid line in panel c), and PA $= -65$ (red line; dashed line in panel c). Figure adapted from J19.

case, one can see both the front cone and back cone (i.e., counter-cone) which is behind the host galaxy from our line-of-sight. The dearth of [O III] emission along the star forming knots is consistent with metal-rich gas.

In the full analysis by J19, the combination of those three strong emission lines with the addition of Hβ and [S II] $\lambda\lambda6717, 6731$ are used to produce line ratio maps in order to apply the BPT and VO87 diagnostic diagrams. The results indicate that the star forming disk and bar are consistent with metal-enriched star forming dominated regions surrounded by diffuse medium (possibly warm-hot interstellar medium, or WHIM). The cones were found to have Seyfert-like line ratios indicating AGN photoionization. The areas on either side of the cones are characterized by LINER-like ratios, which could potentially indicate shock-excitation. This possible interpretation is consistent with the outflowing kinematics of the cones, which we describe in the next section.

3.3. *Star & Ionized Gas Kinematics*

The star and gas kinematics, respectively derived from the continuum fitting, and emission line fitting with LZIFU, are shown in the top row of Figure 5. The velocity maps show a clear large-scale rotation of the host galaxy for both the stellar component (panel a) and the first gas component (panel b). By definition, the second gas component was attributed to spaxels which had a significant second (or offset) component in the [O III] $\lambda5007$ emission line. It clearly shows a biconical outflows originating from the center of the galaxy, with a blue-shifted front cone and a redshifted back cone. The projected velocities of the outflowing ionized gas reach around ± 200 km s^{-1}. While the first gas component is characterized by motion more similar to the stellar component, there are also some notable differences. For instance, the velocity field is generally less regular, suggesting motion beyond a pure simple rotating disk. In particular, there appear to be mild deviations parallel to the PA, which is also the angle of the bar, and which could suggest gas streaming along the leading edge of the bar.

The bottom row displays the velocity profiles along the galaxy PA (157 degrees East of North) for the stellar and first gas component, and the velocity profiles along the edges of the cones for the second gas component (angles of 15 and -65 degrees, as shown in panel c). The velocity profiles for the stellar component clearly shows enhanced features near the center corresponding to a kinematically distinct core (KDC) with respect to the larger scale rotation. The KDC is rotating in the same direction as the galactic disk but with a differential, elevated velocity (by approximately 40 km s^{-1}). The KDC shows as two spots on the velocity field from panel a, and likely corresponds to a disk or ring of material. As described in J19, a ring interpretation is favored due to the likely spatial coincidence with the inner Lindblad resonance of the bar, and furthermore the presence of the AGN and associated outflows, which are possibly driving material away from the inner portion of the ring. The velocity profile for the first gas component is similar to that of the stars except for presenting a systematic offset along the redshifted side by about 40 km s^{-1}. It is possible that this difference arises due to the known presence of a large-scale bar in this galaxy that could affect the motion of the gas along the PA traced in panel b. Lastly, the velocity profiles along the edges of the biconical outflows suggest comparable velocities between the front and back cones along PA $= 15$, where we can see both components. We lack the bulk of the redshifted component along PA $= -65$, possibly due to the presence of foreground dust lanes in the host galaxy obscuring the optical light emission.

3.4. *Implications & Lessons Learnt*

Putting together the results on the gas ionization, gas kinematics, and stellar kinematics for the central ~ 8 kpc of NGC 7582, we paint an overall picture where nuclear ionized gas outflows have a symmetrical and biconical shape, and are predominantly AGN-photoionized. Their opening angle and the spatial coincidence of the base of the cones with the central KDC suggest that the latter may be responsible for collimating them. We thus propose that the outflows are AGN-driven but (re-)collimated by the host galaxy substructure, similarly to what Prieto *et al.* (2014) previously postulated (also see Ricci *et al.* 2018). If that were the case, then the host galaxy substructure would play a role in how the feedback from the central AGN affects the host galaxy. Here, the central KDC may be protecting or shielding the host galaxy material from the effects of AGN outflows. More work is needed to assess how common this scenario may be among AGN hosts. Overall, this interpretation implies that we may need to revisit the interplay between the central AGN and the host galaxy substructure in order to fully understand the connection between supermassive black holes and galaxies. We also conclude that integral field spectroscopy – also known as 3D spectroscopy – is a useful tool to investigate this question (see also, e.g., Husemann *et al.* 2019).

References

Abazajian, K. N., Adelman-McCarthy, J. K., Agüeros, M. A., *et al.* 2009, *ApJS*, 182, 543
Alexander, D. M. & Hickox, R. C. 2012, *NewAR*, 56, 93
Bacon, R., Accardo, M., Adjali, L., *et al.* 2010, *SPIE Conference Series*, Vol. 7735, id. 773508

Baldwin, J. A., Phillips, M. M., & Terlevich, R. 1981, *PASP*, 93, 5
Bassani, L., Dadina, M., Maiolino, R., *et al.* 1999, *ApJS*, 121, 473
Beckmann, R. S., Devriendt, J., Slyz, A., Peirani, S., *et al.* 2017, *MNRAS*, 472, 949
Bongiorno, A., Merloni, A., Brusa, M., Magnelli, B., *et al.* 2012, *MNRAS*, 427, 3103
Bower, R. G., Benson, A. J., Malbon, R., Helly, J. C., *et al.* 2006, *MNRAS*, 370, 645
Brinchmann, J., Charlot, S., White, S. D. M., *et al.* 2004, *MNRAS*, 351, 1151
Calzetti, D., Armus, L., Bohlin, R. C., *et al.* 2000, *ApJ*, 533, 682
Cappellari, M. & Emsellem, E., 2004 *PASP*, 116, 138
Chen, C.-T. J., Hickox, R. C., Alberts, S., Brodwin, M., *et al.* 2013, *ApJ*, 773, 3
Cid Fernandes, R., Stasínska, G., Mateus, A., & Vale Asari, N. 2011, *MNRAS*, 413, 1687
Ciotti, L., Ostriker, J. P., & Proga, D. 2010, *ApJ*, 717, 708
Cisternas, M., Jahnke, K., Inskip, K. J., *et al.* 2011, *ApJ*, 726, 57
Croton, D. J., Springel, V., White, S. D. M., *et al.* 2006, *MNRAS*, 365, 11
Davies, R. L., Groves, B., Kewley, L. J., *et al.* 2016, *MNRAS*, 462, 1616
Delvecchio, I., Lutz, D., Berta, S., Rosario, D. J., *et al.* 2015, *MNRAS*, 449, 373
Dey, A., Schlegel, D. J., Lang, D., Blum, R., *et al.* 2019, *AJ*, 157, 168
Diamond-Stanic, A. M. & Rieke, G. H. 2012, *ApJ*, 746, 168
Ellison, S. L., Teimoorinia, H., Rosario, D. J., & Mendel, T. 2016, *MNRAS*, 458, L34
Fabian, A. C. 2012, *ARA&A*, 50, 455
Ferrarese, L. & Merritt, D. 2000, *ApJ*, 539, L9
Freudling, W., Romaniello, M., Bramich, D. M., *et al.* 2013, *A&A*, 559, A96
Gabor, J. M. & Bournaud, F. 2013, *MNRAS*, 434, 606
Gabor, J. M. & Davé, R. 2012, *MNRAS*, 427, 1816
Gavazzi, *et al.* 2015, *A&A*, 580, 13
Grogin N. A., Conselice, C. J., Chatzichristou, E., Alexander, D. M., *et al.* 2005, *ApJ*, 627, L97
Groves, B., Brinchmann, J., & Walcher, C. J. 2012, *MNRAS*, 419, 1402
Harrison, C. M., Costa, T., Tadhunter, C. N., Flütsch, A., *et al.* 2018, *NatAs*, 2, 198
Harrison, C. M. 2017, *NatAs*, 1, 16
Hickox, R. C., Jones, C., Forman, W. R., *et al.* 2009, *ApJ*, 696, 891
Hirschmann, M., Charlot, S., Feltre, A., *et al.* 2017, *MNRAS*, 472, 2468
Hirschmann, M., Dolag, K., Saro, A., *et al.* 2014, *MNRAS*, 442, 2304
Ho, I.-T., Medling, A. M., Groves, B., *et al.* 2016, *Ap&SS*, 361, 280
Ho, L. C. 2008, *ARA&A*, 46, 475
Husemann, B., Scharwächter, J., Davis, T. A., *et al.* 2019, *arXiv e-prints*, arXiv:1905.10385
Jahnke, K. & Maccío, A. V. 2011, *ApJ*, 734, 92
Juneau, S., Bournaud, F., Charlot, S., *et al.* 2014, *ApJ*, 788, 88
Juneau, S., Dickinson, M., Bournaud, F., Alexander, D. M., *et al.* 2013, *ApJ*, 764, 176
Juneau, S., Dickinson, M., Alexander, D. M., & Salim, S. 2011, *ApJ*, 736, 104
Karouzos, M., Im, M., Trichas, M., Goto, T., *et al.* 2014, *ApJ*, 784, 137
Kartaltepe, J. S., Sanders, D. B., Le Floc'h, E., Frayer, D. T., *et al.* 2010, *ApJ*, 721, 98
Kauffmann, G., Heckman, T. M., White, S. D. M., *et al.* 2003a, *MNRAS*, 341, 33.
Kauffmann, G., Heckman, T. M., Tremonti, C., *et al.* 2003b, *MNRAS*, 346, 1055
Kewley, L. J., Dopita, M. A., Leitherer, C., *et al.* 2013, *ApJ*, 774, 100
Kewley, L. J., Groves, B., Kauffmann, G., & Heckman, T. 2006, *MNRAS*, 372, 961
Kewley, L. J., Dopita, M. A., Sutherland, R. S., *et al.* 2001, *ApJ*, 556, 121
Lamastra, A., Bianchi, S., Matt, G. *et al.* 2009, *A&A*, 504, 73
Leslie S. K., Kewley L. J., Sanders, D. B., & Lee, N. 2015 *MNRAS*, 455, 82
Lusso, *et al.* 2013, *ApJ*, 777, 28
Magorrian J. *et al.* 1998, *AJ*, 115, 2285
Morris, S., Ward, M., Whittle, M., Wilson, A. S., & Taylor, K. 1985, *MNRAS*, 216, 193
Mullaney, J. R., Alexander, D. M., Aird, J., Bernhard, E., *et al.* 2015, *MNRAS*, 453, 83
Novak G. S., Ostriker J. P., & Ciotti L. 2011, *ApJ*, 737, 26
Osterbrock, D. E. & Ferland, G. J. 2006, *U*niversity Science Books, Astrophysics of Gaseous Nebulae and Active Galactic Nuclei, 2edn.

Prieto, M. A., Mezcua, M., Fernández-Ontiveros, J. A., & Schartmann, M. 2014, *MNRAS*, 442, 2145
Ricci, T. V., Steiner, J. E., May, D., Garcia-Rissmann, A., & Menezes, R. B. 2018, *MNRAS*, 473, 5334
Riffel, R. A., Storchi-Bergmann, T., Dors, O. L., & Winge, C. 2009, *MNRAS*, 393, 783
Roos, O., Juneau, S., Bournaud, F., & Gabor, J. M. 2015, *ApJ*, 800, 19
Rosario, D. J., Santini, P., Lutz, D., Netzer, H., *et al.* 2013, *ApJ*, 771, 63
Salim, *et al.* *ApJS*, 173, 267
Schawinski, K., Koss, M., Berney, S., & Sartori, L. F. 2015, *MNRAS*, 451, 2517
Schawinski, K., Urry, C. M., Simmons, B. D., *et al.* 2014, *MNRAS*, 440, 889
Schawinski, K., Thomas, D., Sarzi, M., *et al.* 2007, *MNRAS*, 382, 1415
Schaye, J., Crain, R. A., Bower, R. G.; Furlong, M., *et al.* 2015, *MNRAS*, 446, 521
Silverman, J. D., Lamareille, F., Maier, C., Lilly, S. J., *et al.* 2009, *ApJ*, 696, 396
Simpson, C. 2005, *MNRAS*, 360, 565
Steinborn, L. K., Hirschmann, M., Dolag, K., Shankar, F., *et al.* 2018, *MNRAS*, 481, 341
Tanaka, M. 2012, *PASJ*, 64, 37
Treister, E., Schawinski, K., Urry, C. M., & Simmons, B. D. 2012, *ApJ*, 758, 39
Tremonti, C. A., Heckman, T. M., Kauffmann, G., *et al.* 2004, *ApJ*, 613, 898
Trouille, L., Barger, A. J., & Tremonti, C. 2011, *ApJ*, 742, 46
Trump, J. R., Weiner, B. J., Scarlata, C., *et al.* 2011, *ApJ*, 743, 144
Trump, J. R., Sun, M., Zeimann, J. R., *et al.* 2015, *ApJ*, 811, 26
Veilleux, S. & Osterbrock, D. E. 1987, *ApJS*, 63, 295
Yan, R. & Blanton, M. R. 2012, *ApJ*, 747, 61

Uncovering Early Galaxy Evolution in the ALMA and JWST Era
Proceedings IAU Symposium No. 352, 2019
E. da Cunha, J. Hodge, J. Afonso, L. Pentericci & D. Sobral, eds.

doi:10.1017/S1743921319009591

ALMA Deep Field in the SSA22 proto-cluster at z = 3

Hideki Umehata[1,2]

[1]RIKEN Cluster for Pioneering Research, 2-1 Hirosawa, Wako-shi, Saitama 351-0198, Japan
email: hideki.umehata@riken.jp

[2]Institute of Astronomy, School of Science, The University of Tokyo, 2-21-1 Osawa, Mitaka, Tokyo 181-0015, Japan

Abstract. Galaxies and nuclei in dense environment at high redshift provide a good laboratory to investigate accelerated, most extreme evolution of galaxies. The SSA22 proto-cluster at $z = 3.1$ is known to have a three-dimensional 50 (comoving) Mpc-scale filamentary structure, traced by Lyα emitters, which makes the field a suitable target in this regard. To identify dust-obscured star-formation, a contiguous 20 arcmin^2 region at the node of the cosmic structure was observed in ALMA band 6. In total 57 ALMA sources have been identified above 5σ, which makes the field one of the richest field in ALMA-identified (sub)millimeter galaxies. The follow-up spectroscopy confirmed about 20 sources as exact proto-cluster members so far. Together with high X-ray AGN fraction, our results suggest that the vigorous star formation activity and the growth of super massive black holes occurred simultaneously in the densest regions at $z \sim 3$.

Keywords. galaxies: evolution, galaxies: high-redshift

---

1. Introduction

Models of cosmological structure formation in a cold dark matter universe predict that the matter in the intergalactic medium (IGM) forms a "cosmic web" of sheets and filaments (e.g., Bond *et al.* 1996). The most massive dark matter halos lie at the intersections of filaments (e.g., Kauffmann *et al.* 1999), and high over densities in the dark matter distribution would make cradles in which galaxies form and thrive, eventually. Streams of cold gas flowing along the IGM filaments can provide fuel for the mass growth of galaxies and super massive black holes (SMBHs) through the circumgalactic medium (CGM, e.g., Dekel *et al.* 2009). Consequently galaxies and nuclei in the densest environments at high redshift are a nice laboratory for understanding of galaxy formation and evolution in their forming era.

In the last decades, such galaxy overdensities have been found preferentially tracing rest-UV-selected star-forming galaxies such as Lyα emitters (LAEs) and Lyman break galaxies (LBGs) (e.g., Steidel *et al.* 1998), (sometimes) in conjunction with 'beacons' of the over densities like high-redshift radio galaxies (HzRGs) (e.g., Venemans *et al.* 2002) (see Overzier *et al.* 2016 as a review). While such UV-bright populations indicates the existence of proto-clusters and give us some fundamental insights such as structures and overdensity, accumulating evidence now suggests that they are just part of star-forming galaxies and a complementary approach is essential to comprehend the formation of galaxies, especially for massive populations.

This is because actively star-forming galaxies generate significant amount of dust as they evolve and increase stellar mass, which absorbs the stellar lights significantly. Eventually their spectral energy distribution (SEDs) are dominated by rest-frame Far-IR to (sub)mm emission (for reviews; Blain *et al.* 2002; Casey *et al.* 2014). Therefore it is crucial to uncover not only the 'visible' part (i.e., UV/Optical views) but also the 'obscured' part (i.e., FIR views) of various galaxy populations (e.g., Dunlop *et al.* 2017).

Motivated by this, a number of extragalactic surveys were carried out at submm/mm using bolometer cameras onboard single-dish telescopes (e.g., SCUBA, SCUBA2/JCMT, AzTEC/ASTE) (e.g., Tamura *et al.* 2009; Umehata *et al.* 2014). Together with follow-up observations using an interferometer such as the SMA and (J)VLA, association between submillimeter galaxies (SMGs) and proto-clusters at high redshift was reported in some works (e.g., Tamura *et al.* 2010). However, the identification has been hard for years, and our knowledge had been limited to several exceptionally bright SMGs. As a reasonable successive step, astronomers have sought for an ALMA follow-up, which provide us with unprecedented sensitivity and angular resolution.

While snapshot surveys of single-dish selected submm/mm sources is one cost-effective way, such observations suffer from selection bias and completeness issue, which would result in a highly biased view. Therefore, ALMA deep surveys - mapping a significant area contiguously - are essential to obtain an unbiased submm/mm view. Such surveys are now on-going in general fields (e.g., Franco *et al.* 2018; Hatsukade *et al.* 2018), and a proto-cluster, which is the theme of this paper.

An ALMA deep survey is especially vital to understand massive galaxy formation, although it is helpful for understanding various aspects of galaxy formation. In the local Universe, the centers of cluster are known to be occupied by massive elliptical galaxies (Dressler *et al.* 1980). Simulations and galactic archeology suggest that such massive ellipticals experienced an episodic starburst phase at $z > 2-3$ in an ancestor of cluster - proto-cluster- (e.g., De Lucia *et al.* 2006). As we described above, the massive galaxies in the forming phase, possibly also forming a central SMBH, would be heavily obscured by dust. ALMA deep surveys are expected to identify and characterize such galaxies in the early Universe.

2. Survey Description

Our target field is a proto-cluster at $z \approx 3$. The SSA22 field is known to harbor a >100-Mpc-scale huge structure traced by $z \approx 3.09$ LAEs (Yamada *et al.* 2012), which includes a three-dimensionally-confirmed 50-Mpc-scale filamentary structure in the densest, proto-cluster region (Matsuda *et al.* 2005). This is the most spectacular known proto-cluster amongst all the known over-densities at $z > 2$ (Yamada *et al.* 2012), which provides a unique laboratory to investigate the co-evolution of galaxies and cosmic structures.

In ALMA Cycle-2, we have started a series of ALMA projects in the SSA22 proto-cluster field to uncover the cold regime - the universe of gas and dust-. The details of the observations are reported in previous papers (Umehata *et al.* 2015, 2017, 2018, 2019) and to be reported in an upcoming paper (H. Umehata *et al.*, in prep), and hence here we just briefly presented our observation. We mapped a central area of the proto-cluster contiguously and we name the project ALMA Deep Field in SSA22 or ADF22. So far we mapped 7 arcmin^2 area in band 6 down to 25 μJy Beam^{-1}. We also observed an adjacent 13 arcmin^2 field with 1σ sensitivity of 60 μJy Beam^{-1} (Umehata *et al.* 2018). The former 7 arcmin^2 area is also covered by band 3 observations which are designed to detect $z = 3.1$ CO(3-2) emission line (Umehata *et al.* 2019). The representative noise level per a 100 km s^{-1} channel is $\sigma = 90$ μJy beam^{-1} at around 84-86 GHz.

3. The ADF22 Results

Identification of SMGs in the proto-cluster: The deep and wide 1.1 mm map uncovers 57 SMGs above 5σ (note that the noise level is heterogenous within the map, as we explained

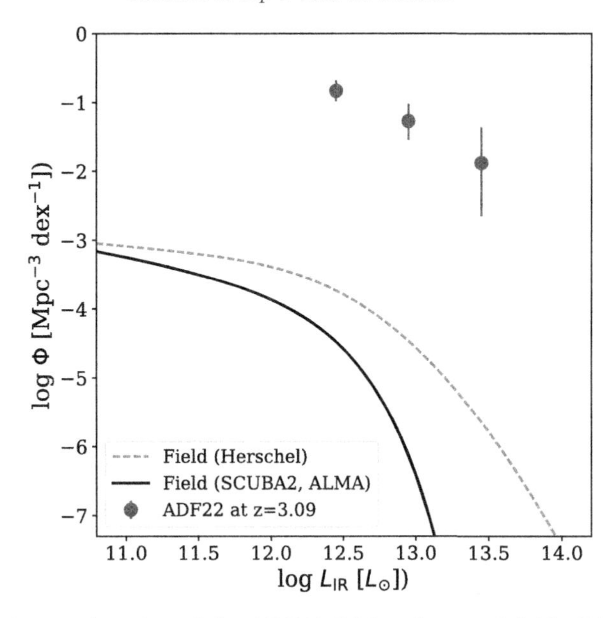

Figure 1. IR luminosity function of the ADF22 field and general fields. Blue points shows the ADF22 results. For comparison, the luminosity function at similar epoch, derived from *Herschel* (Gruppioni *et al.* 2013) and the combination of SCUBA2 and ALMA (Hatsukade *et al.* 2018) in general fields are shown. The excess in ADF22 is easily recognizable compared to both. The difference is larger for the SCUBA2 and ALMA results, which do not suffer from source confusion so much and hence more reliable. The excess shows that the dusty, intense star-formation seen in rapidly growing massive galaxies preferentially occurs in such a dense environment at $z \sim 3$.

above.) The negative map shows no *negative* 5σ peak, which demonstrates that our threshold is conservative enough not to count artificial peak due to noise fluctuations (e.g., Umehata *et al.* 2018). If we relax the criteria, the source number increases significantly. We perform various spectroscopic follow-up observations for the detected ALMA sources like NearIR spectroscopy using MOSFIRE on Keck and mm spectroscopy using ALMA as above (Umehata *et al.* 2019). So far 20 SMGs are confirmed to be the exact proto-cluster members through mult-wavelength spectroscopic follow-ups, which results in clear excess on 1 mm number counts (e.g., Umehata *et al.* 2018).

Luminosity Function at the proto-cluster: Here we further investigate the excess, focusing on the ADF22A (i.e., the deepest 7 arcmin2 region), which has the richest ancillary data. Among the top brightest 18 SMGs with $S_{1.1\mathrm{mm}} \geqslant 0.5$ mJy, 16 SMGs are found to be at $z = 3.09$ (2 SMGs are confirmed to be at $z = 3.991$ and $z = 2.05$, respectively; e.g., Umehata *et al.* 2018). Then the accurate redshifts allows us to perform SED fitting and derive total IR luminosity (rest-frame 8-1000 μm). Since the 16 SMGs are 'flux-limited' samples, then we can evaluate IR luminosity function at the core of the proto-cluster. The results are shown in Fig. 1). As shown in the figure, the excess is quite remarkable, showing more than 3 dex, while the excess of LBGs is ~6 times than general field (Steidel *et al.* 1998). This further demonstrates that intense star-forming activity in massive galaxies is accelerated in the proto-cluster core.

High X-ray AGN fraction: The SSA22 proto-cluster has been observed by *Chandra* X-ray observatory (e.g., Lehmer *et al.* 2009a,b). Thanks to the relatively deep integration time (about 400 ks), 8 X-ray luminous AGNs at $z = 3.09$ with X-ray Luminosity $L_x \approx 10^{44}$ erg s^{-1} have been found in ADF22 (e.g., Umehata *et al.* 2015, 2019). Furthermore, six of the eight are also detected in ALMA and included in the 16 brightest SMGs in the field. The volume density of the X-ray AGN is three orders of magnitudes higher than the

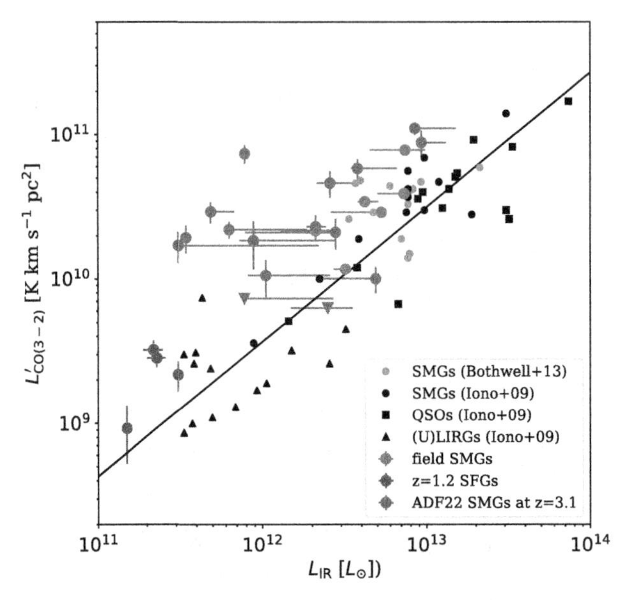

Figure 2. The relation between IR luminosity ($L_{\rm IR}$) and CO(3-2) line luminosity ($L'_{\rm CO(3-2)}$) for the ADF22 SMGs. For comparison, we plot the results for SMGs in general fields (e.g., Iono *et al.* 2009; Bothwell *et al.* 2013; Wardlow *et al.* 2018). The solid line shows best-fit function from Iono *et al.* (2009). The digram suggests that some of the ADF22 SMGs have relatively molecular gas-rich nature, compared to field SMGs.

general field (Umehata *et al.* 2019). Together with the high X-ray AGN fraction, this indicates that there is not only the star-bursting activity but also accelerated accretion activity at the node of the cosmic structure.

Molecular gas reservoirs The co-spatial band 3 survey provides a census of molecular gas for all the 16 SMGs at $z = 3.09$ in ADF22A. We show the relation between the IR luminosity ($L_{\rm IR}$) and the CO(3-2) line luminosity ($L'_{\rm CO(3-2)}$) as a proxy of the SFR-$M_{\rm gas}$ relation (where SFR shows star-formation rate and $M_{\rm gas}$ shows molecular gas mass, Fig. 2). There is certain diversity among the ADF22 SMGs (as we can easily recognize it from the fact that some of SMGs are not detected in CO(3-2)), which would somehow reflect the evolutionary stage. Another interesting feature is that some of the SMGs show relatively large CO(3-2) line luminosity at a given IR luminosity, compared to field SMGs. The ADF22 SMGs are located at the proto-cluster core, where the potential well is expected to be deep. Hence one interpretation is that the SMGs are fueled by cosmic web filaments and abundant gas supply leads the starbursts and large molecular gas reservoir. But we also note that we assume a brightness temperature ratio which is derived for SMGs in general field (Bothwell *et al.* 2013), which may not be appropriate. More data are required to make a final conclusion.

4. Summary and Prospects

We present the background, survey strategy, and some key results of our ongoing ALMA survey towards the SSA22 proto-cluster at $z = 3$. We have uncovered unusual active growth of massive galaxies and SMBHs at the center of the proto-cluster on the basis of a contiguous mm mapping, utilizing the unprecedented capability of ALMA. Further ALMA observations such as fine structure lines and/or very high angular resolution imaging would give us further insights. Moreover, synergy with other wavelengths, such as optical IFU (e.g., MUSE), would be of great importance to uncover various aspects of galaxy formation at the proto-cluster.

References

Blain, A. W. *et al.* 2002, *PhR*, 369, 111
Bond, J. R. *et al.* 1996, *MNRAS*, 380, 603
Bothwell, M. S. *et al.* 2013, *MNRAS*, 429, 3047
Casey, C. M. *et al.* 2014, *PhR*, 541, 45
Dekel, A. *et al.* 2009, *Nature*, 457, 451
De Lucia, G. *et al.* 2006, *MNRAS*, 366, 499
Dressler, A. 1980, *ApJ*, 236, 351
Dunlop, J. S. *et al.* 2017, *MNRAS*, 466, 861
Franco, M. *et al.* 2018, *A&A*, 620, 152
Gruppioni, C. *et al.* 2013, *MNRAS*, 432, 23
Hatsukade, B. *et al.* 2018, *PASJ*, 70, 105
Iono, D. *et al.* 2009, *ApJ*, 695, 1537
Kauffmann, G. *et al.* 1999, *MNRAS*, 307, 529
Lehmer, B. D. *et al.* 2009, *ApJ.*, 691, 687
Lehmer, B. D. *et al.* 2009, *MNRAS.*, 400, 299
Matsuda, Y. *et al.* 2005, *ApJ*, 634, 125
Overzier, R. A. *et al.* 2016, *A&ARv*, 24, 14
Steidel, C. C. *et al.* 1998, *ApJ*, 492, 428
Tamura, Y. *et al.* 2009, *Nature*, 459, 61
Tamura, Y. *et al.* 2010, *ApJ*, 724, 1270
Umehata, H. *et al.* 2014, *MNRAS*, 440, 3462
Umehata, H. *et al.* 2015, *ApJ*, 815, L8
Umehata, H. *et al.* 2017, *ApJ*, 835, 98
Umehata, H. *et al.* 2018, *PASJ*, 70, 65
Umehata, H. *et al.* 2019, submitted to *Science*
Venemans, B. P. *et al.* 2002, *ApJ*, 569, 11
Wardlow, J. L. *et al.* 2018, *MNRAS*, 479, 3879
Yamada, T. *et al.* 2012, *AJ*, 143, 79

Uncovering Early Galaxy Evolution in the ALMA and JWST Era
Proceedings IAU Symposium No. 352, 2019
E. da Cunha, J. Hodge, J. Afonso, L. Pentericci & D. Sobral, eds.

doi:10.1017/S1743921319008895

More than star formation: High-J CO SLEDs of high-z galaxies

Chelsea E. Sharon[1], Reni Chng[1], Kebron K. Gurara[1], Axel Weiß[2], Jeremy Darling[3], Dominik Riechers[4] and Carl Ferkinhoff[5]

[1]Yale-NUS College, Singapore, 138527, Singapore
email: chelsea.sharon@yale-nus.edu.sg

[2]Max-Planck-Institut für Radioastronomie, Auf dem Hügel 69 D-53121 Bonn, Germany

[3]Center for Astrophysics and Space Astronomy, Department of Astrophysical and Planetary Sciences, University of Colorado, Boulder, CO 80309-0389, USA

[4]Department of Astronomy, Cornell University, Ithaca, NY 14853, USA

[5]Physics Department, Winona State University, Winona, MN 55987, USA

Abstract. Theoretical work suggests that AGNs play an important role in quenching star formation in massive galaxies. In addition to molecular outflows observed in the local universe, emission from very high-J CO rotational transitions have been a key piece of evidence for AGN directly affecting the molecular gas reservoirs that fuel star formation. However, very few observations exist of CO rotational lines past the peak of the CO spectral line energy distribution (SLED) for galaxies in the early universe. Here we present new ALMA observations of high-J CO rotational lines (from CO(5–4) to CO(16–15)) in six $z > 2$ IR-bright systems, including several sources not known to contain a strong AGN for comparison. We detect significant amounts of high-excitation CO emission that suggests the presence of energy sources beyond UV-heating.

Keywords. Galaxies: high-redshift, Galaxies: active, Galaxies: star formation, ISM: molecules

1. Background

Explaining the mismatch between the observed number of massive galaxies and the number predicted by ΛCDM cosmology has largely relied on feedback from active galactic nuclei (AGNs) to suppress star formation (for a review, see Fabian 2012). However, the exact AGN feedback mechanisms and whether they are sufficient to quench star formation are still debated. Modes of AGN feedback can largely be classified as either indirect, where the radio jet prevents accretion of new gas onto the galaxy (i.e. kinetic, jet, radio, or maintenance mode feedback; e.g., Kereš *et al.* 2005), or direct, where the AGN removes gas already present in the galaxy or otherwise prevents that gas from forming stars (i.e. radiative, wind, or quasar mode feedback; e.g., Silk & Rees 1998). While evidence for indirect AGN feedback exists in hot X-ray gas halos (e.g., Bîrzan *et al.* 2012), concerns about the duty cycles of AGNs' radio phase remain (e.g., Shabala *et al.* 2008). Indirect AGN feedback also does not suppress star formation in galaxies with existing large molecular gas reservoirs (e.g., submillimeter galaxies; SMGs; e.g., Casey *et al.* 2014). Studies of direct AGN feedback have largely focussed on the ubiquity of observed outflows (e.g., Veilleux *et al.* 2005), although it is challenging to distinguish between AGN-driven and starburst-driven winds (e.g., Diamond-Stanic *et al.* 2012; Gowardhan *et al.* 2018), and it is unclear what fraction of winds' mass eventually return to the galaxy. Molecular gas outflow rates have been recently confirmed to correlate with AGN luminosity at low redshift (Cicone *et al.* 2014), but similar observations are too observationally expensive

Table 1. Summary of observations & spectral line fits

Source	Class[a]	$z_{\rm fiducial}$	Spectral Line	$z_{\rm fit}$	FWHM ($\rm km\,s^{-1}$)	$S\Delta v^b$ $\rm Jy\,km\,s^{-1}$
BR 1202–0725 North	SMG	4.695	CO(13–12)	4.6932 ± 0.0005	1314 ± 64	3.0 ± 0.3
			CO(14–13)	4.6931 ± 0.0008	1301 ± 98	1.9 ± 0.2
BR 1202–0725 South	AGN	4.695	CO(13–12)	4.6944 ± 0.0004	348 ± 47	0.61 ± 0.08
			CO(14–13)	4.6948 ± 0.0005	423 ± 70	0.54 ± 0.07
Cloverleaf	AGN+L	2.560	CO(11–10)	2.55785 ± 0.00005	412 ± 9	33 ± 3
SMM J04431+0210	SMG	2.510	CO(5–4)	2.5082 ± 0.0008	360 ± 160	0.83 ± 0.23
			CO(8–7)	2.5084 ± 0.0003	287 ± 59	0.82 ± 0.13
SMM J14011+0252	SMG	2.565	CO(11–10)	2.5651 ± 0.0002	136 ± 33	0.43 ± 0.08
MG 0751+2716	AGN+L	3.200	CO(10–9)	3.1992 ± 0.0005	670 ± 84	7.3 ± 0.9
			CO(11–10)	3.1996 ± 0.0003	560 ± 52	4.1 ± 0.5
PSS J2322+1944[c]	AGN+L	4.119	CO(15–14)	4.1206 ± 0.0003	225 ± 41	0.74 ± 0.11
			CO(16–15)	4.1199 ± 0.0001	227 ± 55	0.80 ± 0.13
			OH 1835 GHz		314 ± 15	2.6 ± 0.3
			OH 1838 GHz			2.9 ± 0.3

[a] Indicates whether the source contains an AGN or is a SMG. "L" denotes sources that are strongly lensed.
[b] Uncertainties include statistical and assumed 10% flux calibration uncertainties combined in quadrature.
[c] The CO(16–15) line and two OH lines were fit simultaneously. The line separations were held fixed while fitting a global redshift. The two CO lines were constrained to have the same FWHM.

near the peak epoch of cosmic star formation (e.g., Madau & Dickinson 2014) and black hole accretion (e.g., Delvecchio *et al.* 2014), where most AGN feedback might be expected. At high redshift, detections of molecular outflows are often detected in less commonly studied species (such as CH^+ or OH in absorption; e.g., Falgarone *et al.* 2017; Spilker *et al.* 2018) that are challenging to convert into mass outflow rates.

It may be possible that correlations between CO excitation and AGN strength will reveal direct modes of AGN feedback. IR-bright AGN are known to produce CO SLEDs that peak near $J_{\rm upper} \sim 7$ (e.g., Weiß *et al.* 2007), while starburst-dominated galaxies like SMGs have CO SLEDs that peak near $J_{\rm upper} \sim 5$ (e.g., Carilli & Walter 2013). While initial systematic differences in low-J CO line ratios between SMGs and AGN (e.g., Swinbank *et al.* 2010; Harris *et al.* 2010; Ivison *et al.* 2011; Danielson *et al.* 2011; Riechers *et al.* 2011; Thomson *et al.* 2012) were not statistically significant in larger samples (Sharon *et al.* 2016), a more quantitative classification of AGN strength based on mid-IR spectral features suggests systematically higher excitation CO line ratios across the entire SLED for AGN (albeit with limited statistical power; Kirkpatrick *et al.* 2019). Looking for evidence of AGN feedback via CO excitation not only probes the fraction of the molecular gas affected by the AGN, but can also constrain how the AGN interacts with the gas (for example, via X-ray or shock heating; e.g., van der Werf *et al.* 2010; Meijerink *et al.* 2013). Constraining X-ray and shock heating requires observing very high-J CO lines, and there are few published observations of $J_{\rm upper} > 8$ lines at high redshift (Weiß *et al.* 2007; Riechers *et al.* 2013; Gallerani *et al.* 2014; Tuan-Anh *et al.* 2017; Oteo *et al.* 2017; Yang *et al.* 2017; Cañameras *et al.* 2018; Stacey & McKean 2018).

2. Observations & Results

In order to explore the prevalence of AGN-affected molecular gas and look for evidence of AGN feedback, we observed several high-J CO lines in a sample of $z > 2$ galaxies. We selected high-z galaxies accessible from Atacama Large Millimeter/submillimeter Array (ALMA) that either (a) already have a $J_{upper} > 9$ detection indicating that there is a high-excitation component that needs to be characterized more accurately, or (b) have an existing CO SLED that includes both the CO(1–0) and CO(7–6) line to aid

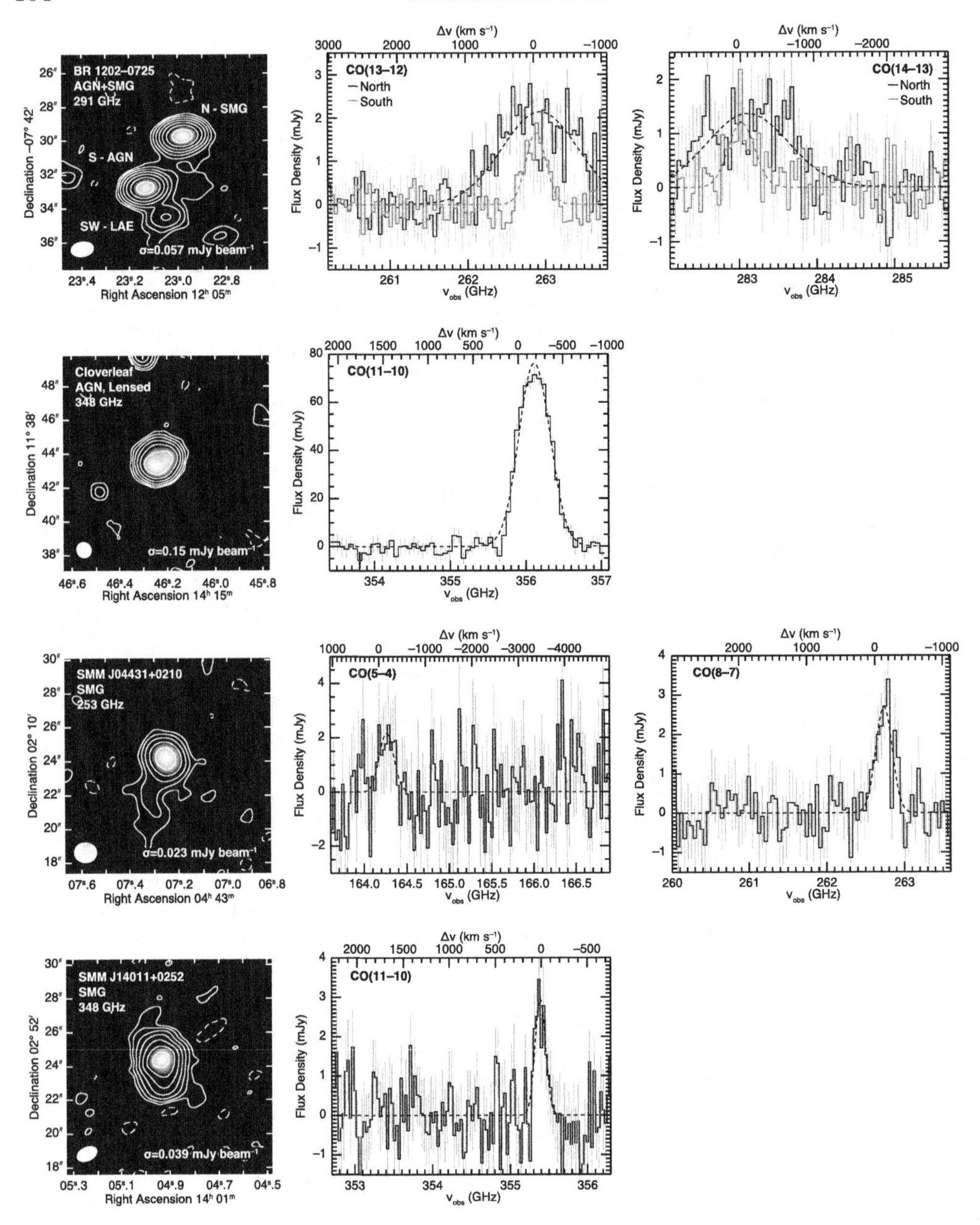

Figure 1. Continuum maps (left column) and continuum-subtracted spectra (middle and right columns) for our ALMA observations. We show the continuum maps from the higher frequency observations in cases where we observe multiple CO lines. Contours are powers of two times the map noise (i.e., $\pm 2\sigma$, $\pm 4\sigma$, $\pm 8\sigma$, etc.); negative values are dashed and the map noise is given in the lower right corner. The beam FWHM is shown in the lower left corner. For the spectra, the vertical bars show the $\pm 1\sigma$ statistical uncertainty in each channel. Dashed lines show the best-fit Gaussian line profile. The velocity axis is calculated relative to the fiducial redshift listed in Table 1. For BR 1202–0725, we show the spectra for the northern SMG in black and the southern AGN in red; since we do not detect significant line emission from the SW Lyman-α emitter we do not show its spectra. Note that SMM J14011+0252 is only weakly lensed.

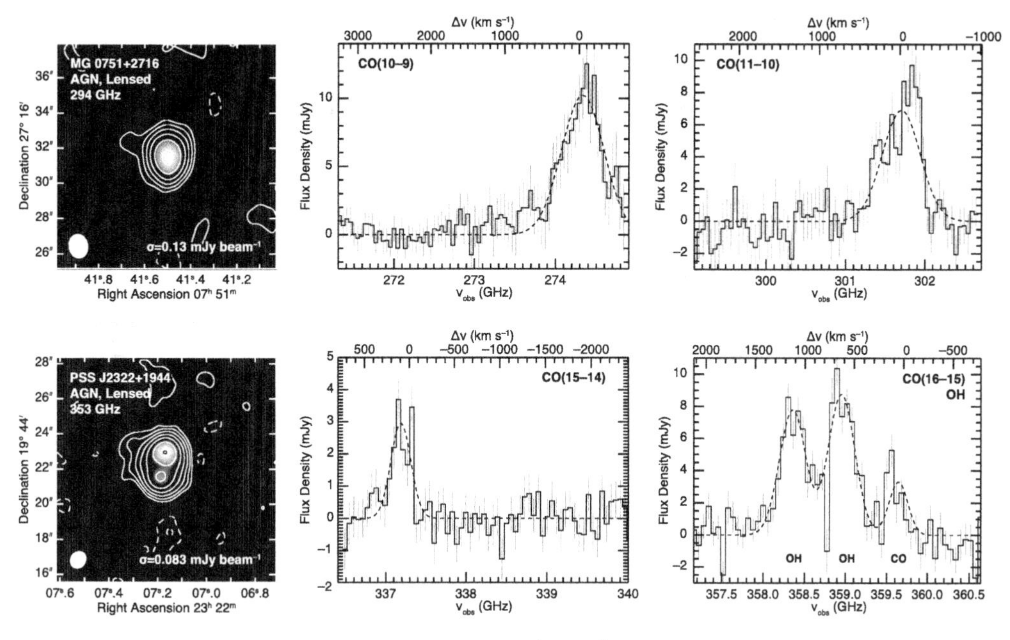

Figure 1. (Cont.)

estimates of the higher-J line strengths. We observed the next 1–2 higher-J CO lines during ALMA Cycle 5 (PID: 2017.1.00963.S). In addition to known high-z AGN, we also targeted several SMGs that are expected to be starburst-dominated in order to provide a baseline comparison sample for the effects of star formation alone on gas heating. For one target, we also observed a more moderate-excitation CO line in order to help fill in the SLED. For a complete list of targets and observed CO lines, see Table 1.

We show continuum maps and continuum-subtracted spectra in Figure 1. We detected all observed CO lines, although the detections of the CO(5–4) line for SMM J04431+0210 and the CO(16–15) line for PSS J2322+1944 are marginal ($\sim 5\sigma$ peak detections in the integrated line maps). In addition, for PSS J2322+1944, we detected the OH $^2\Pi_{1/2}$ $(3/2-1/2)$ doublet at 1835/1838 GHz. We also detected continuum from the Lyman-alpha emitter closely aligned with BR 1202–0725, but did not detect any significant CO emission.

3. Implications

While we have not yet compared the CO observations to radiative transfer models, the significant high-excitation CO emission in the AGN-dominated sources suggests there are additional heating mechanisms that must be considered (Fig. 2; however, one must be wary of differential lensing; e.g., Serjeant 2012). Many of these sources' SLEDs fall above the envelope for sources from the local *Herschel* Comprehensive (U)LIRG Emission Survey sample that *requires* an additional excitation mechanism, such as mechanical shocks and/or heating from X-rays or cosmic rays (Rosenberg *et al.* 2015). However, the differences in line ratios may be partially explained by the cosmic microwave background, particularly for the lower excitation SMGs and $z\sim5$ sources (e.g., da Cunha *et al.* 2013). Only two sources (BR 1202–0725 North from this sample, and MG 0414 from Stacey & McKean 2018) show larger line FWHMs at higher-J (their lower-J CO FWHMs are $\sim$300–700 km s^{-1}; Omont *et al.* 1996; Barvainis *et al.* 1998; Carilli *et al.* 2002; Salomé *et al.* 2012) suggesting they trace a dynamically distinct region.

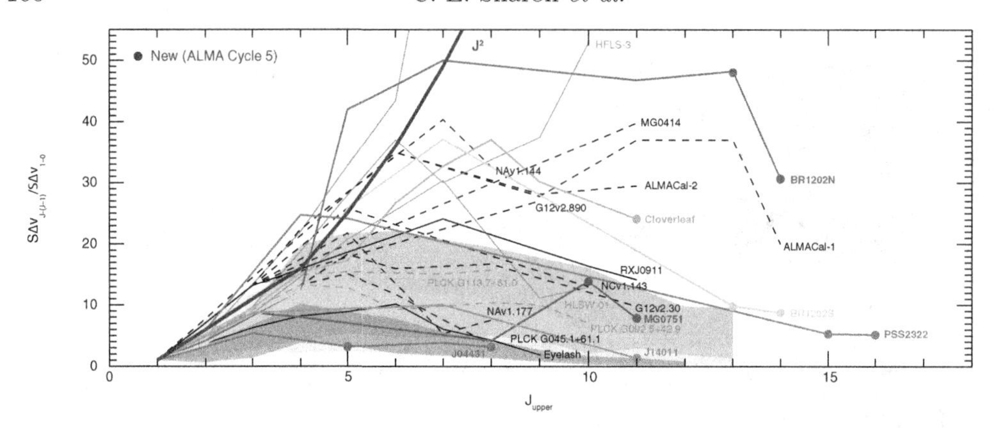

Figure 2. CO SLEDs for our sample (colored lines with symbols for new ALMA observations) normalized by the CO(1–0) flux. Literature SLEDs with at least one detection at $J_{\rm upper} \geq 7$ are in thin gray/black lines (black for ALMA-observable sources; gray for northern sources). Sources lacking CO(1–0) have dashed lines; we extrapolate a CO(1–0) flux from the lowest-J measured line using average HerCULES ratios (Rosenberg *et al.* 2015). Shaded regions show envelopes of CO SLED classes identified with HerCULES: SLEDs that peak near $J_{\rm upper} \sim 4$–6 then swiftly decline which only require UV-heating from young stars (red), SLEDs that peak near $J_{\rm upper} \sim 6$–7 but do not decline as quickly which have a mixture of energy sources (yellow), and SLEDs that peak near $J_{\rm upper} \sim 9$ which *require* mechanical shock, X-ray, and/or cosmic ray heating (blue). Thermalized line ratios ($\propto J^2_{\rm upper}$) are shown in the thick black line.

The results for starburst-dominated galaxies are more mixed. While the SLEDs of SMM J14011+0252 and SMM J04431+0210 are certainly declining at higher-J (similar to the Eyelash; Danielson *et al.* 2011), BR 1202–0725 North is more highly excited than its AGN companion, perhaps due to an AGN too deeply buried in dust to have been detected by previous observations.

References

Barvainis, R., Alloin, D., Guilloteau, S., & Antonucci, R. 1998, *ApJL*, 492, L13
Bîrzan, L., Rafferty, D. A., Nulsen, P. E. J., *et al.* 2012, *MNRAS*, 427, 3468
Cañameras, R., Yang, C., Nesvadba, N. P. H., *et al.* 2018, *A&A*, 620, A61
Carilli, C. L. & Walter, F. 2013, *ARA&A*, 51, 105
Carilli, C. L., Cox, P., Bertoldi, F., *et al.* 2002, *ApJ*, 575, 145
Casey, C. M., Narayanan, D., & Cooray, A. 2014, *PhR*, 541, 45
Cicone, C., Maiolino, R., Sturm, E., *et al.* 2014, *A&A*, 562, A21
da Cunha, E., Groves, B., Walter, F., *et al.* 2013, *ApJ*, 766, 13
Danielson, A. L. R., Swinbank, A. M., Smail, I., *et al.* 2011, *MNRAS*, 410, 1687
Delvecchio, I., Gruppioni, C., Pozzi, F., *et al.* 2014, *MNRAS*, 439, 2736
Diamond-Stanic, A. M., Moustakas, J., Tremonti, C. A., *et al.* 2012, *ApJL*, 755, L26
Fabian, A. C. 2012, *ARA&A*, 50, 455
Falgarone, E., Zwaan, M. A., Godard, B., *et al.* 2017, *Nature*, 548, 430
Gallerani, S., Ferrara, A., Neri, R., & Maiolino, R. 2014, *MNRAS*, 445, 2848
Gowardhan, A., Spoon, H., Riechers, D. A., *et al.* 2018, *ApJ*, 859, 35
Harris, A. I., Baker, A. J., Zonak, S. G., *et al.* 2010, *ApJ*, 723, 1130
Ivison, R. J., Papadopoulos, P. P., Smail, I., *et al.* 2011, *MNRAS*, 412, 1913
Kereš, D., Katz, N., Weinberg, D. H., & Davé, R. 2005, *MNRAS*, 363, 2
Kirkpatrick, A., Pope, A., Sajina, A., *et al.* 2015, *ApJ*, 814, 9
Kirkpatrick, A., Sharon, C., Keller, E., & Pope, A. 2019, *ApJ*, 879, 41
Madau, P. & Dickinson, M. 2014, *ARA&A*, 52, 415
Meijerink, R., Kristensen, L. E., Weiß, A., *et al.* 2013, *ApJL*, 762, L16

Omont, A., Petitjean, P., Guilloteau, S., *et al.* 1996, *Nature*, 382, 428
Oteo, I., Zwaan, M. A., Ivison, R. J., Smail, I., & Biggs, A. D. 2017, *ApJ*, 837, 182
Riechers, D. A., Carilli, C. L., Maddalena, R. J., *et al.* 2011, *ApJL*, 739, L32
Riechers, D. A., Bradford, C. M., Clements, D. L., *et al.* 2013, *Nature*, 496, 329
Rosenberg, M. J. F., van der Werf, P. P., Aalto, S., *et al.* 2015, *ApJ*, 801, 72
Salomé, P., Guélin, M., Downes, D., *et al.* 2012, *A&A*, 545, A57
Serjeant, S. 2012, *MNRAS*, 424, 2429
Shabala, S. S., Ash, S., Alexander, P., & Riley, J. M. 2008, *MNRAS*, 388, 625
Sharon, C. E., Riechers, D. A., Hodge, J., *et al.* 2016, *ApJ*, 827, 18
Silk, J. & Rees, M. J. 1998, *A&A*, 331, L1
Spilker, J. S., Aravena, M., Béthermin, M., *et al.* 2018, Science, 361, 1016
Stacey, H. R. & McKean, J. P. 2018, *MNRAS*, 481, L40
Swinbank, A. M., Smail, I., Longmore, S., *et al.* 2010, *Nature*, 464, 733
Thomson, A. P., Ivison, R. J., Smail, I., *et al.* 2012, *MNRAS*, 425, 2203
Tuan-Anh, P., Hoai, D. T., Nhung, P. T., *et al.* 2017, *MNRAS*, 467, 3513
van der Werf, P. P., Isaak, K. G., Meijerink, R., *et al.* 2010, *A&A*, 518, L42
Veilleux, S., Cecil, G., & Bland-Hawthorn, J. 2005, *ARA&A*, 43, 769
Weiß, A., Downes, D., Walter, F., & Henkel, C. 2007, in Astronomical Society of the Pacific Conference Series, Vol. 375, From Z-Machines to ALMA: (Sub)Millimeter Spectroscopy of Galaxies, ed. A. J. Baker, J. Glenn, A. I. Harris, J. G. Mangum, & M. S. Yun, 25
Yang, C., Omont, A., Beelen, A., *et al.* 2017, *A&A*, 608, A144

Uncovering Early Galaxy Evolution in the ALMA and JWST Era
Proceedings IAU Symposium No. 352, 2019
E. da Cunha, J. Hodge, J. Afonso, L. Pentericci & D. Sobral, eds.

doi:10.1017/S1743921319009475

The molecular gas content in a protocluster at z = 1.7: Star formation and AGN feedback

Q. D'Amato[1,2], I. Prandoni[1], R. Gilli[3], M. Massardi[1], E. Liuzzo[1], M. Mignoli[3], C. Vignali[2], A. Peca[3] and R. Nanni[2,3]

[1]INAF/IRA, Istituto di Radioastronomia, Via Piero Gobetti 101, 40129, Bologna, Italy
email: quirino.damato2@unibo.it

[2]Dipartimento di Fisica e Astronomia dell'Università degli Studi di Bologna, via P. Gobetti 93/2, 40129 Bologna, Italy

[3]INAF/OAS, Osservatorio di Astrofisica e Scienza dello Spazio di Bologna, via P. Gobetti 93/3, 40129 Bologna, Italy

Abstract. A large-scale structure has been recently discovered at z = 1.7, around a powerful FRII radio galaxy. Eight Star Forming Galaxies (SFGs) have been discovered within $\Delta z \approx 0.0095$ and at <1 Mpc from the FRII, indicating that this is a signpost of a protocluster. Furthermore, a significant X-ray diffuse emission overlapping the Eastern lobe of the FRII has been detected. Protoclusters are the ideal targets to investigate the complex assembly processes leading to the formation of local galaxy clusters. We will exploit new ALMA CO(2-1) observations (PI: R. Gilli) of the entire region around the FRII galaxy to trace the molecular gas content, in order to discover new protocluster members. Coupling these measurements with the multi-wavelength data coverage available for this field, we aim at placing constrains on the physical conditions in which star formation occurs, and ultimately infer the role of the radio jets in triggering it.

Keywords. galaxies clusters, quasars, supermassive black holes, shock waves, high-redshift

---

VLT/MUSE data show an overdensity of 6 SFGs in the redshift range 1.6871–1.6967 (Fig. 1, green circles). From a subsequent spectroscopic follow-up carried out with LBT/LUCI we measured the spectroscopic redshift z = 1.6987 for the FRII optical host, and serendipitously discovered an additional protocluster member candidate, at z = 1.6966 (Fig. 1, blue circles). We also estimated the photometric redshift of the additional radio source North of the Eastern lobe of the FRII, obtaining a redshift estimate between z = 1.4 and z = 1.9 at 68% confidence level. We obtained new ALMA observation of the region covered by the FRII (Fig. 1, cyan rectangle) down to $L'_{CO(2-1)} = 2 \times 10^{10}$ K km/s pc^{-2} in order to detect new members of the protocluster and measure their position, gas mass, gas fraction and Star Formation Rate (SFR). Deep Chandra observations reveal a diffuse X-ray emission (Fig. 2), in particular around the FRII Eastern lobe. Four out of six the VLT/MUSE sources are distributed at the edge of such emission, with three of them in an arc-like shape. We speculate that the diffuse X-rays trace shock-heated gas by positive AGN feedback, promoting star formation by compression of the cold gas. Based on the number density of field CO emitters with $L'_{CO} \geqslant 3 \times 10^9$ K km/s pc^{-2} (Decarli+16, ApJ, 833, 69), from the CO (2-1) observation we expect to detect ~10 new galaxies of the protocluster. Based on their location we will infer whether the SF is related to the FRII activity. In Fig. 3, right-top corner,

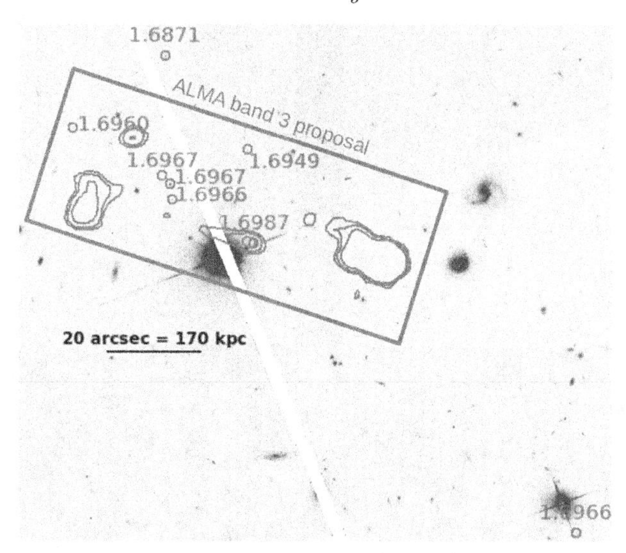

Figure 1. HST-ACS F850LP image of the field around the FRII. North is up and East is to left. The magenta contours show the 1.4 GHz emission at the 3σ, 5σ, 10σ level, respectively. An additional radio source (possible part of the protocluster) is also labeled in magenta. The green circles mark the VLT/MUSE protocluster candidates, labeled with their spectroscopic redshift. The two LBT/LUCI possible members, including the FRII are marked with the blue circles. The cyan rectangle covers the ALMA proposal region.

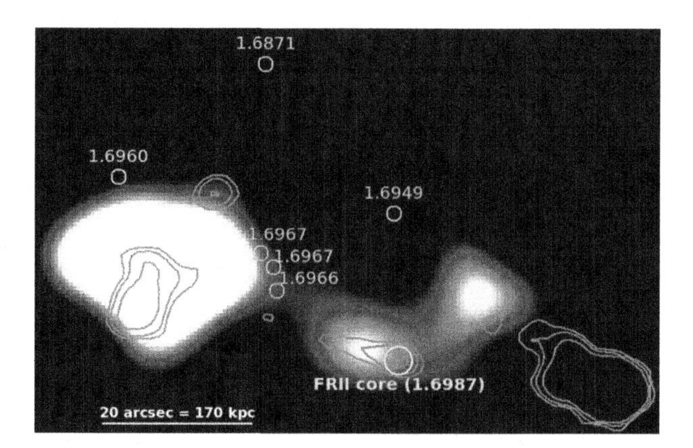

Figure 2. Point-source subtracted and smoothed Chandra image in the 0.5-7 KeV band, overlaid with 1.4 GHz contours (magenta). We mark the protocluster galaxies with green circles, labeled with their redshift. The FRII core position is indicated with a cyan circle.

we report the serendipitous discovery of a CO emitter located close (projected distance <50 kpc, spectral shift <300 km/s) to the FRII core, which is one of the best candidates as new protocluster member. The host of the FRII was found to be a strong CO emitter with a complex kinematic structure, as visible from the double-peaked shape of the line (Fig. 3, left-top corner), that could be ascribed either to a rotating system or even to two separated, but spatially unresolved, galaxies.

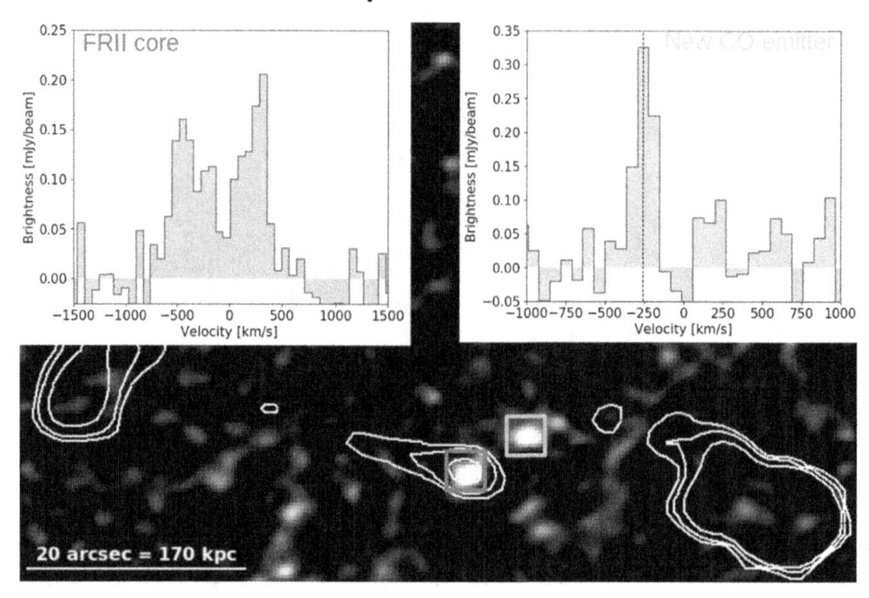

Figure 3. Continuum-subtracted image of the CO(2-1) emission, overlaid with 1.4 GHz contours in white. The spectrum of the serendipitous source (highlighted by the green box) is reported in the right-top corner, in which the black dashed line indicates the channel corresponding to the image (~-255 km/s). In the left-top corner we report the spectrum of the FRII host (in the red box). The spectrum velocity is centered at the FRII rest-frame, the channel width is 70 km/s.

Uncovering Early Galaxy Evolution in the ALMA and JWST Era
Proceedings IAU Symposium No. 352, 2019
E. da Cunha, J. Hodge, J. Afonso, L. Pentericci & D. Sobral, eds.

doi:10.1017/S1743921319009396

Clustering of galaxies around quasars at $z \sim 4$

Cristina García-Vergara[1,2,3], Joseph F. Hennawi[3,4], L. Felipe Barrientos[2] and Fabrizio Arrigoni Battaia[5]

[1]Leiden Observatory, Leiden University, P.O. Box 9513, 2300 RA Leiden, The Netherlands
[2]Instituto de Astrofísica, Pontificia Universidad Católica de Chile, Avenida Vicuña Mackenna 4860, Santiago, Chile
[3]Max-Planck Institut für Astronomie (MPIA), Königstuhl 17, D-69117 Heidelberg, Germany
[4]Department of Physics, University of California, Santa Barbara, CA 93106, USA
[5]Max-Planck Institut für Astrophysik (MPA), Karl-Schwarzschild-Str. 1, 85741 Garching bei München, Germany

Abstract. We conduct a survey for Lyman break galaxies (LBGs) and Lyman alpha emitters (LAEs) in the environs of six and 17 $z \sim 4$ quasars respectively, probing scales of $R \lesssim 9\,h^{-1}$ Mpc. We detect an enhancement of galaxies (both LBGs and LAEs) in quasar fields, a positive and strong quasar-galaxy cross-correlation function, consistent with a power-law shape, and a strong galaxy auto-correlation function in quasar fields. The three mentioned results are all indicators that quasars trace massive dark matter halos in the early universe.

Keywords. galaxies: active, galaxies: high-redshift, galaxies: quasars: general, cosmology: large-scale structure of universe, cosmology: early universe.

---

1. Introduction

The strong observed clustering of $z > 3.5$ quasars indicates they are hosted by massive ($M_{\rm halo} \gtrsim 10^{12}\,h^{-1}\,{\rm M}_\odot$) dark matter halos (Shen *et al.* 2007). This should manifest as strong clustering of galaxies around quasars. Previous works on high-redshift quasar environments, mostly focused at $z > 5$, have failed to find convincing evidence for these overdensities. Most of previous works aim to detect overdensities of galaxies around individual or at most a handful of quasars, and the large statistical fluctuations expected from cosmic variance could explain why they have been inconclusive. One strategy for overcoming this complication is to target a large sample of quasars, and focus on measuring the quasar-galaxy cross-correlation function.

2. Galaxy overdensity and quasar-galaxy cross-correlation function

We observed six quasar fields with VLT/FORS1 to search for LBGs at $z = 3.8$. We detected 44 LBGs in quasar fields, while only 28.6 LBGs are expected in the same volume in blank fields (computed using the galaxy luminosity function (LF) at $z \sim 4$ from Ouchi *et al.* 2004b). This implies a LBG overdensity of 1.5 in quasar fields (Garcia-Vergara *et al.* 2017). Additionally, we observed 17 quasar fields with VLT/FORS2 to search for LAEs at $z = 3.9$. We detected 25 LAEs, while only 17.3 LAEs are expected in the same volume in blank fields (computed using the galaxy LF at $z \sim 4$ from Ouchi *et al.* 2008). This implies a LAE overdensity of 1.4 in quasar fields (Garcia-Vergara *et al.* subm.).

We also measure the volume-averaged quasar-galaxy cross-correlation function using both the LBG and LAE samples. For both LBG and LAE, we do detect a strong

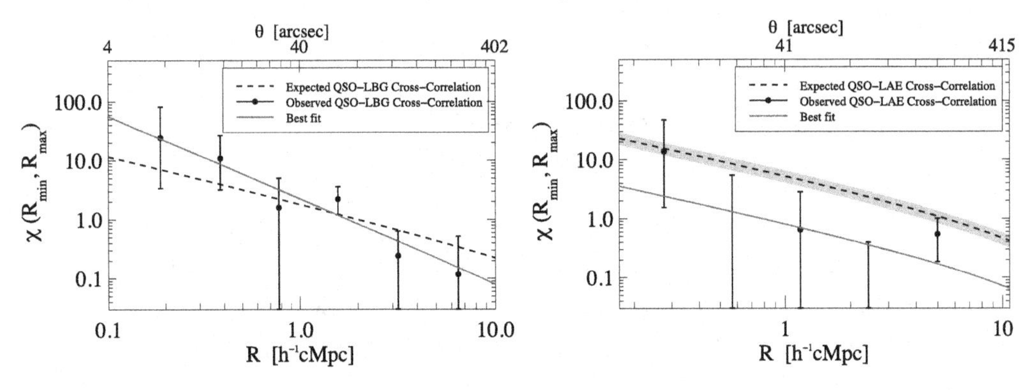

Figure 1. Quasar-galaxy cross-correlation function computed using the LBG sample (left panel) and the LAE sample (right panel). We show our measurement (filled circles) with 1σ Poisson error bars and its best fit (red curve). The dashed black line shows the theoretical expectation computed assuming a deterministic bias model. The gray shaded region in the right panel indicates the 1σ error on the theoretical expectation.

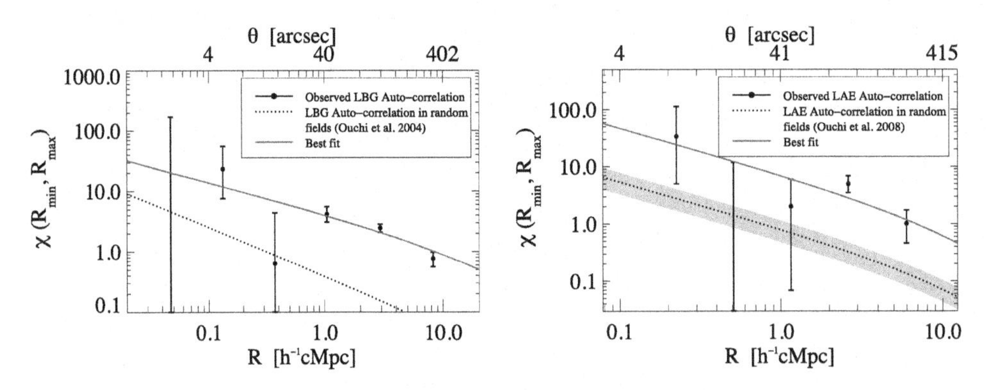

Figure 2. Galaxy auto-correlation function in quasar fields, computed using the LBG sample (left panel) and the LAE sample (right panel). We show our measurement (data points), its best fit (red curve), and the galaxy clustering in blank fields at $z \sim 4$ (dotted black curve).

quasar-galaxy cross-correlation function, consistent with a power-law shape indicative of a concentration of galaxies centered on quasars (see Fig. 1). We compare the observed clustering with the expectation from a deterministic bias model, and find that our measurements are in good agreement in the case of LBG, but fall short of the predicted overdensities by a factor of 2.1 in the case of LAEs. Some possible explanations for this last discrepancy are related with i) the possibility that galaxies in the Mpc-scale quasar environments are on average significantly more dusty, or ii) the possibility of a larger ($R \gtrsim 9\,h^{-1}$ Mpc) scale overdensity in quasar fields (for details see Garcia-Vergara *et al.* subm.).

3. Galaxy auto-correlation function in quasar fields

If quasars reside in overdensities of galaxies, then we expect the galaxy auto-correlation to be enhanced compared with blank fields. We measure the galaxy auto-correlation function in our fields, and compare it with the galaxy clustering in blank fields. For both, LBGs and LAEs we find that galaxies in quasar fields are significantly more clustered compared with their clustering in blank fields (see Fig. 2).

References

Garcia-Vergara, C. *et al.* 2017, *ApJ*, 848, 7
Garcia-Vergara, C. *et al.* 2019, *subm. to the ApJ*, arXiv:1904.05894
Ouchi, M. *et al.* 2004, *ApJ*, 611, 685
Ouchi, M. *et al.* 2004b, *ApJ*, 611, 660
Ouchi, M. *et al.* 2008, *ApJ*, 176, 301
Shen, Y. *et al.* 2007, *AJ*, 133, 2222

SESSION 5: The interstellar medium of high-redshift galaxies

Uncovering Early Galaxy Evolution in the ALMA and JWST Era
Proceedings IAU Symposium No. 352, 2019
E. da Cunha, J. Hodge, J. Afonso, L. Pentericci & D. Sobral, eds.

doi:10.1017/S1743921319009244

The interstellar medium content of galaxies in the ALMA era

Manuel Aravena

Núcleo de Astronomía, Facultad de Ingeniería y Ciencias, Universidad Diego Portales, 8370191 Santiago, Chile
email: manuel.aravenaa@mail.udp.cl

Abstract. The advent of the Atacama Large Millimeter/submillimeter Array (ALMA) has enabled a new era for studies of the formation and assembly of distant galaxies. Cosmological deep field surveys with ALMA and other interferometers have flourished in the last few years covering wide ranges of galaxy properties and redshift, and allowing us to gain critical insights into the physical mechanisms behind the galaxy growth. Here, we present a brief review of recent studies that aim to characterize the interstellar medium properties of galaxies at high redshift ($z > 1$), focusing on blank-field ALMA surveys of dust continuum and molecular line emission. In particular, we show recent results from the ALMA Spectroscopic Survey in the Hubble Ultra Deep Field (ASPECS) large program.

Keywords. galaxies: high-redshift, galaxies: evolution, galaxies: ISM

1. Introduction

One of the most outstanding challenges in galaxy evolution today is to understand how galaxies obtain their cold gas to sustain star formation activity through cosmic time. A significant advance in the last 20 years has been the determination of the cosmic star formation rate (SFR) density out to the end of the Epoch of Reionization ($z \sim 6 - 8$). From this epoch on, the cosmic SFR density rose gradually up to a peak level ($1 < z < 3$), before smoothly declining by an order of magnitude towards the present day ($0 < z < 1$). The period of peak star-formation constitutes the main epoch of galaxy assembly, when roughly half the stars in the Universe were formed (Fig. 1; Madau & Dickinson 2014). However, why the cosmic SFR density follows this evolution is a matter of increasingly active investigation, with various studies aiming to measure the evolution of the cosmic density of molecular gas, as the latter is the fuel for active star formation. Critical questions are related to whether there are fundamental variations in average molecular gas content in galaxies through cosmic time that would produce the observed shape in the cosmic SFR density, or whether this is produced by changes in the molecular gas efficiency (e.g., Decarli *et al.* 2014).

A key finding has been that since early times, $z < 6$, most galaxies display a linear relation between their stellar masses and SFRs, forming what is usually referred to as the main sequence of star-forming galaxies (Fig. 1). Galaxies with SFRs above and below this sequence are respectively termed 'starburst', and 'quiescent' (e.g., Elbaz *et al.* 2007). Most main sequence galaxies resemble clumpy rotating disks, whereas starburst galaxies are typically associated to galaxy collisions and mergers (Daddi *et al.* 2010; Tacconi *et al.* 2013). Main sequence galaxies constitute the population that dominates the cosmic SFR density, and thus they are critical to understand the transformation of galaxies through cosmic time (e.g. Magnelli *et al.* 2012).

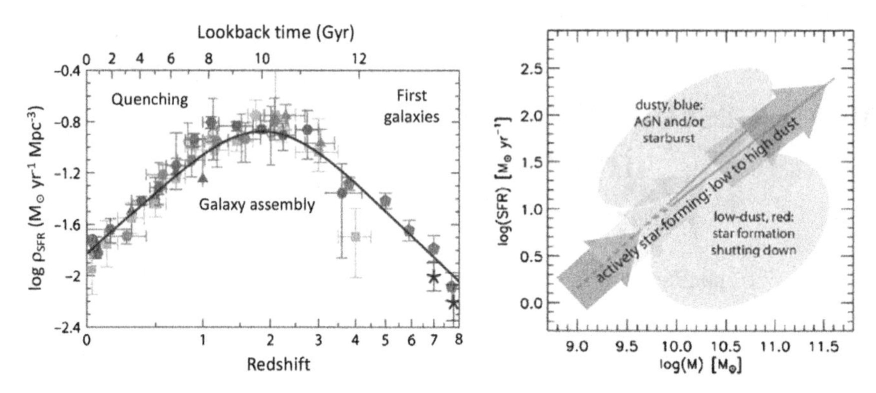

Figure 1. *(Left:)* Evolution of the cosmic SFR density with redshift adapted from Madau & Dickinson (2004). The main epochs for galaxy formation are highlighted. While the rise and fall of star formation with redshift has been known for several years, little is known about the reasons behind this evolution. *(Right:)* Cartoon view of the location of MS galaxies, starburst and quenched galaxies in the SFR vs stellar mass plot (Whitaker *et al.* 2014). Targeted studies have so far focused in follow-up observations of highly star forming galaxies in top part of this diagram.

Efforts to measure the molecular gas content in galaxies at high-redshift have focused on detection of the ^{12}CO line and dust continuum emission, as these are the most direct tracers of the total molecular gas (e.g., Carilli & Walter *et al.* 2013). Dust continuum emission has received particular interest since it is easier to detect observationally and benefits from a negative K-correction out to high redshifts (e.g. Scoville *et al.* 2014). These tracers, however, are not free from uncertainties, since CO and dust measurements rely on assumptions on the CO-to-gas mass conversion factor and the gas-to-dust ratio, respectively, which are dependent on metallicity and environment and thus can vary by factors of a few (even though they can be calibrated under certain conditions). Recent studies have proposed the use of [CI] and even [CII] line emission as tracers for molecular gas mass (Valentino *et al.* 2018; Zanella *et al.* 2018). However, similarly to CO and dust, [CI] determinations require assumptions on the [CI]/H_2 abundance ratio and the use of [CII] as a molecular gas tracer is not so direct.

2. Approaches to detect the ISM at high-z

Significant efforts have been devoted in the last 20 years to conduct blank-field submillimeter surveys over large areas of the sky with bolometer cameras in single dish telescopes (e.g. Smail *et al.* 1997; Hughes *et al.* 1998; Weiss *et al.* 2009). These surveys discovered an important population of dust-rich starburst galaxies at high redshift that contributed a significant fraction of the cosmic SFR density (e.g. Casey *et al.* 2014). These sources, called "submillimeter galaxies" (SMGs), turned out to be inconspicuous at optical wavelengths due to heavy dust obscuration. Although detailed characterization has been difficult, it is relatively clear that SMGs are massive galaxies undergoing episodes of enhanced star formation, typically catalogued as starburst galaxies in the stellar mass versus SFR diagram, with SFRs above 200 $M_\odot$ yr^{-1} and redshifts in the range 1–4.

Since these (sub)millimeter surveys are only able to grasp the brightest dusty star-forming galaxies in the sky, pinpointing the typical population of galaxies at $z \sim 1-3$ with typical SFRs $\sim$50 – 200 $M_\odot$ yr^{-1}, has required targeted observations of dust continuum and molecular gas emission in galaxies that have been pre-selected through their optical colors and/or faint IR emission detected in *Herschel* IR surveys (e.g. Daddi *et al.* 2008, 2010; Dannerbauer *et al.* 2009; Aravena *et al.* 2010, 2012; Saintonge *et al.* 2013;

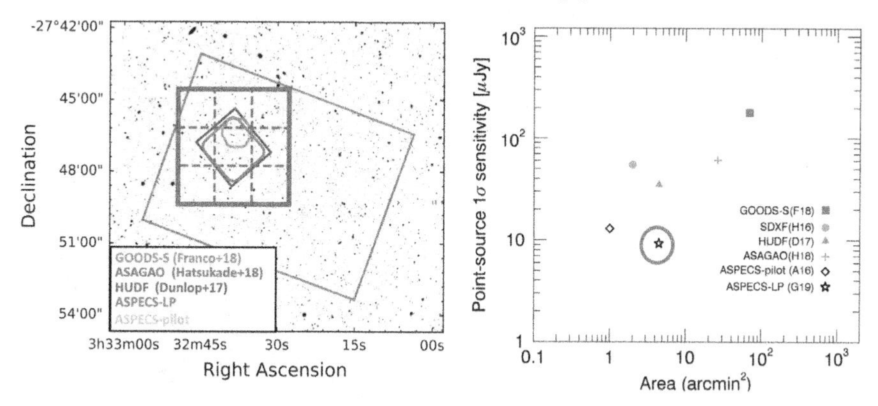

Figure 2. *(Left:)* Relative location of the different ALMA millimeter deep surveys in the GOODS-S field (adapted from Hatsukade et al. (2018). *(Right:)* Comparison of the area and point-source sensitivities achieved by the various ALMA deep field efforts at 1mm. Current surveys have been following a "wedding cake" approach. The red circle highlights the great sensitivity and area covered by the ASPECS large program.

Tacconi *et al.* 2010, 2013, 2018; Bolatto *et al.* 2015; Scoville *et al.* 2017; Freundlich *et al.* 2018). These follow-up programs have been extremely successful to measure molecular gas masses of a large number of galaxies, enabling the determination of observational scaling relations between various interstellar medium (ISM) parameters (gas fractions, efficiencies, distance to the MS, etc) and the establishment of a framework to understand the evolution of galaxies (e.g. Genzel *et al.* 2015; Tacconi *et al.* 2018).

A complementary approach to reach the fainter dusty galaxies has been the use of archival data toward an increasing number of extragalactic pointings and calibrator fields observed with the Atacama Large Millimeter/submillimeter Array (ALMA) (Oteo *et al.* 2015, 2017; Carniani *et al.* 2014; Fujimoto *et al.* 2016, 2017). The main motivation of these projects is to take advantage of the already available data to look for galaxies in the field of the main target of the ALMA observations, which due to the large number of visits of those fields can reach great depths. These surveys have been successful in putting constraints on the (sub)millimeter number count distribution, however, they typically do not have enough multi-wavelength ancillary data to measure the galaxies' properties.

3. Millimeter continuum deep fields

Recent deep millimeter continuum surveys over increasingly large contiguous areas with ALMA toward cosmological deep fields are revolutionizing galaxy evolution studies by accessing the dust emission from the faint star-forming galaxy population.

These surveys have been roughly following a "wedding cake" approach, covering large areas at shallower depths and narrow areas deeper, and concentrating in the 1-mm band due to the efficiency to reach good depth and area coverage, while targeting a sufficiently high part of the Rayleigh-Jeans tail of the dust spectral energy distribution of galaxies at $z > 1$. Figure 2 compares the depth and area achieved by the ALMA millimeter continuum surveys to date. In order of decreasing area, current millimeter continuum surveys include: The ALMA survey of the GOODS-S field (Franco *et al.* 2018), the ALMA twenty-Six Arcmin^2 Survey of GOODS-S at One-millimeter (ASAGAO; Hatsukade *et al.* 2018), the ALMA deep field in SSA22 (Umehata *et al.* 2018), the ALMA 1.3-mm continuum survey of the Hubble Ultra Deep Field (HUDF; Dunlop *et al.* 2017) and The ALMA Subaru-XMM Deep field (SXDF) ALMA 2-arcmin^2 deep field survey (Hatsukade *et al.* 2016).

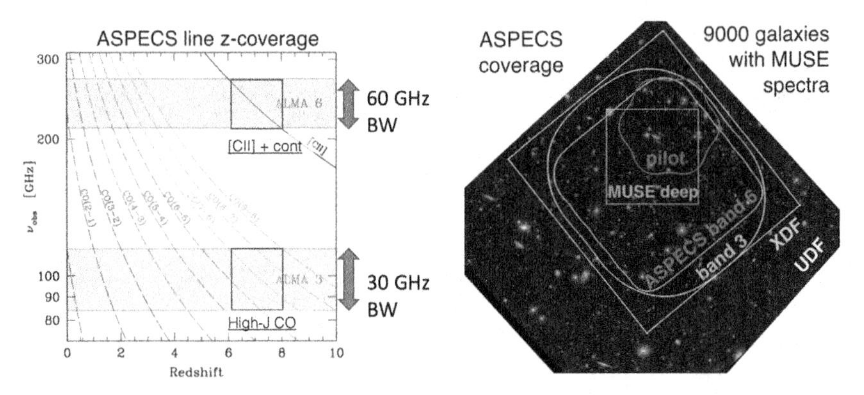

Figure 3. *(Left:)* Frequency and redshift coverage for CO and [CII] line emission for the ASPECS program. The cyan and green bands highlight the frequency coverage in ALMA bands 3 and 6, representing bandwidths (BW) of 30 and 60 GHz, respectively. The yellow boxes represent the coverage for CO/[CII] at $z = 6 - 8$. *(Right:)* Sky coverage of the ASPECS pilot and large programs with respect to the location of the HUDF, XDF and MUSE coverage in the field.

The availability of ancillary data has been key to provide a detailed characterization of the identified galaxies. For example, the galaxies detected by the ALMA 1.3-mm survey of the HUDF, with $S_{1.3\mathrm{mm}} > 120\mu$Jy ($3.5\sigma$), are shown to be massive galaxies mostly located in the main-sequence of star formation with stellar masses $M_{\mathrm{stars}} > 2 \times 10^{10}\ M_{\odot}$ and typical redshifts $z \sim 2$ (Dunlop *et al.* 2017). Similar results are found in other ALMA surveys (Hatsukade *et al.* 2016). The sample of galaxies drawn from these ALMA surveys can thus be seen as galaxies as massive as SMGs, but that are not necessarily undergoing a starburst phase with enhanced SFR. This is exemplified by the $5 - 10\times$ lower 1-mm fluxes observed. Figure 2 shows the location of the various ALMA 1mm deep field surveys so far conducted in the GOODS-S field, and compares the area and point source sensitivities achieved so far.

4. The ALMA Spectroscopic Survey in the HUDF

The ALMA Spectroscopic Survey in the HUDF (ASPECS) project pioneered a parallel, complementary approach to pure continuum millimeter deep field observations, following previous results using the Plateau de Bureau Interferometer (PdBI; Decarli *et al.* 2014; Walter *et al.* 2014) and the Very Large Array (VLA; Aravena *et al.* 2012). By scanning the full ALMA bands 3 and 6 over the ranges 84-115 GHz and 212-275 GHz, the ASPECS observations are able to access the CO and [CII] emission lines in the redshift ranges $z = 0 - 6$ and $6 - 8$, respectively. By collapsing along the frequency axis, great sensitivity in the 3mm and 1mm continuum can be achieved. Figure 3 shows the ASPECS survey design, including the frequency coverage and location with respect to the HUDF area.

4.1. *ASPECS pilot*

The ASPECS pilot program (ALMA cycle-2) covered a 1 arcmin^2 in the northern part of the HUDF, using a single pointing in band 3 and 7-pointing mosaic pattern in band 6. The ASPECS pilot region overlaps significantly with an ultra-deep integration obtained with the VLT/MUSE instrument, providing unique spectroscopic coverage over the redshift range $z = 0 - 6$. The ASPECS pilot 1.2-mm continuum image reached a roughly uniform depth of $\sim 13\mu$Jy over this region. The ASPECS pilot program, while covering a modest area of the sky, probed to be extremely useful to test the techniques necessary and explore que science that could be achieved through these observations. The results from

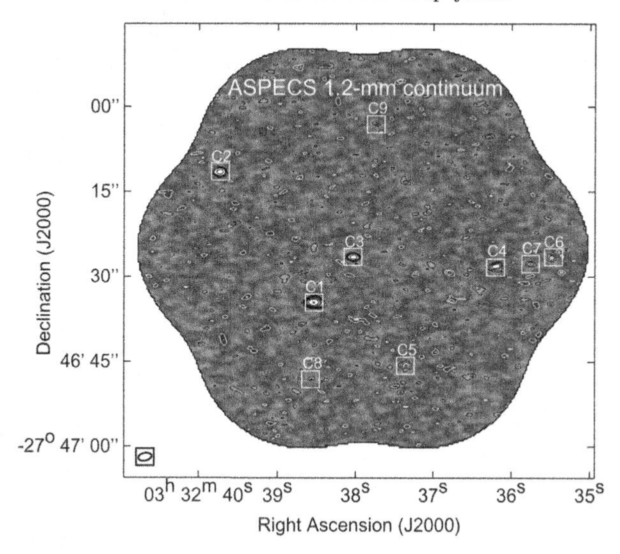

Figure 4. ALMA 1.2-mm S/N image obtained by the ASPECS pilot project in the HUDF (from Aravena *et al.* 2016). Black nd white contours show positive and negative emission, respectively. The boxes show the position of the sources detected with our extraction procedure at S/N$\sim$3.5.

this survey are presented in a series of seven papers, covering a variety of topics related to galaxy evolution including: the line catalog and survey design (Walter *et al.* 2014); the millimeter continuum imaging, number counts and characterization of the faint dusty galaxies (Aravena *et al.* 2016a); determination of the CO luminosity function and evolution of the cosmic molecular gas density (Decarli *et al.* 2016a); characterization of the ISM properties of CO line emitters (Decarli *et al.* 2016b); [CII] line emitting candidates at $z = 6 - 8$ (Aravena *et al.* 2016b); measurements of the infrared excess relation at high redshift (Bouwens *et al.* 2016); and constraints on CO intensity mapping experiments (Carilli *et al.* 2016). Figure 4 shows the ALMA 1.2-mm map obtained by the ASPECS pilot project.

4.2. *ASPECS LP: results from ALMA band 3*

The ASPECS large program (LP; ALMA cycles 4 and 5) built on the pilot observations, expanding the covered area to $\sim$4.5 arcmin^2, comprising roughly the area known as the eXtremely Deep Field (XDF) within the HUDF. This region is fully covered by deep VLT/MUSE observations and the deepest HST observations ever obtained. The ASPECS LP thus complements greatly the legacy value of this field, which will also be covered by Granted Time Observations (GTO) with the James Webb Space Telescope (JWST).

The first results of this survey, based on the ALMA band 3 observations are presented in 5 recent papers (González-López *et al.* 2019; Decarli *et al.* 2019; Aravena *et al.* 2019; Boogard *et al.* 2019; Popping *et al.* 2019).

In González-López *et al.* (2019), the full suite of techniques for line detection, cube search, and fidelity and completeness measurements were scrutinized in detail through comparison of different tools/algorithms, yielding a sample of 16 statistically significant CO line emitters with reliable counterparts. The ASPECS 3mm continuum map yielded a sample of 6 sources, also detected at 1mm, providing the tightest contraints on the 3mm number counts to date. Interestingly, the identified counterparts to the 3mm sources seem to lie at significantly high redshifts than the population of 1mm selected sources (e.g. from Aravena *et al.* 2016).

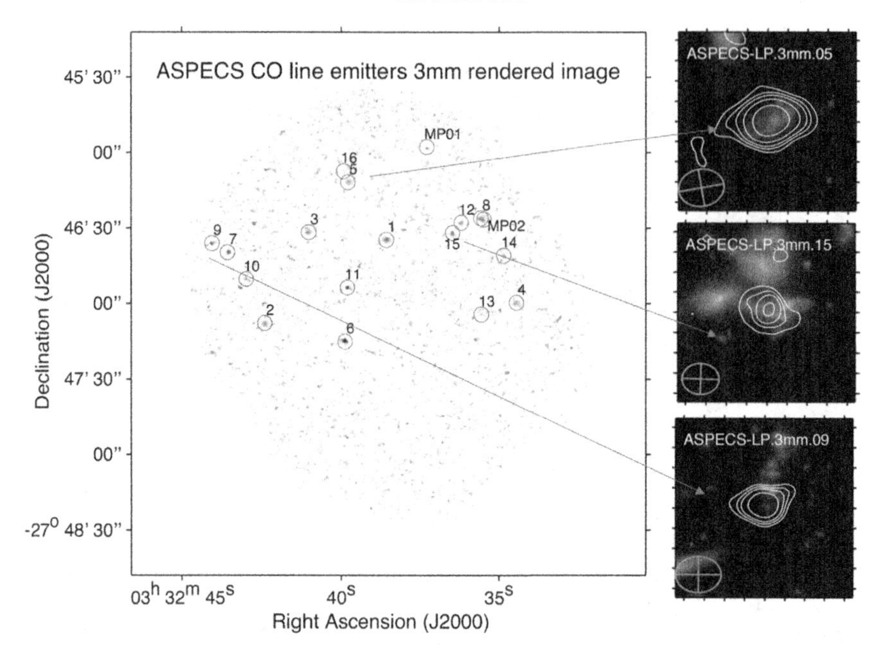

Figure 5. *Left:)* Rendered CO image obtained by the ASPECS LP toward the HUDF, obtained by co-adding the individual average CO line maps around the bright CO-selected galaxies and the 2 lower significance MUSE-based CO sources (adapted from Aravena *et al.* 2019). (*Right:*) HST cutouts with CO contours overlaid on three of the ASPECS sources. These are shown to exemplify the variety of optical morphologies and brightness of the sources being identified. Source ID9 would not have been pre-selected by targeted follow-up programs.

The properties of the CO-selected galaxies discovered by the ASPECS survey are presented in Boogaard *et al.* (2019) and (Aravena *et al.* 2019). Results from the former are discussed in a paper in this proceeding series. Figure 5 shows the blank-field CO line detections in the ASPECS LP field, and highlight the optical properties observed in a few examples. In short, Boogaard *et al.* (2019) report the search for optical/near-IR counterparts in the HST images, determination of MUSE/VLT spectroscopic redshifts, and derivation of physical properties based on spectral energy distribution (SED) fitting using the MAGPHYS code (da Cunha *et al.* 2008). Where available, optical emission lines are used to measure metallicities for the ASPECS galaxies. Using the measured SFRs and stellar masses, Boogaard *et al.* (2019) study in detail the location of the ASPECS CO-selected galaxies with respect to the main-sequence of star formation. Furthermore, they find that a significant fraction of the CO selected galaxies are X-ray sources, suggesting an important presence of AGN activity.

Through these line search measurements, it is possible to quantify the abundance of CO line emitters as a function of CO luminosity. Decarli *et al.* (2019) provided the key constraints on the CO luminosity function at different redshifts, which can thus be converted into measurements of the cosmic molecular gas density as a function of redshift (Fig. 6). These measurements were found to be in excellent agreement with the parallel results from the CO Luminosity Density z (COLDz) survey based on VLA low-J CO observations in the COSMOS field (Pavesi *et al.* 2018; Riechers *et al.* 2019), and provided the tightest constraints to date on the evolution of the molecular gas density. Most importantly, the shape of the cosmic molecular gas density is found to decrease by a factor of $\sim$6.5$\times$ from $z=1$ to $z\sim 0$, in overall consistency with the observed decline in the cosmic SFR density in this redshift range. This implies that the shape of the cosmic

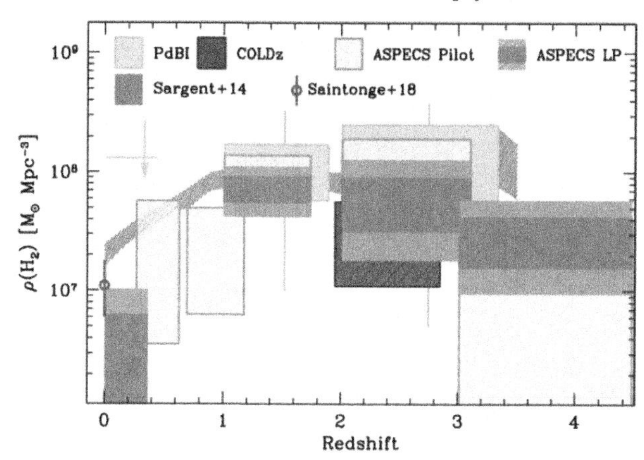

Figure 6. Cosmic density of molecular gas (H_2) derived by the ASPECS LP survey, compared to previous measurements (from Decarli *et al.* 2019). A rise and fall of the molecular gas content in galaxies is measured, decreasing by a factor of 6.5 from $z = 2.5$ to 0, similar to the evolution of the cosmic SFR density.

SFR density is likely produced by a similar evolution in the molecular gas reservoirs, and very likely not due to changes in the star formation efficiencies. These results are found to be in agreement with the previously observed changes in the molecular gas fractions with cosmic time (e.g. Daddi *et al.* 2008, 2010; Tacconi *et al.* 2010, 2013; 2016; Saintonge *et al.* 2013; Schinnerer *et al.* 2016; Scoville *et al.* 2017.

Popping *et al.* (2019) compare the results from the ASPECS CO properties with that of two cosmological galaxy formation models, from the IllustrisTNG hidrodynamical simulations and the Santa Cruz semi-analytical model. This study finds that the predicted molecular gas mass of galaxies at $z > 1$ as a function of stellar mass is $2 - 3\times$ lower than the observations, with the models not being able to reproduce the number of molecular gas rich galaxies detected with ASPECS. As such the ASPECS observations are found to be in tension with these models.

In complement to this study, Aravena *et al.* (2019) studied the molecular gas properties from the CO-selected galaxies. They find that the ASPECS galaxies generally follow the ISM scaling relations in terms of the evolution of the molecular gas fraction and gas depletion timescales as a function with redshift (Fig. 7). They find that a good fraction of the CO-galaxies are located above and below the MS of star formation given the CO-selected nature of the survey. Finally, they find that main sequence galaxies contribute the dominant contribution to the cosmic molecular gas density at all redshifts proven, with an increasing/decreasing contribution from starburst/passive galaxies at increasing redshift (from $z \sim 1$ to $z \sim 2.5$).

4.3. *ASPECS LP: results from ALMA band 6*

The ALMA band 6 observations reached a uniform depth of 9.3μJy per beam in the 1.2-mm continuum map, which is unprecedented for an area of ~4.5 arcmin^2. Many of the results of this part of the ASPECS program, mostly based on the ultra-deep 1.2-mm continuum map, have been submitted for publication in Journals as of this writing.

González-López *et al.* (2019b) finds a sample of 32 significant continuum sources in this area, with additional 26 sources found using a optical counterpart priors. Measurements of the number counts as a function of 1.2-mm flux density show that there is a clear flattening at faint fluxes (below 100μJy), which indicates that the observations are able to resolve most (~90%) of the Extragalactic Background Light at 1.2-mm in the HUDF.

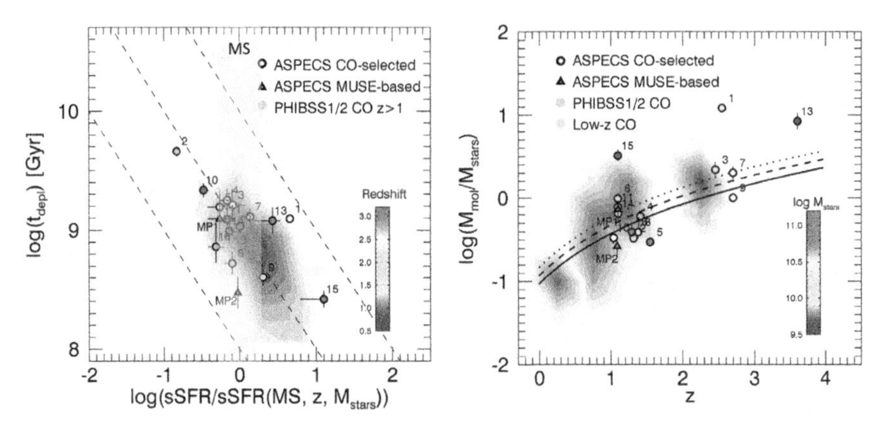

Figure 7. *(Left:)* Relationship between the molecular gas depletion timescales and normalized sSFR (offset from the MS) for the ASPECS CO galaxies compared to literature galaxies at $z > 1$ (from Aravena *et al.* 2019). The location of the MS is highlighted. ASPECS galaxies seem to broadly follow the well established scaling relations, yet several galaxies fall above and below the MS. *(Right:)* Evolution of the molecular gas fraction with redshift for the ASPECS galaxies compared to literature galaxies (from Aravena *et al.* 2019). The steady decrease of the molecular gas fraction by a factor of roughly an order of magnitude from $z = 2$ to 0 is well supported by the CO-selected ASPECS sample.

Interestingly, splitting the number counts in bins of stellar mass, SFR, dust mass, and redshift allowed for measurements of the population that dominates the actual aggregate cosmic dust emission (Gonzalez-Lopez *et al.* 2019b; Popping *et al.* 2019b). The comparison of these observed number counts with models that use simple assumptions about scaling relations and ISM properties (from Popping et al., 2019b) shows a remarkable agreement. Characterization of the physical properties of the dust continuum detected sources shows again a variety, from starburts to passive galaxies and confirm previous ASPECS pilot measurements, yielding a median redshift for this population of $z \sim 1.5$ (Aravena *et al.* 2019b).

Using stacking analysis on the 1.2-mm map, Magnelli *et al.* (2019) measured the evolution of the dust mass function with redshift for galaxies in the ASPECS footprint. They find that the observations are able to probe deep, below the "knee" of the dust mass function. By combining the aggregate contribution from different stellar mass ranges, Magnelli et al. are able to recover the evolution of the cosmic density of dust mass. After assuming a metallicity-dependent gas-to-mass ratio, they furthermore measure the cosmic density of molecular gas. Their results are found to be in good agreement with the CO-based estimates of the cosmic molecular gas density.

5. Concluding remarks

• In the last 20 years, we have learnt significantly from blank-field deep submm surveys and targeted observations of star-forming galaxies in dust and CO line emission.

• While these studies have been fundamental to provide a foundational framework for galaxy evolution and the transformation of gas to stars in galaxies, they might not be encapsulating the full picture.

• A complementary approach is to reach faint SFGs by conducting deep ALMA of surveys of dust continuum and molecular line emission in cosmological deep fields, with an increasing number of these surveys to date and on going.

• Among these surveys, the ASPECS pilot and large programs are proving to provide unique legacy value in the HUDF, in the era of JWST.

• The observed ASPECS CO-based sources are found to have a diversity of ISM properties, but consistent with "scaling relations".
• Molecular gas density measurements based on ASPECS observations are found to be consistent with the evolution of cosmic SFR density.
• Ultra deep 1mm continuum ASPECS observations are able to place tight constraints on 1-mm number counts, and confirm a flattening of the 1-mm number counts at faint fluxes.

References

Aravena, M., Carilli, C., Daddi, D. A., *et al.* 2010, *ApJ*, 718, 177
Aravena, M., Carilli, C. L., Salvato, M., *et al.* 2012, *MNRAS*, 426, 258
Aravena, M., Hodge, J. A., Wagg, J., *et al.* 2014, *MNRAS*, 442, 558
Aravena, M., Decarli, R., Walter, F., *et al.* 2016, *ApJ*, 833, 68
Aravena, M., Decarli, R., Walter, F., *et al.* 2016, *ApJ*, 833, 71
Aravena, M., Decarli, R., Gónzalez-López, F., *et al.* 2019, *ApJ* accepted, arxiv:1903.09162
Bolatto, A. D., Warren, S. R., Leroy, A. K., *et al.* 2015, *ApJ*, 809, 175
Boogaard, L. A., Decarli, R. González-López, J., *et al.* 2019, *ApJ* accepted, arxiv:1903.09167
Bouwens, R. J., Aravena, M., Decarli, R., *et al.* 2016, *ApJ*, 833, 72
Carilli, C. L. & Walter, F. 2013, *ARA&A*, 51, 105
Carilli, C. L., Chluba, J., Decarli, R., *et al.* 2016, *ApJ*, 833, 73
Carniani, S., Maiolino, R., De Zotti, G., *et al.* 2014, *A&A*, 584, A78
da Cunha, E. Charlot, S., Elbaz, D., *et al.* 2008, *MNRAS*, 388, 1595
Daddi, E., Dannerbauer, H., Elbaz, D., *et al.* 2008, *ApJ*, 673, 21
Daddi, E., Bournaud, F., Walter, F., *et al.* 2010 2010, *ApJ*, 713, 686
Dannerbauer, H., Daddi, E., Riechers, D. A., *et al.* 2009, *ApJ*, 698, 178
Decarli, R., Walter, F., Carilli, C., *et al.* 2014, *ApJ*, 782, 68
Decarli, R., Walter, F., Aravena, M., *et al.* 2016, *ApJ*, 833, 69
Decarli, R., Walter, F., Aravena, M., *et al.* 2016, *ApJ*, 833, 70
Decarli, R., Walter, F., Gónzalez-L'opez, J., *et al.* 2019, *ApJ* accepted, arXiv:1903.09164
Dunlop, J. S., McLure, R. J., Biggs, A. D., *et al.* 2017, *MNRAS*, 466, 861
Elbaz, D., Daddi, E., Le Borgne, D., *et al.* 2007 , *A&A*, 468, 33
Franco, M., Elbaz, D., Béthermin, M., *et al.* 2018, *A&A*, 620, 152
Freundlich, J., Combes, F., Tacconi, L. J., *et al.* 2019, *A&A*, 622, A105
Fujimoto, S., Ouchi, M., Ono, Y., *et al.* 2016, *ApJS*, 222, 1
Fujimoto, S., Ouchi M., Shibuya, T., *et al.* 2017, *ApJ*, 850, 83
González-López, J., Decarli, R., Pavesi, R., *et al.* 2019, *ApJ* accepted, arxiv:1903.09161
Genzel, R., Tacconi, L. J., Lutz, D., *et al.* 2015, *ApJ*, 800, 20
Hatsukade, B., Kohno, K., Umehata, H., *et al.* 2016, *PASJ*, 68, 36
Hatsukade, B., Kohno, K., Yamaguchi, Y., *et al.* 2018, *PASJ*, 70, 105
Hughes, D. H., Serjeant, S., Dunlop, J., *et al.* 1998, *Nature*, 394, 241
Madau, P. & Dickinson, M. 2004 , *ARA&A*, 52, 415
Magnelli, B., Saintonge, A., Lutz, D., *et al.* 2012 , *A&A*, 548, A22
Oteo, I., Zwaan, M. A., Ivison, R. J., Smail, I., & Biggs, A. D. 2016, *ApJ*, 822, 36
Pavesi, R. Sharon, C. E. Riechers, D. A., *et al.* 2018, *ApJ*, 864, 49
Popping, G., Pillepich, A., Somerville, R. S., *et al.* 2019, *ApJ* accepted, arxiv:1903.09158
Riechers, D. A., Pavesi, R. Sharon, Chelsea E., *et al.* 2019, *ApJ*, 872, 7
Saintonge, A., Lutz, D., Genzel, R., *et al.* 2013, *ApJ*, 778, 2
Scoville, N., Lee, N., Vanden Bout, P., *et al.* 2015, *ApJ*, 837, 150
Smail, I., Ivison, R. J., & Blain, A. W. 1997, *ApJL*, 490, 5
Tacconi, L. J., Neri, R., Genzel, R., *et al.* 2013, *ApJ*, 768, 74
Tacconi, L. J., Genzel, R., Neri, R., *et al.* 2010, *Nature*, 463, 781
Tacconi, L. J., Genzel, R., Saintonge, A., *et al.* 2018, *ApJ*, 853, 179
Umehata, H. Hatsukade, B., Smail, I., *et al.* 2018, *PASJ*, 70, 65

Valentino, F., Magdis, G. E., Daddi, E., *et al.* 2018, *ApJ*, 869, 27
Walter, F., Decarli, R., Sargent, M., *et al.* 2014, *ApJ*, 782, 79
Walter, F., Decarli, R., Aravena, M., *et al.* 2016, *ApJ*, 833, 67
Weiss, A., Kovács, A., Coppin, K., *et al.* 2009, *ApJ*, 707, 1201
Whitaker, *et al.* 2014, *ApJ*, 795, 20
Zanella, A., Daddi, E., Magdis, G., *et al.* 2018, *MNRAS*, 481, 2

Uncovering Early Galaxy Evolution in the ALMA and JWST Era
Proceedings IAU Symposium No. 352, 2019
E. da Cunha, J. Hodge, J. Afonso, L. Pentericci & D. Sobral, eds.

doi:10.1017/S1743921319009232

Galactic outflows at high spatial resolution via gravitational lensing

Justin Spilker

University of Texas at Austin, 2515 Speedway Stop C1400, Austin, TX 78712, USA
email: `spilkerj@gmail.com`

Abstract. The completion of the Atacama Large Millimeter/submillimeter Array (ALMA) has led to the ability to make observations with unprecedented resolution at sub-millimeter wavelengths, allowing novel probes of the ISM and kinematics of high-redshift galaxies. Because they are magnified by foreground galaxies or clusters, gravitationally lensed galaxies allow the highest possible spatial resolution to be obtained, and/or a sharp reduction in the observing time required to detect faint objects or spectral lines. These benefits have made lensed galaxies useful benchmark systems for ALMA, enabling a wide variety of science cases. Here I focus in particular on spatially-resolved observations of massive galactic outflows in the very distant $z > 4$ universe, summarizing plausible tracers of the cold molecular phase of these outflows. The prospects of joint *JWST* and ALMA observations will be revolutionary, including the chance to take a full census of galactic outflows in multiple gas phases at matched spatial resolution.

Keywords. galaxies: high-redshift, galaxies: formation, galaxies: ISM

1. Galactic Feedback and Outflows

Self-regulating galactic feedback is a critical component of galaxy evolution, yet one that has remained difficult to probe observationally. For simple models in which only gravitation plays an important role in the growth of galaxies, virtually all baryonic mass collapses into stars very early in the history of the universe, yielding a population of galaxies far more massive and older than those observed in the real universe. Some energetic processes must be invoked to limit the growth rate of galaxies, by heating, depleting, expelling, or increasing the turbulence in the gas from which stars form. Energy and momentum from active galactic nuclei (AGN), Type II supernovae, and/or massive star winds and radiation pressure are in general capable of providing the necessary inputs to prevent the over-production of extremely massive galaxies.

Aside from limiting the overall mass of galaxies, feedback is commonly invoked to explain a wide variety of observed galaxy properties, ranging from the creation and maintenance of the red sequence, the stellar mass to halo mass relation, the correlation between supermassive black hole mass and stellar velocity dispersion, the mass-metallicity relation, the normalization of the Kennicutt-Schmidt star formation 'law', and the presence of vast quantities of metal-enriched gas in the circumgalactic medium surrounding galaxies (e.g. Fabian 2012; Hopkins *et al.* 2014; Kormendy & Ho 2013; Werk *et al.* 2014; Kennicutt & Evans 2012).

From a theoretical standpoint, feedback has been difficult to implement in hydrodynamical simulations as the spatial and temporal scales of the energy/momentum deposition are much smaller than the typical resolution limit of simulations. At the high-mass end, high-resolution hydrodynamical simulations currently favor dramatic processes such as merger-induced starbursts and subsequent supermassive black hole accretion to provide enough energy to heat and expel the gas supply and stave off further star formation (e.g. Hopkins *et al.* 2008; Ceverino *et al.* 2015; Wellons *et al.* 2015). Feedback also plays an important role for less extreme galaxies. Recent simulation work with novel prescriptions for feedback yields galaxies with highly time-variable star formation rates (SFRs) as the star-forming gas is temporarily disrupted before being able to cool and resume forming stars (e.g. Hopkins *et al.* 2014).

Massive galactic winds or outflows are a nearly-ubiquitous signature of feedback in nearby and intermediate-redshift galaxies. One challenge is that outflows are inherently multiphase in structure, spanning five or more orders of magnitude in gas temperature and density even within a single galactic wind (e.g., Scannapieco 2013; Leroy *et al.* 2015; Schneider & Robertson 2017). Powerful outflows have been observed in hot ionized plasma, 'warm' ($T \sim 10^4$ K) ionized gas (e.g. Förster Schreiber *et al.* 2014; Genzel *et al.* 2014), and cooler neutral atomic gas (e.g. Shapley *et al.* 2003; Rupke *et al.* 2005). Most intriguing, however, has been the discovery of ubiquitous massive *molecular* winds in low-redshift galaxies, as this may indicate that feedback is capable of suppressing star formation by acting directly on the gas from which stars form (e.g., Feruglio *et al.* 2010; Veilleux *et al.* 2013; Cicone *et al.* 2014).

Because the spectral signatures of outflows are very faint, few detections of cold atomic/molecular outflows exist at high redshifts (e.g., Cicone *et al.* 2015; Spilker *et al.* 2018). Nevertheless, the $z > 1$ era represents an important epoch in the history of the universe, probing the early and rapid formation of the first massive galaxies, the creation of the first passive 'red-and-dead' objects, and the peak epoch of cosmic star formation (e.g. Straatman *et al.* 2014; Madau & Dickinson 2014). The completion of ALMA now allows for cold galactic winds to be detected and in some cases spatially resolved into the distant universe, while JWST will enable detections of the warm ionized phase of outflows beyond $z = 2$.

2. Plausible Long-Wavelength Outflow Tracers for the High-Redshift Universe

Based on lessons learned from extensive observational investment in the nearby universe, a number of plausible outflow tracers are accessible by ALMA in the distant universe. In most cases it is the vastly increased sensitivity of ALMA (rather than spatial resolution or receiver frequency coverage) that makes these tracers observable. I also note that most of these tracers are also accessible to other sub-mm/mm interferometers such as the SMA and NOEMA, although the lower sensitivity of these facilities makes detections more difficult. This is not an exhaustive list of possible outflow tracers, instead only including those most commonly observed.

Hydroxyl (OH). Far-infrared lines of OH are perhaps the best and most reliable tracer of molecular outflows available. Compared to the other available tracers of molecular outflows, OH offers several advantages (as well as some disadvantages). First, the presence of winds is manifest through blueshifted absorption features against the bright dust continuum of the host galaxy, an unambiguous signature of outflowing material. This is particularly relevant in the high-redshift universe, where the violent processes of galaxy accretion and assembly can lead to emission at high velocities relative to the systemic that is unrelated to galactic outflows (e.g. accreting subhalos or major and minor mergers; Narayanan *et al.* 2015). Second, its ground-state transitions that trace the bulk of the molecular material lie at rest-frame wavelengths near the peak of the dust spectral energy distribution, $\sim$100 μm. This allows for observations of a given duration to reach better levels of contrast with the continuum compared to longer-wavelength molecular lines, the relevant quantity for absorption line work. Finally, due to its high dipole moment, the absorption line transitions of OH are strong and sensitive even to relatively small amounts of molecular material. For example, the timescale for spontaneous emission of the ground-state 119 μm transitions of OH is ~100 s, compared to ~10^7 s for carbon monoxide. This ensures that the vast majority of the OH molecules are in the ground state and available to absorb continuum photons.

The strength of its transitions is also one of the primary drawbacks of OH – the lines become optically thick even for low column densities of gas, requiring additional lower-opacity lines to be observed to improve estimates of, e.g., the mass outflow rate. Another drawback of OH is that the transitions most useful for observing outflows are at short rest-frame wavelengths (<120 μm), and so can only be observed by ground-based interferometers at redshifts $z > 2$ (and practically only $z > 4$ given weather requirements for high-frequency work). Future far-IR facilities such as the *Origins Space Telescope* will be able to detect these lines in virtually every star-forming galaxy out to $z \sim 3$.

The *Herschel*/PACS instrument observed multiple transitions of OH in nearby quasar host galaxies and Ultra/Luminous Infrared Galaxies (U/LIRGs), finding evidence for ubiquitous outflows (e.g. Veilleux *et al.* 2013; González-Alfonso *et al.* 2017). The high detection rate, ~70% of sources with detected outflows, implies that both the intrinsic occurrence rate *and* the covering fraction / opening angle of these outflows must be very high (or they could not be detected in absorption). Currently the most distant unambiguous outflow is presented by Spilker *et al.* (2018) in a highly dust-obscured star-forming galaxy at $z = 5.3$, discussed in more detail below.

Carbon Monoxide (CO). Lines of CO have long been used as tracers of molecular gas in galaxies because it is both a very abundant molecule and its transitions are easily accessible in the local universe with ground-based millimeter telescopes (e.g. Wilson *et al.* 1970). For CO the outflow signature manifests as excess emission at high relative velocities to the host systemic velocity. In the nearby universe where it is possible to resolve the CO emission from galaxies at high resolution, it is also possible to distinguish the emission associated with outflows kinematically, as the outflowing material is not expected to follow the predominant galactic kinematics (e.g. launched from the minor axis, perpendicular to the overall rotation curve). The CO emission associated with outflows is typically >10× fainter than the galaxy itself, requiring very deep observations in order to detect and characterize outflows this way. A compilation of local-universe CO outflow detections can be found in Fluetsch *et al.* (2019). Reassuringly, the results from CO observations of outflows generally agree with those from OH despite very different assumptions and modeling approaches (Lutz *et al.* in prep.).

At high redshift CO remains a difficult tracer of outflows. The ground-state CO(1–0) transition is generally not accessible, requiring observations to target a more highly-excited transition that may not trace the bulk of the molecular material. Like OH, transitions of CO are also optically thick, requiring an estimate of a conversion factor to translate the observed luminosity to an outflowing mass (the infamous CO-H_2 conversion factor, see Bolatto *et al.* 2013 for a review). Nevertheless, some recent studies suggest detections of high-velocity CO emission possibly signaling molecular outflows (e.g. Brusa *et al.* 2018; Herrera-Camus *et al.* 2019). Future facilities such as the next generation Very Large Array (ngVLA) should be able to detect low-excitation transitions of CO in outflows for 'normal' star-forming galaxies out to $z > 3$ (Spilker & Nyland 2018).

Ionized Carbon (C^+). As typically the brightest far-IR or millimeter line in galaxies, the 158 μm transition of [CII] has become a workhorse for high-redshift observations, detectable by ALMA even in 'normal' galaxies with star formation rates <100 $M_\odot$/yr in a few hours or less. [CII] emission arises in gas with a wide variety of temperatures and densities, including [HII] regions, the largely-neutral ISM, and the outer parts of molecular clouds. This also means that it is difficult to disentangle the contributions to the total luminosity from any of these individual sources; outflows detected in [CII] likely trace some mixture of both cool atomic and molecular gas in the winds. As with CO, the signature of outflows from [CII] is excess emission in high-velocity wings of the line profile that is much fainter than the host galaxy. Janssen *et al.* (2016) present an analysis of *Herschel* observations, comparing outflow properties from [CII] with those from OH absorption in the same galaxies, finding that nearly all galaxies would be characterized as showing outflow in both tracers and with a consistent outflow mass from both tracers.

At high redshift the use of [CII] as an outflow tracer remains controversial. Cicone *et al.* (2015) present PdBI observations of a $z = 6.4$ quasar that shows tentative evidence of [CII] wings extending up to $1400\,\mathrm{km\,s^{-1}}$ relative to the galaxy, although this line emission therefore fills nearly the entire PdBI bandpass making continuum level estimation more difficult. The advent of far more sensitive ALMA observations has not settled this debate, with different groups finding evidence either for or against [CII] outflows from stacked spectra of the same sample of galaxies (Decarli *et al.* 2018; Gallerani *et al.* 2018), where the former strongly rules out an outflow with the same brightness as that claimed by Cicone *et al.* (2015); none of the sources individually show evidence for outflows. It is also interesting to note that we do not detect an outflow in [CII] in any of the objects discussed further below despite very high signal-to-noise and clear and obvious outflows seen in OH. At present, then, the evidence that [CII] can or should be used as an outflow tracer is mixed at best – [CII] does not appear to be a reliable outflow tracer.

3. Spatially-Resolved Molecular Winds from the South Pole Telescope Survey

The South Pole Telescope collaboration has been conducting a survey of molecular outflows traced by OH in a sample of $z > 4$ gravitationally lensed dusty, star-forming galaxies. Due to its long selection wavelength and flux cutoff that mostly selects lensed sources, the redshifts of this sample are typically much higher than 'traditional' single-dish $870\,\mu$m-selected sources, with $z_{\mathrm{median}} \sim 4$ extending to $z \approx 7$ (Weiß *et al.* 2013; Béthermin *et al.* 2015; Strandet *et al.* 2016; Marrone *et al.* 2018). The brightness of these sources and the availability of pre-existing gravitational lens models (Spilker *et al.* 2016) makes this sample ideal for detections of molecular outflows via OH absorption.

The first object from this sample was recently published (Figure 1, Spilker *et al.* 2018), and this detection currently stands as both the highest-redshift and only spatially-resolved molecular outflow in the distant universe. In that work, we found an outflow that reaches speeds up to $800\,\mathrm{km\,s^{-1}}$ relative to the galaxy's systemic redshift (determined from [CII]). A lensing reconstruction showed that the outflow covers a large fraction of the source $\approx$80%, in rough agreement with the overall (spatially unresolved) detection fraction of low-redshift U/LIRGs from *Herschel*. The lensing analysis also showed that while the continuum emission, corresponding to obscured star formation in the galaxy, was fairly smooth, the outflowing material showed signs of clumpiness on kiloparsec scales. From simple assumptions about the OH optical depth and outflow geometry, we determined that the outflow rate of molecular material alone was within a factor of 2 of the star formation rate, even without including any contributions from warmer neutral or ionized gas. This implies that outflows can be efficient at removing the star-forming gas from high-redshift galaxies, and possibly links these highly star-forming galaxies to early massive quiescent galaxies observed by $z \sim 4$ (Straatman *et al.* 2014).

We have subsequently been working to extend this work to a large sample of $z > 4$ galaxies with spatially-resolved outflows (Spilker *et al.* in prep.). Even before any lensing analysis it is already clear that outflows must be nearly ubiquitous among dusty star-forming galaxies at high redshift – we have detected an unambiguous outflow in nearly

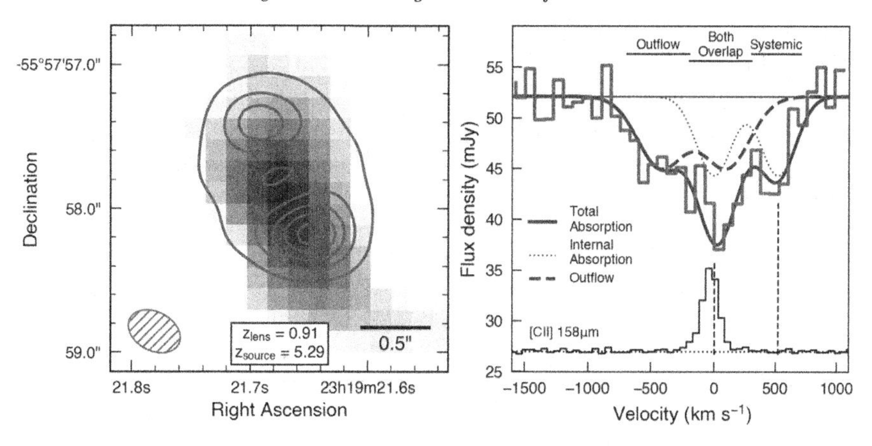

Figure 1. ALMA observations of the molecular outflow in the $z = 5.3$ galaxy SPT2319-55. *Left:* ALMA rest-frame 119 μm dust continuum observations (red contours) on a K-band image of the foreground lensing galaxy (grayscale). *Right:* The apparent spectrum (not corrected for lensing magnification) of the OH 119 μm doublet transitions, with vertical dashed lines indicating the velocities of the two doublet components. A [CII] spectrum is also shown with arbitrary scaling as an indication of the line width of this galaxy; we seen no evidence for [CII] emission associated with the outflow. The spectrum is fit with two components representing systemic absorption from gas within the galaxy as well as a strongly blueshifted component representing the molecular outflow, which reaches velocities up to $\approx$800 km s^{-1}. Figure adapted and reproduced with permission from Spilker *et al.* (2018).

every source thus far observed. As with the low-redshift *Herschel* results, this again must imply that the occurrence rate of molecular winds must be nearly unity. The fact that the outflows are detected in absorption also implies that either the opening angle of these outflows must also be near unity, or that the outflows typically exhibit some preferred geometry such that they are capable of absorbing photons along nearly all lines of sight (e.g., spherically expanding from the nuclear regions). Lensing reconstructions should be able to determine which of these possibilities is correct, if not both.

4. Future Prospects: ALMA and JWST

The outflows described in the previous section were nearly trivial to detect and spatially resolve with ALMA – no source was observed for longer than 30min. Thus it should be possible to push the study of outflows from high-redshift sources to either intrinsically less-luminous (but still lensed) sources, and/or to unlensed (but intrinsically very luminous) sources with observation times of a few hours per source or less.

Our observations only reached spatial resolutions of $\approx$0.3–0.5″, nowhere near the maximum possible spatial resolution achievable by ALMA. Future directions could thus include detecting outflows at low spatial resolution and then resolving them on scales of a few hundred parsecs in order to determine the detailed outflow geometry. This is of particular interest for two main reasons. First, while it is possible to directly follow the mass in outflows in simulations, it is very difficult to measure accurate outflow rates from observations – every available tracer requires numerous assumptions in order to translate observed quantities into outflow masses or rates. The structure of outflows, on the other hand, can be determined in both observations and simulations and directly compared. Second, simulations typically find it extremely challenging to produce fast molecular outflows as observed, because the molecular gas is rapidly shredded by hydrodynamical instabilities long before it can be accelerated to the observed velocities (e.g. Schneider *et al.* 2018). This has led some to suggest that the molecular material in outflows has possibly cooled and condensed from a hotter wind at large radii, after it has already been accelerated (e.g. McCourt *et al.* 2018; Richings & Faucher-Giguère 2018). These two scenarios lead to very different predictions for the detailed structure of molecular outflows: in the former case the molecular gas is concentrated in the few small clumps that were dense enough to avoid destruction, while the latter case predicts a far more uniform 'fog' of parsec-scale molecular droplets that would be unresolvable by any observation.

The advent of the *James Webb Space Telescope* will also prove revolutionary for the study of galactic outflows at very high redshifts. *JWST* will allow for the first observations of the multi-phase nature of outflows at $z > 4$ through deep integral field observations of Hα emission. Hα has been shown to be a good tracer of the warm ionized phase of outflows out to $z \sim 2$, but is redshifted beyond the accessible atmospheric windows for the very early universe. The high spatial resolution, sensitivity, and wavelength coverage of the NIRSPEC instrument's integral field mode will allow observations of both the ionized and molecular phases of outflows on matched spatial scales, which has thus far only been possible for very nearby galaxies.

References

Béthermin, M., De Breuck, C., Sargent, M., & Daddi, E. 2015, *A&A*, 576, L9
Bolatto, A. D., Wolfire, M., & Leroy, A. K. 2013, *ARA&A*, 51, 207
Brusa, M., Cresci, G., Daddi, E., *et al.* 2018, *A&A*, 612, A29
Ceverino, D., Dekel, A., Tweed, D., & Primack, J. 2015, *MNRAS*, 447, 3291
Cicone, C., Maiolino, R., Sturm, E., *et al.* 2014, *A&A*, 562, A21
Cicone, C., Maiolino, R., Gallerani, S., *et al.* 2015, *A&A*, 574, A14
Decarli, R., Walter, F., Venemans, B. P., *et al.* 2018, *ApJ*, 854, 97
Fabian, A. C. 2012, *ARA&A*, 50, 455
Feruglio, C., Maiolino, R., Piconcelli, E., *et al.* 2010, *A&A*, 518, L155
Fluetsch, A., Maiolino, R., Carniani, S., *et al.* 2019, *MNRAS*, 483, 4586
Förster Schreiber, N. M., Genzel, R., Newman, S. F., *et al.* 2014, *ApJ*, 787, 38
Gallerani, S., Pallottini, A., Feruglio, C., *et al.* 2018, *MNRAS*, 473, 1909
Genzel, R., Förster Schreiber, N. M., Rosario, D., *et al.* 2014, *ApJ*, 796, 7
González-Alfonso, E., Fischer, J., Spoon, H. W. W., *et al.* 2017, *ApJ*, 836, 11
Herrera-Camus, R., Tacconi, L., Genzel, R., *et al.* 2019, *ApJ*, 871, 37
Hopkins, P. F., Hernquist, L., Cox, T. J., & Kereš, D. 2008, *ApJS*, 175, 356
Hopkins, P. F., Kereš, D., Oñorbe, J., *et al.* 2014, *MNRAS*, 445, 581
Janssen, A. W., Christopher, N., Sturm, E., *et al.* 2016, *ApJ*, 822, 43
Kennicutt, R. C., & Evans, N. J. 2012, *ARA&A*, 50, 531
Kormendy, J., & Ho, L. C. 2013, *ARA&A*, 51, 511
Leroy, A. K., Walter, F., Martini, P., *et al.* 2015, *ApJ*, 814, 83
Madau, P., & Dickinson, M. 2014, *ARA&A*, 52, 415
Marrone, D. P., Spilker, J. S., Hayward, C. C., *et al.* 2018, *Nature*, 553, 51
McCourt, M., Oh, S. P., O'Leary, R., & Madigan, A.-M. 2018, *MNRAS*, 473, 5407
Narayanan, D., Turk, M., Feldmann, R., *et al.* 2015, *Nature*, 525, 496
Richings, A. J., & Faucher-Giguère, C.-A. 2018, *MNRAS*, 474, 3673
Rupke, D. S., Veilleux, S., & Sanders, D. B. 2005, *ApJS*, 160, 115
Scannapieco, E. 2013, *ApJ*, 763, L31
Schneider, E. E., & Robertson, B. E. 2017, *ApJ*, 834, 144
Schneider, E. E., Robertson, B. E., & Thompson, T. A. 2018, *ApJ*, 862, 56
Shapley, A. E., Steidel, C. C., Pettini, M., & Adelberger, K. L. 2003, *ApJ*, 588, 65
Spilker, J., & Nyland, K. 2018, arXiv e-prints, arXiv:1810.06605
Spilker, J. S., Marrone, D. P., Aravena, M., *et al.* 2016, *ApJ*, 826, 112
Spilker, J. S., Aravena, M., Béthermin, M., *et al.* 2018, Science, 361, 1016

Straatman, C. M. S., Labbé, I., Spitler, L. R., *et al.* 2014, *ApJ*, 783, L14
Strandet, M. L., Weiß, A., Vieira, J. D., *et al.* 2016, ArXiv e-prints, arXiv:1603.05094
Veilleux, S., Meléndez, M., Sturm, E., *et al.* 2013, *ApJ*, 776, 27
Weiß, A., De Breuck, C., Marrone, D. P., *et al.* 2013, *ApJ*, 767, 88
Wellons, S., Torrey, P., Ma, C.-P., *et al.* 2015, *MNRAS*, 449, 361
Werk, J. K., Prochaska, J. X., Tumlinson, J., *et al.* 2014, *ApJ*, 792, 8
Wilson, R. W., Jefferts, K. B., & Penzias, A. A. 1970, *ApJ*, 161, L43

Uncovering Early Galaxy Evolution in the ALMA and JWST Era
Proceedings IAU Symposium No. 352, 2019
E. da Cunha, J. Hodge, J. Afonso, L. Pentericci & D. Sobral, eds.

doi:10.1017/S1743921320000605

Serendipitous discovery of an "ALMA-only" galaxy at $5 < z < 6$ in an ALMA 3-mm survey

Christina C. Williams

Steward Observatory, University of Arizona, 933 North Cherry Avenue, Tucson, AZ 85721, USA
email: ccwilliams@email.arizona.edu

Abstract. We discuss the serendipitous discovery of a dusty high-redshift galaxy in a small (8 arcmin^2) ALMA 3-mm survey Williams *et al.* (2019). The galaxy was previously unknown and is absent from existing multi-wavelength catalogs ("ALMA-only"). Using the ALMA position as prior, we perform forced deblended photometry to constrain its spectral energy distribution. The spectral energy distribution is well described by a massive ($M^* = 10^{10.8} M_\odot$) and highly obscured ($A_V \sim 4$) galaxy at redshift $z = 5.5 \pm 1.1$ with star formation rate $\sim$300 $M_\odot yr^{-1}$. Our small survey area implies an uncertain but large contribution to the cosmic star formation rate density, similar to the contribution from all ultraviolet-selected galaxies combined at this redshift. This galaxy likely traces an abundant population of massive galaxies absent from current samples of infrared-selected or sub-millimeter galaxies, but with larger space densities, higher duty cycles, and significant contribution to the cosmic star-formation rate and stellar mass densities.

Keywords. galaxies: formation, galaxies: evolution, galaxies: high-redshift

---

1. Introduction

Single dish sub-millimeter surveys have discovered dust-obscured star-forming galaxies (Casey, Narayanan & Cooray *et al.* 2014) that contribute significantly to the cosmic star formation rate density at $1 < z < 3$ (Madau & Dickinson 2014). However, beyond $z > 3$ our view of the dust-obscured Universe is incomplete, with only the brightest and most extreme galaxies identified at $z > 4$ (e.g. Marrone *et al.* 2018). Gravitational lensing has enabled the discovery of some dusty galaxies beyond $z > 5$ (Spilker *et al.* 2016; Zavala *et al.* 2018a), but the lensing correction and selection effects make it difficult to measure how much they contribute to the cosmic star formation rate density. Thus the complete census of star formation in the early Universe including the fraction that is dust obscured is unknown.

Selection at longer wavelengths ($\lambda > 2$ mm) is thought to optimize for the identification of dust-obscured star formation at redshift $z > 4$ (Béthermin *et al.* 2015; Casey *et al.* 2018), which could help select dust-obscured galaxies for census studies. Relatively few surveys at such long wavelength exist compared to the multitude of large and deep surveys at sub-millimeter wavelengths (850μm-1mm). However, ALMA 3-mm surveys to date have so far predominantly identified low redshift sources ($z < 3$; González-López *et al.* 2019; Zavala *et al.* 2018b), similar redshifts to sub-millimeter surveys. Observations with IRAM/GISMO indicate that $\lambda > $2mm surveys could select higher-redshift sources (Magnelli *et al.* 2019), but counterpart identification to establish redshifts is difficult because of the large beam sizes of single-dish observatories. Nonetheless, relatively few

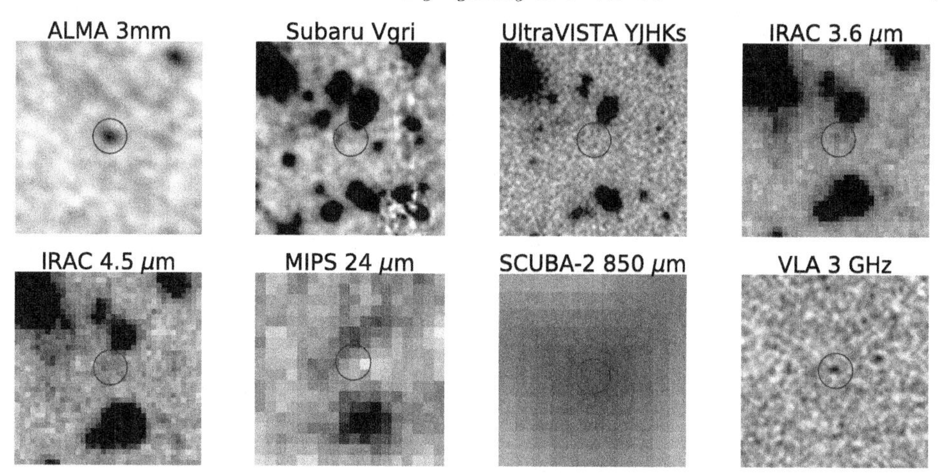

Figure 1. Cutouts (20″x20″) centered at the 3-mm position of the ALMA-only galaxy (blue circle; 3″ diameter). The galaxy is not detected (>3σ) in deep optical and near-IR stacks, or *Spitzer*, *Herschel*, and 850μm. Including the ALMA position as a prior when measuring the photometry results in marginal 2−3σ measurements at 3.6+4.5 and 850μm, and a marginal detection at 3GHz (4σ). Figure adapted from (Williams *et al.* 2019).

dust-obscured candidates exist at z > 4. Therefore the amount of dust-obscured star formation in the early universe is still unconstrained.

In a recent ALMA Band 3 survey targeting CO(2-1) molecular gas emission in unrelated quiescent galaxies at $z \sim 1.5$ (Williams *et al.* in prep.), we serendipitously identified two previously unknown galaxies from our ALMA 3-mm imaging (Williams *et al.* 2019). Our survey size (8 arcmin^2) is comparable to deep field campaigns by ASPECS (González-López *et al.* 2019) and therefore represents an opportunity to investigate the prevalence of high-redshift dust-obscured galaxies. In this proceeding, we summarize the SED properties of one of the blindly selected 3-mm sources and the evidence for high-redshift nature, and discuss the implications for our understanding of massive galaxy evolution at $z > 5$ (published in Williams *et al.* 2019).

2. SED properties and modeling

The serendipitously identified galaxy is in the COSMOS field, and has deep coverage at optical-to-radio wavelengths (0.6μm-1.4 GHz), but does not have any counterpart in multi-wavelength catalogs (i.e. "ALMA-only"; see Williams *et al.* 2019 and references therein). Using the ALMA 3-mm position as prior, we perform deblended forced photometry as described in Williams *et al.* (2019). The galaxy is not significantly detected ($< 3\sigma$) from optical to sub-millimeter, with a marginal 4σ detection in the deep 3GHz imaging (Figure 1). The resulting SED is shown in Figure 2. The deep non-detections between $24 - 500\mu$m and the extreme flux ratios between mid- and far-infrared strongly suggest that the peak of the dust emission is at high redshift ($z > 4$). Similar dark sources with extreme flux ratios have also been seen by Cowie *et al.* 2018; Yamaguchi *et al.* 2019.

We use the Bayesian Analysis of Galaxies for Physical Inference and Parameter EStimation (BAGPIPES) code (Carnall *et al.* 2018) to model the SED. We find that the observations are consistent with the SED of a massive $10^{10.8\pm0.4}$M$_\odot$, star forming $SFR = 309^{+241}_{-149}$M$_\odot$/yr, highly obscured $A_V \sim 4^{+1.4}_{-1.0}$ galaxy at very high redshift $z \sim 5.5^{+1.2}_{-1.1}$. The SED fitting results are shown in Figure 2. The Bayesian posterior probability distribution for redshift and the measured infrared luminosity (Log$_{10}$ L$_{IR}$ = 12.6) are well constrained despite the lack of strong detections at any wavelength other than 3-mm.

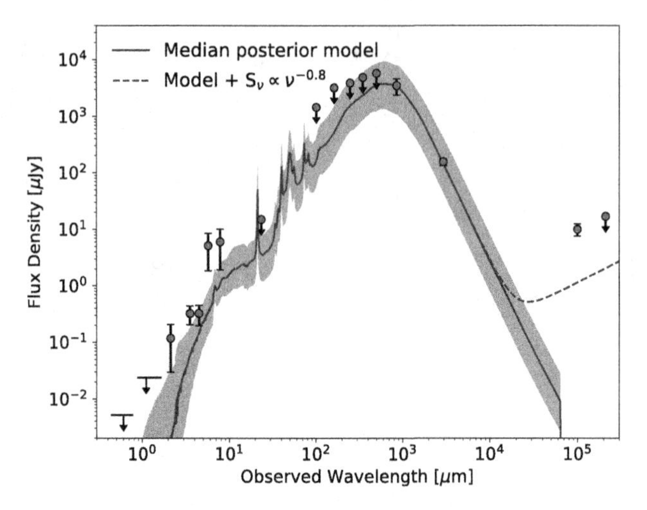

Figure 2. Photometry of the ALMA-only galaxy (red points). Data with signal-to-noise < 1 are shown as upper limits at the 1σ rms value. Photometry with signal-to-noise > 1 is plotted with 1σ uncertainties, but does not necessarily indicate a significant detection. Shown are the median posterior spectrum (dark blue) and 16-84th percentile range (light blue) from BAGPIPES. Dotted line is the radio spectrum predicted from L_{IR} (Tisanić *et al.* 2019), suggesting possible excess radio emission. Figure adapted from (Williams *et al.* 2019).

This result is driven by the deep photometric limits at shorter wavelengths (*Spitzer*/MIPS and *Herschel*/PACS $100 - 160\mu$m) in combination with the high signal-to-noise ALMA measurement. We also estimate a high molecular gas mass of $M_{\rm gas} \sim 0.5 - 1.5 \times 10^{11}$ $M_\odot$, based on the calibration of the sub-millimeter flux density to gas mass (Scoville *et al.* 2016), implying a high inferred gas fraction ($\sim$60%).

3. Contribution to the $z > 4$ galaxy census

This galaxy was identified in a small survey area (8 arcmin^2) which suggests that similar galaxies may be relatively abundant in the early Universe. The implied source density is $0.13^{+0.30}_{-0.10}$ arcmin^{-2}, an order of magnitude higher sub-millimeter galaxies at $z > 4$ which are relatively rare (0.01–0.02 arcmin^{-2}; e.g. Danielson *et al.* 2017; Marrone *et al.* 2018)). Such an abundant population of massive star-forming galaxies would have large impact on our census of star formation rate and stellar mass densities at $z > 4$. Based on our measured star formation rate and the estimated selection volume, we find that the contribution to the cosmic star formation rate density of this one galaxy is $\rho_{\rm SFR}$ $0.9^{+2.0}_{-0.7} \times 10^{-2}$ $M_\odot$ yr^{-1} Mpc^{-3} (Figure 3). Bright sub-millimeter galaxies beyond $z > 4$ in comparison contribute about a factor of 10 less (Swinbank *et al.* 2014; Michałowski *et al.* 2017. If this galaxy is representative, similar galaxies could contribute as much to the cosmic star formation rate density as all known ultraviolet-selected galaxies at similar redshifts combined (Williams *et al.* 2019). Dust-obscured star formation could dominate the cosmic star formation history beyond $z > 4$, however, a larger sample is needed for more certain measurements.

The high stellar mass of the galaxy, and its large inferred space density also suggest significant cosmic stellar mass density in similar galaxies at $z \sim 5$: $\rho^* = 1.9^{+4.4}_{-1.5} \times 10^6\,M_\odot$Mpc^{-3} (Williams *et al.* 2019). This is higher than measurements from the rarer, bright ($S_{850} > 4$ mJy) sub-millimeter galaxies ($\approx 0.5 \times 10^6$ $M_\odot$Mpc^{-3}; Michałowski *et al.* 2017). By comparison, galaxies selected using Hubble Space Telescope contribute $\sim 6.3 \times 10^6 M_\odot$Mpc^{-3} to the stellar mass density (e.g. Song *et al.* 2016), suggesting

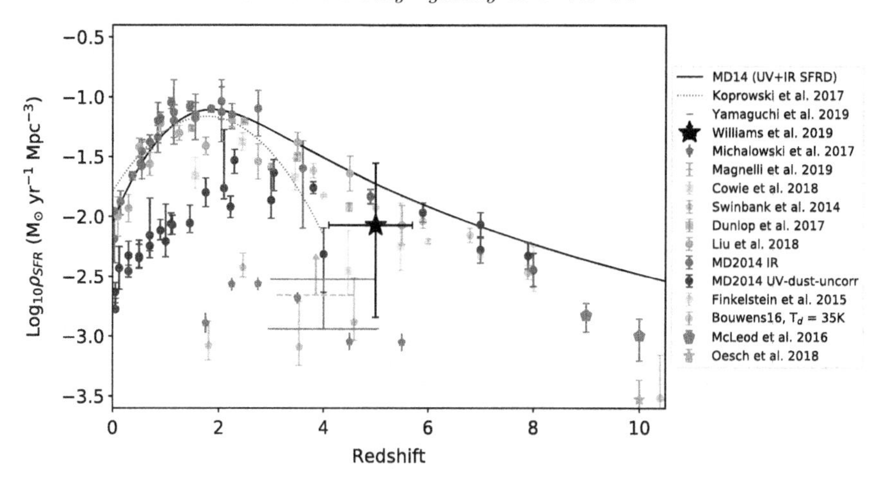

Figure 3. The cosmic star formation history. Blue and green points are measurements based on rest-frame UV (dust un-corrected) and red points are based on star formation rates derived from IR-to-millimeter measurements. In addition to the compilation of Madau & Dickinson (2014; blue circles) we have included later literature measurements $z > 4$ by Finkelstein *et al.* 2015; Bouwens *et al.* 2016; McLeod *et al.* 2016; Oesch *et al.* 2018. We similarly add to the IR compilation of Madau & Dickinson (2014; red circles), with recent constraints at $z > 2$ from Swinbank *et al.* 2014; Koprowski *et al.* 2017; Magnelli *et al.* 2019; Cowie *et al.* 2018; Dunlop *et al.* 2017; Liu *et al.* 2018. The black star indicates the contribution of the "ALMA-only" galaxy Williams *et al.* (2019).

that such galaxies could contribute a significant fraction (22^{+25}_{-16}%) to the total at these redshifts (Williams *et al.* 2019).

Another relevant question is if this galaxy traces a population that could evolve into the earliest known massive quiescent galaxies at $3 < z < 4$ with N$\sim 3 - 5 \times 10^{-5}$ Mpc^{-3} and Log (M/M$_\odot$) $\gtrsim 10.6$ (e.g. Straatman *et al.* 2014). UV-selected galaxies at $z > 4$ are less massive and star forming than necessary to be their progenitors. Similarly, the number densities of bright (> 4 mJy) sub-millimeter galaxies at $z > 4$ are likely too low to be their progenitors ($\sim 0.1 - 3 \times 10^{-6}$ Mpc^{-3} e.g. Michałowski *et al.* 2017). Although sub-millimeter galaxies have large enough star formation rates to rapidly form massive galaxies, their relatively low gas masses indicate very rapid gas depletion timescales ($10 - 100$ Myr; e.g. Aravena *et al.* 2016; Spilker *et al.* 2018). To reconcile rapid gas depletion time with low number density requires large duty cycle corrections. In contrast, the inferred space density of our galaxy is already comparable to the earliest known quiescent galaxies (e.g. Straatman *et al.* 2014). The galaxy has both a large $\sim 10^{11} M_\odot$ gas mass in combination with lower star formation rate, which suggests long depletion timescales compared to sub-millimeter galaxies ($\sim 200 - 500$ Myr). This galaxy may therefore have a longer duty cycle ($\sim 50 - 100$%; Williams *et al.* 2019). Galaxies such as this one could be evidence for a more gradual path to forming massive galaxies, in contrast to the rapid bursts that are associated with sub-millimeter galaxies (e.g. Pavesi *et al.* 2018; Marrone *et al.* 2018).

Until the *James Webb Space Telescope* (*JWST*) launches, ALMA is the only facility that can study infrared-dark galaxies in detail. Future surveys such as the *JWST* Advanced Deep Extragalactic Survey (JADES) will identify $\sim 15 - 30$ galaxies similar to this galaxy (assuming the measured number densities published in Williams *et al.* 2019; Zavala *et al.* 2018a, for the survey specifications described in Williams *et al.* 2018). In Williams *et al.* 2018, we predicted the number of high-redshift galaxies likely to be identified in *JWST* Cycle 1 from the JADES survey. These predictions are based on the rest-frame UV luminosity functions, which above $z > 4$ is currently the only source of a complete galaxy census. The discovery of infrared-dark galaxies therefore indicates these predicted counts are underestimates, in particular at the massive end of the stellar mass function. With *JWST* it will be possible to fully characterize their stellar population properties and redshifts. *JWST* measurements combined with ALMA observations of star formation, molecular gas, and dust properties will provide powerful constraints on the growth of massive galaxies in the early Universe.

Acknowledgements

This proceeding is a summary of work relevant to the symposium previously presented in Williams *et al.* (2018, 2019), to which the reader is referred for further details and the full results. The author acknowledges the important contributions of the co-authors of these publications, without which this proceeding would not be possible.

References

Aravena, M., Decarli, R., Walter, F., *et al.* 2016, *ApJ*, 833, 68
Béthermin, M., De Breuck, C., Sargent, M., & Daddi, E. 2015, *A&A*, 576, L9
Bouwens, R. J., Oesch, P. A., Labbé, I., *et al.* 2016, *ApJ*, 830, 67
Carnall, A. C., McLure, R. J., Dunlop, J. S., & Davé, R. 2018, *MNRAS*, 480, 4379
Casey, C. M., Narayanan, D., & Cooray, A. 2014, *Phys. Rep.*, 541, 45
Casey, C. M., Hodge, J., Zavala, J. A., *et al.* 2018, *ApJ*, 862, 78
Cowie, L. L., González-López, J., Barger, A. J., *et al.* 2018, *ApJ*, 865, 106
Danielson, A. L. R., Swinbank, A. M., Smail, I., *et al.* 2017, *ApJ*, 840, 78
Dunlop, J. S., McLure, R. J., Biggs, A. D., *et al.* 2017, *MNRAS*, 466, 861
Finkelstein, S. L., Ryan, Jr., R. E., Papovich, C., *et al.* 2015, *ApJ*, 810, 71
González-López, J., Decarli, R., Pavesi, R., *et al.* 2019, arXiv e-prints, arXiv:1903.09161
Koprowski, M. P., Dunlop, J. S., Michałowski, M. J., *et al.* 2017, *MNRAS*, 471, 4155
Liu, D., Daddi, E., Dickinson, M., *et al.* 2018, *ApJ*, 853, 172
Madau, P., & Dickinson, M. 2014, *ARA&A*, 52, 415
Magnelli, B., Karim, A., Staguhn, J., *et al.* 2019, arXiv e-prints
Marrone, D. P., Spilker, J. S., Hayward, C. C., *et al.* 2018, *Nature*, 553, 51
McLeod, D. J., McLure, R. J., & Dunlop, J. S. 2016, *MNRAS*, 459, 3812
Michałowski, M. J., Dunlop, J. S., Koprowski, M. P., *et al.* 2017, *MNRAS*, 469, 492
Oesch, P. A., Bouwens, R. J., Illingworth, G. D., Labbé, I., & Stefanon, M. 2018, *ApJ*, 855, 105
Pavesi, R., Riechers, D. A., Sharon, C. E., *et al.* 2018, *ApJ*, 861, 43
Scoville, N., Sheth, K., Aussel, H., *et al.* 2016, *ApJ*, 820, 83
Song, M., Finkelstein, S. L., Ashby, M. L. N., *et al.* 2016, *ApJ*, 825, 5
Spilker, J. S., Marrone, D. P., Aravena, M., *et al.* 2016, *ApJ*, 826, 112
Spilker, J. S., Aravena, M., Béthermin, M., *et al.* 2018, Science, 361, 1016
Straatman, C. M. S., Labbé, I., Spitler, L. R., *et al.* 2014, *ApJ*, 783, L14
Swinbank, A. M., Simpson, J. M., Smail, I., *et al.* 2014, *MNRAS*, 438, 1267
Tisanić, K., Smolčić, V., Delhaize, J., *et al.* 2019, *A&A*, 621, A139
Williams, C. C., Curtis-Lake, E., Hainline, K. N., *et al.* 2018, *ApJS*, 236, 2
Williams, C. C., Labbe, I., Spilker, J., *et al.* 2019, arXiv e-prints; ApJ submitted
Yamaguchi, Y., Kohno, K., Hatsukade, B., *et al.* 2019, arXiv e-prints
Zavala, J. A., Casey, C. M., da Cunha, E., *et al.* 2018a, *ApJ*, 869, 71
Zavala, J. A., Montaña, A., Hughes, D. H., *et al.* 2018b, *Nature Astronomy*, 2, 56

Uncovering Early Galaxy Evolution in the ALMA and JWST Era
Proceedings IAU Symposium No. 352, 2019
E. da Cunha, J. Hodge, J. Afonso, L. Pentericci & D. Sobral, eds.

doi:10.1017/S1743921320001209

ALMA reveals large molecular gas reservoirs in recently-quenched galaxies

Katherine Suess

UC Berkeley, USA

Abstract. We still do not understand the physical mechanisms that are responsible for suppressing star formation in galaxies. Observations of post-starburst galaxies, whose spectra indicate that an intense period of star formation was followed by rapid quenching, are the ideal sample to probe the quenching process. We have conducted an ALMA survey of CO(2-1) in 13 of these recently- quenched galaxies at $z \sim 0.7$ – high enough redshift that these galaxies likely just concluded their primary epoch of star formation, but low enough redshift for follow-up observations to be feasible. Our observations reveal a stunning diversity of molecular gas properties: despite a uniform optical selection and low apparent SFRs, the detected galaxies span a factor of > 30 in CO luminosity and have inferred gas fractions ranging from $< 1\%$ to 20%. These observations indicate that quenching does not require the total removal or depletion of molecular gas. No current models of the quenching process can fully explain our results.

Uncovering Early Galaxy Evolution in the ALMA and JWST Era
Proceedings IAU Symposium No. 352, 2019
E. da Cunha, J. Hodge, J. Afonso, L. Pentericci & D. Sobral, eds.

doi:10.1017/S1743921320000836

Large reservoirs of turbulent diffuse gas around high-z starburst galaxies

E. Falgarone[1], A. Vidal-García[1], B. Godard[2,1], M. A. Zwaan[3], C. Herrera[4], R. J. Ivison[3], E. Bergin[5], P. M. Andreani[3], A. Omont[6] and F. Walter[7]

[1]Laboratoire de Physique de l'ENS, ENS, Université PSL, CNRS, Sorbonne Université, Université de Paris, 24 rue Lhomond, 75005 Paris, France
email: edith.falgarone@ens.fr

[2]LERMA, Observatoire de Paris, CNRS, 61 avenue de l'Observatoire, 75014 Paris, France

[3]European Southern Observatory, Karl-Schwarzschild-Strasse 2, 85748 Garching, Germany

[4]IRAM, 300 rue de la Piscine, Domaine universitaire, 38406 Saint Martin d'Hères, France

[5]University of Michigan, 311 West Hall, 1085 S. University Ave, Ann Arbor, MI 48109, USA

[6]UMPC Université Paris 6 & Institut d'Astrophysique de Paris, CNRS, 75014 Paris, France

[7]Max Planck Institute für Astronomie, Heidelberg, Germany

Abstract. Starburst galaxies at $z \sim 2-4$ are among the most intensely star-forming galaxies in the universe. The way they accrete their gas to form stars at such high rates is still a controversial issue. ALMA has detected the $CH^+(J=1\text{-}0)$ line in emission and/or absorption in all the gravitationally lensed starburst galaxies targeted so far at $z \sim 3$. Its unique spectroscopic and chemical properties enable CH^+ to highlight the sites of most intense dissipation of mechanical energy. The absorption lines reveal highly turbulent, massive reservoirs of low-density molecular gas. The broad emission lines, arising in myriad UV-irradiated molecular shocks, reveal powerful galactic winds. The CH^+ lines therefore probe the fate of prodigious energy releases, due to infall and/or outflows, and primarily stored in turbulence before being radiated by cool molecular gas. The turbulent reservoirs act as mass and energy buffers over the duration of the starburst phase.

Keywords. galaxies: starburst, galaxies: high-redshift, galaxies: ISM, galaxies: formation, intergalactic medium, turbulence, shock waves

---

The power of absorption spectroscopy

In cosmological simulations, the growth of galaxies in the early universe results from the accretion of fresh gas in virialised dark matter halos, modulo ejection of matter by stars and AGNs (Madau & Dickinson 2014). While ejection is observed in ubiquitous ionised and neutral galactic winds (Veilleux *et al.* 2005) and contributes to the self-regulation of cosmic star formation (Schaye *et al.* 2015), evidence for accretion has remained elusive. The dust continuum emission of starburst galaxies at high-redshift makes them extremely bright sub-millimetre sources that provide unique background sources to probe their large scale environment with highly sensitive absorption spectroscopy in the sub-millimetre domain, just as quasars (QSO) do in the visible domain.

The sub-millimetre domain happens to be that of the fundamental transitions of light hydrides, that are the very first steps of chemistry in space, linking hydrogen with heavy elements (Gerin *et al.* 2016). Among all hydrides, CH^+, one of the three first molecular species ever detected in space (Douglas & Herzberg 1941), has unique chemical and

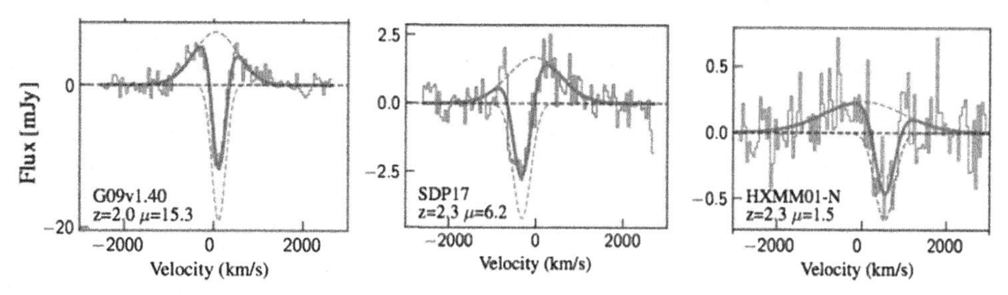

Figure 1. A subset of ALMA continuum-subtracted $CH^+(J=1\text{-}0)$ spectra of starburst galaxies illustrating the three kinds of spectra found in the observations. *Left:* Broad emission and narrow absorption centered at about the same velocity. *Middle:* P-Cygni profile with broad emission and blue-shifted absorption. *Right:* Inverse P-Cygni profile with broad emission and red-shifted absorption. The galaxy name, redshift and lens magnification are given in the lower left corner. Note the factor of 50 between the emission line intensity of the brightest and weakest source.

spectroscopic properties. CH^+ has both a highly endoenergic formation ($E_{\rm form} \sim 0.4\,{\rm eV}$) and a fast destruction rate, by collisions with H and H_2: in low-density and weakly irradiated gas, a warm chemistry activated by dissipation of turbulence in shocks and/or intense velocity shears is able to overcome its fast destruction (Godard *et al.* 2014 and references therein). Once formed, its lifetime is so short, $t \sim 1\,{\rm yr}$, even in low-density gas, that, unlike the CO molecule, CH^+ is always observed where it forms, i.e. it is not transported. Moreover, CH^+ being a light hydride with high dipole moment, the critical density ($\sim 10^7$ cm^{-3}) of its $J=1\text{–}0$ transition is almost $10^5\times$ larger than that of CO($J=1\text{–}0$), so the line appears in absorption in low-density gas ($n_{\rm H} < 10^3$ cm^{-3}), and its opacity is a direct measure of the CH^+ abundance. This abundance in turn provides the turbulent dissipation rate, in the scenario where the supra-thermal energy required for its formation is provided by turbulent dissipation (Godard *et al.* 2014). In emission, the $CH^+(J=1\text{–}0)$ line is detected only in dense and UV-illuminated gas, either photon-dominated regions or UV-irradiated molecular shocks because, unlike most molecules, the abundance of CH^+ is enhanced in intense UV fields (Godard *et al.* 2019). The CH^+ absorption and emission lines therefore both highlight the sites of dissipation of mechanical energy. By doing so, they reveal not only gas components that have never been detected before but also new trails of the gravitational potential energy.

High detection rate of CH^+(1-0) lines in high-z starburst galaxies

With ALMA, the $CH^+(J=1\text{–}0)$ line has been detected in all the 18 lensed submillimetre-selected galaxies (SMGs) targeted so far at redshifts $z = 1.7 - 4.2$ (Falgarone *et al.* 2017, Falgarone *et al.*, in prep.). These are lensed starburst galaxies discovered by *Herschel* (Eales *et al.* 2010; Oliver *et al.* 2010) and SCUBA-2 (Ivison *et al.*, 2010). CH^+ absorption lines are found in 15 of the 18 sources observed. The absorptions are deep (opacities 0.25–1.2) and broad (average FWHM $\sim$ 400 km s^{-1}), blue- or red-shifted with respect to the SMG redshift at almost the same occurrence, (i.e. the average offset is $\sim$ 60 km s^{-1}) with a large scatter (i.e. the variance of the offset velocities is 300 km s^{-1}). The emission lines, when fully visible, are extremely broad (FWHM $\sim$1500 km s^{-1}) and are understood as originating in myriad dense shocks presumably powered by hot galactic winds penetrating the circum-galactic medium (CGM). In Fig. 1, three spectra illustrate the variety of the overall CH^+ lineshapes. In several cases, the absorption centroid is only slightly shifted compared to that of the emission (left panel) but in other cases, the whole profile can be either P-Cygni, revealing outflowing material (central panel) or inverse P-Cygni, revealing infalling gas (right panel).

Since blue- (resp. red-) shifted absorption against a background continuum source *unambiguously* traces outflowing (resp. infalling) gas with respect to the background source, the similar occurrence of blue- and red-shifted absorptions suggests that absorption lines sample gas that is as often infalling towards the starburst galaxy, as outflowing from it. We ascribe this diversity to large scale turbulence of the gas sampled randomly by the line-of-sight towards the SMGs, the width of the absorption lines providing the gas turbulent velocity dispersion. This similar occurrence of blue- and red-shifted absorption lines can also be seen as a manifestation of the co-existence of outflows and infall, with time-dependent domination of either one (i.e. bursty accretion followed by bursty outflows). The sign of the absorption velocity offset may also depend on the orientation of the outflow and infall directions (if any) with respect to the line-of-sight (Zabl *et al.* 2019). Nonetheless, turbulence resulting from a fluid instability unavoidable in large-scale cosmic flows, it has to be ubiquitous in those systems.

Starburst CGM properties

It is well known that absorption lines alone cannot provide the extent of the absorbing medium along the line-of-sight. The method followed to determine the size and mass of the turbulent gas reservoirs (TR) traced by the CH^+ absorption is detailed in Falgarone *et al.*, (2017). It involves the following steps: (1) we infer that the TR lifetime is of the order of the duration of the starburst phase, t_{SB}, given the high occurrence of absorption detections, (2) the CH^+ abundance, proportional to the turbulent energy transfer rate, depends on the gas velocity dispersion σ_{TR} (inferred from the absorption linewidth $\Delta v_{abs} = 2.35\,\sigma_{TR}$) and the unknown size r_{TR}, causing a degeneracy between the CH^+ abundance and r_{TR}, (3) we break this degeneracy with the finding that, in the five cases where the stelllar mass of the starburst is known, the radius at which the escape velocity of the galaxy surrounded by its massive CGM is equal to the mean turbulent velocity is approximately $\sim\Delta v_{abs}\,t_{SB}$, with $t_{SB}\sim 50$Myr, and in spite of very different stellar masses and gas velocity dispersions. We therefore adopt the estimate $r_{TR}\sim\Delta v_{abs}\,t_{SB}$, meaning that the TR is virialized in the potential well of the galaxy and that its lifetime is the dynamical timescale of its turbulence, $t_{dyn}\sim r_{TR}/\sigma_{TR}$. A validation of these unrelated assumptions can only be achieved by direct imaging of the CGM.

Once the size is known, the gas mass is directly inferred from the absorption line and the CH^+ abundance. Massive, M_{TR} up to a few 10^{10} $M_\odot$, and highly turbulent reservoirs, extending far outside ($r_{TR} > 10$ kpc) the compact starburst cores of radii ≈ 1 kpc are therefore unveiled. The resulting gas fraction within tens of kpc therefore approaches $f_{gas}\sim 0.5$. These large-scale gaseous halos are likely multi-phasic, comprising the cool ($T\sim 100$ K), low-density and turbulent gas in which CH^+ forms, probably embedded in a warmer and more dilute gas.

Link of the turbulent luminosity and the star formation rate (SFR)

A remarkable finding, given the many assumptions discussed above and the uncertainties on the SFRs, is that the turbulent luminosity of 11 of the TRs, $L_{turb} = \frac{1}{2}M_{TR}\sigma_{TR}^2/t_{dyn}$ is found to be proportional to the SFR of the galaxies (Fig. 2). The SFRs are derived from the total FIR luminosity (Kennicutt 1998) and could be overestimated by large factors if AGNs contribute, and if the stellar initial mass function (IMF) is not a Salpeter IMF (Zhang *et al.* 2018). This linear relation, however, provides a key insight to the dynamics of the whole CGM.

The broad CH^+ emission lines arise in myriad UV-irradiated molecular shocks at $v_{sh} < 50\,\mathrm{km\,s^{-1}}$. Their width traces the velocity dispersion of the shocks themselves, likely driven by galactic-scale high-velocity winds. The shocks are thought to form at the

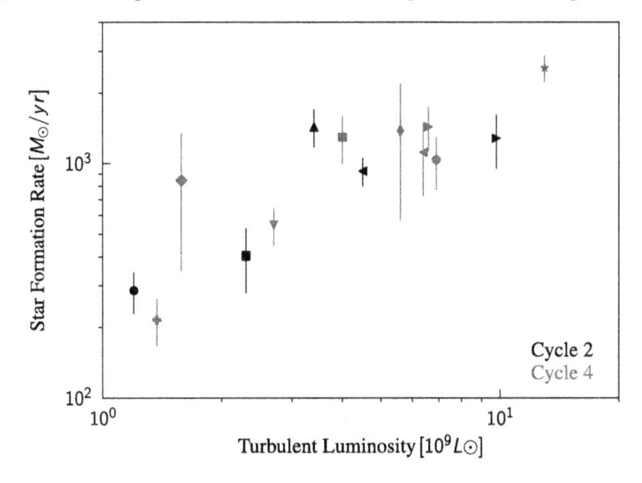

Figure 2. Star formation rates (SFR) of starburst galaxies at $1.7 < z < 4.2$ vs. the turbulent luminosity of their massive reservoirs of diffuse gas unveiled by CH^+(1-0) absorption. Nominal errors bars on $L_{\rm turb}$ are less than a factor of 2, those on the SFRs may be larger than displayed due to possible AGN contributions to the FIR luminosity and top-heavy initial mass function (see text).

interface of the wind and the CGM. The CH^+ emission linewidth, close to 1500 $\rm km\,s^{-1}$ in all the cases where it is fully detected, provides the velocity of the outflowing material, $v_{\rm out} \sim 700$ $\rm km\,s^{-1}$. The outflows trigger large-scale turbulence (and phase transition) in the CGM so that the TR turbulent luminosity is fed by these outflows, $L_{\rm turb} \propto \dot{M}_{\rm out} v_{\rm out}^2$. The mass outflow rates, $\dot{M}_{\rm out}$, are found to be too small to offset the SFRs and we infer from this imbalance that inflows have to contribute to the mass feeding of the TRs to explain their long lifetime around such active starburst galaxies.

Indeed, CH^+ absorption lines in the direction of SMMJ02399, a small group of galaxies comprising a broad absorption line (BAL) QSO and a SMG (Ivison *et al.* 2010), are red-shifted by $\sim$500 $\rm km\,s^{-1}$, indicating an inflow towards the SMG of this magnitude. The size of the TR around these galaxies is inferred to be $r_{\rm TR} \sim 20{\rm kpc}(t_{\rm SB}/50{\rm Myr})$. The CH^+ emission lines have FWZI $\sim$2500 $\rm km\,s^{-1}$. A most extended (> 80 kpc) Lyα nebula has been imaged with the *Keck*/KCWI in that field (Li *et al.* 2019) with remarkable properties: the Lyα lines in the direction of the SMG and the BAL QSO are all extremely broad, with FWZI $\sim$2500 $\rm km\,s^{-1}$. They are asymmetric, with a red-shifted side much weaker than the blue-shifted side, a characteristic of infall (Verhamme *et al.* 2006). Interestingly, the velocity coverage of the CH^+ absorptions in this field is the same as that of the red side of the Lyα lines. The comparison of the CH^+ absorption and emission lines with those of Lyα led to the conclusion that the front part of the Lyα nebula (and possibly the whole nebula) is inflowing towards the galaxies (Falgarone *et al.* 2020). It is therefore a multi-phasic medium in which the warm gas (partially or fully ionized) at 10^4 K that emits Lyα is mixed with much cooler and denser gas in which CH^+ forms because the Lyα emission and the CGM that absorbs CH^+ have similar spatial extent and velocity coverage. Lastly, the similarity of the FWZI of the Lyα and CH^+ emission lines suggests that some contribution to the Lyα line comes from self-irradiated molecular shocks (Lehmann *et al.*, these proceedings).

The emerging picture

This is work in progress. Although the above quantitative estimates have been achieved so far with only a handful of galaxies, a coherent picture may be drawn that relies on the fact that CH^+ is a unique tracer of dissipative processes. The CH^+ lines, in absorption, unveil massive reservoirs of diffuse molecular gas around starburst galaxies at $z \sim 2-4$ that are long-lived, multi-phasic, and extend up to 40 kpc from the starburst cores. In emission, the CH^+ lines trace molecular shocks with prodigious velocity dispersions that are likely driven by galactic winds. Gas accretion is required to sustain the long lifetime of these turbulent massive reservoirs. The origin of this accretion is unknown: it might be due to tidal streams in mergers or cold streams of pristine gas. Turbulence driven by the galactic winds likely processes the circumgalactic matter and allows the outflowing gas to be recycled, mitigating the negative feedback. Turbulence kinematically heats the CGM, making it a long lived buffer of mass and energy.

References

Douglas, A. E. & Herzberg, G. 1941, *ApJ*, 94, 381
Eales, S., Dunne, L., Clements, D., *et al.* 2010, *PASP*, 122, 499
Falgarone, E., Zwaan, M.A., Godard, B., *et al.* 2017, *Nature*, 548, 430
Gerin, M., Neufeld, D. A., & Goicoechea, J. R.. 2016, *ARAA*, 54, 181
Godard, B., Falgarone, E., & Pineau des Forêts, G. 2014, *A&A*, 570, A27
Godard, B., Pineau des Forêts, G. Lesaffre, P., *et al.* 2019, *A&A*, 622, A100
Ivison, R. J., Swinbank, A. M., Swinyard, B., *et al.* 2010, *A&A*, 518, L35
Kennicutt, Jr Robert C. 1998, *ApJ*, 498, 541
Li, Q., Cai, Z., Prochaska, J. X., *et al.* 2019, *ApJ*, 875, 130
Madau, P. & Dickinson, M. 2014, *ARAA*, 52, 415
Oliver, S. J., Wang, L., Smith, A. J., *et al.* 2010, *A&A*, 518, L21
Schaye, J., Crain, R. A., Bower R. G., *et al.* 2015, *MNRAS*, 446, 521
Veilleux, S., Cecil, G., & Bland-Hawthorn, J. 2005, *ARAA*, 43, 769
Verhamme, A., Schaerer, D., & Maselli, A. 2006, *A&A* 460, 397
Zabl, J., Bouché, N. F., Schroetter, L., *et al.* 2019, *MNRAS*, 485, 1961
Zhang, Z.-Y., Ivison, R. J., George, R. D., *et al.* 2018, *MNRAS*, 481, 59

Uncovering Early Galaxy Evolution in the ALMA and JWST Era
Proceedings IAU Symposium No. 352, 2019
E. da Cunha, J. Hodge, J. Afonso, L. Pentericci & D. Sobral, eds.

doi:10.1017/S1743921319008469

Molecular gas across cosmic time

Georgios E. Magdis[1,2,3]

[1]Cosmic Dawn Center (DAWN)
[2]Niels Bohr Institute, University of Copenhagen, Lyngbyvej 2, 2100 Copenhagen, Denmark
[3]Institute for Astronomy, Astrophysics, Space Applications and Remote Sensing, National Observatory of Athens, 15236, Athens, Greece
email: georgios.magdis@nbi.ku.dk

Abstract. We have entered an era where the gas mass estimates of distant galaxies do not rely on a single tracer but rather on an inventory of different and independent methods, much like the case for the determination of the star formation rate (SFR) of the galaxies. This is crucial as the traditional $M_{\rm gas}$ tracers, i.e. low-J CO transition lines and dust continuum emission are becoming highly uncertain as we move to higher redshifts due to metallicity and CMB effects. Here, we present a homogeneous and statistically significant investigation of the use of atomic carbon as an alternative $M_{\rm gas}$ tracer (Valentino *et al.* 2018) and provide evidence of optically thick far-IR emission in high$-z$ starbursts that point towards higher dust temperatures and lower dust and gas mass estimates than previously inferred (Cortzen *et al.* 2019, submitted). Finally, we present direct observations of the effect of the CMB on the far-IR SEDs of high-z SBs, manifested by unphyscally large ($\beta = 2.5 - 3.5$) apparent spectral indexes in R-J tail (Jin *et al.* 2019, submitted).

Keywords. galaxy evolution, ISM, starbursts

---

1. Introduction

A critical property that dictates galaxy evolution is the amount and the physical conditions of the cold, star forming gas primarily in the form of H_2, that is bound within the galaxies. However, since under typical conditions molecular hydrogen cannot be directly observed, in order to measure the amount of gas ($M_{\rm gas}$) in the galaxies, and much like the case of star formation rate (SFR) indicators, we have been are building an inventory of indirect gas tracers and techniques. A collection of such tracers are shown in Figure 1 and summarised below:

- the low-J CO transition lines (e.g. Solomon & Vanden Bout 2005, Carilli & Walter 2013)
- the dust continuum emission, either through the metallicity depended dust to gas mass ratio technique (e.g. Magdis *et al.* 2012) or though the monochromatic flux density in the R-J tail (e.g. Scoville *et al.* 2017)
- the neutral atomic carbon ([CI]) lines (e.g. Valentino *et al.* 2018)
- the single ionised atomic carbon line ([CII]) (e.g. Zanella *et al.* 2018)
- the polycyclic aromatic hydrocarbon (PAHs) emission in the mid-IR (e.g. Cortzen *et al.* 2019)
- the HCN emission line, sensitive to the dense gas reservoir (e.g. Gao & Solomon 2004) and
- dynamical analysis (e.g. Tacconi *et al.* 2008, Daddi *et al.* 2010)

Applying primarily the first two techniques (CO and dust) in large samples of galaxy populations has enable the discovery of various important scaling relations between the

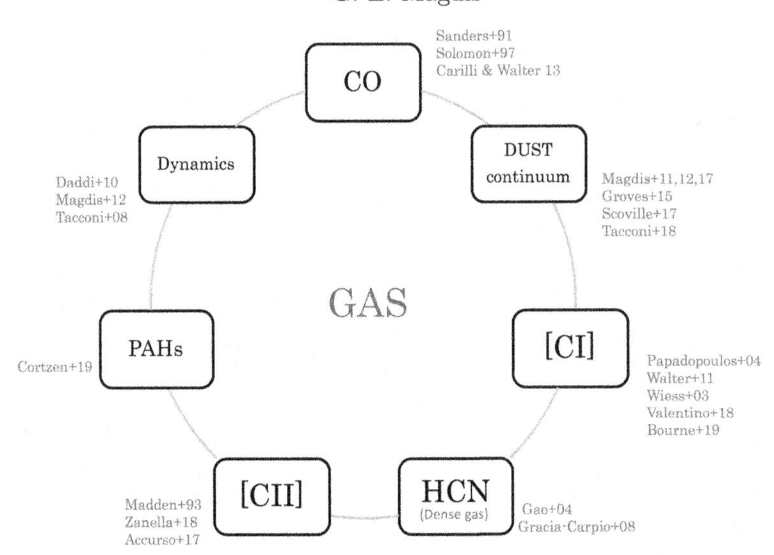

Figure 1. A collection of indirect gas mass tracers that have been employed so far, fo both local and high$-z$ galaxies. The provided references are by far incomplete, but they are indicative of the available literature for each method.

amount of available gas, the stellar mass and the star formation rate of he galaxies that have sed light into the characterisation of the star formation mode, the gas depletion time scales and the star formation efficiency of the galaxies out to $z \sim 3$ and beyond (e.g. Magdis *et al.* 2012, Sargent *et al.* 2014, Scoville *et al.* 2017, Tacconi *et al.* 2018) Furthermore, these studies have traced the evolution of the gas fraction, and more recently the evolution of the molecular gas density, coupling the well established and more *mature* studies that have unveiled the evolution of the specific star formation rate and the evolution of the star formation rate density of the Universe. However, while the various empirical and observationally motivated models of the gas content of the galaxies tend to agree up to $z \sim 3$, the available samples decrease and the uncertainties on how to convert the observable into gas mass estimates increase substantially as we move to higher redshifts, leaving our knowledge of the gas content of the early galaxies poorly constrained.

2. Caveats and the way forward

A main caveat of the CO and dust method to trace gas masses is their dependence on the gas phase metallicity (though the adopted $\alpha_{\rm CO}$ and $\delta_{\rm G/D}$) that varies as a function of environment, of redshift but also as a function of star formation mode. Indeed, critical open question that still remains open and has been the topic of hot debate is the variation of the metallicity (and therefore of the appropriate $\alpha_{\rm CO}$ and $\delta_{\rm G/D}$ conversion factors) between the main sequence (MS) and starbursts (SB) galaxies, a that could lead into a bimodal or alternatively a continuous star formation mode in the galaxies (e.g. Daddi *et al.* 2010, Sargent *et al.* 2014, Scoville *et al.* 2017). Another caveat arises from the increase of the temperature and of the background emission of the CMB that makes the detection of the low$-J$ CO transitions and of the and dust continuum emission in the R-J tail against the CMB background extremely challenging at high redshifts (e.g. da Cunha *et al.* 2013). Finally, until now the dust method has been applied under the implicit assumption that the far-IR emission is optically thin. However, previous studies local ULIRGs and more recent ALMA observations of high$-z$ star-bursting galaxies (Hodge *et al.* 2016, Simpson *et al.* 2017) provide evidence that the emission of the galaxies

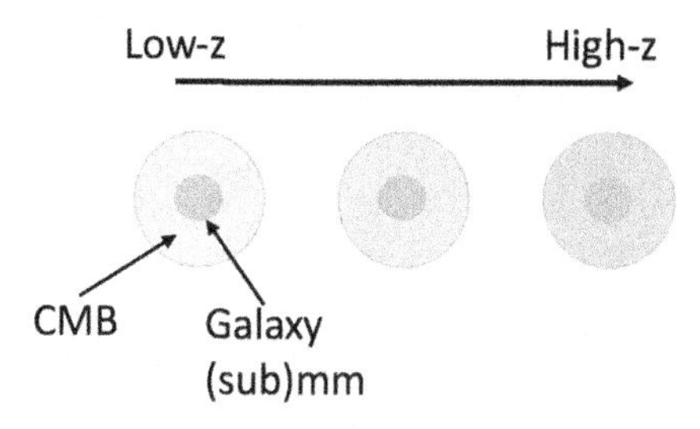

Figure 2. A sketch depicting the effect of the CMB on the measured dust continuum emission as we move to higher redshifts. The CMB background emission becomes progressively more prominent, washing out the contrast between the galaxy emission and the background.

remains optically thick out to $\lambda_{\rm rest}$ $200\mu m$, with important consequences on the derived dust temperature and therefore the dust (and gas) mass of the galaxies.

An alternative method that has been put forward as a promising gas tracer is the [CI] line (e.g. Papadopoulos *et al.* 2004) that come with several advantages with respect to the CO and dust method: 1) it is optically thin, 2) cosmic rays tend to dissociate CO to C leaving CO dark clouds of gas, 3) it is less affected by metallicity and 4) it has a low excitation potential, typical of GMC. However before using [CI] as gas tracer for high$-z$ galaxies we first need to consistently and homogeneously calibrate it among different environments at lower redshifts. In the same direction, it is also important to identify and quantify the effect of the CMB on the dust-continuum emission and also disentangle between optically thin and optically thick far-IR emission and thus yield in more robust dust and gas mass estimates.

3. ALMA [CI] surveys, CMB effect and dust opacity in the early galaxies

To assess the potential of [CI] as a gas mass tracer and to explore how to convert the [CI] line luminosities into $M_{\rm gas}$ we have been carrying out ALMA Band 6 and Band 7 surveys targeting CI[1-0] emission from a sample of ~70 MS and SBs at $z = 0.5 - 1.5$, that also benefit from robust far-IR properties ($T_{\rm d}$ $L_{\rm IR}$ $M_{\rm dust}$) obtained from their well sampled mid to far-IR SEDs from *Spitzer*and *Herschel*. Preliminary results published by Valentino *et al.* 2018, indicate that for fixed $L_{\rm IR}$, MS galaxies have on average higher CI luminosities with respect to local and high$-z$ SBs something that closely resembles the $L'_{\rm CO}-L_{\rm IR}$ relation were again MS galaxies here in green exhibit on average higher $L'_{\rm CO}/L_{\rm IR}$ ratios, indicative of higher $M_{\rm gas}$/SFR therefore of longer gas depletion time scales and lower star formation efficiencies. Most importantly, when the [CI] lines are combined with the traditional CO and dust-based $M_{\rm gas}$ estimates, Valentino *et al.* 2018 reports a [CI] / [H_2] abundance ratio of $\sim 1.6 - 1.9 \times 10^{-5}$ that, for the given set of assumptions is $3 - 8$ times larger than that of SBs galaxies at any redshift (Figure 3 left).

Another diagnostic power of the neutral atomic carbon is that thanks to its simple 3 level structure the CI[2-1]/CI[1-0] line ratio can be used to infer the excitation temperature of the gas ($T_{\rm exc}$, e.g., Weiß*et al.* 2003). Since under LTE conditions $T_{\rm exc} \approx T_{\rm kin} \approx T_{\rm d}$, the line ratio can be used to shed light into the "true" $T_{\rm d}$ of high$-z$ starbursts that might appear to be cold due to the fact the their far-IR emission remains optically thick out to $\lambda_{rest} \sim 200 - 300\mu$m while in fact their ISM is characterised by much a higher $T_{\rm d}$. To test this, Cortzen *et al.* 2019 (submitted), used NOEMA observation to measure the CI[2-1]/CI[1-0] line ratio of GN20, a template star-bursting galaxy at $z = 4.05$ (e.g. Pope *et al.* 2006, Magdis *et al.* 2011, Tan *et al.* 2014) The [CI] data yield a $T_{\rm exc} \sim 47 \pm 8$ which is much higher than the $T_{\rm d}$ of 33$\pm$2 K derived under the assumption of an optically thin

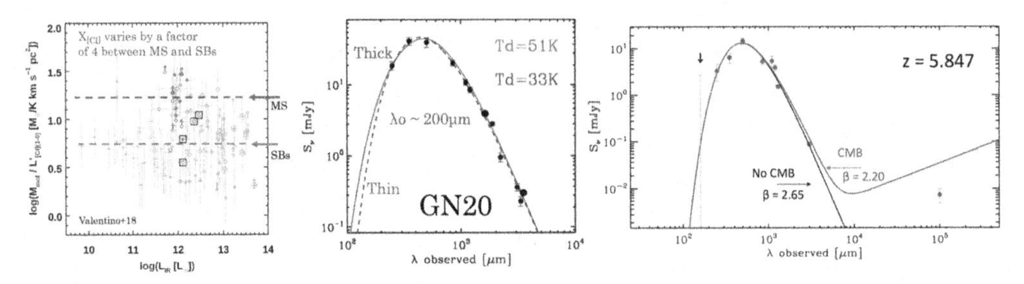

Figure 3. Left) $M_{\rm gas}$/lci vs $L_{\rm IR}$ for local and high-z SBs (grey) as well as z 1.0 MS galaxies from Valentino *et al.* 2018. Middle) Optically thin and optically thick fit to the far−IR continuum data of GN20. Right) The SED of a SB galaxy from Jin *et al.* 2019 with the best fit MBB models with and without the effects of the CMB. The black line corresponds to the apparent SED of the source while the red line to the intrinsic SED when corrected for the CMB effects.

modified black body. Instead, they show that when a general opacity model is adopted, the derived $T_{\rm d}$ ($\sim 52\pm 4\,$K) is much closer the one indicated by the [CI] data, suggesting that GN20 becomes optically thin at $\lambda_{rest} \sim 200\mu$m (Figure 3 middle). This pilot study serves as first evidence that actual $T_{\rm d}$ high$-z$ SB galaxies is much warmer than previously thought, providing an elegant solution to the puzzling observation of SBs becoming colder than MS galaxies at $z > 2$ and mitigates their unphysically high $M_{\rm dust}/M_*$ ratios reported in the literature (e.g. Bethermin *et al.* 2015), since higher $T_{\rm d}$ correspond to lower $M_{\rm dust}$ (and consequently lower $M_{\rm gas}$) estimates.

These findings are very timely with respect to the recent discovery of a population of very cold high$-z$ ($z \sim 4-6$) SBs by Jin *et al.* 2019 (submitted). Using ALMA spectral scanning they confirmed the redshifts of high$-z$, dusty star-bursting candidates in the COSMOS field, which turned out to have (under the opticaly thin assumption) considerably lower dust temperatures with respect to MS galaxies at the same redshifts. An optically thick far-IR emission that would yield higher $T_{\rm d}$ is a very appealing solution to this puzzling observation, that is also in line with the small/compact sizes of the dust emitting region in these galaxies. Furthermore, the most notable characteristic of these sources is that we find for the first time direct observational evidence of the effect of the CMB on the R-J continuum emission of high$-z$ galaxies. This is revealed through the uphysically steep spectral indexes ($\beta = 2.5-3.5$) that are recover when CMB effects are neglected. On the other hand, when the CMB effects are taken into account (following the recipe described in da Cunha *et al.* 2013), we recover reasonable β values in the range of 1.8 − 2.2. The SED along with the fits with and without the inclusion of the CMB effects for one for the sources from Jin *et al.* 2019 is shown in Figure 3 (right).

4. Summary

While the CO and dust methods to trace $M_{\rm gas}$ has provided a robust and coherent picture out to $z \sim 3$, pushing molecular gas studies at higher redhsifts require alternative techniques mainly due to the dependance of the two methods on metallicity dependance and the CMB effects that hamper the detectability of both dust emission in the R-J regime as well as of the low$-J$ CO lines at high redshifts. Our ALMA surveys, combined with literature data, suggest that the [CI] line that has been put forward as an alternative gas tracer also suffers from similar systematics, with a varying [CI] / [H_2] abundance ratio among different galaxy types. On top of the we provide evidence that the far-IR emission of high$-z$ starbursts is optically thick with considerably higher dust temperatures and thus lower $M_{\rm dust}$ and $M_{\rm gas}$ estimates than previously inferred. Similarly, we find first direct observations of the impact of the CMB in the dust continuum emission of a population of "apparently" cold $z > 3$ SBs thanks to the abnormally high β values that we recover when fitting the ALMA continuum data in the R-J tail without taking into account the effects of the CMB. When these effects are taken into consideration the derived β values are more consistent with the well established $\beta = 1.5 - 2.0$ range.

The presented results are based on the works of Valentino *et al.* 2018, Jin *et al.* 2019 (submitted) and Cortzen *et al.* 2019 (submitted) with invaluable contributions from E. Daddi and D. Liu.

References

Accurso, G., Saintonge, A., Catinella, B., *et al.* 2017, *MNRAS*, 470, 4750
Bethermin, M., Daddi, E., Magdis, G., *et al.* 2015, *A&A*, 573, A113
Bourne, N., Dunlop, J. S., Simpson, J. M. *et al.*, 2019, *MNRAS*, 482, 3135
Carilli, C. L., & Walter, F. 2013, *Annual Review of A&A*, 51, 1
Cortzen, I., Garrett, J., Magdis, G., *et al.* 2019, *MNRAS*, 482, 1618
da Cunha, E., Groves, B., Walter, F., *et al.* 2013, *ApJ*, 766, 13
Daddi, E., Bournaud, F., Walter, F., *et al.* 2010, *ApJ*, 713, 686
Daddi, E., Elbaz, D., Walter, F., *et al.* 2010, *ApJ*, 714, L118
Gao, Y. & Solomon, P. M., 2004, *ApJ*, 606, 271
Gracia-Carpio, J., Garcia-Burillo & Santiago, P., *et al.* 2008, *A&A*, 479, 703
Groves, B., A., Schinnerer, E., Leroy, A., *et al.* 2015, *ApJ*, 799, 96
Hodge, J. A., Swinbank, A. M., Simpson, J. M., *et al.* 2016, *ApJ*, 833, 103
Madden, S. C., Geis, N., Genzel, R., *et al.* 1993, *ApJ*, 407, 579
Jin, *et al.* 2019, *ApJ*, 887, 15
Magdis, G., Daddi, E., Elbaz, D., *et al.* 2011, *ApJ*, 740, 15
Magdis, G. E., Daddi, E., Bethermin., *et al.* 2012, *ApJ*, 760, 6
Magdis, G. E., Rigopoulou, D., Daddi, E., *et al.* 2017, *A&A*, 603, A93
Papadopoulos, P. P., Thi, W. F., & Viti, S. 2004, *MNRAS*, 351, 147
Pope, A., Scott, D., Dickinson, M., *et al.* 2006, *MNRAS*, 370, 1185
Sanders, D. B., Scoville, N. Z., & Soifer, B. T., 1991, *ApJ*, 370, 158
Sargent, M., Daddi, E., Bethermin, M., *et al.* 2014, *ApJ*, 793, 19
Scoville, N., Lee, N., Vanden Bout, P., *et al.* 2017, *ApJ*, 837, 150
Simpson, J. M., Smail, I., Swinbank, A. M., *et al.* 2017, *ApJ*, 839, 58
Solomon, P. M., Downes, D., Radford, S. J. E., *et al.* 1997, 478, 144
Solomon, P. M., & Vanden Bout, P. A. 2005, *ARAA*, 43, 677
Tacconi, L. J., Genzel, R., Smail, I., *et al.* 2008, *ApJ*, 680, 246
Tacconi, L. J., Genzel, R., Saintonge, A., *et al.* 2018, *ApJ*, 853, 179
Tan, Q., Daddi, E., Magdis, G., *et al.* 2014, *A&A*, 569, A98
Valentino, F., Magdis, G. E., Daddi, E, *et al.* 2018, *ApJ*, 869, 27
Walter, F., Weiß, A., Downes, D., *et al.* 2011, *ApJ*, 730, 18
Weiß, A., Henkel, C., Downes, D., & Walter, F. 2003, *A&A*, 409, L41
Zanella, A., Daddi, E., Magdis, G., *et al.* 2018, *MNRAS*, 481, 1976

Uncovering Early Galaxy Evolution in the ALMA and JWST Era
Proceedings IAU Symposium No. 352, 2019
E. da Cunha, J. Hodge, J. Afonso, L. Pentericci & D. Sobral, eds.

doi:10.1017/S1743921319009116

ALPINE: The ALMA [*CII*] survey of normal star-forming galaxies at $4 < z < 6$

Olivier Le Fèvre[1], Matthieu Bethermin[1], Andreas Faisst[2], P. Capak[2], P. Cassata[3], J. D. Silverman[4], D. Schaerer[5], L. Yan[2] and the ALPINE team[6]

[1]Aix Marseille Université, CNRS, LAM (Laboratoire d'Astrophysique de Marseille) UMR 7326, 13388, Marseille, France
emails: olivier.lefevre@lam.fr, matthieu.bethermin@lam.fr

[2]IPAC, California Institute of Technology, 1200 East California Boulevard, Pasadena, CA 91125, USA
emails: afaisst@ipac.caltech.edu, capak@ipac.caltech.edu, lyan@ipac.caltech.edu

[3]INAF Osservatorio di Padova and Dipartimento di Fisica e Astronomia, Università di Padova, Vicolo dell'Osservatorio, 3 35122 Padova, Italy
email: paolo.cassata@unipd.it

[4]Kavli Institute for the Physics and Mathematics of the Universe, The University of Tokyo, Kashiwa, Japan 277-8583 (Kavli IPMU, WPI)
email: john.silverman@ipmu.jp

[5]CNRS and Geneva Observatory, University of Geneva, ch. des Maillettes 51, CH-1290 Versoix, Switzerland
email: Daniel.Schaerer@unige.ch

[6]ALPINE team: R. Amorin, S. Bardelli, Boquien, A. Cimatti, M. Dessauges-Zavadsky, Seiji Fujimoto, Ginolfi, M., Hemmati, S., Ibar , G. Jones, A. Koekemoer, G. Lagache , B. Lemaux, R. Maiolino, P. Oesch, Pozzi, Riechers, M. Talia, L. A. M. Tasca, R. Thomas, L. Vallini, D. Vergani, F. Walter, G. Zamorani, and E. Zucca

Abstract. The ALMA-ALPINE [CII] survey (A2C2S) aims at characterizing the properties of normal star-forming galaxies (SFGs) observed in the [CII]-158μm line in the period of rapid mass assembly at redshifts $4 < z < 6$. Here we present the survey and the selection of 118 galaxies observed with ALMA, selected from large samples of galaxies with spectroscopic redshifts derived from UV-rest frame. The observed properties derived from the ALMA data are presented and discussed in terms of the overall detection rate in [CII] and far-IR continuum. The sample is representative of the SFG population at these redshifts. The overall detection rate is 61% down to a flux limit of 0.07 mJy. From a visual inspection of the [CII] data cubes together with the large wealth of ancillary data we find a surprisingly wide range of galaxy types, including 32.4% mergers, 25.7% extended and dispersion dominated, 13.5% rotating discs, and 16.2% compact, the remaining being too faint to be classified. ALPINE sets a reference sample for the gas distribution in normal star-forming galaxies at a key epoch in galaxy assembly, ideally suited for studies with future facilities like JWST and ELTs.

Keywords. Galaxies: high redshift – Galaxies: formation – Galaxies: star formation

1. Introduction

The mass assembly in galaxies at different epochs proceeds from several physical processes which, together, produce the remarkable observed evolution of the star formation rate density (SFRD) with cosmic time (Silk & Mamon (2012); Madau & Dickinson (2014); Dayal & Ferrara (2018)). The SFRD first rises during the reionization epoch, to reach a peak at z$\sim 2-3$ after a $\sim$1 dex increase in $\sim$3 Gyr, then decreases by $\sim$0.8 dex in $\sim$10 Gyr to the current time (Madau & Dickinson (2014); Bouwens *et al.* (2015)). Along with star formation, the total stellar mass density (SMD) in galaxies is observed to rise steeply from early times to z~ 2, followed by a slower increase at z<2 Ilbert *et al.* (2013).

At the root of the SFRD and SMD evolution, the transformation of gas into stars in a hierarchical picture of galaxy assembly is a key element. Two main processes are shown from more and more detailed simulations to drive this evolution: gas accretion and galaxy-galaxy merging . This is expected to be tempered by feedback processes from gas expelled from galaxies by strong AGN and/or stellar jets and winds. While this is appealing from a theoretical and simulation standpoint, there is actually very little observational support towards a comprehensive, consistent, and quantitative picture, particularly at early cosmic epochs when mass assembly is in a major phase. Galaxy mergers are observed at all epochs, with a major merger rate increasing to $z\sim 2$ possibly flattening to $z\sim 4-5$, while gas accretion suffers from weak signatures difficult to identify observationally. On the other hand, feedback processes are directly measured e.g. Le Fèvre *et al.* (2019), and thought to affect both the bright (AGN) and faint end of the galaxy luminosity function (LF) Croton (2006).

To disentangle the relative contributions of these processes, the far infrared (FIR) domain redshifted in the sub-mm for high-z galaxies is proving particularly rich. From the sub-mm it is now possible to investigate the properties of star-forming galaxies up to the epoch of HI reionisation. The [CII]-158μm line is the dominant coolant making it one the strongest FIR lines. The primary [CII] emission is from photo-dissociation regions (PDR) and cold neutral medium (CNM) of molecular clouds. [CII] at high-z has raised considerable interest as it probes the gas from which stars form in normal galaxies, and then broadly traces star formation activity, offering an important window on galaxy formation Carilli & Walter (2013). This led to the detection of strong [CII] emitters, up to very high redshifts Capak *et al.* (2015), an easier measurement than the FIR continuum. Searching for [CII] emission, interpreting and simulating the observations, and comparing with other line emission like Lyman-α, has therefore become a major new way of studying high-z galaxies. The strong UV radiation in high-z galaxies results in a non-negligible fraction of [CII] emission from the extended warm ISM. The evolution of [CII] emission and its resolved velocity profile provide important information on the SFR and ISM properties, setting constraints on the dynamical and gas masses of galaxies. The morphology of the [CII] emission indicates if star formation is compact or extended, an important element to understand high-z star formation . Beyond the line flux, the FIR continuum emission adjacent to [CII] is near the peak of the FIR emission. It constrains the total FIR luminosity, and provides a good measure of the total SFR when combined with UV continuum measurements. The continuum flux can also be combined with UV colors and luminosity to construct the Infrared-Excess (L_FIR/L_UV, so-called IRX) vs. UV color (β) diagnostic, providing insight into the spatial distribution of dust, dust grain properties, and metallicity.

Simulations of forming galaxies during and right after reionization are informing on the possible properties of these galaxies despite the difficulty to take into account early galaxy formation processes during and right after the EoR in a consistent way. Specific predictions related to [CII] emission are useful to guide and compare to observations.

This whole domain opened up at high redshift $z > 4$ with ALMA becoming fully operational, when it was realized from pilot observations that detecting [CII] for normal galaxies was ubiquitous even with short on-source exposure times Capak *et al.* (2015). Galaxies with star formation rates as low as a few $M_{\odot}.yr^{-1}$ have been reported at z~ 5 Capak *et al.* (2015), and [CII] is now detected for galaxies well into the reionisation epoch.

However, existing observations of [CII] in normal galaxies at these epochs are still scarce. While strong sub-mm sources have been primarily targeted, they provide a view biased towards the highly star-forming population with SFR$>1000\ M_{\odot}.yr^{-1}$. Normal galaxies, that is galaxies with SFR~ 10 to a few hundred, lying on the so-called main sequence at these redshifts Khusanova *et al.* (2019), have not been observed in statistically representative numbers. The Capak *et al.* (2015) observations proved that this was feasible and prompted us to submit the ALPINE Large Program, largely designed based on the properties of the Capak *et al.* (2015) sample. A key element was the availability of large samples of these normal galaxies, with accurate *spectroscopic redshifts* to be able to define ALMA observations with a high success rate in detecting [CII].

In these proceedings, we present the sample selection as in the original proposal. We give an overview of the ALMA observations. We then describe the main properties of the sample, including the redshift distribution, detection rates in [CII] and continuum, and observed flux limits. We give examples of maps of sample galaxies in the [CII] line.With these maps, kinematic data, and all ancillary imaging data we perform an empirical visually-based morpho-kinematic classification.

2. Sample selection

The sample is drawn from large spectroscopic survey samples of normal SFGs in the COSMOS and ECDFS fields.

A key element is that galaxies must have a reliable spectroscopic redshift in $4.4 < z_{spec} < 5.8$ ($< zspec > \sim 4.7$, excluding $4.65 < z < 5.05$ where [CII] falls in a low transmission atmospheric window). Galaxies are UV-selected with $L_{UV} > 0.6L^*$ to include most of the star formation traced by the UV, and excluding AGN. Accurate redshifts come from extensive spectroscopic campaigns at the VLT Le Fèvre *et al.* (2015) and Keck Hasinger *et al.* (2018), unbiased against Lyman-α emitters or absorbers. The absolute UV luminosity cut ($M_{UV} < -20.2$) is equivalent to SFR$>$10 $M_{\odot}/yr$. Using the relation for high-z galaxies Capak *et al.* (2015), this SFR limit is equivalent to $L[CII] > 1.2 \times 10^8 L_{\odot}$. This sample is representative of the overall SFG population, rather than ULIRGS, that is with galaxies essentially positioned on or near the so-called main sequence in the SFR versus M_{star} plane observed at these redshifts (Tasca *et al.* (2015), Khusanova *et al.* (2019)).

3. ALMA observations

This program was awarded an ALMA Large Program status under number 2017.1.00428.L for a total of 69 hours. ALMA observations were carried out in Band-7 starting in May 2018 during Cycle 5 and completed in February 2019 in Cycle 6. Each target was observed for about 30 minutes and up to one hour of on-source integration time, pointing at the rest-frame UV positions of the sources. The availability of spectroscopic redshifts allowed to accurately set the main spectral window on the expected [CII] frequencies. The other side-band was then used for FIR continuum measurements.

At these redshifts, the velocity width of one ALMA bandpass in band-7 is as narrow as $\sim$3000 km/s. Samples with photometric redshifts accurate to $\sim 0.05 \times (1+z)$ Ilbert *et al.* (2013), would have added a considerable uncertainty on the detection of [CII] and associated incompleteness, making it more hazardous to build e.g. the [CII] luminosity function. With an accuracy of a few hundred km/s even at low spectral resolution from optical (UV rest-frame) spectroscopy Le Fèvre *et al.* (2015), the availability of z_{spec} is therefore a key element of this program ensuring a high [CII] detection rate and, for those galaxies which would be undetected, setting stringent upper limits.

We use the TDM mode of the ALMA correlator, which offers the largest bandwidth to optimize the continuum sensitivity. The resolution varies with redshift from 26 to 35 km/s. We assumed 235 km/s FWHM line width (or sigma ~100 km/s), which is the average width measured in the Capak *et al.* (2015) sample. Emission lines were thus expected to be spectrally resolved giving the possibility to measure the line width when the SNR is sufficient.

We privilege detection over spatial and spectral resolution, and with the typical size of the Capak *et al.* (2015) sources being 0.5-0.7 arcsec, we elected to use ALMA array configurations offering an angular resolution not larger than 0.7 arcsec. The median beam size of the ALMA observations is then about 0.7 arcsec FWHM.

4. Detection rate

The Signal to noise ratio (SNR) obtained on the [CII] line detection has a median $SNR \simeq 6.2$. See Béthermin *et al.* (2019, in prep.) for more details. Taking 3.5σ as a conservative detection limit, ALPINE detected [CII] in 73 galaxies over 118, hence a success rate of ~62%. In the continuum adjacent to [CII], 25 galaxies, or 21%, are detected. These rates are quite impressive given the redshift of the sources and relatively short integration times. The SNR in [CII] of most other targets varies in the range from 0.5 to 3, providing useful upper limits.

5. [CII] maps and morpho-kinematic classification

Example images in the [CII] line of sources with [CII] detected at more than 3.5σ are presented in Fig. 1.

These images give a first view of the shape of the [CII] emission in normal galaxies at $4 < z < 6$. There are several facts worth noting. Even though the observations were carried out with a beam size providing moderate spatial resolution with FWHM~0.7 arcsec, about two third of the sources are resolved in [CII]. This means that intrinsic (total) sizes must be reaching several kilo-parsecs. By itself this fact gives an indication that physical processes at work in those galaxies are puffing up their sizes beyond being compact. Another striking evidence from these images is the large diversity of [CII] emission morphology. Some objects appear as very extended, some others with double merger-like components while others are compact (unresolved). This diversity must also reflect a diversity in the physical processes at work.

We perform an empirical visual-based morpho-kinematic classification using the [CII] maps the velocity and spectral information in the 3D cubes and all optical and NIR ancillary data. We find 32.4% of galaxies in the merger class 2, 25.7% in the extended and dispersion dominated class 3, 13.5% rotating discs in class 1, and 16.2% compact in class 4, the rest of the sample being too difficult to classify (class 5). We note the high fraction of mergers, indicating that mass assembly through merging is frequent at these redshifts for normal main sequence SFGs. Preliminary examination of spatial and velocity information indicates that most merging systems would merge within 0.5 to 1 Gyr (see the triple merger system presented in Jones *et al.* (2019)), which then means that most of these mergers would end-up forming one single galaxy by $z \sim 2.5$.

6. Conclusions

The ALMA-ALPINE [CII] survey (A2C2S) provides an unprecedented view of a representative sample of 118 star-forming galaxies observed in formation right after the end of HI reionisation at redshifts $4 < z < 6$. Galaxies are selected on the basis of an existing reliable spectroscopic redshift, and using the SED-based SFR to predict the [CII] flux using the De Looze *et al.* (2014) relation and selecting SFR such that $L[CII] > 1.2 \times 10^8 L_{\odot}$. The overall detection rate is 61% for galaxies detected in [CII] 3.5σ above the noise,

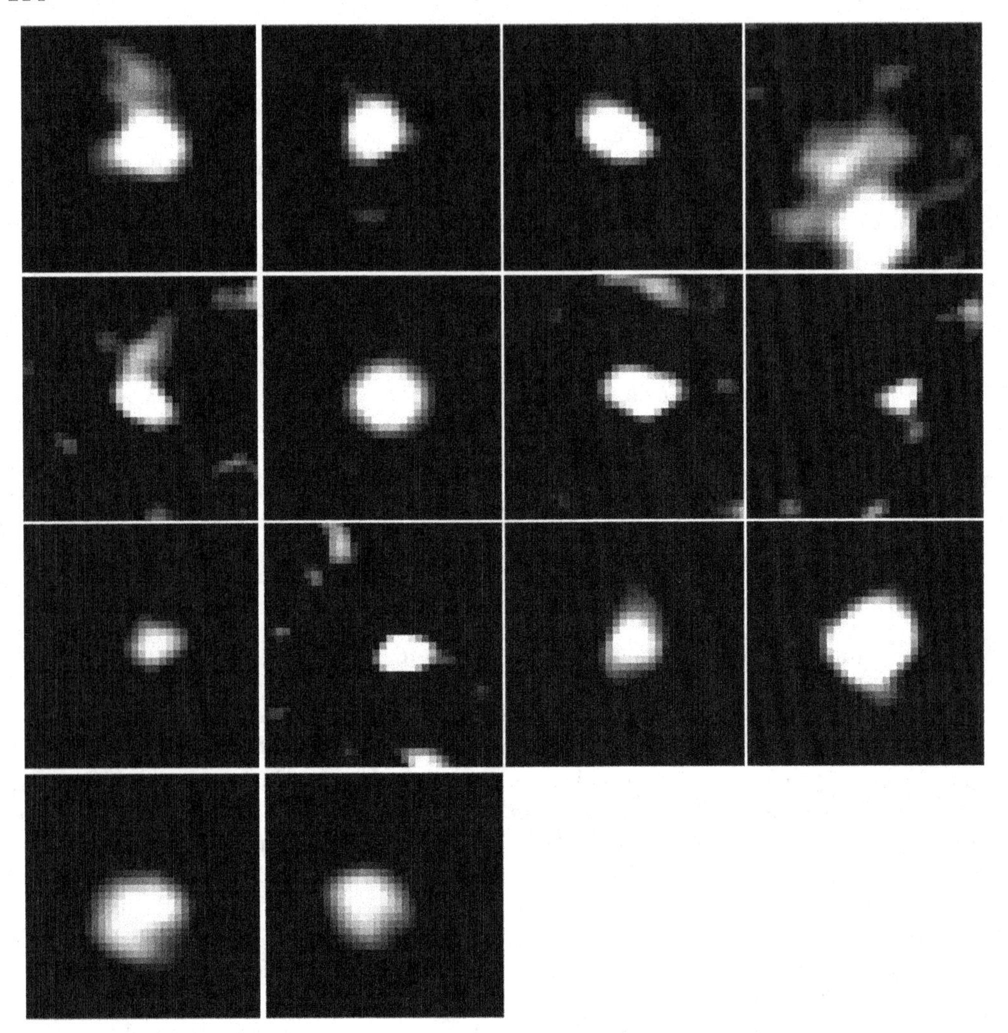

Figure 1. [CII] 'moment 0' maps obtained from projecting the ALPINE cubes along all frequencies corresponding to ±1500 km/s around the peak [CII] flux, unless cut by the edge of the ALMA bandpass (see text). Each panel is 5×5 arcsec^2 or about 6 kpc on a side at the mean redshift of the survey. From top to bottom, left to right, objects vuds-cosmos-5100822662, vuds-cosmos-5100969402, vuds-cosmos-5100994794, vuds-cosmos-5101209780, vuds-cosmos-5101210235, vuds-cosmos-5101218326, vuds-cosmos-5101244930, vuds-cosmos-5101288969, vuds-cosmos-510596653, uds-cosmos-510605533, vuds-cosmos-510786441, uds-cosmos-5110377875, vuds-cosmos-5180966608, vuds-efdcs-530029038. These images show the large range of morpholopy in[CII], from compact to extended and resolved or merger-like.

down to a flux limit of 0.07 mJy. projected. Combining projected [CII] 'moment zero' maps with velocity channel maps, velocity field, and all available ancillary information, we establish a classification scheme. We find a surprisingly wide range of galaxy types, including 32.4% mergers, 25.7% extended and dispersion dominated, 13.5% rotating discs, and 16.2% compact, the remaining being too faint to be classified. This diversity of types indicates that several physical processes are at work to assemble mass in these galaxies, first and foremost galaxy-galaxy merging.

References

Bouwens, R. J., Illingworth, G. D., Oesch, P. A., *et al.* 2015, Ap.J., 803, 34
Capak, P. L., Carilli, C., Jones, G., *et al.* 2015, Nature, 522, 455
Carilli, C. L. & Walter, F. 2013, Annual Review, Astronomy and Astrophysics, 51, 105
Croton, D. J. 2006, MNRAS, 369, 1808
Dayal, P. & Ferrara, A. 2018, Physics Review, 780, 1
De Looze, I., Cormier, D., Lebouteiller, V., *et al.* 2014, A&A, 568, A62
Hasinger, G., Capak, P., Salvato, M., *et al.* 2018, Ap.J., 858, 77
Ilbert, O., McCracken, H. J., Le Fèvre, O., *et al.* 2013, A&A, 556, A55
Jones, G. C., Bethermin, M., Fudamoto, Y., *et al.* 2019, arXiv e-prints, arXiv:1908.07777
Khusanova, Y., Le Fèvre, O., Cassata, P., *et al.* 2019, arXiv e-prints, arXiv:1903.01884
Le Fèvre, O., Tasca, L. A. M., Cassata, P., *et al.* 2015, A&A, 576, A79
Le Fèvre, O., Lemaux, B. C., Nakajima, K., *et al.* 2019, A&A, 625, A51
Madau, P. & Dickinson, M. 2014, Annual Review, Astronomy and Astrophysics, 52, 415
Silk, J. & Mamon, G. A. 2012, Research in Astronomy and Astrophysics, 12, 917
Tasca, L. A. M., Le Fèvre, O., Hathi, N. P., *et al.* 2015, A&A, 581, A54

Uncovering Early Galaxy Evolution in the ALMA and JWST Era
Proceedings IAU Symposium No. 352, 2019
E. da Cunha, J. Hodge, J. Afonso, L. Pentericci & D. Sobral, eds.

doi:10.1017/S1743921319009189

Near-IR spectroscopic studies of galaxies at $z \sim 1-3$

Irene Shivaei

Steward Observatory, University of Arizona, Tucson, AZ 85721, USA, Hubble Fellow
email: ishivaei@email.arizona.edu

Abstract. ISM comprises multiple components, including molecular, neutral, and ionized gas, and dust, which are related to each other mainly through star formation – some are fuel for star formation (molecular gas) while some are the products of it (ionized gas, dust). To fully understand the physics of star formation and its evolution throughout cosmic time, it is crucial to measure and observe different ISM components of galaxies out to high redshifts. I will review the current status of near-IR studies of galaxies during the peak of star formation activity ($z \sim 1-3$). Using rest-frame optical emission lines, we measure dust, star formation, and gaseous properties of galaxies. *JWST* will advance such studies by probing lower luminosities and higher redshifts, owing to its significantly higher sensitivity. Incorporating ALMA observations of cold dust and molecular gas at $z > 1$ will give us a nearly complete picture of the ISM in high-redshift galaxies over a large dynamic range in mass.

Keywords. galaxies: high-redshift, galaxies: abundances, galaxies: evolution, galaxies: general, galaxies: ISM, dust, extinction

1. Introduction

The interstellar medium (ISM) comprises various components, including molecular, neutral, and ionized gas, large dust grains in thermal equilibrium with their surrounding, and small dust grains, such as PAHs, heated by single photons (Figure 1). The different components of ISM are related to each other mainly through star formation – some are fuel for star formation (molecular gas) while some are the products of it (ionized gas, dust). Therefore, to fully understand the physics of star formation, its evolution, and its connection to metal and dust enrichment, it is crucial to trace various ISM components through a range wavelengths. The ionized phase of ISM, which is the focus of this review, is gas photoionized by the energetic photons of hot stars, and, if present, AGN. It has a temperature of $T \sim 10^4$ K and can be in dense or diffuse regions with densities of $n_H \sim 0.3 - 10^4$ cm^{-3} (Draine 2011). Some the main diagnostics of the properties of ionized gas are the nebular emission lines in the rest-frame optical spectra of galaxies, including hydrogen Balmer lines, Hα, Hβ, and atomic fine-structure lines of [OII], [OIII], [NII], and [SII].

In this review, I focus on the advances of our understanding in ISM physics at $z \sim 1-3$, corresponding to ~$2-6$ billion years after Big Bang. Redshifts of $z \sim 1-3$ are an important era to study star formation activity and metal and dust enrichment in the history of universe, as galaxies in that era were in the process of assembling most of their stellar mass and quasar activity was at its peak (Madau & Dickinson 2014). At these redshifts, the optical nebular emission lines are redshifted to near-IR wavelengths. Owing to the high near-IR terrestrial background, obtaining rest-frame optical spectra of large samples of galaxies at $z \sim 2$ has been very challenging prior to the advent of

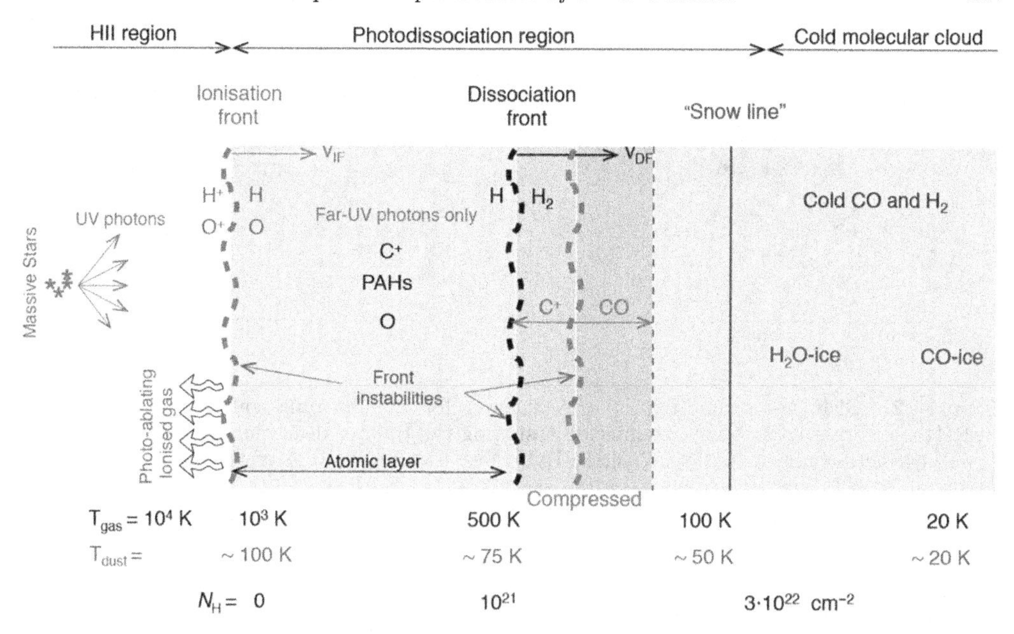

Figure 1. Structure of a UV-irradiated molecular cloud edge (Goicoechea *et al.* 2016)

multi-object near-IR spectrographs on 10-m class telescopes. As a result, the existing near-IR surveys were limited to small samples using the classical single-object long-slit instruments (e.g., Erb *et al.* 2006a) or larger samples of low resolution grism data (e.g., Brammer *et al.* 2012). The multiplexing capability and high sensitivity of revolutionary instruments such as MOSFIRE on the 10-m Keck (first light in 2012; McLean *et al.* 2012), KMOS on the 8.2-m VLT (first light in 2012; Sharples *et al.* 2013), FMOS on the 8.2-m Subaru (first light in 2008; Kimura *et al.* 2010), and LUCI on the 8.4-m LBT (first light in 2008; Ageorges *et al.* 2010), enabled the first statistically large studies of ionized ISM at high redshifts. Among these large spectroscopic surveys of $z \sim 1-3$ galaxies are KBSS-MOSFIRE (Steidel *et al.* 2016; Strom *et al.* 2017b) and MOSDEF (Kriek *et al.* 2015) with respectively ~1100 and ~1500 MOSFIRE spectra of galaxies at $z \sim 1.5-3.5$, KMOS-3D with KMOS near-IR integral field spectra of over 700 galaxies at $z \sim 0.6-2.7$ (Wisnioski *et al.* 2015), and FMOS-COSMOS with ~1900 FMOS spectra at $z \sim 1.6$ (Kashino *et al.* 2019).

In future, with *JWST* and its cutting-edge instruments, we will be able to build upon the current high-redshift near-IR surveys by probing fainter and higher redshift objects. The multi-object NIRSpec spectrograph on board of *JWST* will provide us with near-IR spectra out to 5 μm, in the absence of terrestrial background and sky line contamination. NIRCam camera will obtain high resolution imaging out to 5μm, and with MIRI we will be able to acquire mid-IR photometry out to 28 μm, with much higher sensitivity and spatial resolution compared to its predecessor, *Spitzer*.

2. Optical emission line diagnostics

In the rest-frame optical spectra, there are multiple strong nebular emission lines that can be used as diagnostics of the physical conditions of ionized gas (e.g., excitation and ionization properties, electron densities and temperatures, chemical abundances), stars (star-formation rate and production rate of ionizing photons), and dust (nebular reddening). In this section, I will briefly review the observational analysis and diagnostic

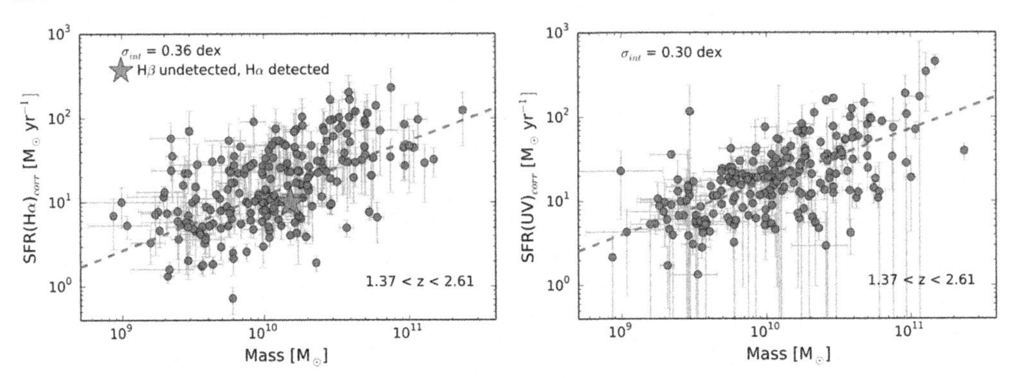

Figure 2. SFR as a function of stellar mass for star-forming galaxies at $z\sim2$. *Left*: SFR(Hα) – corrected for dust attenuation assuming the Balmer decrement and a Cardelli *et al.* (1989) extinction curve – vs. M_*. *Right*: SFR(UV) vs. M_*. The SFR is inferred from the UV luminosity at 1600 Å and is dust corrected by the UV slope, both determined from the multiband photometry. The green lines show the regression lines fitted to the log(SFR)-log(M_*) relation. The measurement-subtracted scatter in SFR (i.e., the scatter after the subtraction of SFR and mass measurement uncertainties) is provided in the upper left corner of each plot. Figure from Shivaei *et al.* (2015b).

measurements of such properties using near-IR (rest-frame optical) spectra of $z\sim2$ galaxies. In each section, the potential synergies with longer wavelength IR/submm facilities, such as *Sptizer*, *Herschel*, ALMA, and future *JWST* are also briefed.

2.1. *Star formation rate and stellar ionizing radiation*

The brightest emission line in the rest-frame optical spectra of galaxies is the Hα Balmer line, which is emitted from the ionized gas around the most massive and hot stars, those with O and early-B spectral types. As the lifetime of these stars on the main sequence is very short ($\sim$10 Myr), hydrogen nebular lines are considered as nearly instantaneous tracers of SFR. The conversion of Hα luminosity to SFR has been studied by numerous authors and is widely used at low and high redshifts (see the review by Kennicutt & Evans 2012). An important correction that needs to be applied to the observed Hα luminosity to recover the total SFR, is dust attenuation correction. The reddening or color excess ($E(B-V)$) can be calculated using various methods, with the ratio of the two Balmer lines (i.e., Balmer decrement, see Section 2.4) being the most direct one for nebular emission lines (Balmer decrement, see Section 2.4).

Figure 2 shows a comparison of the two SFR indicators, the attenuation-corrected Hα and UV SFRs, in the SFR-M_* plane at $z\sim2$. In principle, the two SFR indicators trace SFR on different timescales, and hence the difference in the scatter of the two SFR-M_* relations (the σ values on the corners of the plots in Figure 2) are thought to indicate recent star-formation burstiness. However, there are multiple factors in addition to a recent star formation burst that may cause deviations from the nominal value of the SFR(Hα)-to-SFR(UV) ratio in a galaxy. Figure 5 shows the distribution of the ratio of dust attenuation corrected SFR(Hα)-to-SFR(UV) for a large sample of $z\sim2$ galaxies from the MOSDEF survey (Shivaei *et al.* 2018). As shown, there is a large scatter in the observed values with an intrinsic scatter of 0.28 dex. Part of the scatter and the uncertainty in such calculations are associated with uncertainties in the assumed dust attenuation curve (in this case, the Calzetti starburst curve versus the SMC extinction curve). Such uncertainties are discussed in more detail in Section 2.4. Other than dust correction uncertainties, galaxy-to-galaxy variations in the IMF (the high-mass end slope or cutoff), stellar metallicity, and stellar rotation and binarity may also alter the rate

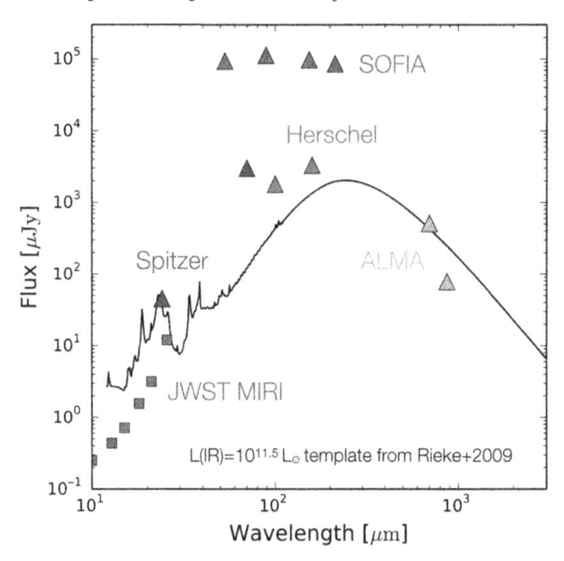

Figure 3. 3σ detection limits for 1-hour integrations with the major previous, current, and future IR and sub-mm facilities. The limits for MIPS 70 and 24μm are calculated based on the confusion limits from Frayer *et al.* (2006). The black SED is the $10^{11.5}\,L_\odot$ template of Rieke *et al.* (2009), represented as the IR emission of a *typical* $z \sim 2$ galaxy (based on the IR luminosity function of Magnelli *et al.* (2011), L(IR)* is $10^{11.83\pm0.13}\,L_\odot$ at $z \sim 2$).

of ionizing photons production at a given SFR. These effects are shown in the lower panel of Figure 5, calculated by varying different parameters in the stellar population synthesis models. The details of the models can be found in Shivaei *et al.* (2018). The important take-away point is that although Balmer lines are tracers of the most recent SFR, their conversion to SFR should be taken with caution, particularly for high-redshift systems, which may have different stellar populations (e.g., lower stellar metallicities and more intense and/or harder ionizing radiation at a given SFR) compared to that of local galaxies.

While dust-corrected Hα SFRs are commonly used, the Balmer lines may miss optically thick star-forming regions (Shivaei *et al.* 2016). This is an important issue at redshifts of $z \sim 1-3$, where galaxies are more star forming (and hence, more dusty) at a given stellar mass compared to $z \sim 0$, and the obscured star formation dominates the total SFR at masses of $M_* \gtrsim 10^{9.5}\,\mathrm{M}_\odot$ (Whitaker *et al.* 2017). Therefore, to gain a full understanding of total star formation activity in galaxies at $z \sim 2$, both the obscured and unobscured components of star formation need to be measured. *Spitzer* and *Herschel* opened a new window into measuring bolometric SFRs by allowing us to directly trace the obscured star forming regions (Reddy *et al.* 2012a; Whitaker *et al.* 2014b; Shivaei *et al.* 2015a). However, due to the low spatial resolution and sensitivity of PACS and SPIRE, *Herschel* studies at high redshifts are limited to dusty and IR-bright galaxies. Meanwhile, the higher sensitivity of *Spitzer*/MIPS 24μm detector enabled us to detect mid-IR emission in *individual typical* galaxies at $z \sim 2$ (Figure 3). MIPS 24μm traces the PAH emission at $z \sim 2$, which can be converted to total IR luminonisity (L(IR)) by using empirical conversions (Wuyts *et al.* 2008; Reddy *et al.* 2012a) or IR templates (Chary & Elbaz 2001; Rieke *et al.* 2009; Elbaz *et al.* 2011). However, caution needs to be taken in using such conversions at low masses ($\mathrm{M}_* < 10^{10}\,\mathrm{M}_\odot$) as the PAH intensity scales with metallicity, such that at low metallicities (and low masses, owing to the mass-metallicity relation), the PAH-to-L(IR) ratio decreases by a factor of $\sim$2 (Shivaei *et al.* 2017). Figure 4 shows the good agreement between the SFRs inferred from dust-corrected Hα luminosity with

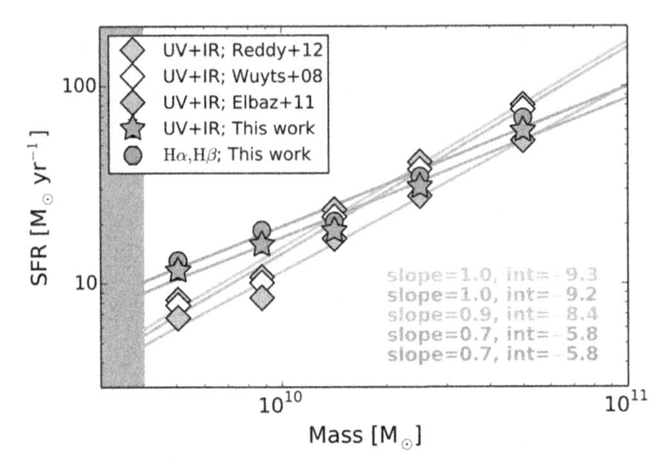

Figure 4. Star-formation rate as a function of stellar mass at $z \sim 2$. Orange circles are averages of dust-corrected Hα SFR in each bin of stellar mass. Other symbols show the sum of SFR(UV) at 1600Å and SFR(IR) derived from 24μm stacks using different conversions. Blue stars adopt the mass-dependent conversion of (Shivaei *et al.* 2017). Cyan, yellow, and purple diamonds are from single-value conversions of Reddy *et al.* (2012a), Wuyts *et al.* (2008), and Elbaz *et al.* (2011), respectively. The best-fit lines, slopes, and intercepts estimated from a simple linear least-squares regression to log(SFR) vs. log(M_*) are shown on the plot with respective colors. Figure from Shivaei *et al.* (2017).

those derived from PAH emission using the mass-dependent calibration of Shivaei *et al.* (2017). In the same figure, it is shown that using the single conversions of PAH-to-L(IR) from literature underestimates SFR at low masses, as such conversions are calibrated based on samples of massive and dusty galaxies.

ALMA bands 6 and 7 trace the Rayleigh-Jeans (RJ) tail of the IR emission at $z \sim 1-3$, which can be used to constrain dust masses and possibly the total IR emission (Aravena *et al.* 2016; Dunlop *et al.* 2017; Franco *et al.* 2018). The disadvantages of these observations are a) due to the small field of view of ALMA, it is very time consuming (and not feasible) to obtain large surveys of IR emission of typical galaxies at high redshifts, and b) the conversion of the RJ emission to L(IR) is highly dependent on the assumed IR template, which is uncertain at high redshifts and low-metallicity regimes, as most of the existing templates are locally calibrated (Casey 2012; De Rossi *et al.* 2018; Schreiber *et al.* 2018).

The higher sensitivity and angular resolution of *JWST*/MIRI compared to *Spitzer*/MIPS 24μm† will enable us, for the first time, to obtain a nearly complete census of the PAH and IR emission at high redshifts, and to detect the 7.7μm feature in *individual* galaxies down to masses as low as 10^9 M$_\odot$ (Figure 3).

2.2. *Electron density and temperature*

The average electron density (n_e) may be estimated by comparing the line fluxes of two lines of the same ion, in which the lines have nearly the same excitation energies but different collision strengths and radiative transition probabilities. In this case, the relative population of the two levels, and hence the relative intensities of the two lines, depends on the density of electrons (Osterbrock & Ferland 2006). The two commonly-used electron density-sensitive doublets are [OII]$\lambda\lambda$3726, 2729 and [SII]$\lambda\lambda$6716, 6731, with their critical densities being comparable to the average ISM electron densities of galaxies.

† *JWST*/MIRI has ~50 times the sensitivity and 7 times the angular resolution of *Spitzer*/MIPS.

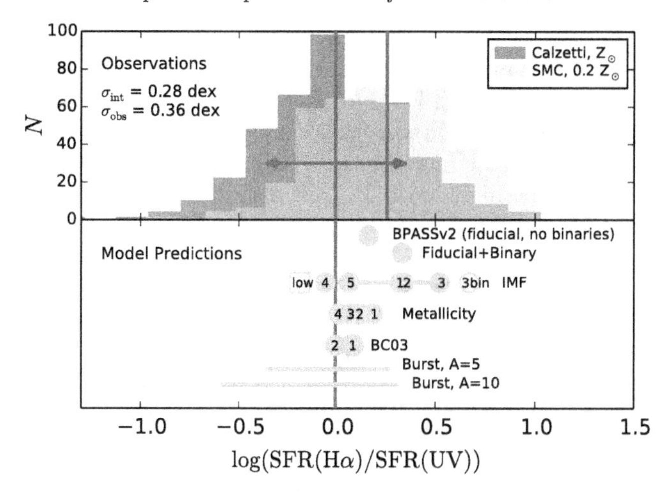

Figure 5. *Top*: distribution of SFR(Hα)-to-SFR(UV) for a sample of $z = 1.4-2.6$ galaxies, assuming the Calzetti attenuation curve for the UV dust correction in violet and an SMC curve in green. The observed and intrinsic (i.e., measurement-subtracted) scatters are 0.36 and 0.28 dex for the Calzetti distribution and are shown with black and violet arrows, respectively. The vertical lines show the averages of the sample. Assuming an SMC curve systematically increases xion by $\sim$0.3 dex, indicating the sensitivity of the SFR ratios to the assumed UV attenuation curve. *Bottom*: SFR(Hα)-to-SFR(UV) predictions from the stellar population models of BPASSv2 and BC03. Each row shows the variation of the SFRs ratio by changing the labeled quantity in the model, described in detail in Table 1 of Shivaei *et al.* (2018). Figure is modified from Shivaei *et al.* (2018).

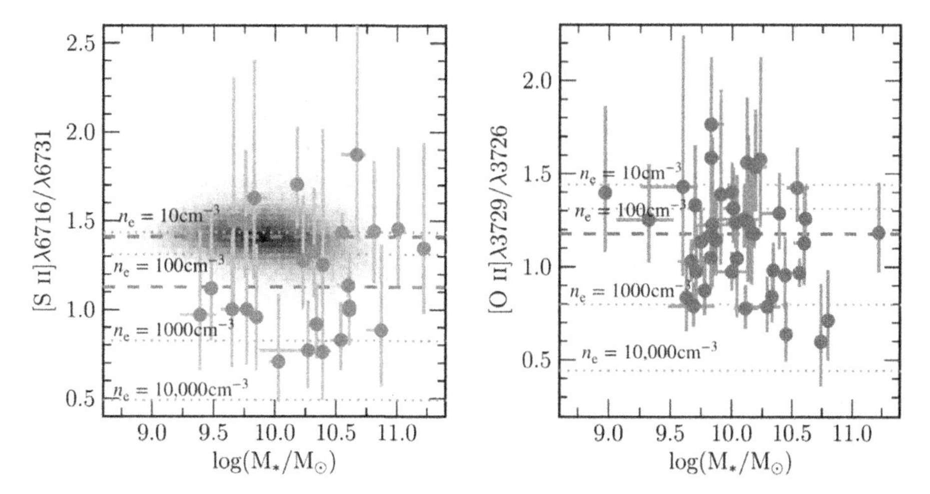

Figure 6. The ratio of [SII]$\lambda\lambda$6716, 6731 doublet (left) and [OII]$\lambda\lambda$3726, 2729 doublet (right) as a function of stellar mass at $z \sim 2.3$ (green and blue points) and $z \sim 0$ (gray histogram). The green, blue, and gray lines show the median line ratios for the corresponding samples. Dotted horizontal lines show the line ratios corresponding to electron densities of 10, 100, 1000, and 10,000 cm$^-$3. Figure modified from Sanders *et al.* (2016a).

Studies at $z \sim 2$ have shown that electron densities inferred from [OII] and [SII] doublets, assuming an electron temperature of 10^4 K, are an order of magnitude higher at $z \sim 2$ compared to the SDSS results at $z \sim 0$ (Sanders *et al.* 2016a; Strom *et al.* 2017b).

Electron temperature (T_e) of the ionized gas may be estimated by using the ratio of the intensities of two lines of the same ion but with considerably different excitation energies, which makes the ratio strongly dependent on T_e. An example of such line ratios in the optical spectra are [OIII] λ4363 auroral line to λ4959 and λ5007, and [NII]λ5755 to λ6583 and λ6548 (Osterbrock & Ferland 2006). The auroral lines are extremely week (e.g., [OII]λ4363 is typically 100 times weaker than [OIII] λ5007 at low metallicities, and even weaker in solar and higher metallicities). The faintness of these lines makes them extremely hard to detect at high redshifts, resulting in fewer than 20 galaxies at $z > 1$ that have T_e measurements (Sanders *et al.* 2019).

2.3. *Metallicity*

The most direct way of getting the abundances of ions is by taking the relative strength of their recombination lines to that of their Balmer lines. The strength of metal recombination lines are mildly dependent on electron temperature but due to the low elemental abundances, the metal recombination lines are very weak compared to hydrogen recombination lines, making this method practically limited to very bright HII regions.

The widely adopted method to determine chemical abundances in galaxies is using the relative ratio of the collisionally excited lines of elements to hydrogen recombination lines. The flux of collisionally excited lines depends on electron temperature and density – once those are determined, one can infer the ionic abundances, and by assuming an ionization correction for the unobserved ions, the elemental abundances (metallicity) can be derived. In the optical spectra, the most commonly used auroral line to derive T_e is [OIII]λ4363, and the corresponding lines to measure the oxygen abundance ($12 + \log(\frac{\mathrm{O}}{\mathrm{H}})$) are [OIII]$\lambda\lambda$4959, 5007 and Balmer lines. However, as mentioned in Section 2.2, auroral lines that are used to estimate T_e are intrinsically faint, making the direct T_e-based methods hard to conduct, particularly at high redshifts. As a result, studies rely on calibrations of strong emission line ratios to derive metallicities. Such calibrations are made empirically based on $z \sim 0$ galaxies or by using photoionization models. The most commonly used strong line diagnostics are R23 ($\frac{[\mathrm{OII}]\lambda3727+[\mathrm{OIII}]\lambda\lambda4959,5007}{\mathrm{H}\beta}$), O3N2 ($\frac{[\mathrm{OIII}]\lambda\lambda4959,5007/\mathrm{H}\beta}{[\mathrm{NII}]\lambda6584/\mathrm{H}\alpha}$), and N2 ([NII]$\lambda$6584/H$\alpha$; e.g., Kewley & Dopita 2002; Pettini & Pagel 2004). The advantage of using the two latter diagnostics is the insensitivity to dust attenuation, as the dust attenuation is wavelength dependent and the constituent lines are close in wavelength.

Studies show that galaxies at $z \sim 2$ have $\sim$0.3 – 0.5 dex lower metallicities at a given stellar mass compared to that at $z \sim 0$, depending on the metallicity indicator used (Strom *et al.* 2017b; Sanders *et al.* 2018; the comparison of the $\frac{[\mathrm{OIII}]/\mathrm{H}\beta}{[\mathrm{NII}]/\mathrm{H}\alpha}$, and [NII]/H$\alpha$ indicators between $z \sim 0$ and 2 are shown in Figure 7). However, a major concern is whether locally-calibrated strong line diagnostics still hold at high redshifts. For example, it is known that $z \sim 2$ galaxies have harder ionizing radiation (Steidel *et al.* 2016; Strom *et al.* 2017b), higher ionization parameters (Kewley *et al.* 2015; Kashino *et al.* 2017), and higher electron density and ISM pressure (Liu *et al.* 2008; Bian *et al.* 2010) at a fixed oxygen abundance, compared to local galaxies. Additionally, using the oxygen abundances may not be representative of the real metallicity of the gas, as the relative abundances of the chemical elements may vary with respect to the solar value at high redshifts. Indeed, studies show that $z \sim 1 - 3$ galaxies have lower Fe/H and higher O/H ratios compared to $z \sim 0$ galaxies (Steidel *et al.* 2016; Strom *et al.* 2017a; Kriek *et al.* 2019).

These uncertainties and complications indicate the need for direct T_e method calibrations at high redshifts, as well as independent methods of inferring metallicity to overcome the T_e dependency. The far-IR ground-state fine structure lines provide independent diagnostics that are insensitive to electron temperature and dust obscuration

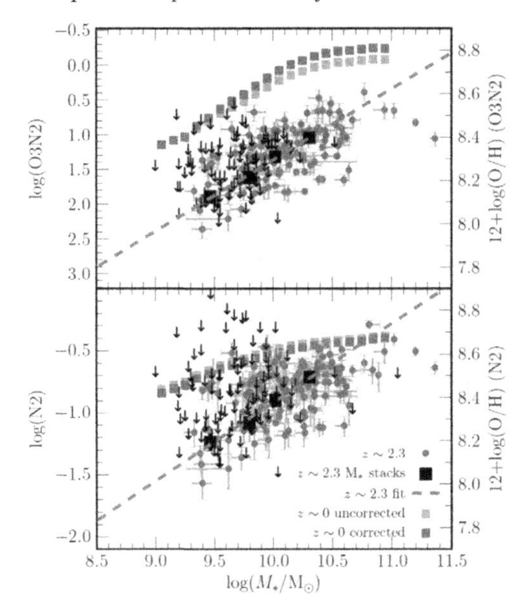

Figure 7. The emission line ratios $\frac{[\text{OIII}]/\text{H}\beta}{[\text{NII}]/\text{H}\alpha}$ (O3N2; top) and [NII]/Hα (N2; bottom) as a function of stellar mass at $z \sim 2.3$ and $z \sim 0$. The right y-axis shows the translation of the emission line ratios into metallicity. The black arrows display 3σ limits in cases where one or more of the required lines is not detected. Black squares show stacks of both detections and non-detections in bins of stellar mass. The red dashed line is the best-fit linear relation to the stacks in each panel. Figure from Sanders *et al.* (2018).

(Nagao *et al.* 2011; Smith *et al.* 2019). ALMA observations of such lines at high redshifts can be joined with optical line diagnostics to study metal enrichment and directly calibrate metallicity diagnostics at high redshifts.

As a complementary tracer of metal enrichment, dust features such as the mid-IR PAH emissions may be valuable. Local and high-redshift studies have shown that the PAH intensity (defined as the strength of PAH 7.7μm feature to IR luminosity or SFR) drops below a certain oxygen abundance ($12 + \log(\frac{\text{O}}{\text{H}}) \sim 8.1 - 8.2$ at $z \sim 0$ and $\sim 8.4 - 8.5$ at $z \sim 2$; Figure 8; Draine *et al.* 2007b; Engelbracht *et al.* 2005; Shivaei *et al.* 2017). The nature of this drop may be due to lower carbon abundances in low-metallicity environment, or harder ionizing radiation that destroys the PAH molecules. The future *JWST*/MIRI will enable us to trace the PAH features in large samples of $z \sim 1 - 2$ galaxies, down to at least an order of magnitude lower masses than have been studied previously at these redshifts (Elbaz *et al.* 2011; Shipley *et al.* 2016; Shivaei *et al.* 2017).

2.4. *Nebular Dust*

Balmer line ratios (e.g., Balmer decrement, Hα/Hβ) are almost insensitive to electron temperature and density, and hence, any deviation of the measured ratio from the theoretical intrinsic line ratio may be attributed to dust attenuation. Balmer optical depth is defined as $\tau_\text{b} \equiv \ln(\frac{\text{H}\alpha/\text{H}\beta}{2.86})$, where 2.86 is the lines ratio for a Case B recombination with $T_e = 10^4$ K and $n_e = 100\,\text{cm}^{-3}$ (Osterbrock & Ferland 2006). Assuming a shape for the dust attenuation/extinction curve, Balmer optical depth indicates the nebular reddening ($E(B-V)_\text{nebular}$), which is the reddening along the line of sight towards the ionized gas. Nebular reddening should be used to correct dust attenuation in nebular emission lines. On the other hand, $E(B-V)_\text{stellar}$, inferred from the continuum indicators such as the UV slope, is advised to be used to correct the stellar continuum emission. Many studies

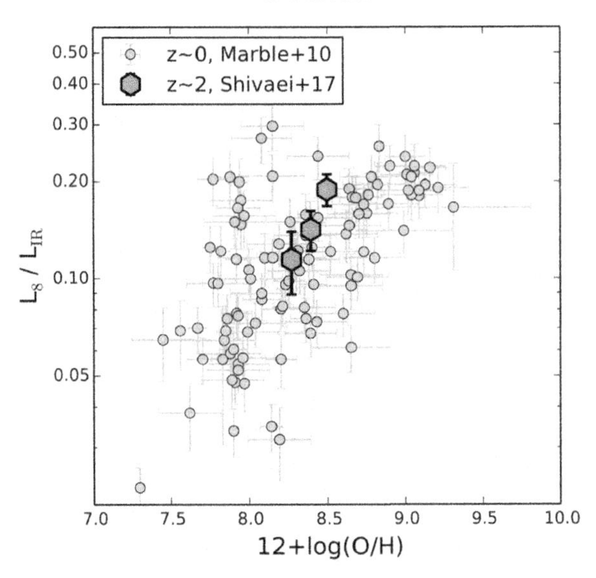

Figure 8. Relative strength of 7.7 μm luminosity to total IR luminosity as a function of metallicity. Orange symbols are stacks of $z \sim 2$ galaxies in bins of metallicity, where the metallicity is derived from $\frac{[\mathrm{OIII}]/\mathrm{H}\beta}{[\mathrm{NII}]/\mathrm{H}\alpha}$ diagnostic (Shivaei *et al.* 2017). The blue symbols are individual measurements at $z \sim 0$ (Marble *et al.* 2010).

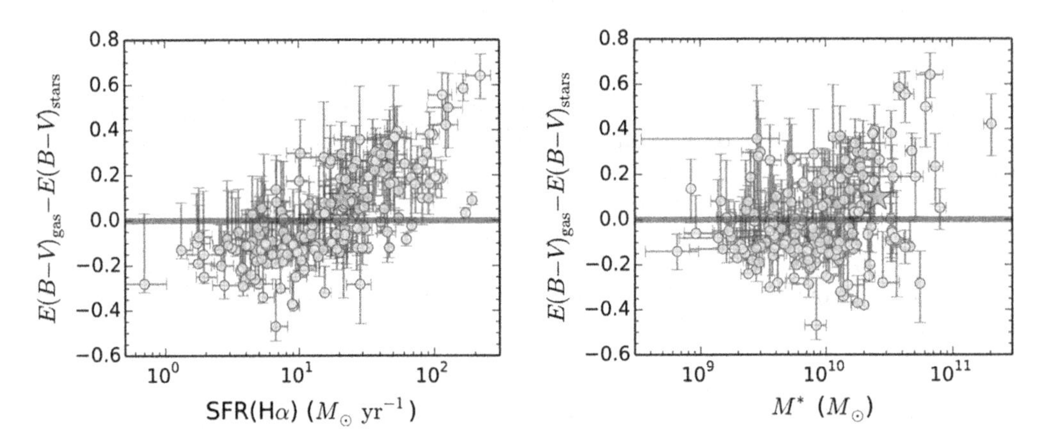

Figure 9. Difference between gas and stellar continuum reddenings as a function of SFR (left) and stellar mass (right) for a sample of ~200 galaxies at $z \sim 2$. The solid lines indicate no difference between the color excess of the nebular regions and the stellar continuum. The yellow stars indicate the values for Hβ-undetected galaxies. Figure from (Reddy *et al.* 2015).

in the literature have explored the relation between the two reddenings (Calzetti *et al.* 2000; Price *et al.* 2014; Shivaei *et al.* 2015a; Reddy *et al.* 2015). Owing to the stellar birth cloud dissipation and/or stars migrating from their parent molecular clouds, the differences between the mean optical depth probed by continuum and line photons may vary with SFR or specific SFR (Reddy *et al.* 2015; Theios *et al.* 2019). Therefore, the general consensus is that the relation between the two reddenings changes with SFR or specific SFR (Figure 9).

At low redshifts, dust *extinction* curves have been determined along multiple sight lines of the Milky Way, LMC, and SMC (Savage & Mathis 1979; Prevot *et al.* 1984, e.g.,). An extinction curve is determined for a simple system of a star with a foreground

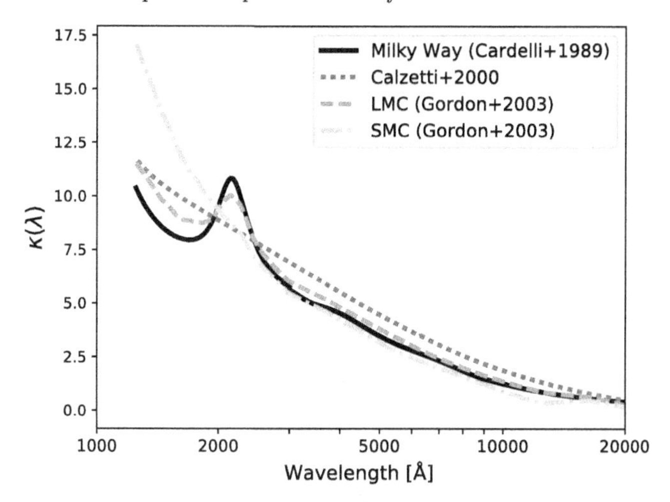

Figure 10. Comparison of the commonly used dust extinction/attenuation curves. As evident, two of the most distinct differences between the curves are the steepness and the strength of a broad absorption feature at 2175Å (the UV bump).

screen of dust. The dust screen absorbs and scatters photons out of the line-of-sight. Therefore, variations in extinction curves reflect differences in dust grain physics (the composition and size distribution). On the other hand, *attenuation* curves are inferred for more complex systems, such as galaxies, in which the geometry of dust with respect to the emitting source also plays a role in determining the shape of the curve. In high redshift studies, the Calzetti *et al.* (2000) attenuation curve is often adopted, which is calculated based on a sample of local starburst galaxies.

The extinction and attenuation curves have profound differences. The two most distinct differences are the strength of the UV extinction bump at 2175 Å and the UV slope of the attenuation curve. As shown in Figure 10, the Milky Way and LMC extinction curves both have strong UV bumps, while the bump is absent in the SMC and the Calzetti curves. Also, the SMC curve has a much steeper slope compared to the other ones. The underlying cause of these differences are not well known; it is speculated that as the SMC has the lowest metallicity of the three galaxies ($12 + \log(\mathrm{O/H}) \sim 8, Z \sim 0.2Z_{\odot}$ Kurt & Dufour 1998), therefore it has a different grain size and composition distribution compared to the other two. These observations emphasize the importance of the galaxy-to-galaxy variations of the attenuation curves at high redshifts.

Using Balmer decrement as an independent tracer of dust reddening, one can directly determine a dust attenuation curve by comparing the average SED of the "more" to "less" dusty galaxies with similar intrinsic (i.e., dust free) SED shapes (e.g., similar star formation histories, ages, etc.). This method requires large rest-optical spectroscopic samples (Calzetti *et al.* 1994; Battisti *et al.* 2016), and hence a challenging task at high redshifts (Reddy *et al.* 2015). Additionally, the normalization of the attenuation curve can be independently determined by incorporating IR data from *Spitzer* and *Herschel* and using an energy balance argument between the UV and IR emission (Calzetti *et al.* 2000). Given the observed variations of the dust emission properties with gas-phase metallicity (e.g., the PAH intensity correlation with metallicity in Section 2.3), it is also expected to see changes in the slope of the attenuation curve and the strength of the UV bump as a function of galaxy properties. Such studies require large enough spectroscopic datasets that can be divided into bins of metallicity, specific SFR, and stellar mass, to compare their inferred attenuation curves with each other. In the future, with *JWST*, the dust attenuation properties of obscured and optically-thick systems can be independently studied by simultaneously observing multiple hydrogen recombination lines, including Paschen and Balmer series, and directly derive the nebular attenuation curve.

References

Ageorges, N., Seifert, W., Jütte, M., *et al.* 2010, in Society of Photo-Optical Instrumentation Engineers (SPIE) Conference Series, Vol. 7735, Ground-based and Airborne Instrumentation for Astronomy III, 77351L

Aravena, M., Decarli, R., Walter, F., *et al.* 2016, *ApJ*, 833, 71

Battisti, A. J., Calzetti, D., & Chary, R.-R. 2016, *ApJ*, 818, 13

Bian, F., Fan, X., Bechtold, J., *et al.* 2010, *ApJ*, 725, 1877

Brammer, G. B., van Dokkum, P. G., Franx, M., *et al.* 2012, *ApJS*, 200, 13

Calzetti, D., Armus, L., Bohlin, R. C., *et al.* 2000, *ApJ*, 533, 682

Calzetti, D., Kinney, A. L., & Storchi-Bergmann, T. 1994, *ApJ*, 429, 582

Casey, C. M. 2012, *MNRAS*, 425, 3094

Chary, R., & Elbaz, D. 2001, *ApJ*, 556, 562

De Rossi, M. E., Rieke, G. H., Shivaei, I., Bromm, V., & Lyu, J. 2018, *ApJ*, 869, 4

Draine, B. T. 2011, Physics of the Interstellar and Intergalactic Medium

Draine, B. T., Dale, D. A., Bendo, G., *et al.* 2007b, *ApJ*, 663, 866

Dunlop, J. S., McLure, R. J., Biggs, A. D., *et al.* 2017, *MNRAS*, 466, 861

Elbaz, D., Dickinson, M., Hwang, H. S., *et al.* 2011, *A&A*, 533, A119

Engelbracht, C. W., Gordon, K. D., Rieke, G. H., *et al.* 2005, *ApJ*, 628, L29

Erb, D. K., Shapley, A. E., Pettini, M., *et al.* 2006a, *ApJ*, 644, 813

Franco, M., Elbaz, D., Béthermin, M., *et al.* 2018, *A&A*, 620, A152

Frayer, D. T., Huynh, M. T., Chary, R., *et al.* 2006, *ApJ*, 647, L9

Goicoechea, J. R., Pety, J., Cuadrado, S., *et al.* 2016, Nature, 537, 207

Kashino, D., Silverman, J. D., Sanders, D., *et al.* 2017, *ApJ*, 835, 88

—. 2019, *ApJS* Series, 241, 10

Kennicutt, R. C., & Evans, N. J. 2012, *ARA&A*, 50, 531

Kewley, L. J., & Dopita, M. A. 2002, *ApJS*, 142, 35

Kewley, L. J., Zahid, H. J., Geller, M. J., *et al.* 2015, *ApJ*, 812, L20

Kimura, M., Maihara, T., Iwamuro, F., *et al.* 2010, *PASJ*, 62, 1135

Kriek, M., Shapley, A. E., Reddy, N. A., *et al.* 2015, *ApJS*, 218, 15

Kriek, M., Price, S. H., Conroy, C., *et al.* 2019, *ApJ*, 880, L31

Kurt, C. M., & Dufour, R. J. 1998, in Revista Mexicana de Astronomia y Astrofisica Conference Series, Vol. 7, Revista Mexicana de Astronomia y Astrofisica Conference Series, ed. R. J. Dufour & S. Torres-Peimbert, 202

Liu, X., Shapley, A. E., Coil, A. L., Brinchmann, J., & Ma, C.-P. 2008, *ApJ*, 678, 758

Madau, P., & Dickinson, M. 2014, *ARA&A*, 52, 415

Magnelli, B., Elbaz, D., Chary, R. R., *et al.* 2011, *A&A*, 528, A35

Marble, A. R., Engelbracht, C. W., van Zee, L., *et al.* 2010, *ApJ*, 715, 506

McLean, I. S., Steidel, C. C., Epps, H. W., *et al.* 2012, in Society of Photo-Optical Instrumentation Engineers (SPIE) Conference Series, Vol. 8446, Society of Photo-Optical Instrumentation Engineers (SPIE) Conference Series, 0

Nagao, T., Maiolino, R., Marconi, A., & Matsuhara, H. 2011, *A&A*, 526, A149

Osterbrock, D. E., & Ferland, G. J. 2006, Astrophysics of gaseous nebulae and active galactic nuclei

Pettini, M., & Pagel, B. E. J. 2004, *MNRAS*, 348, L59

Prevot, M. L., Lequeux, J., Prevot, L., Maurice, E., & Rocca-Volmerange, B. 1984, *A&A*, 132, 389

Price, S. H., Kriek, M., Brammer, G. B., *et al.* 2014, *ApJ*, 788, 86

Reddy, N., Dickinson, M., Elbaz, D., *et al.* 2012a, *ApJ*, 744, 154

Reddy, N. A., Kriek, M., Shapley, A. E., *et al.* 2015, *ApJ*, 806, 259

Rieke, G. H., Alonso-Herrero, A., Weiner, B. J., *et al.* 2009, *ApJ*, 692, 556

Sanders, R. L., Shapley, A. E., Kriek, M., *et al.* 2016a, *ApJ*, 816, 23

—. 2018, *ApJ*, 858, 99
Sanders, R. L., Shapley, A. E., Reddy, N. A., *et al.* 2019, arXiv e-prints, arXiv:1907.00013
Savage, B. D., & Mathis, J. S. 1979, *ARA&A*, 17, 73
Schreiber, C., Elbaz, D., Pannella, M., *et al.* 2018,*A&A*, 609, A30
Sharples, R., Bender, R., Agudo Berbel, A., *et al.* 2013, The Messenger, 151, 21
Shipley, H. V., Papovich, C., Rieke, G. H., Brown, M. J. I., & Moustakas, J. 2016, *ApJ*, 818, 60
Shivaei, I., Reddy, N. A., Steidel, C. C., & Shapley, A. E. 2015a, *ApJ*, 804, 149
Shivaei, I., Reddy, N. A., Shapley, A. E., *et al.* 2015b, *ApJ*, 815, 98
Shivaei, I., Kriek, M., Reddy, N. A., *et al.* 2016, *ApJ*, 820, L23
Shivaei, I., Reddy, N. A., Shapley, A. E., *et al.* 2017, *ApJ*, 837, 157
Shivaei, I., Reddy, N. A., Siana, B., *et al.* 2018, *ApJ*, 855, 42
Smith, J. D., Armus, L., Davé, R., *et al.* 2019, *BAAS*, 51, 400
Steidel, C. C., Strom, A. L., Pettini, M., *et al.* 2016, *ApJ*, 826, 159
Strom, A. L., Steidel, C. C., Rudie, G. C., *et al.* 2017a, *ApJ*, 836, 164
—. 2017b, *ApJ*, 836, 164
Theios, R. L., Steidel, C. C., Strom, A. L., *et al.* 2019, *ApJ*, 871, 128
Whitaker, K. E., Pope, A., Cybulski, R., *et al.* 2017, ArXiv e-prints 1710.06872, arXiv:1710.06872
Whitaker, K. E., Franx, M., Leja, J., *et al.* 2014b, *ApJ*, 795, 104
Wisnioski, E., Förster Schreiber, N. M., Wuyts, S., *et al.* 2015, *ApJ*, 799, 209
Wuyts, S., Labbé, I., Schreiber, N. M. F., *et al.* 2008, *ApJ*, 682, 985

Uncovering Early Galaxy Evolution in the ALMA and JWST Era
Proceedings IAU Symposium No. 352, 2019
E. da Cunha, J. Hodge, J. Afonso, L. Pentericci & D. Sobral, eds.

doi:10.1017/S1743921319008974

Automated mining of the ALMA archive in the COSMOS field (A^3COSMOS): Cold molecular gas evolution

Daizhong Liu[1] and A^3COSMOS Team[2]

[1]Max-Planck-Institut für Astronomie,
Königstuhl 17, D-69117 Heidelberg, Germany
email: dzliu@mpia.de

[2]https://sites.google.com/view/a3cosmos/team

Abstract. We present new constraints on the cosmic cold molecular gas evolution out to redshift 6 based on systematic mining of the public ALMA archive in the COSMOS field (A^3COSMOS). Our A^3COSMOS dataset contains ~700 galaxies ($0.3 \lesssim z \lesssim 6$) with high-confidence ALMA detection and multi-wavelength SEDs. Combining with ~1,200 CO-observed galaxies at $0 \lesssim z \lesssim 4$ (75% at $z < 0.1$) in the literature, we parameterize galaxies' molecular gas depletion time and gas fraction each as a function of stellar mass, offset from the star-forming main-sequence and cosmic age. We propose a new functional form which provides a better fit and implies a "downsizing" effect and "mass-quenching". By adopting galaxy stellar mass functions and applying our gas fraction function, we obtain a cosmic cold molecular gas density evolution in agreement with recent CO blind field surveys as well as semi-analytic modeling. These together provide us a coherent picture of galaxy cold molecular gas, SFR and stellar mass evolution.

Keywords. galaxies: evolution — galaxies: high-redshift — galaxies: ISM — submillimeter

1. Introduction

Cold molecular gas is the fuel of star formation activity in galaxies. In recent years, our knowledge of the cosmic evolution of star formation and stellar mass growth has been obtained out to redshift ~5, however, the cosmic evolution of the cold molecular gas is much less well constrained, and the validity of different methods are debated.

Briefly, there are three most widely used methods for probing gas mass in high-redshift galaxies: *(1)* blind deep field CO luminosity function and CO-to-H_2 conversion method (e.g., Walter *et al.* 2014; Decarli *et al.* 2016, 2019; Riechers *et al.* 2019); *(2)* IR-to-millimeter(mm) spectral energy distribution (SED) fitting dust mass to gas mass method (e.g., Santini *et al.* 2010, 2014; Magdis *et al.* 2011, 2012; Magnelli *et al.* 2012, 2014; Genzel *et al.* 2015); and *(3)* Rayleigh-Jeans(RJ)-tail dust continuum to gas mass calibration method (e.g., Scoville *et al.* 2014, 2016, 2017; Groves *et al.* 2015; Schinnerer *et al.* 2016; Hughes *et al.* 2017).

Earlier works have led to a coevolution picture of dust, gas and star formation from redshift 3 to present where: *(a)* the molecular gas to total baryon mass fraction, $f_{\mathrm{mol\,gas}} \equiv M_{\mathrm{mol\,gas}}/(M_{\mathrm{mol\,gas}} + M_{\star})$, or $\mu_{\mathrm{molgas}} \equiv M_{\mathrm{mol\,gas}}/M_{\star}$, increases with redshift and depends slightly on the SFR and stellar mass; *(b)* the molecular gas depletion time, $\tau_{\mathrm{depl}} \equiv M_{\mathrm{mol\,gas}}/\mathrm{SFR}$, decreases with redshift and also depends on SFR and stellar mass; *(c)* whereas τ_{depl} and μ_{molgas} more primely correlated with the $M_{\star} - \mathrm{SFR}$ main-sequence (MS; e.g., Speagle *et al.* 2014), denoted as $\delta\mathrm{MS} \equiv \mathrm{SFR}/\mathrm{SFR}_{\mathrm{MS}}$, or $\Delta\mathrm{MS} \equiv \log_{10}(\mathrm{SFR}/\mathrm{SFR}_{\mathrm{MS}})$, than the absolute SFR or specific-SFR ($\equiv \mathrm{SFR}/M_{\star}$).

While the aforementioned trends have been widely studied at local and high redshift, an accurate parametrization ($\mu_{\rm molgas}$ and $\tau_{\rm depl}$ functions) across the parameter space is still under debate. Utilizing the RJ-tail dust continuum method (at rest-frame $850\,\mu$m), Scoville *et al.* (2017, hereafter S17) studied a sample of 708 high-redshift *Herschel*-selected galaxies ($0.3 < z < 4.5$) which have detection with Atacama Large Millimeter/submillimeter Array (ALMA), and characterized the $\mu_{\rm molgas}$ and $\tau_{\rm depl}$ functions as:

$$\mu_{\rm molgas} = 0.71 \times (\delta{\rm MS})^{+0.32} \times M_{\star,10}^{-0.70} \times (1+z)^{+1.84},$$
$$\tau_{\rm depl} = 3.23\,{\rm Gyr} \times (\delta{\rm MS})^{-0.70} \times M_{\star,10}^{-0.01} \times (1+z)^{-1.04}, \tag{1.1}$$

where $M_{\star,10} \equiv M_{\star}/(10^{10}\,{\rm M_{\odot}})$. With the same method but at rest-frame 250–500 μm, Schinnerer *et al.* (2016) studied a smaller, optically-selected sample at $z = 2.8 - 3.6$. However, discrepancies exist due to the different wavelengths and samples. Tacconi *et al.* (2018, hereafter T18) expanded the work of Genzel *et al.* (2015) using all three methods to obtain the molecular gas masses for 1,444 galaxies at $0 < z < 4$ (including stacking), and fitted them all together to derive the $\mu_{\rm molgas}$ and $\tau_{\rm depl}$ functions:

$$\mu_{\rm molgas} = 0.75 \times (\delta{\rm MS})^{+0.53} \times M_{\star,10}^{-0.35} \times 10^{-3.62\times(\log_{10}(1+z)-0.66)^2},$$
$$\tau_{\rm depl} = 1.06\ {\rm Gyr} \times (\delta{\rm MS})^{-0.44} \times M_{\star,10}^{+0.09} \times (1+z)^{-0.62}, \tag{1.2}$$

where we adapted their equation to $M_{\star,10}$ (same as in Eq. 1.1) and ignored galaxy size dependency. Comparing Eqs. 1.1 and 1.2 at redshift 3 and $M_{\star} = 5 \times 10^{10}\,{\rm M_{\odot}}$ reveals a factor of 6 difference in $\mu_{\rm molgas}$ and a factor of 1.5 in $\tau_{\rm depl}$. Such noticeable differences exist for other parameter values as well, raising concerns on the validity and predictability of the $\mu_{\rm molgas}$ and $\tau_{\rm depl}$ functions. In addition, previous works have constraints only for $z \lesssim 3$.

To solve the discrepancies and understand systematic biases, a large, robust galaxy sample from local to high redshift is needed to carry out the comprehensive analysis. Therefore, here we briefly present our A^{3}COSMOS gas evolution study, where we: *(a)* developed automated pipelines for public ALMA archive data reduction, imaging, photometry, statistical corrections and extensive analyses (see Liu *et al.* 2019 ApJS in press; hereafter paper I), *(b)* combined with 1,200+ CO-detected galaxies at $0 \lesssim z \lesssim 4$ (75% at $z < 0.1$) in the literature to build a largest, most robust sample ($\sim 2,000$), *(c)* obtained new parameterization of $\mu_{\rm molgas}$ and $\tau_{\rm depl}$ by fitting the new data, and *(d)* applied galaxy stellar mass functions and/or galaxy modelings to obtain the final cosmic molecular gas mass density evolution. The full content of this work is presented in Liu *et al.* (to be submitted; hereafter paper II). Here we only highlight the main results.

2. Method

We developed automated pipelines to query and reduce ALMA public archival data (in the COSMOS 2 sq. deg. field), and pipelines for blind-extraction and prior-based photometry, Monte Carlo simulations and photometry quality assessments. This results into nearly 1,900 ALMA imaging products, 1,200+ robust ALMA detections (version 20180801). Fig. 1 shows the spatial distribution of all these ALMA images (we used all ALMA bands). We select the robust ALMA detections with a relatively high signal-to-noise ratio which correspond to about 10% spurious fraction according to our statistical analysis. Then we performed various quality assessment steps to identify most plausible spurious sources (e.g., boosted by noise or blended with other sources) and excluded them from our final robust galaxy catalog (about 700 galaxies with multi-wavelength SEDs, some have multiple-ALMA-band detections).

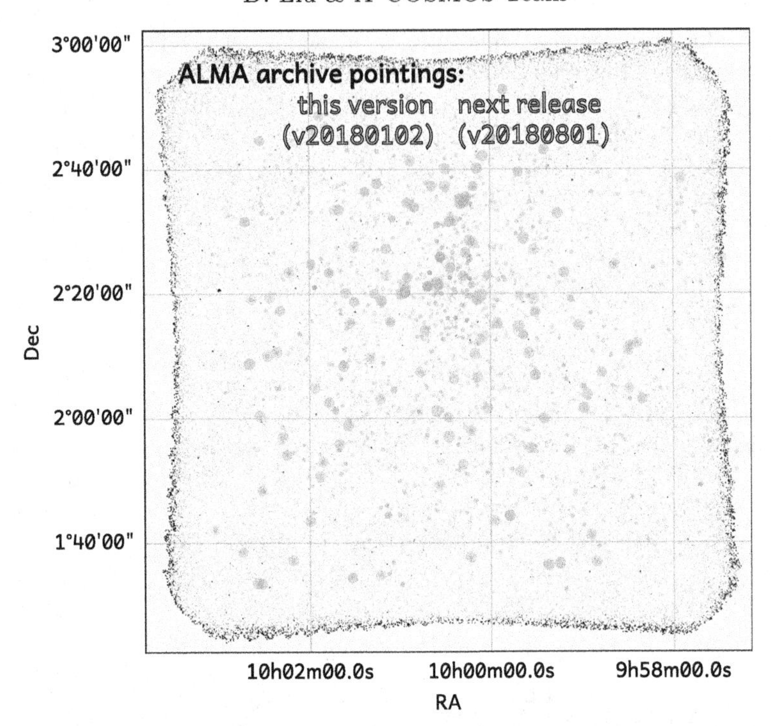

Figure 1. ALMA pointings in the COSMOS field that are publicly accessible. Green and magenta circles represent ALMA pointings which became public before Jan. 2nd, 2018 and Aug. 1st, 2018 respectively. Circle sizes represent the FWHM of the ALMA 12m antennas' primary beam, and the shading reflects the on-source integration time (dark referring to longer integration times). The background image is the *Herschel* PACS 100 μm data from the PACS Evolutionary Probe survey (PEP; Lutz *et al.* 2011).

We included 16 samples of galaxies with CO observations in the literature as complimentary information to our analysis. The full list is presented in our paper II. It encompasses most of the CO-observed samples analyzed by T18. Most of these samples are galaxies in the local Universe, and the largest sample is from the xCOLD GASS survey (Saintonge *et al.* 2017).

In our paper II, we also compared various dust-based gas mass calculation methods, i.e., MAGPHYS (da Cunha *et al.* 2008) SED-fitted dust mass method, RJ 850μm (Scoville *et al.* 2014, 2016, Hughes *et al.* 2017); and RJ 250–500μm method (Groves *et al.* 2015).

We analyzed the complicated dependencies of $\tau_{\rm depl}$ and $\mu_{\rm molgas}$ (respectively) on galaxies' redshift, stellar mass and ΔMS properties. Motivated by the high-dimensional data fitting, we propose a new functional form which accounts for the different behaviors of galaxies due to their stellar masses:

$$\begin{aligned}\log_{10}\tau_{\rm depl} &= (\mathsf{a} + \mathsf{ak} \times \log_{10}(M_\star/10^{10})) \times \Delta{\rm MS} + \mathsf{b} \times \log_{10}(M_\star/10^{10}) \\ &+ (\mathsf{c} + \mathsf{ck} \times \log_{10}(M_\star/10^{10})) \times t_{\rm cosmic\,age} + \mathsf{d} \\ \log_{10}\mu_{\rm molgas} &= (\mathsf{a} + \mathsf{ak} \times \log_{10}(M_\star/10^{10})) \times \Delta{\rm MS} + \mathsf{b} \times \log_{10}(M_\star/10^{10}) \\ &+ (\mathsf{c} + \mathsf{ck} \times \log_{10}(M_\star/10^{10})) \times t_{\rm cosmic\,age} + \mathsf{d} \end{aligned} \tag{2.1}$$

The coefficients a, b, c, d, ak and ck for either $\tau_{\rm depl}$ and $\mu_{\rm molgas}$ (respectively) are provided in our paper II. In Fig. 2 we present the data, fitted functions in this work and from T18 and S17. The parameterizations of $\tau_{\rm depl}$ and $\mu_{\rm molgas}$ with our, T18 and S17 functions are roughly consistent within the overlapping regions of the parameter space, i.e., $z \sim 1-3$,

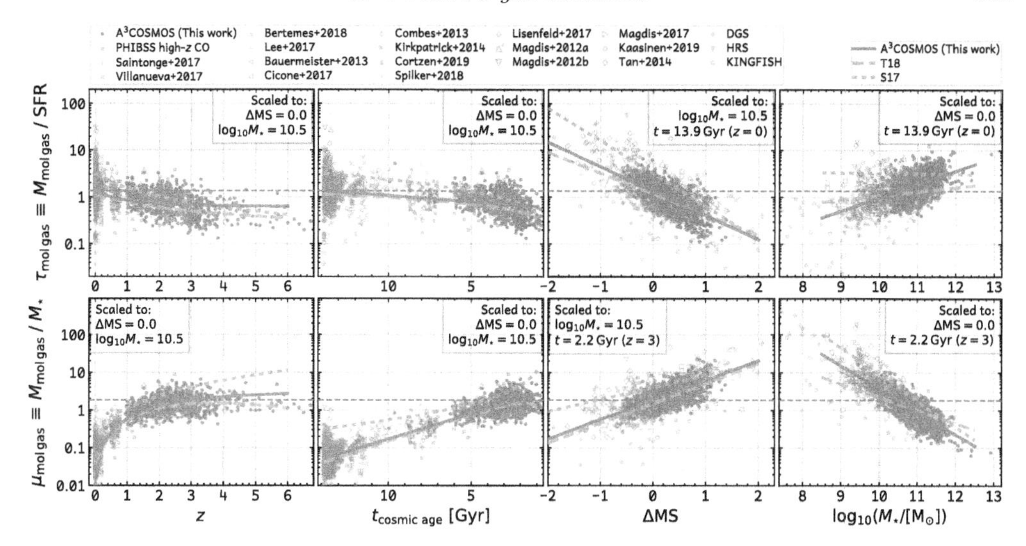

Figure 2. Characterizing molecular gas depletion time $\tau_{\rm depl}$ (*upper panels*) and molecular gas to stellar mass ratio $\mu_{\rm molgas}$ (*lower panels*) in the functional form of Eq. 2.1. From *left* to *right*, we show $\tau_{\rm depl}$ versus redshift, $t_{\rm cosmic\ age}$, ΔMS and $M_\star$, respectively. Data points in each panel are re-scaled using the best-fit function so as to remove the dependency on other parameters and leave only the correlation with the current X-axis parameter (with coefficient(s) labeled at the bottom of each panel). Orangered data points are from this work, while green ones are from PHIBSS 1&2 surveys (T18) and gray ones are from the literature as listed at the top. Our best-fit function is shown as the solid orangered line in each panel, while the functions from T18 (Eq. 1.2) and S17 (Eq. 1.1) are shown as the dashed green and dotted gray lines, respectively.

ΔMS > 0 and log $M_\star > 10.5$. They differ significantly for low-mass and/or main-sequence or below-main-sequence galaxies, which however cannot be verified with current dataset.

Our function implies a "downsizing" effect, i.e., more-massive galaxies evolve earlier than less-massive ones, and "mass-quenching", i.e., gas depletion slows down with cosmic time for more-massive galaxies but speeds up for less-massive ones (see our paper II).

3. Results

The $\mu_{\rm molgas}$ function allows us to infer a galaxy's molecular gas mass with its known z, SFR and $M_\star$. Although this is very uncertain for an individual galaxy, it statistically works for large samples of galaxies. Therefore, if we know galaxy populations and their number densities, e.g., stellar mass functions (SMFs), then by integrating all star-forming galaxies at each redshift we are able to obtain the cosmic molecular gas mass density.

For the first time we provide such an estimation (from gas fraction scaling relations) as detailed in our paper II. Fig. 3 shows three cosmic molecular gas density evolution curves by applying our, T18 and S17 $\mu_{\rm molgas}$ functions, respectively, to the SIDES mock galaxy modeling of Béthermin *et al.* (2017) who fully simulated galaxies within a 2 sq. deg. COSMOS-like field with best-known SMFs, galaxy number counts, infrared/sub-mm luminosity functions and clustering. The differences are mainly driven by the extrapolation of these functions to the lower-mass/below-MS and/or $z > 3$ galaxies. CO blind deep field results from Decarli *et al.* (2016, 2019) and Riechers *et al.* (2019), and semi-analytic modeling from Popping *et al.* (2019) are shown for comparison.

This strongly supports that we now have a coherent picture of galaxies' star formation, stellar mass and molecular gas evolution, which can be well-parameterized by the main sequence correlation (e.g., Speagle *et al.* 2014 #49 with cosmic time), stellar mass

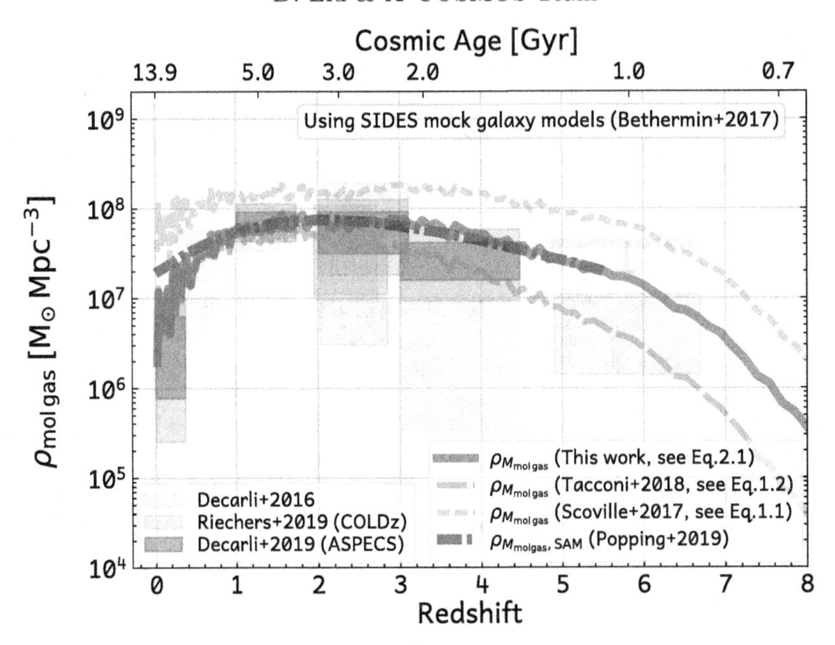

Figure 3. Cosmic evolution of cold molecular gas mass density. Results from high-z CO blind deep field studies from Decarli *et al.* (2016), Riechers *et al.* (2019) and Decarli *et al.* (2019) are shown as green, blue and red boxes, with the X-sides (Y-sides) indicating the observed redshift range (5th and 95th percentiles) (the $z \sim 6$ arrow is an upper limit from Riechers *et al.* 2019). The red solid, green long-dashed and purple short-dashed lines are the integrated molecular gas mass density based on the mock galaxy modeling by Béthermin *et al.* (2017) (integrated down to $M_{\star} = 10^{9.0}$ $M_{\odot}$) and gas fraction function from this work (Eq. 2.1), T18 and S17, respectively. The black dash-dot line is from the Semi-Analytic Model (SAM) simulation of Popping *et al.* (2019). More details in our paper II.

functions (e.g., see references in Béthermin *et al.* 2017) and gas scaling relations (e.g., with our new functional form in Eq. 2.1), respectively.

References

Béthermin, M., *et al.* 2017, *A&A*, 607, A89
da Cunha, E., Charlot, S., & Elbaz, D. 2008, *MNRAS*, 388, 1595
Decarli, R., *et al.* 2016, *ApJ*, 833, 69
—. 2019, arXiv e-prints, arXiv:1903.09164
Genzel, R., *et al.* 2015, *ApJ*, 800, 20
Groves, B. A., *et al.* 2015, *ApJ*, 799, 96
Hughes, T. M., *et al.* 2017, *MNRAS*, 468, L103
Liu, *et al.* 2019, *ApJS*, 244, 42
Lutz, D., *et al.* 2011, *A&A*, 532, A90
Magdis, G. E., *et al.* 2011, *ApJL*, 740, L15
—. 2012, *ApJ*, 760, 6
Magnelli, B., *et al.* 2012, *A&A*, 548, A22
—. 2014, *A&A*, 561, A86
Popping, G., *et al.* 2019, arXiv e-prints, arXiv:1903.09158
Riechers, D. A., *et al.* 2019, *ApJ*, 872, 7
Saintonge, A., *et al.* 2017, *ApJS*, 233, 22
Santini, P., *et al.* 2010, *A&A*, 518, L154
—. 2014, *A&A*, 562, A30
Schinnerer, E., *et al.* 2016, *ApJ*, 833, 112

Scoville, N., *et al.* 2014, *ApJ*, 783, 84
—. 2016, *ApJ*, 820, 83
—. 2017, *ApJ*, 837, 150
Speagle, J. S., *et al.* 2014, *ApJS*, 214, 15
Tacconi, L. J., *et al.* 2018, *ApJ*, 853, 179
Walter, F., *et al.* 2014, *ApJ*, 782, 79

Uncovering Early Galaxy Evolution in the ALMA and JWST Era
Proceedings IAU Symposium No. 352, 2019
E. da Cunha, J. Hodge, J. Afonso, L. Pentericci & D. Sobral, eds.

doi:10.1017/S174392131900886X

$^{13}C/^{18}O$ ratio as a litmus test of stellar IMF variations in high-redshift starbursts

Donatella Romano[1], Zhi-Yu Zhang[2,3], Francesca Matteucci[4,5], Rob J. Ivison[2,3] and Padelis P. Papadopoulos[6]

[1]INAF, Astrophysics and Space Science Observatory, Via Gobetti 93/3, I-40129 Bologna, Italy
email: donatella.romano@inaf.it

[2]Institute for Astronomy, University of Edinburgh, Royal Observatory, Blackford Hill, Edinburgh, EH9 3HJ, UK

[3]European Southern Observatory, Karl-Schwarzschild-Strasse 2, D-85748, Garching bei München, Germany

[4]Physics Department, Section of Astronomy, University of Trieste, Via Tiepolo 11, I-34131, Trieste, Italy

[5]INAF, Astronomical Observatory of Trieste, Via Tiepolo 11, I-34131, Trieste, Italy

[6]Physics Department, Section of Astrophysics, Astronomy and Mechanics, Aristotle University of Thessaloniki, 54124 Thessaloniki, Greece

Abstract. Determining the shape of the stellar initial mass function (IMF) and whether it is constant or varies in space and time is the Holy Grail of modern astrophysics, with profound implications for all theories of star and galaxy formation. On a theoretical ground, the extreme conditions for star formation (SF) encountered in the most powerful starbursts in the Universe are expected to favour the formation of massive stars. Direct methods of IMF determination, however, cannot probe such systems, because of the severe dust obscuration affecting their starlight. The next best option is to observe CNO bearing molecules in the interstellar medium at millimetre/ submillimetre wavelengths, which, in principle, provides the best indirect evidence for IMF variations. In this contribution, we present our recent findings on this issue. First, we reassess the roles of different types of stars in the production of CNO isotopes. Then, we calibrate a proprietary chemical evolution code using Milky Way data from the literature, and extend it to discuss extragalactic data. We show that, though significant uncertainties still hamper our knowledge of the evolution of CNO isotopes in galaxies, compelling evidence for an IMF skewed towards high-mass stars can be found for galaxy-wide starbursts. In particular, we analyse a sample of submillimetre galaxies observed by us with the Atacama Large Millimetre Array at the peak of the SF activity of the Universe, for which we measure $^{13}C/^{18}O \simeq 1$. This isotope ratio is especially sensitive to IMF variations, and is little affected by observational uncertainties. At the end, ongoing developments of our work are briefly outlined.

Keywords. nuclear reactions, nucleosynthesis, abundances, galaxies: abundances, galaxies: evolution, galaxies: ISM, galaxies: starburst, stars: luminosity function, mass function

---

1. Introduction

More than sixty years have passed since Edwin Salpeter first introduced the concept of stellar initial mass function (IMF) in its seminal paper (Salpeter 1955), providing a convenient way to calculate the number of stars within a logarithmic mass interval. Since then, many important questions related to the fundamental nature of the IMF continue to obsess unanswered the astronomers, among which the long-standing issue of whether the IMF is universal or, rather, varies according to the environmental conditions

(Bastian *et al.* 2010). Since the initial mass of a star is the main driver of its evolution, it goes without saying how crucial the IMF is to studies of galaxy formation and evolution†.

Stars of different initial masses pollute the interstellar medium (ISM) on different timescales, with various chemicals. The net yields (matter restored to the ISM in the form of newly-produced elements per stellar generation, normalized to the total mass locked up in low-mass stars and stellar remnants) are strong functions of the adopted IMF. Therefore, in principle, the IMF shape can be recovered from its chemical imprints (Matteucci 2012). This possibility opens new windows on the IMF determination in environments where other methods (e.g., stellar counts or integrated properties, including absorption line indices and mass-to-light ratios) are not applicable (Henkel & Mauersberger 1993; Papadopoulos *et al.* 2014; Romano *et al.* 2017). In this contribution, we will deal in particular with the interpretation of $^{13}C/^{18}O$ ratios measured by us thanks to the unprecedented sensitivity of the Atacama Large Millimeter/submillimeter Array (ALMA) for a sample of four strongly-lensed submillimetre galaxies (SMGs) at redshift $z \simeq 2$–3 (Zhang *et al.* 2018). These systems are converting gas into stars very efficiently (SFRs $\sim$ 100–1000 $M_\odot$ yr^{-1}, Ivison *et al.* 1998), but their starlight is heavily obscured by dust (Simpson *et al.* 2017), making any attempt to estimate their IMF via classical methods unfeasible. Because of their high densities, temperatures and pressures, and owing to the presence of intense radiation fields, however, these extreme starbursts are the best places where to search for clear signs of IMF variation (Papadopoulos 2010).

2. Calibration of the chemical evolution model: the Milky Way as a benchmark

Before we move to the interpretation of the abundance data for our four strong CO emitters at high*(ish)* redshifts, however, we have to calibrate our chemical evolution model against the body of observational data for the Milky Way. A careful preliminary chemical evolution study considering all the major chemical species from H to Zn (Romano *et al.* 2010), allowed the selection of the 'best stellar yield set' (among many available in the literature), namely, the one reproducing at best the majority of the chemical properties observed for a large sample of Galactic stars. As can be seen from Fig. 1, the adopted yield set, in particular, reproduces satisfactorily well the evolution of the CNO isotope ratios in the solar neighbourhood, as well as their gradients along the Galactic disc (see Romano *et al.* 2017, for details about the nucleosynthesis prescriptions and the data sources).

It is worth emphasizing that, while the origin of the main CNO isotopes is quite well known, both qualitatively and quantitatively, the yields (and, sometimes, also the production sites) of the minor isotopes are still very uncertain. The main isotopes of C and O are forged as primary elements (i.e., starting from a mixture of hydrogen and helium) in stars of all masses (^{12}C) and only in massive stars (^{16}O), respectively. The secondary isotope of C, ^{13}C, is produced, similarly to ^{14}N, mostly in intermediate-mass stars, partly as a primary and partly as a secondary product (when starting from ^{12}C seeds already present at star's birth). Low-metallicities massive stars, however, may provide huge amounts of primary ^{13}C and ^{14}N, if they rotate fast (Meynet & Maeder 2002; Limongi & Chieffi 2018). As for ^{15}N, its main production site is highly debated, with either massive stars (Pignatari *et al.* 2015) or novae (Romano & Matteucci 2003; Romano *et al.* 2017) being preferred. The minor isotopes of O, ^{17}O and ^{18}O, have a different origin, being synthesised in intermediate-mass stars, high-mass stars and novae the first, and in massive stars the second.

† We make clear at this point that in this contribution 'IMF' stands for 'galaxy-wide IMF' (see Jeřábková *et al.* 2018 for more detailed explanation; see also contribution by A. Hopkins, this volume).

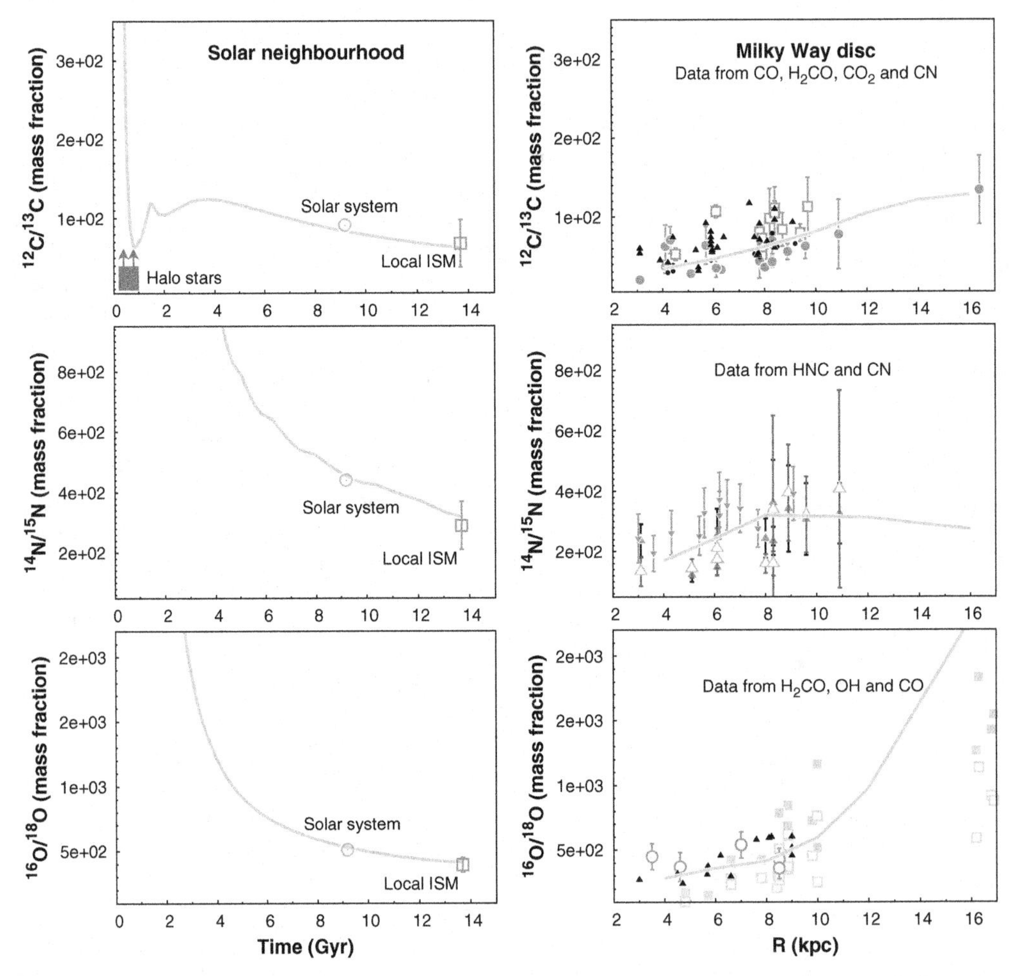

Figure 1. *Left-hand panels:* evolution of $^{12}C/^{13}C$ (top), $^{14}N/^{15}N$ (middle) and $^{16}O/^{18}O$ (bottom) in the solar neighbourhood. *Right-hand panels:* gradients of $^{12}C/^{13}C$ (top), $^{14}N/^{15}N$ (middle) and $^{16}O/^{18}O$ (bottom) across the Milky Way disc. In all panels, the predictions (solid lines) are from our best-fitting Milky Way model, while the symbols refer to the data (see Romano *et al.* 2017, for details about models and data sources).

3. An extremely low $^{13}C/^{18}O$ ratio in SMGs: a signature of a top-heavy IMF?

Once the stellar yields have been calibrated against the rich dataset available for our Galaxy, we can build up models for the prototype SMG. We want to explain the low $^{13}C/^{18}O \simeq 1$ ratio measured from simultaneous observations of ^{13}CO and $C^{18}O$ emission lines in a sample of four SMGs. This ratio has the advantage of being highly sensitive to IMF variations and little biased by differential astro-chemical and lensing effects for the bulk of the molecular gas in the target galaxies (Zhang *et al.* 2018).

We first run some models assuming the same stellar IMF adopted for the Milky Way (namely, a Kroupa 2002 IMF, with a slope $x = 1.7$ for high-mass stars), and vary the strength and duration of the starburst until a value of $^{13}C/^{18}O \simeq 1$ is obtained (we always end up with a final stellar mass of 2×10^{11} $M_\odot$ in the model SMG). In this case, we find that a ratio of $^{13}C/^{18}O$ around unity can be obtained only with a very short burst (lasting less than 100 Myr) and unrealistically high SF rates (up to 15,000 $M_\odot$ yr^{-1}). The

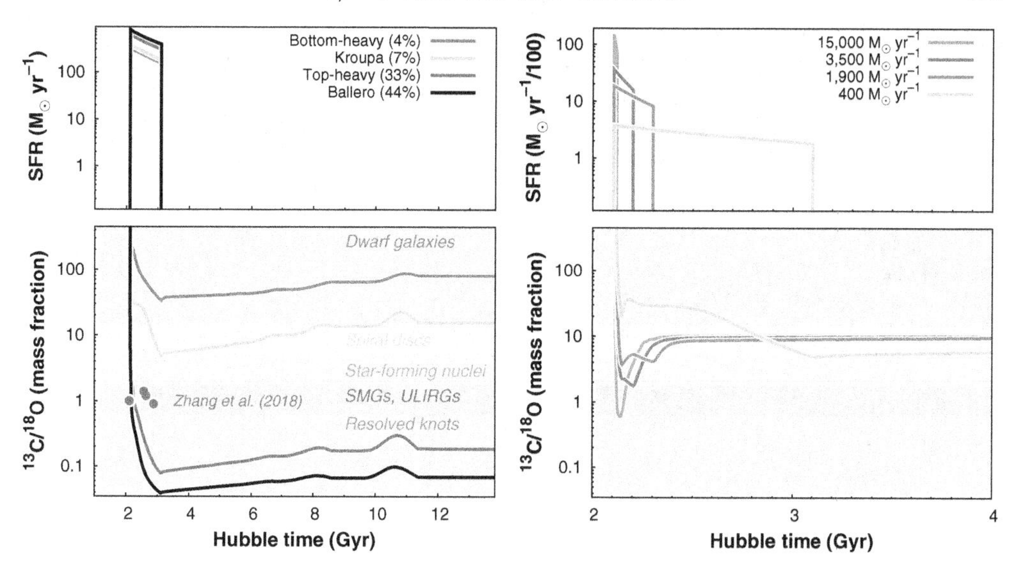

Figure 2. SF rates *(upper panels)* and evolution of $^{13}C/^{18}O$ ratio *(lower panels)* for representative SMG models, where we modify either the SF intensity and duration or the IMF high-mass slope (see text). The fraction of high-mass stars that is obtained with different IMF choices is reported in the top-right corner of the top-left panel. The red dots in the bottom-left panels show the $^{13}C/^{18}O$ ratios of the four SMGs targeted by Zhang *et al.* (2018).

ratio, moreover, suddenly increases to standard Galactic values as soon as the SF ceases, owing to the delayed release of ^{13}C from intermediate-mass stars (see Fig. 2, right-hand panels). According to our models, the only way to obtain a stable, low $^{13}C/^{18}O$ ratio in SMGs is via the assumption of an IMF skewed towards massive stars (see Fig. 2, left-hand panels, and further discussion in Zhang *et al.* 2018).

4. Conclusions and outlook

Our results that the IMF must be top-heavy in high-redshift starbursts is based on a novel approach –we measure the $^{13}C/^{18}O$ ratio in these systems in a regime free from the pernicious effects of dust and use detailed chemical evolution models to disentangle the effects of changes in the IMF or in the SF history on the predicted ratio. It must be cautioned, however, that new developments in stellar evolution and nucleosynthesis theory may still challenge our conclusions. For instance, we must consider the effects that stellar rotation may have on the yields. We are currently working on this.

Acknowledgements

We are grateful to the organizers for the opportunity to attend such a lively and inspiring conference.

References

Bastian, N., Covey, K. R., & Michael R. M. 2010, *ARAA*, 48, 339

Henkel, C. & Mauersberger, R. 1993, *A&A*, 274, 730

Ivison, R. J. et al. 1998, *MNRAS*, 298, 583

Jeřábková, T., Hasani Zonoozi, A., Kroupa, P., Beccari, G., Yan, Z., Vazdekis, A., & Zhang, Z.-Y. 2018, *A&A*, 620, A39

Limongi, M. & Chieffi, A. 2018, *ApJS*, 237, 13

Matteucci, F. 2012, *Chemical Evolution of Galaxies* (Berlin: Springer-Verlag)

Meynet, G. & Maeder, A. 2002, *A&A*, 390, 561

Papadopoulos, P. P. 2010, *ApJ*, 720, 226
Papadopoulos, P. P., Zhang, Z.-Y., Xilouris, E. M., Weiss, A., van der Werf, P., Israel, F. P., Greve, T. R., Isaak, K. G., & Gao, Y. 2014, *ApJ*, 788, 153
Pignatari, M. et al. 2015, *ApJ*, 808, L43
Romano, D. & Matteucci, F. 2003, *MNRAS*, 342, 185
Romano, D., Karakas, A. I., Tosi, M., & Matteucci, F. 2010, *A&A*, 522, A32
Romano, D., Matteucci, F., Zhang, Z.-Y., Papadopoulos, P. P., & Ivison, R. 2017, *MNRAS*, 470, 401
Salpeter, E. E. 1955, *ApJ*, 121, 161
Simpson, J. M., et al. 2017, *ApJ*, 844, L10
Zhang, Z.-Y., Romano, D., Ivison, R. J., Papadopoulos, P. P., & Matteucci, F. 2018, *Nature*, 558, 260

Uncovering Early Galaxy Evolution in the ALMA and JWST Era
Proceedings IAU Symposium No. 352, 2019
E. da Cunha, J. Hodge, J. Afonso, L. Pentericci & D. Sobral, eds.

doi:10.1017/S1743921319009542

ALMA twenty-six arcmin^2 survey of GOODS-S at one millimeter (ASAGAO)

B. Hatsukade[1], K. Kohno[1,2], Y. Yamaguchi[1], H. Umehata[1,3], Y. Ao[4], I. Aretxaga[5], K. I. Caputi[6], J. S. Dunlop[7], E. Egami[8], D. Espada[9,10], S. Fujimoto[11], N. Hayatsu[12,13], D. H. Hughes[5], S. Ikarashi[6], D. Iono[9,10], R. J. Ivison[13,7], R. Kawabe[9,10], T. Kodama[14], M. Lee[15], Y. Matsuda[9,10], K. Nakanishi[9,10], K. Ohta[16], M. Ouchi[11,17], W. Rujopakarn[17,18,19], T. Suzuki[9], Y. Tamura[15], Y. Ueda[16], T. Wang[1,9], W.-H. Wang[20], G. W. Wilson[21], Y. Yoshimura[1], M. S. Yun[21] and ASAGAO team

[1]Institute of Astronomy, Graduate School of Science, The University of Tokyo, 2-21-1 Osawa, Mitaka, Tokyo 181-0015, Japan
email: `hatsukade@ioa.s.u-tokyo.ac.jp`

[2]Research Center for the Early Universe, The University of Tokyo, 7-3-1 Hongo, Bunkyo, Tokyo 113-0033, Japan

[3]RIKEN Cluster for Pioneering Research, 2-1 Hirosawa, Wako-shi, Saitama 351-0198, Japan

[4]Purple Mountain Observatory & Key Laboratory for Radio Astronomy, Chinese Academy of Sciences, 8 Yuanhua Road, Nanjing 210034, China

[5]Instituto Nacional de Astrofísica, Óptica y Electrónica (INAOE), Luis Enrique Erro 1, Sta. Ma. Tonantzintla, Puebla, Mexico

[6]Kapteyn Astronomical Institute, University of Groningen, P.O. Box 800, 9700AV Groningen, The Netherlands

[7]Institute for Astronomy, University of Edinburgh, Royal Observatory, Edinburgh EH9 3HJ UK

[8]Steward Observatory, University of Arizona, 933 N. Cherry Ave, Tucson, AZ 85721, USA

[9]National Astronomical Observatory of Japan, 2-21-1 Osawa, Mitaka, Tokyo 181-8588, Japan

[10]SOKENDAI (The Graduate University for Advanced Studies), 2-21-1 Osawa, Mitaka, Tokyo 181-8588, Japan

[11]Institute for Cosmic Ray Research, The University of Tokyo, Kashiwa, Chiba 277-8582, Japan

[12]Department of Physics, Graduate School of Science, The University of Tokyo, 7-3-1 Hongo, Bunkyo, Tokyo, 113-0033, Japan

[13]European Southern Observatory, Karl-Schwarzschild-Str. 2, D-85748 Garching, Germany

[14]Astronomical Institute, Tohoku University, Aramaki, Aoba-ku, Sendai, Miyagi 980-8578, Japan

[15]Department of Physics, Nagoya University, Furo-cho, Chikusa-ku, Nagoya 464-8601, Japan

[16]Department of Astronomy, Kyoto University, Kyoto 606-8502, Japan

[17]Kavli Institute for the Physics and Mathematics of the Universe, Todai Institutes for Advanced Study, the University of Tokyo, Kashiwa, Japan 277-8583 (Kavli IPMU, WPI)

[18]Department of Physics, Faculty of Science, Chulalongkorn University, 254 Phayathai Road, Pathumwan, Bangkok 10330, Thailand

[19]National Astronomical Research Institute of Thailand (Public Organization), Don Kaeo, Mae Rim, Chiang Mai 50180, Thailand

[20]Institute of Astronomy and Astrophysics, Academia Sinica, Taipei, Taiwan

[21]Department of astronomy, University of Massachusetts, Amherst, MA 01003, USA

Abstract. The ALMA twenty-six arcmin^2 survey of GOODS-S at one millimeter (ASAGAO) is a deep ($1\sigma \sim 61\mu$Jy/beam) and wide area (26 arcmin^2) survey on a contiguous field at 1.2 mm. By combining with archival data, we obtained a deeper map in the same region ($1\sigma \sim 30\mu$Jy beam^{-1}, synthesized beam size $0.59'' \times 0.53''$), providing the largest sample of sources (25 sources at 5σ, 45 sources at 4.5σ) among ALMA blank-field surveys. The median redshift of the 4.5σ sources is 2.4. The number counts shows that 52% of the extragalactic background light at 1.2 mm is resolved into discrete sources. We create IR luminosity functions (LFs) at $z = 1$–3, and constrain the faintest luminosity of the LF at $2 < z < 3$. The LFs are consistent with previous results based on other ALMA and SCUBA-2 observations, which suggests a positive luminosity evolution and negative density evolution.

Keywords. cosmology: observations, galaxies: evolution, galaxies: formation

---

We performed the ALMA twenty-six arcmin^2 survey of GOODS-S at one-millimeter (ASAGAO; Hatsukade *et al.* (2018); Kohno *et al.* (2016); Ueda *et al.* (2018); Fujimoto *et al.* (2018); Yamaguchi *et al.* (2019)). The central 26 arcmin^2 area of GOODS-S was observed at 1.2 mm, providing a map with $1\sigma \sim 61$ μJy beam^{-1} (250 kλ-taper) and a synthesized beam size of $0.51'' \times 0.45''$. By combining the ALMA archival data available in GOODS-S, we obtained a deeper map for the 26 arcmin^2 area, which has a rms noise level of $1\sigma \sim 30$ μJy beam^{-1} for the central region with a 250 kλ-taper and a synthesized beam size of $0.59'' \times 0.53''$. We find 25 sources at 5σ and 45 sources at 4.5σ in the combined ASAGAO map, providing the largest source catalog among ALMA blank field surveys.

The larger sample allow us to construct 1.2 mm number counts with smaller uncertainties from Poisson statistics. The flux coverage of the number counts connects the fainter range probed by ALMA deep observations and the brighter range constrained by ALMA follow-up observations of single-dish detected sources. We find that our number counts are consistent with previous ALMA studies (e.g., Hatsukade *et al.* (2016)). By integrating the derived differential number counts, we find that 52^{+11}_{-8}% of the extragalactic background light (EBL) at 1.2 mm is revolved into the discrete sources. The integration of the best-fitting function reaches 100% at $S_{1.2\text{mm}} \sim 20$ μJy, although there is a large uncertainty to extend the function to the fainter flux range.

By using the 5σ sources, we construct IR luminosity functions (LFs) in the ranges of $1 < z < 2$, $1.5 < z < 2.5$, and $2 < z < 3$. Our study constrains the faintest luminosity end of the LF at $2 < z < 3$ among other studies. We find that the ASAGAO LFs are consistent with those of Koprowski *et al.* (2017), supporting the evolution of LFs (positive luminosity evolution and negative density evolution with increasing redshift). The integration of the best-fitting LF down to the lowest luminosity of the sources ($\log(L_{\text{IR}}/L_\odot) = 11.78$) gives a star-formation rate density (SFRD) of $7.2^{+3.0}_{-1.9} \times 10^{-2}$ $M_\odot$ yr^{-1} Mpc^{-3}. We find that the IR-based star formation of ASAGAO sources contribute to ≈60–90% of the SFRD at $z \sim 2$ derived from UV–IR observation, indicating that the major portion of $z \sim 2$ SFRD is composed of sources with $\log(L_{\text{IR}}/L_\odot) \gtrsim 11.8$.

References

Fujimoto, S., Ouchi, M., Kohno, K., *et al.* 2018, *ApJ*, 861, 7
Hatsukade, B., Kohno, K., Umehata, H., *et al.* 2016, *PASJ*, 68, 36
Hatsukade, B., Kohno, K., Yamaguchi, Y., *et al.* 2018, *PASJ*, 70, 105
Kohno, K., Yamaguchi, Y., Tamura, Y., *et al.* 2016, *IAUS*, 319, 92
Koprowski, M. P., Dunlop, J. S., Michałowski, M. J., *et al.* 2017, *MNRAS*, 471, 4155
Ueda, Y., Hatsukade, B., Kohno, K., *et al.* 2018, *ApJ*, 853, 24
Yamaguchi, Y., Kohno, K., Hatsukade, B., *et al.* 2019, *ApJ*, 878, 73

Uncovering Early Galaxy Evolution in the ALMA and JWST Era
Proceedings IAU Symposium No. 352, 2019
E. da Cunha, J. Hodge, J. Afonso, L. Pentericci & D. Sobral, eds.

doi:10.1017/S1743921319009128

PAHs and star formation in ELAIS N1 as seen by AKARI

Tímea Kovács[1], Denis Burgarella[2], Hidehiro Kaneda[3], Dániel Cs. Molnár[4], Shinki Oyabu[3], Sandor Pinter[1] and L. Viktor Toth[1,5]

[1]Department of Astronomy of the Eötvös Loránd University, Pázmány Péter sétány 1/A, H-1117 Budapest, Hungary

[2]Aix-Marseille Université, CNRS, LAM (Laboratoire d'Astrophysique de Marseille), UMR 7326, 13388 Marseille, France

[3]Graduate School of Science, Nagoya University, Japan

[4]INAF - Osservatorio Astronomico di Cagliari, Via della Scienza 5, I-09047 Selargius (CA), Italy

[5]Konkoly Observatory of the Hungarian Academy of Sciences, Konkoly Thege Miklós út 15-17., H-1121 Budapest, Hungary

Abstract. We have examined the relationship between star formation and polycyclic aromatic hydrocarbons (PAHs) by fitting the spectral energy distributions (SED) of AKARI selected galaxies. PAHs are excited by the ultraviolet (UV) photons of young stars and can trace star formation in galaxies, but they are disassociated by the strong UV radiation in starbursts. AKARI covered the mid-infrared, where the PAHs emit their radiation, with a high density of photometric bands. These observations allow us to estimate the star formation rate and the PAH mass fraction of the dust in galaxies. In the future the James Webb Space Telescope (JWST) will also make measurements in this wavelength range. This research can therefore be considered as a pathfinder to similar studies that will come later from JWST observations.

Keywords. galaxies: ISM, galaxies: star formation, infrared: galaxies

---

1. Data and Methods

We selected sources with Herschel and AKARI detections, and known redshifts in the ELAIS (European Large Area ISO Survey) N1 field (which was observed from the optical to the radio). We used photometric data from the Herschel Point Source Catalogs (Marton *et al.* 2017; Schulz *et al.* 2017) to constrain the far-infrared radiation, and from the AKARI Deep Field Catalogs (Davidge *et al.* 2017, 4 − 18 microns) to estimate the PAH emission. Photometric data at other wavelengths was also collected from different databases (for example WISE, SDSS).

We fitted the spectral energy distributions (SED) of galaxies with CIGALE (Noll *et al.* 2009; Roehlly *et al.* 2014; Boquien *et al.* 2019), and derived their physical properties like stellar mass, star formation rate (SFR), dust mass and the mass fraction of PAH molecules (q_{PAH}). We used the Bruzual & Charlot (2003) stellar emission models with a Chabrier (2003) initial mass function, a delayed star formation history, the Draine & Li (2007) dust emission model, and a double powerlaw for the attenuation.

2. Results

Using the AKARI photometric data in the SED fitting lowered the error of the q_{PAH} parameter on average: the mean relative error of q_{PAH} dropped from 37% to 26%. The models often under or overestimated the PAH emission without the AKARI data. When using AKARI points in the SEDs the $q_{\rm PAH}$ parameter became significantly lower in some cases, and higher in others.

The $q_{\rm PAH}$ in the selected sample is lower than the values usually found in previous works. This could be due to various reasons, such as low metallicity, ongoing active galactic nucleus or starburst activity. The spectrum of one of our sources has line ratios characteristic of AGNs, and another one has low metallicity. Based on the results of the SED fitting, the majority of the sample is on the main sequence of galaxies (Schreiber *et al.* 2015), only 8% of them could be starbursts. We found a decreasing $q_{\rm PAH}$ trend with increasing $A_{\rm V}$ and consequently LIR, suggesting the possible presence of compact star-forming regions.

We investigated different known relations, for example the PAH luminosities were calculated from the fitted SEDs, and compared to the SFR derived by CIGALE: the two methods give a similar SFR (Shipley *et al.* 2016).

3. Discussion and summary

To find the reason behind the low $q_{\rm PAH}$ values (e. g. determining the metallicity) follow-up spectroscopy would be needed.

The James Webb Space Telescope will have spectroscopic and photometric instruments which are sensitive in the near and mid-infrared (0.6 – 28.3 micron). It will be able to measure the equivalent widths of the 3.3 and the 7.7 μm PAH features up to z $\sim$ 6 and z $\sim$ 2.5, respectively. Similarly to the AKARI filters, the MIRI instrument will have 9 broadband filters at wavelengths from 4.6 to 28.8 μm (Gardner *et al.* 2006). The results based on the AKARI selected galaxies are summarized in a paper: Kovács *et al.* (2019).

Acknowledgements

This research was supported by the Hungarian OTKA grant NN111016, and by the ÚNKP-18-2 New National Excellence Program of the Ministry of Human Capacities. This research has made use of the SIMBAD database, and the VizieR catalog access tool, CDS, Strasbourg, France. We made use of data products from the Wide-field Infrared Survey Explorer, and the research is based on observations with AKARI, a JAXA project with the participation of ESA. The conference participation was supported by the IAU.

References

Boquien, M., Burgarella, D., Roehlly, Y., *et al.* 2019, *A&A*, 622, A103
Bruzual, G. & Charlot, S. 2003, *MNRAS*, 344, 1000
Chabrier, G. 2003, *PASP*, 115, 763
Davidge, H., Serjeant, S., Pearson, C., *et al.* 2017, *MNRAS*, 472, 4259
Draine, B. T. & Li, A. 2007, *ApJ*, 657, 810
Gardner, J. P., Mather, J. C., Clampin, M., *et al.* 2006, *SSR*, 123, 485
Kovács, T. O., Burgarella, D., Kaneda, H., *et al.* 2019, *PASJ*, 71, 27
Marton, G., Calzoletti, L., Perez Garcia, A. M., *et al.* 2017, arXiv:1705.05693
Noll, S., Burgarella, D., Giovannoli, E., *et al.* 2009, *A&A*, 507, 1793
Roehlly, Y., Burgarella, D., Buat, V., *et al.* 2014, Astronomical Data Analysis Software and Systems XXIII, 485, 347
Schulz, B., Marton, G., Valtchanov, I., *et al.* 2017, arXiv:1706.00448
Schreiber, C., Pannella, M., Elbaz, D., *et al.* 2015, *A&A*, 575, A74
Shipley, H. V., Papovich, C., Rieke, G. H., *et al.* 2016, *ApJ*, 818, 60

Uncovering Early Galaxy Evolution in the ALMA and JWST Era
Proceedings IAU Symposium No. 352, 2019
E. da Cunha, J. Hodge, J. Afonso, L. Pentericci & D. Sobral, eds.

doi:10.1017/S1743921319008925

An evolving photoelectric efficiency at cosmic noon?

Jed McKinney[1], Alexandra Pope[1], Lee Armus[2], Ranga Chary[3], Mark Dickinson[4] and Allison Kirkpatrick[5]

[1]Department of Astronomy, University of Massachusetts, Amherst, MA 01003, USA

[2]Infrared Processing and Analysis Center, MC 314-6, Caltech, 1200 E. California Blvd., Pasadena, CA 91125, USA

[3]Division of Physics, Math & Astronomy, California Institute of Technology, Pasadena, CA 91125

[4]National Optical Astronomy Observatory, 950 North Cherry Avenue, Tucson, AZ 85719, USA

[5]Department of Physics & Astronomy, University of Kansas, Lawrence, KS 66045, USA

Abstract. To sustain star formation rates (SFRs) of hundreds to thousands of solar masses per year over millions of years, a galaxy must efficiently cool its gas. At $z \sim 2$, the peak epoch for stellar mass assembly, tracers of gas heating and cooling remain largely unexplored. For one $z \sim 2$ starburst galaxy GS IRS20, we present *Spitzer IRS* spectroscopy of Polycyclic Aromatic Hydrocarbon (PAH) emission, and ALMA observations of [C II] 158μm fine-structure emission which we use to probe ISM heating/cooling. Coupled with an unusually warm dust component, the ratio of [C II]/PAH emission suggests a low photolelectric efficiency, and/or the importance of cooling from other far-IR lines in this galaxy. A low photoelectric efficiency at $z \sim 2$ could be key for the peak in the SFR density of the universe by decoupling stellar radiation from ISM gas temperatures.

Keywords. galaxies: high-redshift, galaxies: individual, galaxies: ISM, galaxies: starburst

---

1. Introduction

In galaxies, interstellar gas heats and cools during the cycle of star-formation and feedback processes. Photoelectrons ejected from polycyclic aromatic hydrocarbons (PAHs) are key for converting stellar radiation to thermal energy in photodissociation regions (PDRs), where large gas densities are illuminated by far-UV fluxes from young stars (Tielens & Hollenbach 1985). PDRs cool predominantly via far-infrared (IR) fine-structure lines such as [C II], which can contain as much as $0.1 - 1\%$ of a galaxy's total IR emission (L_{IR}, $8 - 1000$ μm). The ratio of [C II] to mid-IR PAH emission is therefore a powerful diagnostic of PDR regions of high-z galaxies (Helou *et al.* 2001).

[C II] observations of galaxies at cosmic noon are limited to only a handful (Zanella *et al.* 2018), and even fewer have mid-IR PAH spectra from *Spitzer* (e.g., Brisbin *et al.* 2015). Thus, key ISM line-ratios at $z \sim 2$ have yet to be characterized. To this point, we present [C II] and PAH observations of GS IRS20, a starburst galaxy at $z = 1.924$ with the highest signal-to-noise ALMA detection of [C II] at $z \sim 2$ to date (Fig. 1, *Left inset*). GS IRS20 has an unusually warm dust component, and is a unique opportunity to probe the ISM conditions of extreme star formation.

2. Overview

The luminosity ratio of [C II]/L_{IR} and PAH/L_{IR} show a deficit towards higher L_{IR}, with significant scatter related to, amongst other factors, dust temperatures, geometry,

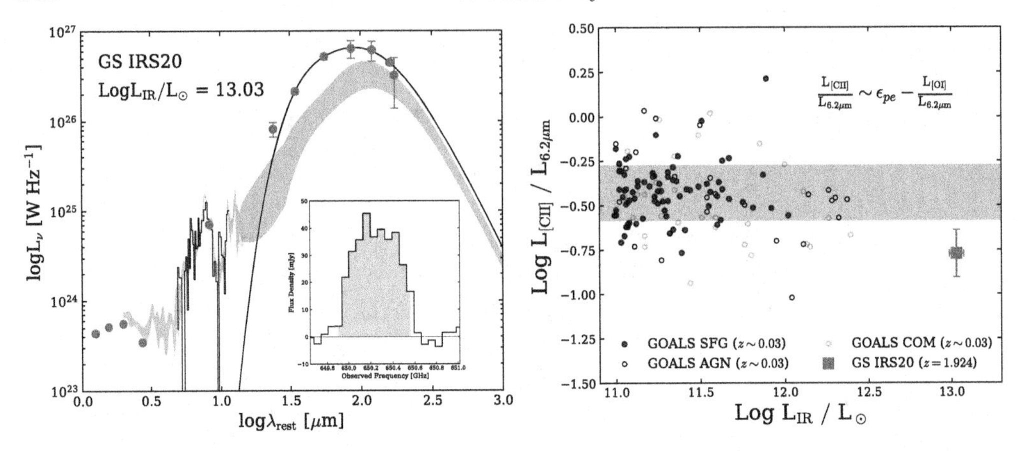

Figure 1. (*Left*) GS IRS20's SED with the ALMA Band 9 [C II] detection shown in the inset panel. Multi-wavelength photometry is shown with red and blue points, and its *IRS* spectrum is shown in black. The shaded gray region corresponds to an empirical SED from Kirkpatrick *et al.* (2015) normalized to GS IRS20's PAH spectrum. The thin black line is a two-temperature graybody fit to photometry past 25μm. (*Right*) The ratio of [C II] luminosity to the 6.2μm PAH feature luminosity. The gray shaded region contains the 1σ scatter about the mean of local star-forming IR galaxies from GOALS. GS IRS20 is shown in red.

and fractional contributions from different gas phases (e.g., Pope *et al.* 2013; Díaz-Santos *et al.* 2017). As noted in Helou *et al.* (2001), PAH emission is a more appropriate normalization factor for $L_{[C\,II]}$ as the collisional excitation of C^+ is proportional to the density of PAH photoelectrons. Indeed, Figure 1 (*Right*) shows that $L_{[C\,II]}/L_{6.2\mu m}$ remains relatively constant as a function of L_{IR} in low-z IR galaxies, particularly those that are purely star-forming. GS IRS20 falls below this average by $\sim 3\sigma$. $L_{[C\,II]}/L_{6.2\mu m}$ traces the balance of gas cooling through the [C II] channel and photoelectric gas heating, and is therefore proportional to the difference in photoelectric efficiency (ϵ_{pe}), and cooling via [O I] assuming other far-IR emission lines are negligible (see Eq. 7 in Croxall *et al.* 2012).

3. Implications

The photoelectric efficiency is a measure of total gas heating to total dust heating, and can fall at high ratios of far-UV flux to gas density as PAHs lose more of their surface electrons, reducing their photoelectric yields (e.g., Helou *et al.* 2001). As a result, PDRs become less efficient at converting stellar radiation to thermal energy in atomic gas. Empirically, Croxall *et al.* (2012) find lower values of ϵ_{pe} in warmer dust environments, consistent with harder and more intense far-UV radiation fields.

GS IRS20 has an unusually warm dust component compared to other $z \sim 2$ IR galaxies (Fig. 1 *Left*), is a starburst galaxy, and has a low $L_{[C\,II]}/L_{6.2\mu m}$ (Fig. 1 *Right*). For these reasons, ϵ_{pe} may be low in GS IRS20, and/or cooling via [O I] plays an important role. We tentatively speculate that a decrease in ϵ_{pe} in high$-z$ star-forming galaxies could suppress stellar feedback and contribute to the peak in the IR SFR density. If this is the case, atomic gas may remain cool to form more stars, yielding higher SFRs.

References

Helou, G., Malhotra, S., Hollenbach, D. J., Dale, D. A., & Contursi, A. 2001, *ApJL*, 548, L73
Tielens, A. G. G. M., & Hollenbach, D. 1985, *ApJ*, 291, 722
Pope, A., Wagg, J., Frayer, D., *et al.* 2013, ApJ, 772, 92
Zanella, A., Daddi, E., Magdis, G., *et al.* 2018, MNRAS, 481, 1976

Brisbin, D., Ferkinhoff, C., Nikola, T., *et al.* 2015, ApJ, 799, 13
Kirkpatrick, A., Pope, A., Sajina, A., *et al.* 2015, ApJ, 814, 9
Díaz-Santos, T., Armus, L., Charmandaris, V., *et al.* 2017, ApJ, 846, 32
Croxall, K. V., Smith, J. D., Wolfire, M. G., *et al.* 2012, ApJ, 747, 81

Uncovering Early Galaxy Evolution in the ALMA and JWST Era
Proceedings IAU Symposium No. 352, 2019
E. da Cunha, J. Hodge, J. Afonso, L. Pentericci & D. Sobral, eds.

doi:10.1017/S1743921319007385

The Far-Infrared emission of the first (z ∼ 6) massive galaxies

George H. Rieke[1], Maria Emilia De Rossi[2], Irene Shivaei[1], Volker Bromm[3] and Jianwei Lyu[1]

[1]Steward Observatory, 533 N. Cherry Ave., The University of Arizona, Tucson, AZ 85721, USA
email: grieke@as.arizona.edu

[2]Universidad de Buenos Aires, Facultad de Ciencias Exactas y Naturales y Ciclo Básico Común. Buenos Aires, Argentina and CONICET-Universidad de Buenos Aires, Instituto de Astronomía y Física del Espacio (IAFE). Buenos Aires, Argentina

[3]Department of Astronomy, The University of Texas, 2515 Speedway, Stop C1400, Austin, Texas 78712-1205, USA

Abstract. The first massive galaxies (z ∼ 6) have (1) very high energy density due to their small diameters and extreme luminosities in young stars and (2) interstellar dust relatively deficient in carbon compared with silicates. Both of these attributes should raise their interstellar dust temperatures compared with lower redshift galaxies. Not only is this temperature trend observed, but the high-z spectral energy distributions (SEDs) are very broad due to very warm dust. As a result total infrared luminosities – and star formation rates – at the highest redshifts estimated by fitting blackbodies to submm- and mm-wave observations can be low by a factor of ∼2.

Keywords. galaxies: high-redshift – evolution – infrared: galaxies

---

1. Introduction

Theoretical models predict that the dust in high redshift galaxies ($z \gtrsim 5$) should be significantly warmer than that in galaxies at modest redshift ($z \lesssim 4$) (De Rossi *et al.* (2018); Ma *et al.* (2019)). The higher temperatures are predicted to result from the compact sizes (diameters of $1-2$ kpc) containing huge luminosities from young stars ($\sim 10^{13} L_{\odot}$), combined with the relatively silicate-rich composition of the dust with poor radiative efficiency at long far infrared wavelengths. We show that the available observations confirm these predictions. In addition, the broader SEDs (i.e., more radiation in the 10 - 40 μm range) at very high z can result in underestimates of the total infrared luminosities of these galaxies, by a factor of ∼2 if the estimates are based on fitting a modified blackbody ($\beta = 1.6$) to submm- and mm-wave measurements. Without taking correct account of the non-blackbody nature of the SEDs, the interpretation of the ALMA measurements will underestimate the star formation rates in these galaxies substantially.

2. Galaxy FIR SEDs at z = 3 and 6

Figure 1 shows the average far-infrared SED of luminous galaxies at $z \sim 3$ as represented by the best-fitting template (De Rossi *et al.* (2018)). The units are νf_{ν}, to indicate directly relative luminosity in logarithmic wavelength intervals. We have shown the average behavior of these galaxies as a fitted template because showing the full data (42 individual galaxies, multiple results from stacks that total several hundreds of galaxies) would be complex; see De Rossi *et al.* (2018) for these details. We have fitted this template with a modified blackbody with T = 34K and $\beta = 1.6$, reflecting common practice

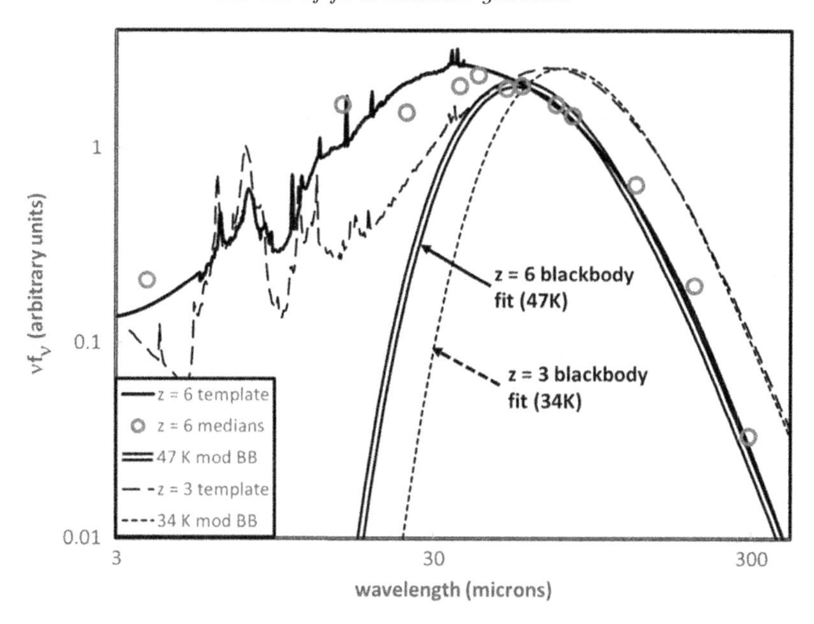

Figure 1. Average far infrared SEDs of galaxies at z ∼ 3 (dashed line) and z ∼ 6 (solid line). The open red circles are medians of the observations of galaxies with $5 \leqslant z < 7$; SEDs and medians from De Rossi *et al.* (2018). Modified blackbodies ($\beta = 1.6$) have been fitted to the parts of the SEDs accessible to ALMA, with a temperature of 34K for z = 3 and 47K for z = 6.

with submm- and mm-wave observations. Over the range of rest wavelengths probed at, e.g., 850μm observed (i.e., for $z < 4$, wavelengths $\geqslant 170$ μm), the correspondence is nearly perfect. However, at wavelengths < 50 μm, the emission from the galaxies substantially exceeds this fit and estimates of the total infrared luminosity based on the blackbody fit will be too small by a significant amount.

Figure 1 also includes (1) median relative flux densities vs. wavelength for galaxies at $5 \leqslant z < 7$; (2) the best-fitting template for these galaxies; and (3) a modified blackbody fit to the template. The medians include those in De Rossi *et al.* (2018) plus one that was missed (Strandet *et al.* (2017)), for a total of 19 (7 from the South Pole Telescope, 12 from other sources). The number of galaxies with useful measurements at rest $\lambda < 30\mu$m is small, hence the scatter, but nonetheless, the figure demonstrates that the template, based on the SED of the local galaxy Haro 11, is a reasonably good representation. The modified blackbody fit is at 47K, $\beta = 1.6$. Again, it is an excellent fit over the range of rest wavelengths accessible with ALMA at 850 μm ($\lambda \geqslant 40\mu$m), but only captures about half of the luminosity. The lower temperature blackbody that fits the data at $z \sim 3$ is clearly not a reasonable fit to the $z \sim 6$ points, and the equivalent temperature of the higher redshift galaxies would be pushed even higher with a metric that accounted for its broader SED caused by a larger contribution by very warm dust.

References

De Rossi, Maria Emilia, Rieke, G. H., Shivaei, Irene, Bromm, Volker, & Lyu, Jianwei 2018, *ApJ*, 869, 4

Ma, Xiangcheng, Hayward, C. C., Casey, C. M., *et al.* 2019, *MNRAS*, 487, 1844
692, 556

Strandet, M. L., Wiess, A., De Breuck, *et al.* 2017, *ApJL*, 842, 15

Uncovering Early Galaxy Evolution in the ALMA and JWST Era
Proceedings IAU Symposium No. 352, 2019
E. da Cunha, J. Hodge, J. Afonso, L. Pentericci & D. Sobral, eds.

doi:10.1017/S1743921319009220

Preliminary results from prebiotic molecules with ALMA in the era of artificial intelligence

Villicana-Pedraza, I.[1], Walterbos, R.[2], Carreto-Parra, F.[3], Ott. J.[4], Momjian, E.[4], Thelen, A.[5], Ginsburg, A.[4], Zapata, L.[6], Gonzalez, M.[7], Floyd, J.[1], Saucedo, J.[8], Binette, L.[9] and Prugh, S.[1]

[1]DACC, New Mexico State University, Las Cruces, NM 88003, USA
email: `ilhui7@nmsu.edu`

[2]Dept. of Astronomy, New Mexico State University, Las Cruces, NM, USA

[3]Dept. of Physics, New Mexico State University, Las Cruces, NM, USA

[4]National Radio Astronomy Observatory, Socorro, NM, USA

[5]NASA Goddard Space Flight Center, USA

[6]Instituto de Radioastronomia,Universidad Nacional Autonoma de Mexico, Mexico

[7]Universidad Autonoma de Ciudad Juarez, Juarez Chihuahua, Mexico

[8]Universidad de Sonora, Hermosillo Sonora, Mexico

[9]Instituto de Astronomia, Universidad Nacional Autonoma de Mexico, Mexico

Abstract. Study of the composition from diverse sources of the Universe helps to us to understand their evolution. Molecular spectroscopy provides detailed information of the observed objects. We present a small study of the starburst NGC 253 with ALMA at 1mm. We detect the prebiotic molecules NH_2CHO, and CNCHO. We obtain the integrated intensity maps and abundances of HNCO, CH_3OH, H_3O+ and CH_3C_2H. We propose the use of Artificial Intelligence for big data to find prebiotic molecules in galaxies.

Keywords. ALMA, Molecules, Spectroscopy, Artificial Intelligence.

---

1. Introduction

The starburst NGC 253 is at a distance of 3.5 Mpc (Rekola 2005) and has a velocity of 258.8km/s. It is one of the brightest extragalactic molecular line sources. Sakamoto *et al.* (2011) studied NGC 253 and identified five peaks (Clumps) at 1.3mm of molecular line and continuum emission over 300pc. Meier *et al.* (2015) did an ALMA multi-line study of NGC 253 which confirmed molecules previously detected at 3 mm as well as reporting the new tentative detection of molecules previously not observed. Ando *et al.* (2017) et al. observed the same source but in the range centered at 0.85mm. They resolved the nuclear starburst in 8 clumps separated 10 pc in scale, detected complex organic molecules, and found that the hot and chemically rich environments are located within 10pc of the nucleus. Villicana-Pedraza *et al.* (2017) studied the same galaxy at the same frequency but using a single dish.

We report here some results centered at 307GHz observed with ALMA (PI. Villicana-Pedraza No. 2013.1.00973).

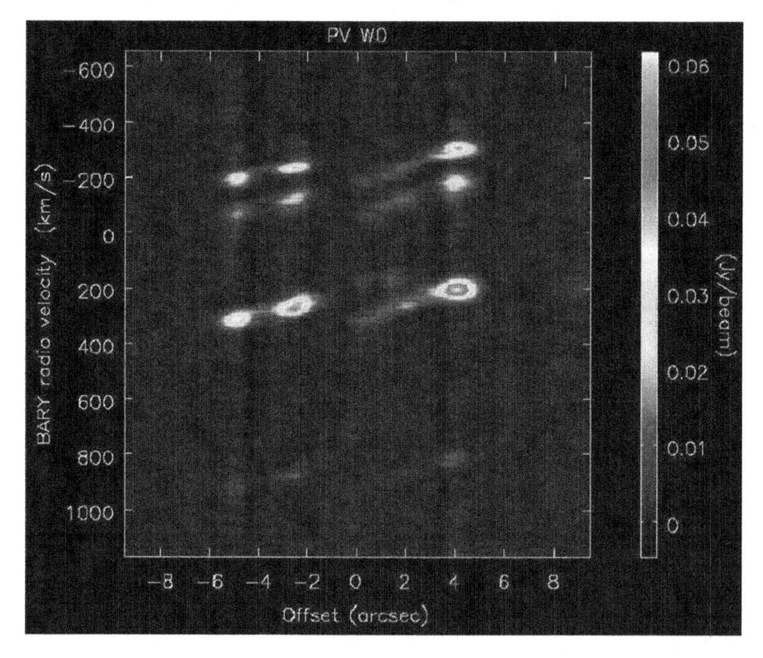

Figure 1. Position-Velocity diagram for the spectral window from 306-308GHz.

2. Analysis and Results

We re-reduced all the observations with the Common Astronomy Software Applications package (CASA). A calibration and the subsequent image extraction of the data cubes were performed. The analysis was carried out using CASA, with GILDAS being preferred for some images. The continuum emission from NGC253 has a flux density of approximately 1 Jy at 294.96GHz. The primary beam has a FWHM of 17.6 arcsec. We identified four clumps in the continuum.

We made Position Velocity maps (P-V) to identify the molecular features of the region and identify blended lines. From the P-V maps we established that the region studied contains features of 4 molecular lines that belong to HNCO, CH_3OH, H_3O+ and CH_3C_2H.We found CH_3C_2H blended, therefore we used a mask of HNCO to separate them (Fig. 1). We obtained integrated intensity maps, column densities and abundances for these molecules. We found one transition for the prebiotic molecules NH_2CHO, and CNCHO; we confirm the tentative detections from Meier *et al.* (2015). We also observed CH_3CH_2OH. The tables with the abundances and maps can be found in Villicana-Pedraza *et al.* in prep.

3. Artificial Intelligence and Machine Learning applications

In this work we report the detection of 2 prebiotic molecules toward one starburst galaxy. Villicana-Pedraza (2017) reported one more detected in a Seyfert galaxy. We will create templates of all molecular lines likely present in galaxies to help the search for prebiotic molecules. The ultimate goal is to establish the environments in which these lines are formed and obtain their abundances. To achieve this goal we will use Artificial Intelligence (AI) and Machine learning (ML) to analyze data for hundreds of galaxies. AI and ML can analyze information much faster than conventional techniques. We can implement an ML algorithm in Python in two stages: The first one is using a Decision Tree Algorithm with the task DecisionTreeClassifier(); The second is to train data sets for the fitting process for the abundances.

References

Rekola, R., Richer, M. G., McCall Marshall, L., Valtonen, M. J., *et al.* 2005, *MNRAS*, 361, 330
Sakamoto, K., Mao, R. Q., Matsushita, S., *et al.* 2011, *ApJ*, 19, 735
Meier, D., Walter, F., Bolatto, A., Leroy, A., Ott, J., *et al.* 2015, *ApJ*, 63, 801
Ando, R., Nakanishi, K., Knohno, K., Izumi, T., *et al.* 2017, *ApJ*, 81, 849
Villicana-Pedraza, I., Martin, S., *et al.* 2017, *IAU*, S321, 305, 11

SESSION 6: Spatially-resolved analyses of $z > 2$ galaxies

Uncovering Early Galaxy Evolution in the ALMA and JWST Era
Proceedings IAU Symposium No. 352, 2019
E. da Cunha, J. Hodge, J. Afonso, L. Pentericci & D. Sobral, eds.

doi:10.1017/S1743921319008421

Resolved views on early galaxy evolution

Stijn Wuyts[1] **and Natascha M. Förster Schreiber**[2]

[1]Dept. of Physics, University of Bath, Claverton Down, Bath, BA2 7AY, UK
email: s.wuyts@bath.ac.uk

[2]Max-Planck-Institut für extraterrestrische Physik, Giessenbachstr. 1, Garching, D-85737, Germany
email: forster@mpe.mpg.de

Abstract. Resolved observations of star-forming galaxies at cosmic noon with the Hubble Space Telescope and large ground-based facilities provide a view on the spatial distribution of stars, gas and dust, and probe gaseous motions revealing the central gravitational potential and local feedback processes at play. In this paper, we review recent insights gained from such observations, with an emphasis on results obtained through optical/near-infrared imaging and imaging spectroscopy. Their context and implications are documented more fully in a forthcoming review article by Förster Schreiber & Wuyts (in prep).

Keywords. Galaxies: structure, Galaxies: kinematics, Galaxies: high-redshift

---

1. Introduction

Census. Accumulating over more than a decade, a large cross section of the galaxy evolution community has contributed to assembling a wedding cake of deep legacy fields from which consistent multi-wavelength photometric catalogs have been extracted and made publicly available. Ranging from ultra-deep pencil-beam surveys such as the Hubble XDF to areas of 2 deg^2 (COSMOS) and beyond (e.g., Hyper-SuprimeCam, VIDEO, KIDS+Viking) they complement each other in a balancing of area versus depth.

Another dimension along which legacy imaging surveys have advanced and augmented the potential to infer physical properties of galaxies is wavelength. The spectral coverage in the aforementioned legacy deep fields now routinely extends from X-ray diagnostics of AGN activity to a bolometric census of star formation by pairing the rest-UV emission as observed with HST with far infrared constraints on the amount of obscured and dust-reprocessed emission from young stars as observed with Spitzer/MIPS and Herschel. Importantly, it also encompasses increasingly a finer sampling of galaxy Spectral Energy Distributions (SEDs) through the use of medium-band filters (e.g., zFOURGE) and slitless grism spectroscopy (e.g., 3D-HST). The intermediate $R \sim 13 - 130$ SED sampling these surveys provide enables a sensitive improvement in photometric redshift estimates, which continue to be of great importance as high-resolution spectroscopic follow-up campaigns struggle to reach high levels of completeness over the same dimensions in area and depth. The enhancement in redshift quality can further aid spectroscopic follow-up campaigns, for example by allowing to select specifically those targets that have their emission lines of interest redshifted away from OH sky lines that render significant swathes of the near-infrared wavelength regime unsuitable for detailed line profile analyses as discussed in Section 3 (Wisnioski *et al.* 2015). Finally, by more accurately pinpointing the location and strength of continuum breaks they yield improved constraints on the mass-to-light ratios of distant galaxies and the stellar population (age) or dust (attenuation) properties responsible for it.

It is this richness of multi-wavelength data that has facilitated a mass-complete census of star-forming and quiescent galaxies over more than 90% of cosmic history (e.g., Davidzon *et al.* 2017).

Scaling relations. Alongside an improved census of the number of galaxies as a function of mass, redshift and type (star-forming/quiescent) multi-wavelength lookback surveys have helped establish the regularity of galaxy properties across cosmic time. This regularity is captured by a set of scaling relations aimed at describing the star-forming galaxy (SFG) population as a one-parameter family. At any given epoch, physical properties of SFGs such as their star formation rate (SFR), gas content, metallicity, rotational velocity and size can be parameterized as a function of galaxy stellar mass. The change in zero point, slope and/or shape of such scaling relations then encodes their evolution in a population averaged sense (Speagle *et al.* 2014; Tacconi *et al.* 2018; Sanders *et al.* 2018; Übler *et al.* 2017; van der Wel *et al.* 2014).

Evolution. The evolution of population-averaged physical properties evaluated at fixed mass does not equate to the evolution of individual galaxies, which grow in mass over time. In recent years, considerable efforts have therefore gone into reconstructing the growth history of individual systems. Perhaps the most common approach of connecting progenitor and descendant galaxies across cosmic time consists of assuming that their relative mass ranking remains preserved (i.e., the most massive galaxy today was also the most massive at earlier epochs and so on), in which case progenitors and descendants can be considered to live at the same comoving number density. Curves of stellar mass growth as a function of redshift can then be combined with the aforementioned scaling relations to establish how an individual galaxy consumes its gas, enriches its ISM, grows its size and builds up angular momentum.

Caveats to this methodology apply. Modifications to the prescription of constructing progenitor-descendant sequences have been proposed by Torrey *et al.* (2017) on the basis of empirical galaxy stellar mass functions and an assessment of the significant galaxy-to-galaxy scatter in growth rates seen in cosmological simulations, even in the absence of frequent major mergers. The primary notion that SFGs can be treated as a one-parameter family can also be questioned. On the grounds of cosmological simulations, Matthee & Schaye (2019) argued that the scatter around the star-forming main sequence features contributions both from short-term stochasticity (so-called 'breathing' due to the cycle of star formation feedback) and a long-term differentiation in growth histories tied to halo formation times. Along similar lines, taking a more empirical view at the evolving galaxy population, Abramson *et al.* (2016) describe the main sequence and size-mass scaling relations as emerging from a population moving through them rather than being composed of the same systems at all times.

Despite the above caveats which prompt further study in this area, specifically regarding the physical origin of scatter around scaling relations, a few basic conclusions can be drawn from the above line of reasoning. For a galaxy of Milky Way mass today, it reveals the so-called era of cosmic noon ($1 < z < 3$) as a key epoch during which the star formation activity peaked, the gas richness dropped most dramatically (despite continuing yet declining gaseous accretion onto the galaxy) and the chemical enrichment proceeded most quickly. In relative terms, the rate of size growth was maintained over a much more extended period of time down to the present day (see Figure 1 of Förster Schreiber & Wuyts in prep).

Toward a multi-wavelength resolved view of galaxies at cosmic noon. While highlighting the importance of the cosmic noon era to understand the growth history of Milky Way and higher mass galaxies, the above galaxy-integrated quantities need to be paired with a resolved view of their structure and kinematics if we are to understand the physical drivers behind their evolution. In fact, as we illustrate below, it is a multi-wavelength resolved

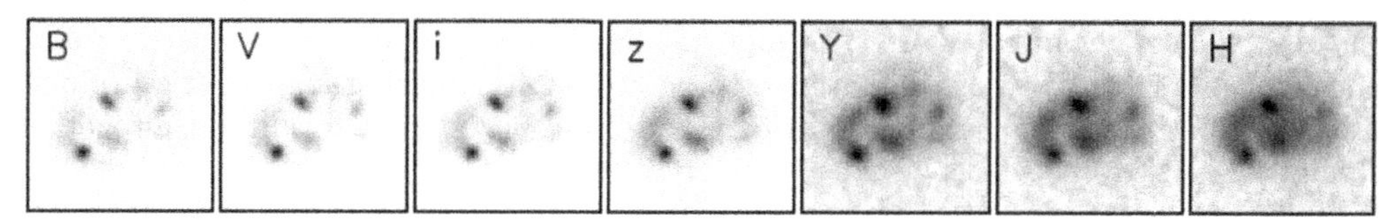

Figure 1. Example of a $z = 2.13$ main sequence galaxy in the HUDF as imaged in the F435W, F606W, F775W, F850LP, F105W, F125W and F160W bands. Postage stamps measure $3'' \times 3''$. This galaxy features a clumpy structure that is most prominent in the rest-frame UV and becomes progressively less pronounced toward redder wavelengths.

view that is needed to construct a comprehensive picture of the physical processes at play. With multi-band resolved maps of thousands of distant galaxies in the rest-UV to optical, resolved observations of ionized gas line emission in hundreds of them (thousands if relying on stacking of low-S/N Hα maps from HST grism spectroscopy out to $z < 1.5$), and far-infrared continuum and molecular gas emission line maps for dozens, the field is increasingly moving towards such spatially resolved, multi-tracer analyses. That said, the number of normal SFGs for which the full complement of multi-wavelength diagnostics is available and sensitively mapped down to kiloparsec scales remains modest. This is in part due to AO-assisted Integral Field Unit (IFU) instrumentation (e.g., SINFONI and its upgraded incarnation ERIS) lacking the multiplexing capabilities of seeing limited IFU instruments such as KMOS, and the emphasis on bright sources during the early cycles of ALMA observations.

In this paper, we review key lessons learned from high-resolution imaging campaigns, starting with monochromatic measures of galaxy structure before delving into a multi-wavelength view (Section 2). We next touch upon a number of insights gained from imaging spectroscopy, with dynamics offering a probe of the gravitational potential and non-gravitational motions complementing metallicity gradients as diagnostics of the feedback processes at play (Section 3). Finally, we present an outlook to new instrumentation and facilities projected to come on-line in the next decade and enabling the tackling of a number of remaining open questions (Section 4).

2. Lessons from high-resolution imaging

Monochromatic structure. Deep HST/WFC3 imaging of the CANDELS fields and more recent wide-area extensions over as much as 0.6 square degrees with COSMOS-DASH provide a means to characterize the rest-optical structure of galaxies spanning a large dynamic range in mass out to cosmic noon. Analyzing the 0th order structural measure of size (parameterized as the half-light radius of a single Sersic model fit to the 2D surface brightness distribution), van der Wel *et al.* (2014) characterize the size growth of SFGs at fixed mass to proceed as $\propto (1+z)^{-0.75}$ and of quiescent galaxies as $\propto (1+z)^{-1.48}$, with no appreciable evolution in the slope of the size - mass relation for either of the two types. Parameterized as a function of the Hubble parameter ($R_e(z) \propto H(z)^{-2/3}$), the size growth at fixed mass is consistent with the halo size evolution at fixed halo mass, validating a basic assumption underpinning virtually all semi-analytical models for galaxy formation, namely that the baryons accreted onto a galaxy disk inherit the specific angular momentum of their dark matter haloes (Mo, Mao & White (1998)). A follow-up analysis by Huang *et al.* (2017) of the galaxy size - halo size relation, inferred from tying the empirical size - mass relation with abundance matching results, arrives at similar conclusions, which are also echoed by direct, kinematic measurements of the angular momentum distribution of galaxies out to $z \sim 2.5$ (Burkert *et al.* 2016). In the same Mo, Mao & White (1998) formalism for disk formation, the distribution of halo spin parameters found in dark matter simulations is by itself sufficient to account for the observed scatter in galaxy sizes at a given mass (Bullock *et al.* 2001; Kravtsov 2013) and has more recently even been argued to be too large, requiring an alteration of disk growth in extreme spin parameter halos or an entirely different formalism altogether (Zanisi *et al.* 2019).

In passing, we note that the above conclusions reflect on the full, mass-complete population of SFGs (and quiescent galaxies). Individual sub-populations of SFGs, such as Lyman Break Galaxies (LBGs), are known to feature different (in the case of LBGs faster) size evolution (e.g., Allen *et al.* 2017). Moreover, as one considers higher redshift SFGs it is well established that their appearance becomes less regular, raising questions about the metric used to quantify size. Ribeiro *et al.* (2016) for example argue that no size evolution is apparent for SFGs at $2 < z < 4.5$ when a non-parametric count of pixels above a given surface brightness threshold is used as the basis for measuring sizes, rather than fitting of a smooth Sersic model.

Even in the context of Sersic fitting, the half-light radius of a galaxy does not capture its structure fully. A 1st order additional specification considers the profile shape, as parameterized by the Sersic index n. It is well known that at all times star-forming and quiescent galaxies differ in their Sersic index distributions, with quiescent galaxies featuring cuspier, more centrally concentrated surface brightness distributions (e.g., Wuyts *et al.* 2011; Whitaker *et al.* 2015). As a consequence, the relative distribution of SFGs and quiescent galaxies in the size - mass plane is altered when defining galaxy size based on an aperture containing a different percentage than 50% of the light. This is explored in more depth by Mowla *et al.* (2019), who leverage the improved number statistics at the massive end provided by COSMOS-DASH, and find SFGs and quiescent galaxies to occupy the same size - mass relation when adopting R_{80} (the radius within which 80% of the light is enclosed) as size definition. They argue it is this size, comprising the bulk of the stars, that relates most tightly to the virial radius of the parent halo ($R_{80} = 0.047\ R_{\rm vir}$), for both galaxy types and across mass and redshift.

Panchromatic structure: rest-UV to rest-optical broad-band colors. With HST legacy surveys such as CANDELS providing up to 7 bands of ACS+WFC3 imaging, sampling the rest-UV to rest-optical SEDs of galaxies at cosmic noon, it was soon realised that galaxies feature internal color variations, encoding spatial variations in the mass-to-light (M/L) ratio (see, e.g., Figure 1).

UV-selected star-forming clumps feature less prominently at longer wavelengths and largely filter out as regions of reduced M/L when reconstructing the stellar mass distributions, which tend to be more centrally concentrated (Wuyts *et al.* 2012). Even in the rest-UV, it should be noted that the underlying diffuse component, and not the giant star-forming clumps, contribute the bulk of the blue light emitted by SFGs at cosmic noon. Taking the census over a mass-complete sample of galaxies at $0.5 < z < 2.5$ and focusing on the azimuthally averaged trends, Figure 2 presents the strength of radial color variations as a function of mass, redshift and galaxy type. Here, we parameterize this in terms of the so-called red centeredness: the difference between an inner and outer rest-optical $(U-V)_{\rm rest}$ color, as interpolated using EAZY (Brammer *et al.* 2008) from the resolved 7-band imaging in the CANDELS-GOODS fields. The boundary between inner and outer regions is taken to be 2 kpc, but similar trends are observed when evaluating red centeredness calculated from the color within and outside the galaxies' half-stellar mass radii. It is apparent that star-forming galaxies feature a more pronounced red-centeredness, especially toward higher masses and increasingly so at later times. While any mass or redshift dependence is reduced for quiescent galaxies, it is clear that they too feature redder centers than outskirts (i.e., $(U-V)_{\rm in} - (U-V)_{\rm out} > 0$), in line with predictions from simulations by Wuyts *et al.* (2010) where these trends come about through a superposition of radial age, metallicity and dust extinction gradients.

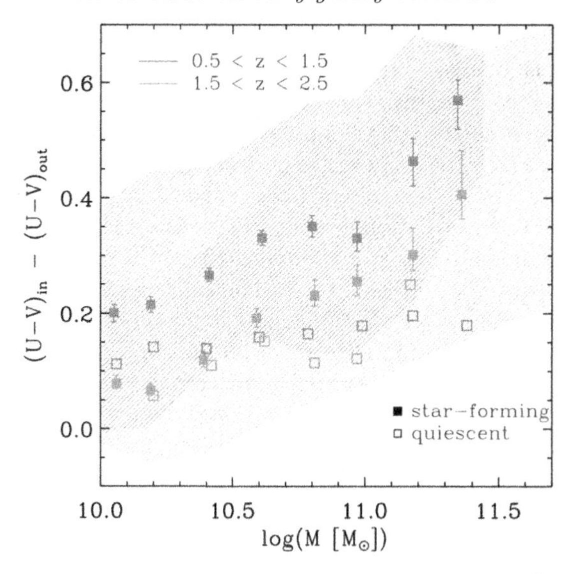

Figure 2. Red centeredness of a mass-complete sample of $M_{\rm star} > 10^{10}$ $M_\odot$ galaxies extracted from the CANDELS-GOODS fields. SFGs and quiescent galaxies both feature negative color gradients, with their inner rest-optical colors being redder than those measured outside a fixed aperture of 2 kpc. The red centeredness of SFGs increases progressively toward higher masses, and is more enhanced at $z \sim 1$ than at $z \sim 2$. Similar trends are found when defining inner/outer colors within/outside an aperture that scales with galaxy size. Quiescent galaxies are also red centered, but exhibit a weaker mass and redshift dependence.

Translated to stellar mass maps, the inferred spatially resolved color profiles imply that SFGs build up significant stellar bulges, comprising over 40% of the total stellar mass, as they grow beyond the Schechter mass, already prior to their eventual quenching (Lang *et al.* 2014). The implications for the (shallower) size - mass relation, quantified on stellar mass rather than light distributions, were recently presented by Suess *et al.* (2019).

Panchromatic structure: Hα maps. A single rest-optical color is insufficient to reveal the origin of the observed spatial M/L ratio variations, as age tracks from stellar population models and dust vectors are closely aligned along a similar M/L - $(U-V)_{\rm rest}$ relation. Owing to HST slitless grism spectroscopy, near-IR IFU and far-IR interferometric observations from the ground, other tracers can increasingly be tied into the analysis of resolved stellar populations and dust distributions. At present, Hα maps on kiloparsec scales are available for statistical stacking analyses for several thousands of galaxies from 3D-HST (Nelson *et al.* 2012, 2013, 2016a; Wuyts *et al.* 2013). For several dozen individual cases the Hα surface brightness distribution has been mapped at the same resolution to a greater sensitivity (and without blended [NII] emission) using adaptive optics assisted IFU spectroscopy (Förster Schreiber *et al.* 2018). More recently, Wilman *et al.* (2019) demonstrated that basic Hα size measurements can also be recovered from seeing-limited IFU observations. The multiplexing capabilities of KMOS on VLT allow for hundreds of such individual size measurements. In common between these studies is the finding that the Hα emission traces an exponential disk distribution, with a scale length that exceeds that of the rest-optical continuum by a factor ~ 1.2. This implies a centrally depressed Hα equivalent width (EW), with the slope of the Hα EW profiles getting steeper as one considers more massive galaxies.

While it is tempting to interpret the Hα EW profiles as probing variations in the local star formation activity ($sSFR \equiv SFR/M_{\rm star}$) and the observed trends therefore as being in line with expectations from inside-out growth and/or quenching scenarios, it is important to note that dust extinction can also affect the observed Hα EW. This is specifically the case as the galaxy's nebular emission may suffer from enhanced levels of attenuation compared to the underlying stellar continuum, due to dust embedded in the birth clouds from which the ionized gas line emission emerges. Various methods have been pursued to characterize the presence of spatial variations, and particularly radial gradients, in dust attenuation within distant SFGs: using HST broad-band colors (e.g., spaxel-by-spaxel SED modeling by Wuyts *et al.* 2012; rest-frame UVI color-color diagrams by Liu *et al.* (2017); rest-UV slopes β by Tacchella *et al.* 2018) or more directly using the Balmer decrement (Nelson *et al.* 2016b, albeit based on stacking of galaxies within relatively large mass bins). All methods agree on finding significant negative gradients, commonly reaching 1.5 to 2 magnitudes of attenuation in the centers of massive galaxies. Quantitative differences in the inferred slopes are seen, which may in part be attributed to differences in sample selection, but more likely also stem from the different methodologies employed. Propagating the estimated dust corrections to quantify the radial sSFR profiles of SFGs at $z \sim 2$, Tacchella *et al.* (2018) conclude that they are generally flat, implying an even build-up of stellar mass by star formation across all radii, with the exception of the most massive objects ($M_{\rm star} > 10^{11}\ M_{\odot}$) where a central depression in the (dust-corrected) Hα EW remains apparent.

Panchromatic structure: dust continuum maps. At the same time, new insights on the nature of the most massive ($M_{\rm star} > 10^{11}\ M_{\odot}$) $z \sim 2$ SFGs are emerging from ALMA. These offer a more bolometric view on the (dust-reprocessed) emission from young stars. Carrying out observations at 870 μm in a compact and extended configuration, Tadaki *et al.* (2017) recovers dust continuum sizes for 12 massive galaxies, nine of which are associated with extremely compact dust emission less than 1.5 kpc in size. Such sizes are more than a factor of 2 smaller than the characteristic rest-optical sizes. Exploiting a purely stellar mass selected sample that is 6 times as large, Tadaki *et al.* (in prep) place these results on a more robust statistical footing. Barring gradients in dust temperature, which could be revealed by multi-band ALMA observations, the compact dust continuum sizes would imply half-SFR sizes that are of order and often smaller than half-stellar mass sizes, suggesting that we are witnessing the build-up of central stellar bulge components. If not attributed to differences in sample selection, it may signal that tracers of dust attenuation may saturate in the centers of the most massive $z \sim 2$ SFGs. It would further imply that within a time span of merely a few hundred Myr these objects can accumulate central stellar mass densities equivalent to those observed in nearby early-type galaxies.

3. Lessons from integral-field spectroscopy

The three-dimensional data cubes produced by IFU spectrographs offer a wealth of information extending beyond the valuable Hα maps discussed in Section 2. Here, we focus on aspects of the internal dynamics and probes of galactic-scale feedback. The results are largely drawn from the KMOS^{3D} programme (PI N.M. Förster Schreiber, co-PI D. Wilman), the survey design and data release of which is documented by Wisnioski *et al.* (2015, 2019). KMOS^{3D} combines deep integrations (from 5 hours at $z \sim 0.9$ to a median of 8.7 hours at $z \sim 2$) for a large number of galaxies (740 in total), spanning a wide dynamic range in star formation rate and mass, and targeting a consistent range of rest-wavelengths across the $0.6 < z < 2.7$ portion of cosmic history.

The mass budget of galaxies since cosmic noon. The ubiquity of ordered rotational motions observed in distant SFGs allows the gaseous kinematic moment maps to be employed to dynamically place constraints on the (total) amount of enclosed mass within the disk regions of distant galaxies. To this end, two aspects are of particular concern regarding the analysis of velocity and velocity dispersion profiles of galaxies in the early

Universe relative to those observed nearby. First, beam smearing effects can often be severe in seeing limited observations, with a typical galaxy size to beam size ratio of 1.7 in the KMOS^{3D} sample. Different approaches have been employed in the literature, ranging from a simple rescaling of the galactocentric radius axis accounting for an addition of the PSF size in quadrature (Tiley *et al.* 2019), to the application of a lookup table with correction factors or a scaling relation based on toy model galaxies to the convert observed to intrinsic rotational velocities and derive an intrinsic velocity dispersion (e.g., Burkert *et al.* 2016; Johnson *et al.* 2018), to finally carrying out a forward modeling of the observed velocity and dispersion profiles simultaneously with rotating disk models that are mock observed under the appropriate inclination and beam smearing conditions (e.g., Cresci *et al.* 2009; Wuyts *et al.* 2016).

While beam smearing reduces the intrinsic rotational velocity leaving only an imprint in the form of an artificially enhanced observed central velocity dispersion, the presence of local random motions (often referred to as the intrinsic velocity dispersion σ_0) revealed as a plateau to the dispersion profile can on its turn impact the observed rotation curve, irrespective of beam smearing effects. Such a component of pressure support has the net effect of reducing the rotational velocity with respect to the circular velocity needed to balance a dynamically cold and hence thin disk against gravity ($v_{\rm rot}^2 = v_{\rm circ}^2 - 2\sigma_0^2 \left(\frac{r}{R_d}\right)$ following Burkert *et al.* 2010).

Accounting for both effects (beam smearing and contributions from pressure support) when modeling the dynamics of high-redshift gas-rich disks, Wuyts *et al.* (2016) found higher redshift SFGs, and most notably those featuring the highest surface densities, to be more baryon dominated (with $\langle f_{\rm bar}(< R_e)\rangle \approx 90\%$) than their more extended and lower redshift counterparts. A comparison to the Illustris cosmological hydro simulation and to a much simpler bath tub model in which disk growth is tied to typical halo growth histories through a set of analytical equations, illustrates that similar trends with galaxy surface density emerge naturally within a ΛCDM context, at least at a qualitative level. Those galaxies in which the baryons managed to assemble within a compact region in the center of the parent dark matter halo naturally feature low contributions from dark matter when quantified within the disk region. Efforts to push measurements of rotation curve shapes out to more than 3 R_e through stacking (Lang *et al.* 2017) or yet deeper integrations (Genzel *et al.* 2017) confirm the baryon dominated nature of $z \sim 2$ SFGs, without reliance on assumptions and systematics that may affect the stellar and gas mass estimates in Figure 3 (IMF, $\alpha_{\rm CO}$, ...).

Returning to the reconstruction of evolutionary sequences discussed in Section 1, we recall that the bulk of $z \sim 2$ galaxies in the kinematic samples from KMOS^{3D} will not evolve into present-day Milky Way-mass systems but will rather end up as higher mass galaxies, predominantly featuring early-type morphologies. If we do not contrast high-z disks to typical spirals in the local Universe, but instead to their more likely early-type descendants, it is noteworthy that those too feature very low dark matter fractions within the confines of their stellar extent (Courteau & Dutton 2015).

Equivalent comparisons between dynamical and stellar masses for galaxies that are already quiescent by cosmic noon have been pursued based on their stellar dynamics. Given their compact sizes, this is most commonly done by means of a virial mass estimator applied to the galaxy-integrated velocity dispersion. In recent years, however, it has become increasingly clear that the quiescent population cannot be treated as purely pressure supported systems, but instead also features substantial amounts of rotational support, as anticipated from a highly dissipational formation process (e.g., Robertson *et al.* 2006; Wuyts *et al.* 2010; Wellons *et al.* 2015). First empirical clues to this end came from a statistical analysis of the axial distribution of early quiescent galaxies,

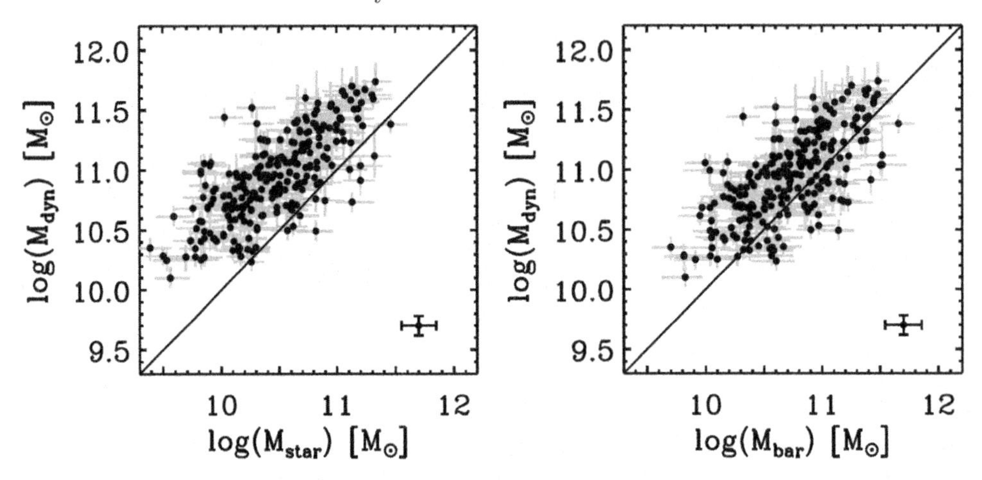

Figure 3. Dynamical mass contrasted to the stellar mass (*left*) and to the baryonic (i.e., stellar + gas) mass (*right*) for 240 KMOS^{3D} galaxies at $0.6 < z < 2.6$ (reproduced from Wuyts *et al.* 2016). While the different mass estimates show a clear correlation, they are offset from the one-to-one line and the scatter around the relation is larger than what can be accounted for by measurement uncertainties. The considerable range in redshifts contributes to part of the scatter, with $z \gtrsim 2$ galaxies being more baryon dominated in their disk regions than their lower redshift counterparts. More predictive of the observed stellar-to-dynamical and baryonic-to-dynamical mass ratio still are measures of the galaxies' surface densities. Extended galaxies probe further into their parent halos and hence contain larger dark matter contributions within R_e.

exhibiting a larger number of elongated projected shapes relative to nearby quiescent galaxies (van der Wel *et al.* 2011; Chang *et al.* 2013). More recently, the presence of rotational motions and hence need for inclination corrections in deriving dynamical masses was alluded to based on galaxy-integrated dynamical measurements by Belli *et al.* (2017). Finally, owing to fortuitous lensing magnifications, Newman *et al.* (2015, 2018) and Toft *et al.* (2017) were able to spatially resolve the rotation curves of a few $z \sim 2$ quiescent galaxies, placing them on the upper envelope of the distribution of so-called 'fast rotators' among nearby early-type galaxies in angular momentum - ellipticity space. Bezanson *et al.* (2018) leverage the larger number statistics from LEGA-C to demonstrate that already at $z \sim 0.8$ a shift toward higher angular momentum quiescent galaxies is notable.

Turbulence in the ISM. Aside from its impact on the inferred dynamical mass budget and the observed rotation curve shapes at cosmic noon, the enhanced levels of turbulence in the ISM of distant galaxies, signaled by a floor to the dispersion profile and parameterized by σ_0, are interesting in their own right. This because it poses the question which process(es) is/are responsible to driving the turbulence and because they can potentially offer an empirical angle on the yet elusive but essential gaseous accretion flows required to continuously replenish the gas-rich high-z disks that feature relatively low depletion times ($t_{\rm dep} \equiv \frac{M_{\rm gas}}{SFR} < 1$ Gyr).

Übler *et al.* (2019) present the latest compilation of intrinsic velocity dispersion measurements. In the ionized gas phase the level of turbulence progressively increases with redshift, reaching $\sigma_0 \approx 45$ km s^{-1} at $z \sim 2.3$, roughly twice of what is typical for nearby disks. In the cold gas phase, the number of measurements at high redshift is still modest, but early ALMA and NOEMA results suggest a similar increase with redshift, albeit offset downward by 10 - 15 km s^{-1}. In other words, it is not just the ionised gas at 10^4 K that is more easily stirred by the enhanced levels of star formation at cosmic noon. In fact, contrasting their measurements to a theoretical model by Krumholz *et al.* (2018) they conclude that stellar feedback alone is not capable to explain the full range of σ_0

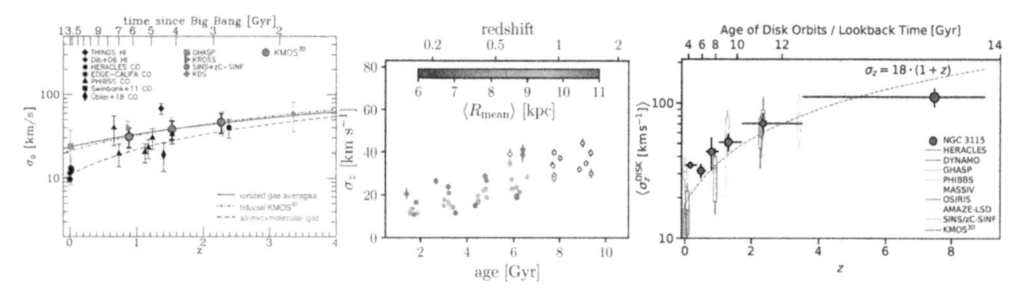

Figure 4. *From left to right:* Intrinsic velocity dispersion of ionized and cold gas as a function of redshift (reproduced from Übler *et al.* 2019); Galactic vertical velocity dispersion as a function of stellar age (reproduced from Mackereth *et al.* 2019); Vertical velocity dispersion as a function of age of disk orbits in NGC3115 (reproduced from Poci *et al.* 2019).

values observed, and specifically their relation to the galaxies' observed SFR, rotational velocity and gas fraction. Instead, a significant source of powering the turbulent velocities in the marginally stable high-z disks is attributed to gravitational instabilities.

Figure 4 illustrates that the observed increase in the intrinsic velocity dispersion with redshift is reminiscent of recent results obtained from Galactic star-by-star archeology and fossil record studies exploiting 3D imaging spectroscopy of a well-resolved individual nearby disk galaxy. In the latter two cases, older stellar populations are associated with progressively thicker disk structures, with a continuum in stellar ages correlating with the vertical velocity dispersion of the respective mono-age stellar population.

It is tempting to associate the similarity to the lookback survey results as an indication that the thicker disk structures were formed as such, from the more gas-rich and turbulent early disks, rather than formed thin and puffed up at a later stage by disturbances such as merger events. Here, again, the caveat applies that in order to consolidate such an interpretation progenitor and descendant galaxies need to be linked up properly. For the high-redshift studies, this generally requires a push to more adaptive optics assisted observations of low-mass galaxies, enabled by ERIS on VLT and in future years JWST as well as ELT.

Indirect probes of feedback: metallicity gradients. Theoretical models of galaxy formation require strong feedback to resolve the long-standing angular momentum problem (Navarro & Steinmetz 2000), prevent the overproduction of stars and reproduce realistic quenched fractions as a function of galaxy mass and cosmic time. Seeing the imprint of such strong feedback observationally is an area where IFU spectrographs can make unique contributions. In an indirect manner, the fact that the metallicity gradients inferred from ionized gas line ratios are found to be weak, if not flat, on average for SFGs at cosmic noon suggests a continuous mixing of ISM material (Wuyts *et al.* 2016b; Förster Schreiber *et al.* 2018). In the absence of injections of energy and momentum, the inside-out growth of stellar disks combined with the chemical enrichment they bring about would otherwise soon lead to the emergence of declining radial profiles in the gas-phase metallicity. Zooming out beyond the disk regions into the Circum-Galactic Medium (CGM), sight-line studies by Steidel *et al.* (2010) have demonstrated the presence of metals at impact parameters beyond 60 kpc. By lack of in situ enrichment so far into the halo the observed metal columns can only be explained by powerful galactic-scale outflows.

Direct probes of feedback: galactic winds. Capturing the launching of such winds can be achieved by decomposing emission line profiles into a systemic and high-velocity component (e.g., Förster Schreiber *et al.* 2019), a technique complementary to that tracing velocity offsets of interstellar absorption lines which shed light on the neutral phase of the outflow (e.g., Shapley *et al.* 2003). To this end, IFU data sets offer the unique advantage that the resolved velocity field can be taken out, thereby reducing the width of the systemic component and hence enhancing the contrast with the broad velocity component representing non-gravitational motions. Numbers, depth and dynamic range in galaxy properties sampled by surveys such as KMOS^{3D} have recently brought studies in this area to a stage where not only the occurence of this phenomenon can be established, but its demographics can be mapped across the SFR - mass plane and its energetics and scaling relations can be compared to those predicted by theoretical models. Förster Schreiber *et al.* (2019) make the case that the outflow phenomenology is best studied for galaxies featuring/lacking diagnostics of AGN activity separately, as their respective broad components differ in characteristic width (FWHM $\sim 400 - 600$ km s^{-1} for star formation driven winds versus FWHM $\sim 1000 - 2500$ km s^{-1} for those emerging from galaxies hosting AGN) and in their demographics. Star formation driven winds are most prevalent in galaxies residing above the star-forming main sequence, where the highest star formation surface densities (and hence highest energy and momentum injection rates from supernovae) are found. The AGN driven winds on the other hand show a steep mass dependence, becoming increasingly common as one considers galaxies above the Schechter mass, irrespective of their level of star formation activity. Moreover, the inferred energetics are such that much of the material launched by stellar feedback may not make it out of the parent halo and is anticipated to rain back in the form of a galactic fountain. Conversely, the strong outflows emerging from AGN-hosting galaxies at the high-mass end may do more (long-term) damage to the gas reservoirs.

4. Outlook and open questions

Deep multi-wavelength lookback surveys with HST mapping the rest-UV to rest-optical emission from distant SFGs on kiloparsec scales, combined with integral-field spectroscopy and accumulating resolved dust continuum and CO data sets are providing a rich multi-tracer view on the nature of gas-rich disk galaxies at cosmic noon. The highly multiplexed nature of new near-infrared (and particularly IFU) spectrographs has enabled placing early results gathered painstakingly by observing of order 1 galaxy per observing night on an 8-meter telescope on a more robust statistical footing. In addition, recent results have also prompted a new array of questions, many of which are to be tackled with the next generation of instruments and observing facilities (Figure 5).

A non-exhaustive list building on the aforedescribed findings include:

• *Where do massive $z \sim 2$ SFGs form their last stars before they get quenched?* Balmer decrement maps for individual galaxies and bolometric UV+IR SFR maps accounting for potential gradients in dust temperature will be required to address whether half-SFR sizes at the tip of the main sequence are smaller than, equal to or larger than the half-stellar mass sizes inferred from multi-wavelength HST imagery.

• *What is the origin of scatter in galaxy scaling relations (SFR - $M_{\rm star}$, $R_e - M_{\rm star}$, $Z - M_{\rm star}$, ...)?* Is any scatter around the observed relations attributed to short-term stochasticity (i.e., the equivalent of 'weather') or an imprint of a long-term differentiation in growth histories among SFGs of the same mass at a given epoch. If the latter, what (halo) property other than mass is most appropriate to describe the SFG population as a two-parameter family? And which observable best serves as an empirical proxy?

• *What is the origin of 'unphysical' baryon fractions ($M_{\rm bar}/M_{\rm dyn} > 1$)?* A robust trend of increasing baryon fractions with increasing surface density is emerging from disk modeling of IFU kinematics. At its extremes, the number of galaxies for which the baryonic mass enclosed within R_e exceeds the dynamical constraint $M_{\rm dyn}(< R_e)$ appears larger than what can be accounted for by random uncertainties. Are these compact SFGs closer to being gas depleted than conventional gas scaling relations, established for the bulk of

Figure 5. *Top:* A new generation of wide-area survey telescopes (LSST, Euclid) on the one hand, and the ultra-sensitive and high-resolution capabilities of JWST and a new class of Extremely Large Telescopes on the other hand, promises to revolutionize the study of galaxies at cosmic noon. These facilities will open a window on an array of outstanding questions inspired by the recent studies reviewed in this paper. *Bottom:* Complementing these rest-UV, rest-optical and near/mid-IR instruments, high-resolution observations with interferometers such as ALMA and NOEMA in the far infrared will remain indispensible to shed full light on the physics at play, by complementing stellar and ionized gas tracers with probes of dust and molecular gas emission.

the galaxy population, suggest? Are uncertainties on their dynamical masses underestimated (e.g., because in the most extreme cases axial ratio based inclinations are hard to determine accurately for marginally resolved compact nuggets)? Or do they represent a subpopulation for which treatment with thick disk models becomes inadequate?

• *What is the physics responsible for setting σ_0?* The redshift evolution of σ_0 can be understood in the framework of marginally stable disks with gas fractions that are dwindling with cosmic time (Wisnioski *et al.* 2015; Übler *et al.* 2019). Yet, at fixed redshift no statistically significant relations between σ_0 and for example the gas fraction within galaxies can be discerned within current data sets. Is this because the dynamic range sampled at a given epoch is modest, and measurement uncertainties in the relevant quantities (σ_0, $f_{\rm gas}$) comparatively large? Or are we missing physics? Furthermore, nearby galaxies are known to feature velocity anisotropies with radial velocity dispersions σ_R typically exceeding those measured along the axis orthogonal to the disk plane. In contrast, the assumption of an isotropic velocity ellipsoid (i.e., $\sigma_0 = \sigma_R = \sigma_z$) is commonly adopted in the analysis of high-z kinematics.

• *Which processes set the local gas-phase metallicity?* The largest samples of resolved (with AO) and marginally resolved (in seeing limited mode) metallicity gradients of

SFGs at cosmic noon to date are based on the N2 (i.e., [NII]/Hα) diagnostic. Its known caveats leave room for improvement, e.g., by employing multi-line diagnostics calibrated for the evolving ISM conditions within distant SFGs compared to normal nearby spirals. KLEVER, a VLT Large Programme piggybacking on KMOS^{3D} and extending it with a sample of lensed galaxies that benefit from enhanced magnification, serves as an example where resolved Hβ, [OIII], Hα and [NII] maps can offer a more comprehensive view on the local ISM conditions.

• *What are the total mass loading and energetics of galactic-scale winds and how does it break down into multi-phase components?* Much of what was discussed above in terms of wind properties and demographics was based on the ionized phase only. A more holistic view on wind properties and their impact on a galaxy's evolutionary path requires the combination of multi-phase tracers. To date, the number of individual normal main sequence SFGs (as opposed to luminous starbursting outliers or extreme quasars) with wind properties mapped in the neutral phase through interstellar absorption line shifts, as well as in the ionized and molecular phase through broad components to the [OIII] or Hα and CO emission lines, remains extremely small. A pilot programme by Herrera-Camus *et al.* (2019) suggests that, equivalent to what is seen in nearby starbursts, the bulk of the mass flow may be in the molecular phase, implying that ionized gas observations alone are insufficient to fully capture their impact on galaxy evolution.

New facilities that will come online this decade promise to advance our knowledge on most of the above outstanding questions. They will yield basic structural measurements for samples orders of magnitude larger than presently available, copious strong lenses to carry out higher resolution studies of internal galaxy structure and kinematics, and significantly enhanced sensitivities and spatial resolution to dissect the in situ physical processes within high-z disks, even for blank field galaxies that lack the systematic uncertainties associated with lens modeling (differential magnification etc.). Among many other areas, this will open a window on the fragmentation processes within marginally stable disks, on not only gas but also stellar kinematics within distant SFGs, and on anticipated deviations from circular motions enabling tests of in situ bulge formation scenarios in which radial gas flows result from violent disk instabilities.

References

Abramson, L. E., Gladders, M. D., Dressler, A., Oemler, A. Jr., Poggianti, B., & Vulcani, B. 2016, *ApJ*, 832, 7
Allen, R. J., Kacprzak, G. G., Glazebrook, K., *et al.* 2017, *ApJ*, 834, 11
Belli, S., Newman, A. B., & Ellis, R. S. 2017, *ApJ*, 834, 18
Bezanson, R., van der Wel, A., Pacifici, C., *et al.* 2018, *ApJ*, 858, 60
Brammer, G. B., van Dokkum, P. G., & Coppi, P. 2008, *ApJ*, 686, 1503
Bullock, J. S., Dekel, A., Kolatt, T. S., *et al.* 2001, *ApJ*, 555, 240
Burkert, A., Förster Schreiber, N. M., Genzel, R., *et al.* 2016, *ApJ*, 826, 214
Burkert, A., Genzel, R., Bouché, N., *et al.* 2010, *ApJ*, 725, 2324
Chang, Y.-Y., van der Wel, A., Rix, H.-W., *et al.* 2013, *ApJ*, 762, 83
Courteau, S. & Dutton, A. A. 2015, *ApJ*, 801, 20
Cresci, G., Hicks, E. K. S., Genzel, R., *et al.* 2009, *ApJ*, 697, 115
Davidzon, I., Ilbert, O., Laigle, C., *et al.* 2017, *A&A*, 605, 70
Förster Schreiber, N. M., Renzini, A., Mancini, C., *et al.* 2018, *ApJS*, 238, 21
Förster Schreiber, N. M., Übler, H., Davies, R. L., *et al.* 2019, *ApJ*, 875, 21
Genzel, R., Förster Schreiber, N. M., Übler, H., *et al.* 2017, *Nature*, 543, 397
Herrera-Camus, R., Tacconi, L., Genzel, R., *et al.* 2019, *ApJ*, 871, 37
Huang, K.-H., Fall, S. M., Ferguson, H. C., *et al.* 2017, *ApJ*, 838, 6
Johnson, H. L., Harrison, C. M., Swinbank, A. M., *et al.* 2018, *MNRAS*, 474, 5076
Kravtsov, A. V. 2013, *ApJ*, 764, 31

Krumholz, M. R., Burkhart, B., Forbes, J. C., & Crocker, R. M. 2018, *MNRAS*, 477, 2716
Lang, P., Wuyts, S., Somerville, R. S., *et al.* 2014, *ApJ*, 788, 11
Lang, P., Förster Schreiber, N. M., Genzel, R., *et al.* 2017, *ApJ*, 840, 92
Liu, F. S., Dongfei, J., Faber, S. M., *et al.* 2017, *ApJ*, 844, 2
Mackereth, J. T., Bovy, J., Leung, H. W., *et al.* 2019, *MNRAS*, in press (arXiv:1901.04502)
Matthee, J. & Schaye, J. 2019, *MNRAS*, 484, 915
Poci, A., McDermid, R. M., Zhu, L., & van de Ven, G. 2019, *MNRAS*, 487, 3776
Mo, H. J., Mao, S., & White, S. D. M. 1998, *MNRAS*, 295, 319
Mowla, L., van der Wel, A., van Dokkum, P. G., & Miller, T. B. 2019, *ApJ*, 872, 13
Navarro, J. F. & Steinmetz, M. 2000, *ApJ*, 538, 477
Nelson, E. J., van Dokkum, P. G., Brammer, G., *et al.* 2012, *ApJ*, 7474, 28
Nelson, E. J., van Dokkum, P. G., Momcheva, I., *et al.* 2013, *ApJ*, 763, 16
Nelson, E. J., van Dokkum, P. G., Förster Schreiber, N. M., *et al.* 2016a, *ApJ*, 828, 27
Nelson, E. J., van Dokkum, P. G., Momcheva, I., *et al.* 2016b, *ApJ*, 817, 9
Newman, A. B., Belli, S., & Ellis, R. S. 2015, *ApJ*, 813, 7
Newman, A. B., Belli, S., Ellis, R. S., & Patel, S. G. 2018, *ApJ*, 862, 126
Ribeiro, B., Le Fèvre, O., Tasca, L. A. M., *et al.* 2016, *A&A*, 593, 22
Robertson, B., Bullock, J. S., Cox, T. J., *et al.* 2006, *ApJ*, 645, 986
Sanders, R. L., Shapley, A. E., Kriek, M., *et al.* 2018, *ApJ*, 858, 99
Shapley, A. E., Steidel, C. C., Pettini, M., & Adelberger, K. L. 2003, *ApJ*, 588, 65
Speagle, J. S., Steinhardt, C. L., Capak, P. L., & Silverman, J. D. 2014, *ApJS*, 214, 15
Steidel, C. C., Erb, D. K., Shapley, A. E., *et al.* 2010, *ApJ*, 717, 289
Suess, K. A., Kriek, M., Price, S., & Barro, G. 2019, *ApJ*, 877, 103
Tacchella, S., Carollo, C. M., Förster Schreiber, N. M., *et al.* 2018, *ApJ*, 859, 56
Tacconi, L. J., Genzel, R., Saintonge, A., *et al.* 2018, *ApJ*, 853, 179
Tadaki, K., Genzel, R., Kodama, T., *et al.* 2017, *ApJ*, 834, 135
Tiley, A. L., Swinbank, A. M., Harrison, C. M., *et al.* 2019, *MNRAS*, 485, 934
Toft, S., Zabl, J., Richard, J., *et al.* 2017, *Nature*, 546, 510
Torrey, P., Wellons, S., Ma, C.-P., Hopkins, P. F., & Vogelsberger, M. 2017, *MNRAS*, 467, 4872
Übler, H., Förster Schreiber, N. M., Genzel, R., *et al.* 2017, *ApJ*, 842, 121
Übler, H., Genzel, R., Wisnioski, E., *et al.* 2019, *ApJ*, 880, 48
van der Wel, A., Rix, H.-W., Wuyts, S., *et al.* 2011, *ApJ*, 730, 38
van der Wel, A., Franx, M., van Dokkum, P. G., *et al.* 2014, *ApJ*, 788, 28
Wellons, S., Torrey, P., Ma, C.-P., *et al.* 2015, *MNRAS*, 449, 361
Whitaker, K. E., Franx, M., Bezanson, R., *et al.* 2015, *ApJ*, 811, 12
Wilman, D. J., Fossati, M., Mendel, J. T., *et al.* 2019, *ApJ*, submitted
Wisnioski, E., Förster Schreiber, N. M., Wuyts, S., *et al.* 2015, *ApJ*, 799, 209
Wisnioski, E., Förster Schreiber, N. M., Fossati, M., *et al.* 2019, *ApJ*, submitted
Wuyts, S., Cox, T. J., Hayward, C., *et al.* 2010, *ApJ*, 722, 1666
Wuyts, S., Förster Schreiber, N. M., van der Wel, A., *et al.* 2011, *ApJ*, 742, 96
Wuyts, S., Förster Schreiber, N. M., Genzel, R., *et al.* 2012, *ApJ*, 753, 114
Wuyts, S., Förster Schreiber, N. M., Nelson, E. J., *et al.* 2013, *ApJ*, 779, 135
Wuyts, S., Förster Schreiber, N. M., Wisnioski, E., *et al.* 2016, *ApJ*, 831, 149
Wuyts, E., Wisnioski, E., Fossati, M., *et al.* 2016, *ApJ*, 827, 74
Zanisi, L., Shankar, F., Lapi, A., *et al.* 2019, *MNRAS*, submitted

Uncovering Early Galaxy Evolution in the ALMA and JWST Era
Proceedings IAU Symposium No. 352, 2019
E. da Cunha, J. Hodge, J. Afonso, L. Pentericci & D. Sobral, eds.

doi:10.1017/S1743921320001210

Sub-galactic views of cold gas and dust in distant star-forming galaxies: Pushing the ~100 pc frontier at $z \sim 3$

Wiphu Rujopakarn

Chulalongkorn University, Thailand

Abstract. While the evolution of spatially-integrated properties of galaxies are relatively well constrained across cosmic time, many of the most fundamental processes are not well understood, especially down to the sub-galactic scales, where frontier questions in galaxy evolution lie: How did galactic spheroids form? How did galaxies and their supermassive black holes co-evolve? With the angular resolution capability of ~tens of milliarcseconds, ALMA has conferred extinction-independent views of cold gas and dust distributions within individual $z \sim 1-4$ galaxies at resolutions approaching ~ 100 pc, thereby opening new avenues to study sub-galactic properties of galaxies at the peak of their assembly. In this talk, I will review recent findings and ongoing challenges enabled by ALMA's extinction-independent, spatially-resolved views of star forming galaxies, particularly the galactic substructures, e.g., clumps (or the lack thereof) from both field and gravitationally-lensed galaxies, and their implications on the bulge assembly scenario. I will also discuss a new synergistic approach between radio and millimeter observations (using, e.g., VLA and ALMA) to independently pinpoint the locations of star-forming region and AGN down to < 100 pc at $z \sim 3$. Lastly, I will discuss the planned surveys with JWST in the first year of operation, and ways that the first datasets can be combined with ALMA to provide new breakthroughs and plan future observations to utilize Webb to the fullest.

Uncovering Early Galaxy Evolution in the ALMA and JWST Era
Proceedings IAU Symposium No. 352, 2019
E. da Cunha, J. Hodge, J. Afonso, L. Pentericci & D. Sobral, eds.

doi:10.1017/S1743921320001222

Spatially resolving the relics: The inferring the physics driving the quenching of massive galaxies from kinematics at $z \sim 1$ and beyond

Rachel Bezanson

University of Pittsburgh, USA

Abstract. Today's massive elliptical galaxies are primarily red-and-dead, dispersion supported ellipticals. The physical process(es) driving the shutdown or 'quenching' of star formation in these galaxies remains one of the least understood aspects of galaxy formation and evolution. Although today's spiral and elliptical galaxies exhibit a clear bimodality in their structures, kinematics, and stellar populations, it may be that the quenching and structural transformation do no occur simultaneously. In this talk I will present evidence that early quiescent galaxies, observed much closer to their quenching epoch at $z \sim 1$, retain significant rotational support ($\sim$twice as much as local ellipticals). This suggests that the mechanisms responsible for shutting down star formation do not also have to destroy ordered motion in massive galaxies; the increased dispersion support could occur subsequently via hierarchical growth and minor merging. I will discuss this evidence in conjunction with recent ALMA studies of the dramatic range in molecular gas reservoirs of recently quenched high redshift galaxies to constrain quenching models. Finally, I will discuss prospects for extending spatially resolved spectroscopic studies of galaxies immediately following quenching with JWST and eventually 30-m class telescopes.

Uncovering Early Galaxy Evolution in the ALMA and JWST Era
Proceedings IAU Symposium No. 352, 2019
E. da Cunha, J. Hodge, J. Afonso, L. Pentericci & D. Sobral, eds.

doi:10.1017/S1743921320001234

Mapping the structure and source of outflows from star-forming galaxies at $z = 2 - 3$

Bethan James

Space Telescope Science Institute, USA

Abstract. As we enter the era of JWST our need to characterise the rest-frame UV spectra of star-forming galaxies becomes essential. By combining the NIR capabilities of JWST with our understanding of UV wavelength science, we have the opportunity to explore fundamental properties of the gas, such as its metallicity and density, as well as the extent, velocity, and magnitude of their outflowing gas, in galaxies out to $z \sim 6$. Galaxy outflows in particular play a fundamental role in the evolution of young galaxies at high redshifts, but their properties remain largely unknown as it is difficult to spatially resolve the outflowing gas. To-date, only two attempts to resolve outflows at redshift ~ 2 have been made using lensing magnification, producing contradictory results on the origin of the outflows. In this talk I will present results from one such groundbreaking study where we combine gravitational lensing with VLT-MUSE to perform one of the first spatially resolved absorption line studies of a galaxy at $z = 2 - 3$. I will discuss how the the distinct kinematical structure and uniform column densities obtained from the outflowing gas maps reveal 'global' rather than 'locally' sourced outflows. I will also present preliminary results from our latest attempt to accurately constrain the structure and source of outflows in star-forming galaxies by observing the brightest galaxy-scale lens known with KCWI. I will conclude with the benefits and limitations of spatially resolved observations in this wavelength range, and possible implications on NIRSpec observations of the high-z Universe.

Uncovering Early Galaxy Evolution in the ALMA and JWST Era
Proceedings IAU Symposium No. 352, 2019
E. da Cunha, J. Hodge, J. Afonso, L. Pentericci & D. Sobral, eds.

doi:10.1017/S1743921319008949

Molecular clouds in a Milky Way progenitor at $z = 1$

Miroslava Dessauges-Zavadsky[1], Johan Richard[2], Françoise Combes[3,4], Daniel Schaerer[1,5], Wiphu Rujopakarn[6,7,8] and Lucio Mayer[9,10]

[1]Observatoire de Genève, Université de Genève, Versoix, Switzerland
email: miroslava.dessauges@unige.ch

[2]Université Lyon, Université Lyon1, ENS de Lyon, CNRS, Centre de Recherche Astrophysique de Lyon UMR5574, Saint-Genis-Laval, France

[3]LERMA, Observatoire de Paris, PSL Research Université, CNRS, Sorbonne Université, UPMC Paris, France

[4]Collège de France, Paris, France

[5]CNRS, IRAP, Toulouse, France

[6]Department of Physics, Faculty of Science, Chulalongkorn University, Bangkok, Thailand

[7]National Astronomical Research Institute of Thailand (Public Organization), Chiang Mai, Thailand

[8]Kavli Institute for the Physics and Mathematics of the Universe (WPI), The University of Tokyo Institutes for Advanced Study, University of Tokyo, Kashiwa, Japan

[9]Center for Theoretical Astrophysics and Cosmology, Institute for Computational Science, University of Zurich, Zurich, Switzerland

[10]Physik-Institut, University of Zurich, Zurich, Switzerland

Abstract. Thanks to the remarkable ALMA capabilities and the unique configuration of the Cosmic Snake galaxy behind a massive galaxy cluster, we could resolve molecular clouds down to 30 pc linear physical scales in a typical Milky Way progenitor at $z = 1.036$, through CO(4–3) observations performed at the $\sim 0.2''$ angular resolution. We identified 17 individual giant molecular clouds. These high-redshift molecular clouds are clearly different from their local analogues, with $10 - 100$ times higher masses, densities, and internal turbulence. They are offset from the Larson scaling relations. We argue that the molecular cloud physical properties are dependent on the ambient interstellar conditions particular to the host galaxy. We find these high-redshift clouds in virial equilibrium, and derive, for the first time, the CO-to-H_2 conversion factor from the kinematics of independent molecular clouds at $z = 1$. The measured large clouds gas masses demonstrate the existence of parent gas clouds with masses high enough to allow the in-situ formation of similarly massive stellar clumps seen in the Cosmic Snake galaxy in comparable numbers. Our results support the formation of molecular clouds by fragmentation of turbulent galactic gas disks, which then become the stellar clumps observed in distant galaxies.

Keywords. gravitational lensing, galaxies: high-redshift, galaxies: ISM, ISM: clouds

1. Introduction

The motivation to search for molecular clouds at high-redshift is two-fold. *First*, the molecular gas in present-day galaxies is known to be structured in discrete cloud complexes, the giant molecular clouds (GMCs). They are the seeds of star formation, as star clusters form by condensation of these GMCs. Typically, local GMCs have gas masses between $10^4\ M_\odot$ to $10^7\ M_\odot$ and radii as small as 5 pc to 100 pc (e.g., Bolatto *et al.* 2008; Columbo *et al.* 2014; Sun *et al.* 2018). Achieving such small scales at $z>1$ is observationally challenging. The higher resolution ALMA observations of high-redshift galaxies were performed at $0.03'' - 0.04''$, that is still ~ 200 pc at $z=1-2$. At this resolution, sub-mm galaxies do show structures in the rest-frame far-IR continuum emission (Hodge *et al.* 2019), but normal star-forming galaxies do not (Rujopakarn *et al.* 2019). The only way we may beat this spatial resolution limitation is with the help of gravitational lensing. Observations of a strongly lensed sub-mm galaxy at $z=3.042$, reading down to $50-100$ pc, revealed massive CO clouds (Swinbank *et al.* 2015).

Second, about 60% of galaxies at the peak of the cosmic star formation have perturbed morphologies, characterized by UV-bright star-forming clumps (e.g., Cowie *et al.* 1995; Elmegreen *et al.* 2013; Guo *et al.* 2018). These clumps have stellar masses between $10^6\ M_\odot$ to $10^9\ M_\odot$, that is, they are, on average, two orders of magnitude more massive than the star cluster complexes found in nearby galaxies (Dessauges-Zavadsky *et al.* 2017). What is the origin of these stellar clumps? Are they accreted satellites following a merger event, or are they formed 'in situ' in host galaxies? In the latter case, we should see the molecular clouds which led to the formation of these stellar clumps. Currently, there is a number of observational findings which favour the in-situ clump origin: (i) the redshift evolution of the clumpy galaxy fraction is inconsistent with the evolutionary trends of minor and major mergers (Shibuya *et al.* 2016); (ii) clumpy galaxies are dominated by disk-like systems with Sersic indices of $n\sim 1$ (Shibuya *et al.* 2016); (iii) the kinematics of the majority of star-forming galaxies at $z=1-2$, including clumpy galaxies, is dominated by ordered disk rotation (e.g., Wisnioski *et al.* 2015; Girard *et al.* 2018); (iv) edge-on clumpy galaxies (the dubbed chain-galaxies) have very comparable disk scale-height to edge-on spirals without UV-bright clumps (Elmegreen *et al.* 2017); and (v) the stellar mass function of high-redshift clumps follows a power law with a slope of -2, which is characteristic of local star clusters (Dessauges-Zavadsky & Adamo 2018). Based on numerical simulations, it has been proposed that the observed high-redshift turbulent disks are subject to violent gravitational instability, caused by intense cold gas accretion flows, which triggers their fragmentation, and then the formation of in-situ gravitationally bound gas clouds believed to be the progenitors of the observed stellar clumps (e.g., Dekel *et al.* 2009; Bournaud *et al.* 2014). State-of-the-art high-resolution simulations predict a stellar mass distribution of clumps formed via disk fragmentation in perfect agreement with the observed distribution of high-redshift stellar clumps (Tamburello *et al.* 2015; Behrendt *et al.* 2016; Dessauges-Zavadsky *et al.* 2017). If the proposed scenario is correct, the observed stellar clumps currently provide an important evidence of galaxy mass build-up via cosmic cold gas accretion flows rather than mergers. The detection of the associated molecular clouds will bring an even stronger support to this scenario.

2. Target and molecular cloud identification

We have undertaken the search for molecular clouds in the 'Cosmic Snake' galaxy at $z=1.036$, strongly lensed by the massive galaxy cluster MACS J1206.2–0847. With a stellar mass of $2.4\times 10^{10}\ M_\odot$, a star formation rate of $18\ M_\odot\ \mathrm{yr}^{-1}$, and a molecular gas fraction of 30%, it is a typical main-sequence galaxy in rotation, inclined by 70 degrees (Patricio *et al.* 2018; Girard *et al.* 2019). It is characterized by a clumpy morphology with 21 stellar clumps detected in the HST images (Cava *et al.* 2018). We obtained ALMA imaging in the CO(4–3) emission with a synthesized beam size of $0.22''\times 0.18''$, comparable to the HST rest-frame UV to optical images. Together with the lens model accuracy of $0.15''$ and magnification factors between 10 and >100, we achieve resolutions of $30-70$ pc, similar to those of local galaxy molecular gas studies.

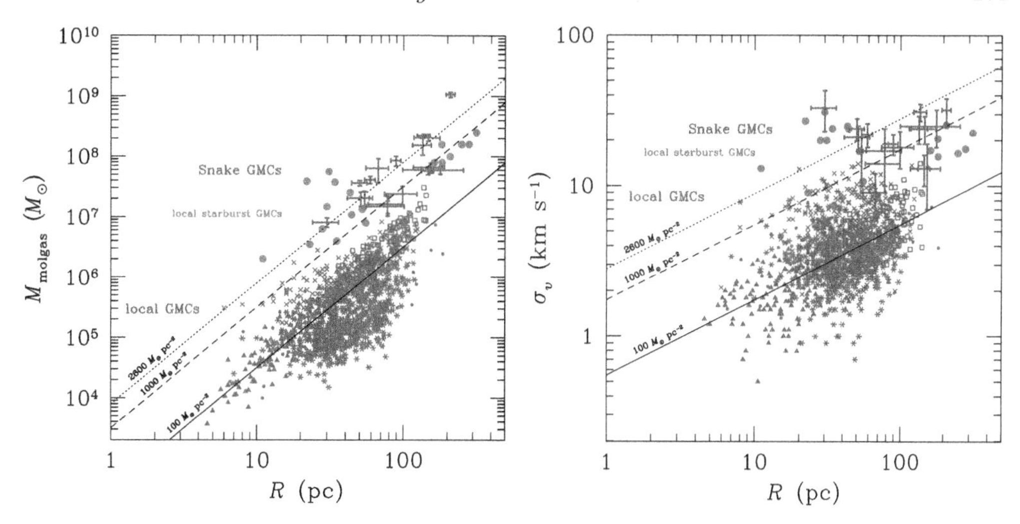

Figure 1. *Left.* Molecular gas masses as a function of radius for the GMCs identified in the Cosmic Snake galaxy (blue data points), the local quiescent galaxies (red symbols from a compilation of various publications), and the nearby starbursting galaxies (magenta filled circles from Wei *et al.* (2012) and Leroy *et al.* (2015)). The black lines show fixed molecular gas mass surface densities. *Right.* Internal velocity dispersions as a function of radius, plotted for the same GMC populations as in the left panel. The black lines show the $\sigma_v \propto R^{0.5}$ relation expected for virialized clouds with fixed gas mass surface density (same values as in the left panel).

We used individual CO(4–3) intensity channel maps of 10.343 km s^{-1} to search for molecular clouds. We first extracted all the emission peaks per channel above our detection threshold set to 100% fidelity (Walter *et al.* 2016), and ewe kept only the emissions detected over at least two adjacent channels. Using the lens model, we associated the emissions with 40 counter-images of 17 molecular clouds. The 17 identified molecular clouds are detected at a significance level $>6-27\sigma$. For each cloud, we measured the radius, the internal velocity dispersion, and the CO(4–3) line-integrated flux, which we converted to the molecular gas mass assuming the CO luminosity correction factor $r_{4,1} = L'_{\rm CO(4-3)}/L'_{\rm CO(1-0)} = 0.33$ (extrapolated from our $r_{4,2}$ measurement in the Cosmic Snake and $r_{2,1}$ obtained in $z \sim 1.5$ BzK galaxies by Daddi *et al.* (2015)) and the Milky Way CO-to-H_2 conversion factor. All the Cosmic Snake molecular clouds are spatially resolved, with measured radii all larger than the equivalent circularised beam detection limit and measured molecular gas masses all larger than the molecular gas mass detection limit (for the 100% fidelity) at the magnification of a given cloud. The radii range between 30 pc to 210 pc, the molecular gas masses between $8 \times 10^6\ M_\odot$ and $1 \times 10^9\ M_\odot$, and the internal velocity dispersions between 9 km s^{-1} and 33 km s^{-1}.

3. Molecular cloud physical properties

To compare the derived physical properties of the Cosmic Snake GMCs with GMCs hosted in the Milky Way and nearby galaxies, we considered the Larson scaling relations used for a long time as the benchmark of local GMC populations and thought to reflect identical GMC physical properties in all galaxies at all times (Larson 1981). As shown in Fig. 1, the Cosmic Snake GMCs are offset from the Larson scaling relations: they have higher molecular gas masses by two orders of magnitude, higher gas mass surface densities by one order of magnitude with a median density as high as 2600 $M_\odot$ pc^{-2}, and larger internal velocity dispersions. This is not so surprising, as genuine variations in the physical properties of GMCs hosted in different interstellar environments have already

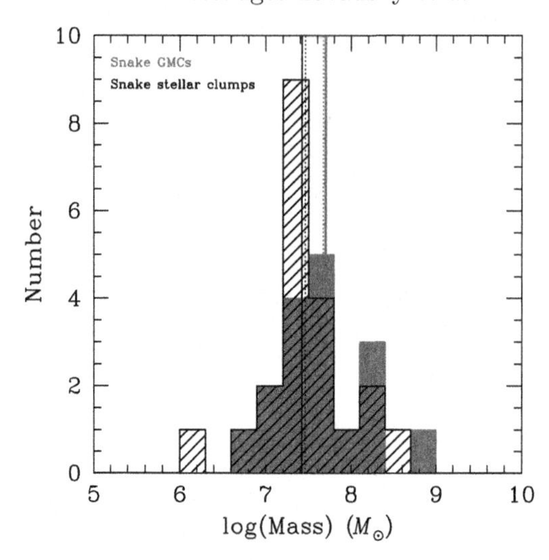

Figure 2. Comparison of the gas mass distribution of the Cosmic Snake molecular clouds (filled blue histogram) with the stellar mass distribution of the stellar clumps identified in the HST UV to near-IR images (hatched black histogram). The medians (solid lines; the dotted lines show the means) are $5.0 \times 10^7\ M_\odot$ for the molecular clouds and $2.6 \times 10^7\ M_\odot$ for the stellar clumps.

been reported in nearby mergers (the Antenna merger; Wei *et al.* 2012) and starbursting nuclear regions (NGC 235; Leroy *et al.* 2015). properties in fact very much resemble those of high-redshift Cosmic Snake GMCs. The universality of GMCs is definitely challenged, the environment must matter. There seems to be a link between the GMC properties and the interstellar medium conditions of the host galaxy, which are more extreme in distant objects in terms of the ambient turbulence and the disk hydrostatic pressure. In the Cosmic Snake disk we measured a 1000 times stronger hydrostatic pressure ($\sim 10^{7.7}$ cm^{-3} K) than in the Milky Way disk.

The enhanced surface density and turbulence of the Cosmic Snake GMCs guarantee their survival and equilibrium. Fourteen out of the 17 identified GMCs are indeed found in virial equilibrium. For the virialized GMCs, we can, for the first time, constrain the CO-to-H_2 conversion factor from the kinematics of individual GMCs at $z = 1$. The derived conversion factor of $3.8 \pm 1.1\ M_\odot/(\text{K km s}^{-1}\ \text{pc}^2)$ is close to the Milky Way value, despite the stronger photodissociating radiation expected in the Cosmic Snake galaxy given its relatively high star formation rate. This implies that the measured gas mass surface densities are high enough to shield the clouds.

The Cosmic Snake GMCs are highly supersonic with 10 times higher Mach numbers than in local GMCs. Does this betray a higher efficiency of star formation? The fact that we have identified GMCs and stellar clumps at the same spatial resolution enables us to provide an estimate of the efficiency of star formation in a similar fashion as it has been done in the local Universe, from the comparison of the mass distributions of the molecular clouds and the stellar clumps (Evans *et al.* 2009). Under the hypothesis that the identified GMCs are representative of the parent gas clouds which gave rise to the observed massive star clusters, we obtained an efficiency of star formation of $26 - 34\%$, fairly much higher than the canonical values smaller than $\sim 10\%$ found in contemporary galaxies. The derived efficiency is, on the other hand, in very good agreement with the FIRE numerical simulations predicting a dependence of the star formation efficiency on the gas mass surface density of molecular clouds (Grudic *et al.* 2018).

4. Conclusion

This is the first time that we detect molecular clouds and stellar clumps in comparable numbers in a high-redshift galaxy. The detection of the molecular clouds in the Cosmic Snake galaxy, moreover, demonstrates the existence of parent gas clouds with masses high enough to allow the in-situ formation of stellar clumps seen in the galaxy. Galactic disk fragmentation can therefore be proposed as the mechanism of formation of massive molecular clouds in distant galaxies.

References

Behrendt, M., Burkert, A., & Schartmann, M. 2016, *ApJ* (Letters), 819, L2
Bolatto, A. D., Leroy, A. K., Rosolowsky, E., Walter, F., & Blitz, L. 2008, *ApJ*, 686, 948
Bournaud, F. *et al.* 2014, *ApJ*, 780, 57
Cava, A., Schaerer, D., Richard, J., Pérez-González, P.G., Dessauges-Zavadsky, M., Mayer, L., & Tamburello, V. 2018, *Nat.As*, 2, 76
Columbo, D. *et al.* 2014, *ApJ*, 784, 3
Cowie, L. L., Hu, E. M., & Songaila, A. 1995, *AJ*, 110, 1576
Daddi, E. *et al.* 2015, *A&A*, 577, A46
Dekel, A. *et al.* 2009, *Nature*, 457, 45
Dessauges-Zavadsky, M., Schaerer, D., Cava, A., Mayer, L., & Tamburello, V. 2017, *ApJ* (Letters), 836, L22
Dessauges-Zavadsky, M. & Adamo, A. 2018, *MNRAS* (Letters), 479, L118
Elmegreen, B. G. *et al.* 2013, *ApJ*, 774, 86
Elmegreen, B. G., Elmegreen, D. M., Tompkins, B., & Jenks, J. G. 2017, *ApJ*, 847, 14
Evans, N. J. II *et al.* 2009, *ApJS*, 181, 321
Girard, M., Dessauges-Zavadsky, M., Schaerer, D., Richard, J., Nakajima, K., & Cava, A. 2018, *A&A*, 619, A15
Girard, M., Dessauges-Zavadsky, M., Combes, F., Chisholm, J., Patricio, V., Richard, J., & Schaerer D. 2019, *A&A*, 631, 10
Grudic, M. Y. *et al.* 2018, *MNRAS*, 475, 3511
Guo, Y. *et al.* 2018, *ApJ*, 853, 108
Hodge, J. A. *et al.* 2019, *ApJ*, 879, 130
Larson, R. B. 1981, *MNRAS*, 194, 809
Leroy, A. K. *et al.* 2015, *ApJ*, 801, 25
Patricio, V. *et al.* 2018, *MNRAS*, 477, 18
Rujopakarn, J. *et al.* 2019, *ApJ*, accepted [arXiv:1904.04507]
Shibuya, T., Ouchi, M., Kubo, M., & Harikane, Y. 2016, *ApJ*, 821, 72
Sun, J. *et al.* 2018, *ApJ*, 860, 172
Swinbank, A. M. *et al.* 2015, *ApJ* (Letters), 806, L17
Tamburello, V., Mayer, L., Shen, S., & Wadsley, J. A. 2015, *MNRAS*, 453, 2490
Walter, F. *et al.* 2016, *ApJ*, 833, 67
Wisnioski, E. *et al.* 2015, *ApJ*, 799, 209
Wei, L. H., Keto, E., & Ho, L. C. 2012, *ApJ*, 750, 136

Uncovering Early Galaxy Evolution in the ALMA and JWST Era
Proceedings IAU Symposium No. 352, 2019
E. da Cunha, J. Hodge, J. Afonso, L. Pentericci & D. Sobral, eds.

doi:10.1017/S1743921319008937

Uncovering the spatial distribution of stars and dust in $z \sim 2$ Submillimeter Galaxies

Philipp Lang[1], Eva Schinnerer[1], Ian Smail[2], U. Dudzevičiūtė[2], A. M. Swinbank[2] and the A3COSMOS, AS2UDS and ALESS Teams

[1]Max-Planck-Institut für Astronomie, Königstuhl 17, D-69117 Heidelberg, Germany

[2]Center for Extragalactic Astronomy, Department of Physics, Durham University, South Road, Durham DH1 3LE, UK

Abstract. The spatial distribution of the dust and stars contains crucial information about the evolutionary pathways of galaxies. We present results of our study combing high-resolution ALMA and HST observations of $z \sim 2$ bright sub-millimeter galaxies (SMGs). We have developed a two-dimensional extinction and age correction technique to obtain accurate stellar mass distributions from HST/CANDELS. For the first time, we can directly compare the spatial distribution of assembled stellar mass and ongoing star formation on kpc scales for distant SMGs, shedding light on their highly debated formation mechanisms. We find that the dust distribution is more compact than the stellar component, regardless if the SMG lies on the main sequence or at the starburst regime. Taking the dust emission as a proxy for dust-obscured star formation, our results imply that high-redshift SMGs are experiencing centrally enhanced star formation. These findings suggests that major galaxy interactions are not necessarily the main formation channel for SMGs with secular disk formation remaining a viable option as suggested by state-of-the-art cosmological simulations. The sizes and stellar densities of our $z \sim 2$ SMGs agree well with the most compact early-type galaxies in the local Universe, strongly supporting the idea that the latter systems are indeed the descendants of massive SMGs at $z \sim 2$.

Keywords. galaxies: evolution, galaxies: high-redshift, galaxies:structure, submillimeter

---

1. Introduction

The brightest class of IR-luminous galaxies, detected with single-dish sub-mm/mm surveys (therefore dubbed sub-mm galaxies, 'SMGs'), are associated with the highest SFRs ($> 300 M_\odot$/yr) and IR luminosities ($L_{\rm IR} > 10^{12} L_\odot$). Since they contribute a significant fraction to the total star formation rate density (~10-30 %; e.g., Swinbank *et al.* 2014) and are thought to be the progenitors of today's massive ellipticals (e.g., Swinbank *et al.* 2006; Ikarashi *et al.* 2015), it is crucial to understand their evolutionary paths in the early Universe. In our canonical picture, the general star-forming population of galaxies on the star-forming main sequence evolves through cold gas accretion (e.g., Dekel *et al.* 2009), while SMGs represent merger-induced and short-lived starbursts (e.g., Narayanan *et al.* 2010). However, theoretical models suggest that distant SMGs are simply the high-mass extension of non-interacting galaxies where star formation is triggered through the slower internal secular processes (e.g. Hayward *et al.* 2011, 2013).

Recent studies based on ALMA (e.g. Simpson *et al.* 2015a; Hodge *et al.* 2016; Tadaki *et al.* 2017a) have revealed compact sub-mm dust cores (with effective radii of $R_{\rm e}$ ~1-2 kpc) in SMGs at $z \sim$ 1-5. Using the dust continuum as a proxy for massive star formation (i.e. tracing the dust heated by young, massive stars) implies that distant

IR-luminous systems host strong central dust-enshrouded star formation. In contrast, the stellar continuum light morphologies at $z \sim 1$-3 – accessible through high-resolution HST observations – are found to be clearly more extended ($R_e \sim$4-5 kpc), irregular, as well as spatially decoupled from the FIR/submm emission (e.g., Chen *et al.* 2015; Hodge *et al.* 2016.

These properties are widely associated with a major-merger scenario, in which dissipative collapse of gas causes central star-formation, while the stellar distribution remains in an extended and perturbed configuration. However, there is yet no consensus towards a secular versus major merger origin of SMGs in the literature since the rest-frame optical emission mostly fails to approximate the underlying stellar distribution due to large spatial variations in age and/or extinction, with the latter being most extreme in SMGs with strong central dust obscuration. In this study, we investigate the formation mechanisms of SMGs by combining information on the resolved distribution of both the dust-obscured star formation and the assembled stellar mass on kpc scales. The latter is inferred from the deep multi-wavelength imaging from the CANDELS survey. The results of this study are publish in Lang *et al.* (2019); below, we briefly summarize the method and results.

2. Sample and Method

For our analysis, we construct a sample of 20 SMGs that have been targeted with deep, high angular-resolution observations from both ALMA and *HST* across three cosmological deep fields. We consider submm sources detected in the following surveys: (1) The ALMA-SCUBA-2 Ultra Deep Survey, henceforth referred to as 'AS2UDS' (Stach *et al.* 2018), (2) ALMA Band 7 follow-up of single-dish submm sources from the LESS survey ('ALESS', Hodge *et al.* 2013, 2016), and (3) ALMA Band 6 (1.3 mm) follow-up of bright AzTEC sources in the COSMOS field (Miettinen *et al.* 2017). Then, we apply the following selection criteria: (a) a redshift range of $1.7 < z < 2.6$, where our method to derive stellar-mass maps is most robust; (b) coverage by *HST*/CANDELS (Koekemoer *et al.* 2011; Grogin *et al.* 2011), and that are detected with sufficient S/N in both J_{125} and H_{160} band filters; (c) no evidence for a central AGN component.

Those yield a final sample of 20 SMGs across a large dynamic range of submm brightness, stellar mass, SFR, and offset from the main sequence. Those properties are further presented in Figure 1, demonstrating that our sample is well representative for the bright and massive SMG population at redshift 2. We note that our selection limits our analysis to systems with bright optical counterparts ($K_s \lesssim 22.9$) due to the requirement of a detection in the HST H and J bands. We derive spatially resolved stellar mass distributions inferred from $J_{125} - H_{160}$ color maps as this filter combination probes rest-frame optical wavelengths at $z \simeq 2$ and can therefore be used to infer stellar M/L ratios. Based synthetic galaxy SEDs from on (Bruzual & Charlot 2003) models, we calibrate a relation between the observed optical M/L ratio and the $J_{125} - H_{160}$ color, that show a robust correlation which is independent of details of the SFH, metallicity and extinction.

3. Results

Figure 2 shows the H_{160}-band cutouts, $J - H$ color maps, and resulting stellar mass maps for a few example cases. Overall, we find that our sources exhibit systematic radial color gradients (i.e. redder centers vs. bluer outskirts). In some cases, off-centered clumps dominating the light distribution in H_{160}-band (such as for ALESS067.1) but only weakly contribute to the stellar mass density. Moreover, the stellar mass distribution of ALESS079.2 appears as a large system with a smooth and strongly centrally peaked mass profile, rather than being comprised of several components as the H_{160}-band image suggests. Those cases highlight the caveat of interpreting highly disturbed rest-frame optical

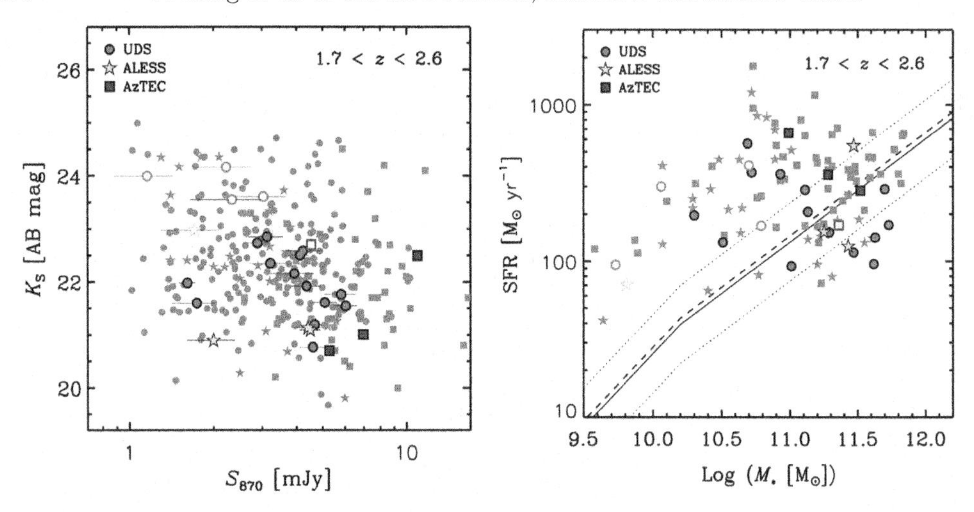

Figure 1. Properties of our SMG sample. Left: K_s-band magnitude versus observed flux density at 870 μm. Right: Our SMG sample shown in the M_*-SFR plane. Both quantities are based on SED-derived values from MAGPHYS. Gray symbols show the parent samples of all AS2UDS, ALESS, and AzTEC SMGs in the redshift range $1.7 < z < 2.6$. Filled colored symbols represent our final SMG sample. In addition, open colored symbols identify targets rejected due to their low surface brightness in the *HST*/WFC3 *J* and *H*-band imaging. The solid and dotted lines represent the main sequence at redshift 2 (Whitaker *et al.* 2014) and its scatter, respectively.

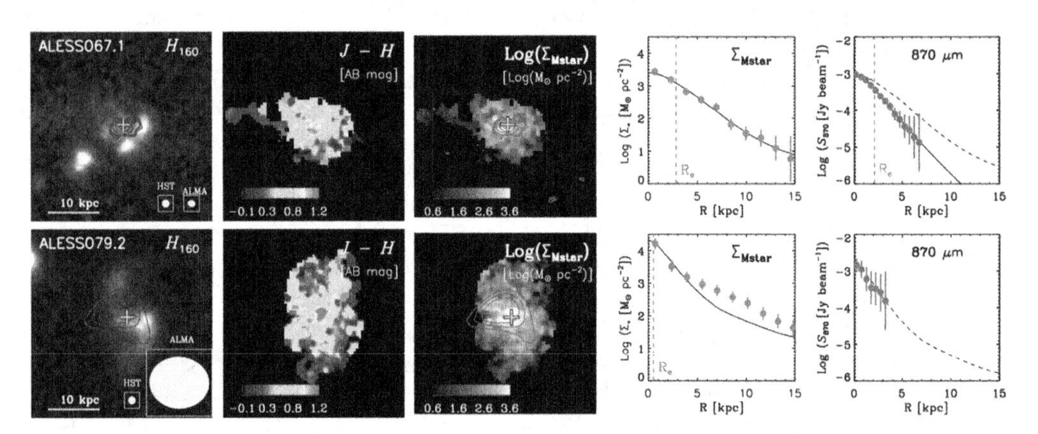

Figure 2. Example cases of our SMG sample; From left to right: H_{160}-band cutouts; $(J_{125} - H_{160})$ color maps; stellar-mass distributions (all with ALMA emission as blue contours); radial stellar mass; and far-infrared ALMA profiles for our SMG sample. Solid lines show our best-fit GALFIT models to the radial profiles. Best-fit mass model re-normalized to the peak of the best-fit submm profile are shown as solid lines.

morphologies commonly seen in SMGs (e.g., Chen *et al.* 2015; Hodge *et al.* 2016). The implied radial trends of M/L within our sources are either caused by variations of stellar age and/or the effects of extinction. With the ALMA/submm emission peak coinciding well with the location of strong color variations, the redder colors are likely the effect of increased extinction towards stronger dust-obscured regions.

As a consequence of the spatial M/L variations, our mass maps show more compact and smoother distributions than the optical images. For the majority of our targets, we confirm that the that stellar mass components are are systematically smaller than those of the H_{160}-band light by measuring a median ratio of $\langle R_{\rm e,mass}/R_{\rm e,H160}\rangle = 0.5 \pm 0.1$.

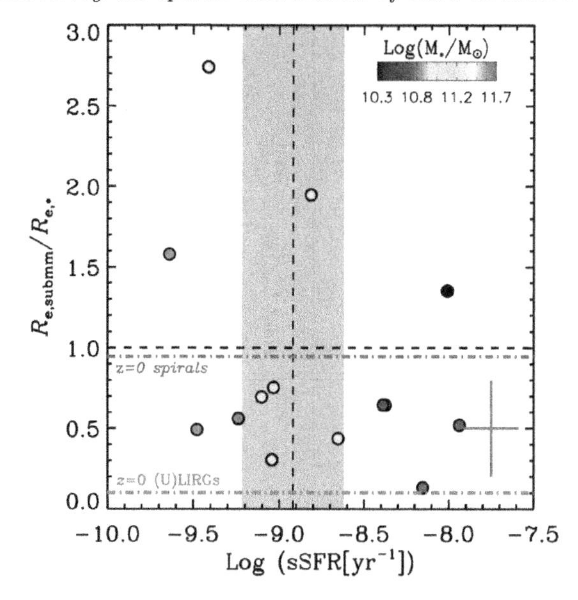

Figure 3. Ratio between intrinsic sizes of the submm and stellar components versus specific star-formation rate, with the color-coding indicating the total stellar mass. The position and scatter of the main sequence, adopted from Whitaker *et al.* 2014, is shown as the dashed vertical line and shaded area, respectively. The colored horizontal lines indicate the median size ratios for local spirals based on KINGFISH sample (Hunt *et al.* 2015), as well as for local (U)LIRGs as based on the GOALS survey (Kim *et al.* 2013). Median uncertainties in the shown properties for our sample are indicated.

We furthermore correlate the inferred size difference between the dust and stellar distributions with the specific star-formation rate, revealing no significant trend (Figure 3). To interpret our observations, we use the dust emission as a tracer of dust-obscured star formation in these systems, relying on the assumption that the dust properties (e.g., dust temperature) do not strongly vary spatially. This points to a picture in which the star formation is clearly more compact than the stellar distribution in high-redshift star-forming galaxies over a large range of specific star-formation rate. Our SMGs exhibit a more compact dust distribution relative to the stars compared to local spiral galaxies, where the dust and stars exhibit an about equal extent. As major interactions can (at least in the local universe) be attributed to the local (U)LIRG population that exhibit very compact dust configurations and large main sequence offsets, our findings indicate that SMGs undergoing strong interactions do not necessarily have more compact star-forming regions than the ones representing secular disks.

Based on our structural measurements, we investigate in the link between SMGs and passive galaxies, which are plausibly connected through the shut-down of star formation (referred to as 'quenching'). In Figure 4, we plot the effective stellar sizes and resulting surface densities of our SMGs as a function of total stellar mass. As a reference sample for the passive galaxy population, we consider quiescent early-type galaxies at $z = 1.5$, as those might represent the 'direct' descendants of SMG once they have undergone quenching and evolved to the passive population within $\sim$1 Gyr. Their sizes and surface densities are measured by van der Wel *et al.* (2014), who have quantified their R_e-M_* relation based on large samples. We further consider nearby massive early-type galaxies as their potential ultimate descendants in the local Universe (e.g. Toft *et al.* 2017).

The median stellar sizes and surface densities of our near-infrared bright SMGs are in good agreement with the quiescent population at $z = 1.5$ at the same stellar mass. Since

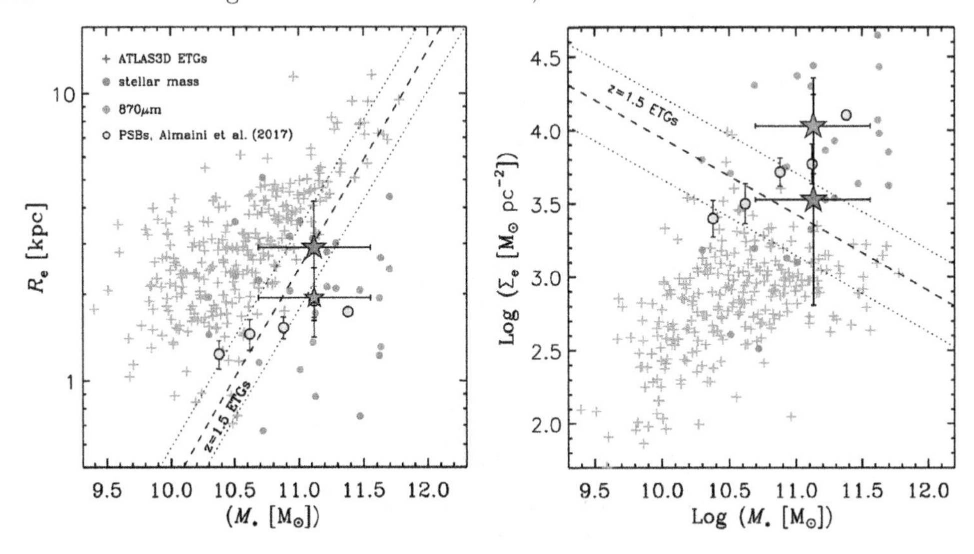

Figure 4. Comparison between the effective radii (left) and surface densities (right) of our SMG sample with local early-type galaxies from the ATLAS3D survey (Cappellari *et al.* 2011). The measurements of the submm and stellar components for our SMGs are shown as blue and red filled circles, respectively. Median values are indicated as stars. The stellar components of the underlying population of ATLAS3D galaxies are shown as gray crosses. The effective stellar sizes and surface densities of quiescent early-types at $z = 1.5$ from van der Wel *et al.* (2014) are shown as dashed lines, together with their scatter shown as dotted lines. Also shown are rest-frame optical sizes and surface densities of post-starburst galaxies (PSBs) at $1 < z < 2$.

the star-forming component is even more compact than the stars for our SMGs, those seem to fade into systems that represent the smaller and denser part of the quiescent galaxies at $z = 1.5$. Figure 4 also plots the effective sizes and resulting surface densities of post-starburst galaxies (i.e., systems selected to represent quiescent systems with very recent episodes of major star formation) at $1 < z < 2$ determined using rest-frame optical imaging by Almaini *et al.* (2017). The post-starburst systems exhibit on average effective sizes of 1.5–2 kpc within the stellar-mass range sampled by our SMGs, in agreement with a decrease in stellar size of SMGs before quenching. Thus, the post-starburst-phase might represent a link between the most immediate descendant of SMGs and the passive population at high redshift. The median stellar sizes and surface densities of our SMGs also occupy the locus of the most compact local ETGs that therefore might represent the ultimate descendants of SMG in the local Universe.

References

Almaini, O., Wild, V., Maltby, D. T., *et al.* 2017, *MNRAS*, 472, 1401
Bruzual, G. & Charlot, S. 2003, *MNRAS*, 344, 1000
Cappellari, M., Emsellem, E., Krajnović, D., *et al.* 2011, *MNRAS*, 416, 1680
Chen, C.-C., Smail, I., Swinbank, A. M., *et al.* 2015, *ApJ*, 799, 194
Dekel, A., Birnboim, Y., Engel, G., *et al.* 2009, *Nature*, 457, 451
Grogin, N. A., Kocevski, D. D., Faber, S. M., *et al.* 2011, *ApJs*, 197, 35
Hayward, C. C., Kereš, D., Jonsson, P., *et al.* 2011, *ApJ*, 743, 159
Hayward, C. C., Narayanan, D., Kereš, D., *et al.* 2013, *MNRAS*, 428, 2529
Hodge, J. A., Karim, A., Smail, I., *et al.* 2013, *ApJ*, 768, 91
Hodge, J. A., Swinbank, A. M., Simpson, J. M., *et al.* 2016, *ApJ*, 833, 103
Hunt, L. K., Draine, B. T., Bianchi, S., *et al.* 2015, *A&A*, 576, A33
Ikarashi, S., Ivison, R. J., Caputi, K. I., *et al.* 2015, *ApJ*, 810, 133
Kim, D.-C., Evans, A. S., Vavilkin, T., *et al.* 2013, *ApJ*, 768, 102

Koekemoer, A. M., Faber, S. M., Ferguson, H. C., *et al.* 2011, *ApJs*, 197, 36
Lang, P., Schinnerer, E., Smail, I., *et al.* 2019, *ApJ*, 879, 54
Miettinen, O., Delvecchio, I., Smolčić, V., *et al.* 2017, *A&A*, 606, A17
Narayanan, D., Dey, A., Hayward, C. C., *et al.* 2010, *MNRAS*, 407, 1701
Simpson, J. M., Smail, I., Swinbank, A. M., *et al.* 2015a, *ApJ*, 799, 81
Stach, S. M., Smail, I., Swinbank, A. M., *et al.* 2018, *ApJ*, 860, 161
Tadaki, K.-i., Genzel, R., Kodama, T., *et al.* 2017a, *ApJ*, 834, 135
Toft, S., Zabl, J., Richard, J., *et al.* 2017, *Nature*, 546, 510
van der Wel, A., Chang, Y.-Y., Bell, E. F., *et al.* 2014, *ApJl*, 792, L6
Swinbank, *et al.* 2006, *MNRAS*, 371, 465
Swinbank, A. M., Simpson, J. M., Smail, I., *et al.* 2014, *MNRAS*, 438, 1267
Whitaker, K. E., Franx, M., Leja, J., *et al.* 2014, *ApJ*, 795, 104

Uncovering Early Galaxy Evolution in the ALMA and JWST Era
Proceedings IAU Symposium No. 352, 2019
E. da Cunha, J. Hodge, J. Afonso, L. Pentericci & D. Sobral, eds.

doi:10.1017/S1743921320001246

Resolving on 100 pc-scales the UV-continuum in Lyman-emitters between redshift 2 to 3 with gravitational lensing

Elisa Ritondale

Max Planck Institute for Astrophysics, Germany

Abstract. Lyman-alpha emitting (LAE) galaxies are thought to be predominantly responsible for the re-ionisation of the Universe and are, as such, one of the most studied star-forming galaxy populations. Current optical and narrow-band studies are limited by the angular resolution of the observations and the considerable investment in telescope time. Strong gravitational lensing is an extremely powerful method that can be used to overcome these limitations. In my talk I will present a study on the first homogeneous sample of 17 lensed Lyman-alpha emitters at redshift $2 < z < 3$. By taking advantage of the lensing magnification, I was able to access the detailed structure of this high redshift star-forming galaxies, finding that they have radii ranging from 0.2 to 1.8 kpc and have a complex and clumpy morphology, with a median ellipticity of 0.49. This is consistent with disk-like structures of star-formation, which would rule out models where the Lyman-alpha emission is only seen perpendicular to the disk, and favours those clumpy models for the escape lines of sight for Lyman-alpha photons. We also find that the star formation rates range from 0.3 to 8.5 $M_\odot$/yr and that these galaxies tend to be very compact. The lower limit to their intrinsic size is about a factor of two smaller than that found for non-lensed LAEs, which highlights the power of gravitational lensing and sophisticated lens modelling techniques for resolving such objects in the high redshift Universe.

Uncovering Early Galaxy Evolution in the ALMA and JWST Era
Proceedings IAU Symposium No. 352, 2019
E. da Cunha, J. Hodge, J. Afonso, L. Pentericci & D. Sobral, eds.

doi:10.1017/S1743921320001258

Lensed quiescent galaxies at $z \sim 2$: What quenched their star formation?

Allison Man

Dunlap Institute, University of Toronto, Canada

Abstract. A key outstanding issue in galaxy evolution studies is how galaxies quench their star formation. I will present new results from our VLT/X-Shooter, ALMA and VLA campaign of a pilot sample of lensed quiescent massive galaxies at $z > 1.5$. Lensing magnification enables us to spatially resolve the stellar structure and kinematics of these compact galaxies, that are otherwise barely resolvable even with HST. Our deep X-Shooter spectra provided multiple absorption lines enabling strong constraints on their stellar populations, namely their star formation rates, ages, dispersions, and in some cases metallicities. Our complementary ALMA+VLA programme probes their molecular gas content through CO emission. All these observations provide unparalleled constraints on their quenching mechanisms. Our results indicate that quiescent galaxies at $z \sim 2$ (1) have short star formation timescales of a few hundred Myrs; (2) have a variety of stellar morphology from exponential disks to bulges; (3) are devoid of molecular gas; and (4) host low-luminosity active galactic nuclei which may be responsible for suppressing star formation. In addition to discussing the insights gained on quenching, I will highlight how these findings bring about new questions that can be addressed with future JWST and ALMA studies.

Uncovering Early Galaxy Evolution in the ALMA and JWST Era
Proceedings IAU Symposium No. 352, 2019
E. da Cunha, J. Hodge, J. Afonso, L. Pentericci & D. Sobral, eds.

doi:10.1017/S1743921319008998

Resolving distant, dusty galaxies using observations and simulations

R. K. Cochrane

SUPA, Institute for Astronomy, Royal Observatory Edinburgh, EH9 3HJ, UK

Abstract. Spatially resolved studies of galaxies in the high-redshift Universe have traditionally been reliant on data at rest-frame optical and UV wavelengths, which can be biased towards the least dust-obscured galaxies. For several years now, we have been able to resolve and probe the morphology of longer-wavelength emission from distant galaxies with ALMA, and a number of recent ALMA studies were presented at the IAU Symposium No. 352. These included our study of the resolved multi-wavelength emission of galaxies at $z \sim 2$. As part of the SHiZELS collaboration, we are mapping the Hα emission line (from SINFONI/VLT), UV continuum (from HST), and the far-infrared (from ALMA) emission from a small sample of Hα-selected galaxies. In this proceedings paper, we showcase the high quality of our data, and the spectacular structures displayed by one of our most dusty sources. We also provide an overview of some highly complementary simulation-based work, using galaxies drawn from the FIRE-2 zoom-in cosmological hydrodynamical simulations. Using sophisticated radiative transfer techniques, we have derived predictions for the spatially-resolved emission of a sample of star-forming galaxies, from rest-frame far-ultraviolet to the far-infrared. For both observed and simulated galaxies, emission maps show striking differences with wavelength, with the same galaxy appearing clumpy and extended in the far-ultraviolet yet compact at far-infrared wavelengths.

Keywords. galaxies: high-redshift, galaxies: starburst, infrared: galaxies, radiative transfer

---

1. Introduction

Recent far-infrared (FIR) observations have shown that $\sim 85\%$ of the total star formation at $z \sim 2$ is enshrouded in dust (Dunlop *et al.* 2017). For high-mass galaxies ($M_* > 2 \times 10^{10} M_\odot$), the SFR derived from long-wavelength FIR emission is an extraordinary 200 times that derived from unobscured light. It is therefore crucial to study the internal properties of massive $z \sim 2$ galaxies at long wavelengths, where dust emits thermally, as well as at shorter wavelengths. In this paper, we present two complementary approaches. The first involves an ambitious observational programme in which we compare several widely-used tracers of star-formation for an individual high-redshift source. The second couples high-resolution zoom-in simulations with radiative transfer to understand the physics at play as well as the observational biases that affect what is measured.

2. Mapping $z \sim 2$ galaxies at multiple wavelengths

The High-Redshift(Z) Emission Line Survey, HiZELS, used a combination of narrow-band and broad-band filters to select thousands of intermediate redshift star-forming galaxies via their emission line fluxes (Sobral *et al.* 2013, 2015) in fields with high-quality multi-wavelength coverage (COSMOS, UDS & SA22). Following a number of studies of the global properties of these galaxies (e.g. Sobral *et al.* 2009, 2010, 2014; Cochrane *et al.* 2017, 2018), the HiZELS collaboration has recently obtained resolved maps of a sample of

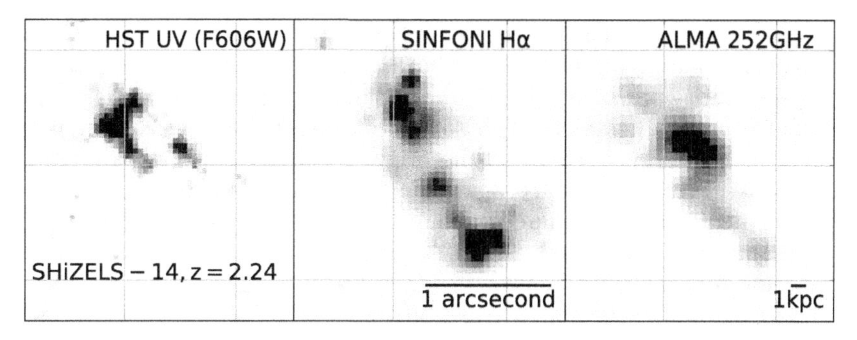

Figure 1. SHiZELS-14, a star-forming galaxy at $z = 2.2$, mapped at matched $\sim$ 1kpc resolution at three wavelengths typically used to probe star formation: rest-frame UV (HST F606W filter), the Hα emission line (SINFONI/VLT), FIR dust continuum emission (ALMA). The three images show strikingly different morphologies, with little short-wavelength emission from the most dusty central region. The distribution of star formation inferred from each of the tracers is distinct. From Cochrane *et al.* (in prep).

Hα-selected galaxies at $z \sim 1.5$ and $z \sim 2.2$. We have obtained $\sim 0.15''$ matched-resolution observations of Hα emission (using adaptive optics with the SINFONI Integral Field Unit on the VLT; see Swinbank *et al.* 2012a,b; Molina *et al.* 2017), dust continuum emission (with ALMA), and in the rest-frame UV and optical (from HST UVIS imaging).

One of our most extensively imaged galaxies is SHiZELS-14 (Figure 1), a $M_*{\sim}10^{11}M_\odot$, highly star-forming galaxy at $z = 2.24$. Although identified via its Hα emission, this galaxy displays a bright sub-millimeter flux density ($S_{\rm obs,\,252GHz} = 2.7 \pm 0.15$mJy), and is not a typical star-forming galaxy, as we might have have inferred from its modest UV and Hα fluxes. Fitting the dust SED with a two grey-body model yields a dust mass of $M_{\rm dust} = 10^{9.1\pm0.1}M_\odot$, a TIR luminosity of $\log_{10}(L_{\rm TIR}/{\rm erg\,s^{-1}}) = 46.39 \pm 0.02$ and a star formation rate of ${\rm SFR} = 950 \pm 50{\rm M_\odot yr^{-1}}$. This bright IR emission places it in the category of a ULIRG. The star formation rates derived from shorter wavelength data using standard luminosity-SFR calibrations (Kennicutt & Evans 2012) are notably lower (${\rm SFR} \sim 10 - 200{\rm M_\odot yr^{-1}}$), highlighting the importance of long-wavelength data for such dusty systems.

Figure 1 shows the impressive quality of our $\sim 0.15''$ data, and highlights the biased view of galaxy morphology provided by UV data. After careful astrometric calibration, the UV emission is spatially offset from the dust emission, and does not represent the position of the bulk of star formation. A number of recent studies have found similar offsets in the multi-wavelength emission of bright sub-millimeter galaxies (e.g. Hodge *et al.* 2016; Chen *et al.* 2017; Calistro Rivera *et al.* 2018).

3. Simulating multi-wavelength emission with FIRE-2

In this section, we describe work undertaken to simulate the emission from galaxies at high spatial resolution. The Feedback In Realistic Environments (FIRE) project (Hopkins *et al.* 2014, 2018) is a set of hydrodynamical cosmological zoom-in simulations that explore the role of stellar feedback in galaxy formation and evolution. FIRE simulations explicitly model multi-channel stellar feedback by resolving the formation of giant molecular clouds and also include models for both energy and momentum return from the main stellar feedback processes, using predictions of stellar population synthesis models without the extensive parameter tuning employed in other simulations. They broadly reproduce many observed galaxy properties, including stellar masses, star-formation histories and the 'main sequence' of star-forming galaxies (see Hopkins *et al.* 2014; Sparre *et al.* 2017), metallicities and abundance ratios (Ma *et al.* 2016; van de Voort *et al.* 2015), as well as morphologies and kinematics of both thin and thick disks (Ma *et al.* 2017).

Here, we summarise our study of the central galaxies of four massive haloes recently published by Cochrane *et al.* (2019). These haloes were originally selected and simulated by Feldmann *et al.* (2016, 2017) with the original FIRE model (Hopkins *et al.* 2014) as part of the MASSIVEFIRE suite. Our sample of central galaxies have dark matter halo masses of $M_{\rm halo} \sim 10^{12.5} M_{\odot}$ and stellar masses of $7 \times 10^{10} - 3 \times 10^{11} M_{\odot}$ at $z = 2$. The mass resolution of our simulations is $3.3 \times 10^{4}\ M_{\odot}$ for gas and star particles and $1.7 \times 10^{5} M_{\odot}$ for dark matter.

To simulate the multi-wavelength emission from these galaxies in a consistent way, we perform radiative transfer. We model the radiation field from stars as a flow of photons through the dusty medium of a galaxy to compute the effects of dust absorption, scattering, and re-emission of the absorbed light, including dust self-absorption. We implement this using the Stellar Kinematics Including Radiative Transfer (SKIRT) Monte Carlo radiative transfer code (Baes *et al.* 2011; Camps & Baes 2015). We extract gas and star particles from the FIRE-2 simulations at a number of time snapshots. For gas particles with temperature $< 10^{6}$ K, we compute dust masses using the metallicity of the gas particles and a dust-to-metals mass ratio of 0.4 (Dwek 1998; James *et al.* 2002). We assume that dust is destroyed in gas particles with temperature $> 10^{6}$ K (Draine & Salpeter 1979; Tielens *et al.* 1994). We use a Weingartner & Draine (2001) Milky Way dust prescription to model a mixture of graphite, silicate and PAH grains. Star particles are assigned Bruzual & Charlot (2003) SEDs according to their ages and metallicities. We model the flux that would be received by an observer on Earth at 100 discrete wavelengths between rest-frame UV and FIR, also considering different inclinations with respect to the disk plane of the galaxy.

Figure 2 shows an example of the output of our radiative transfer for a simulated galaxy at $z = 2.95$. The derived morphologies are notably different in different wavebands, with the same galaxy appearing clumpy in the far-ultraviolet yet regular and spiral-like at far-infrared wavelengths. The observed-frame $870\mu m$ half-light radii of our FIRE-2 galaxies are $\sim 0.5 - 4$kpc, consistent with existing ALMA observations of high redshift galaxies. In these simulated galaxies, the dust continuum emission is more compact than the cold gas, but more extended than the stellar component. The most extreme cases of compact dust emission are driven by particularly compact recent star formation, which can drive steep dust temperature gradients.

4. Summary

In this short paper, we have provided an overview of our ongoing observational and simulation-based work to understand the spatially-resolved, multi-wavelength emission from highly star-forming galaxies around the peak of cosmic star formation. As well as shaping the global SED, dust clearly plays a role in determining the spatial extent of emission observed in different wavebands. For dusty galaxies like SHiZELS-14, short wavelength emission appears to trace little but holes in the dust distribution. Our radiative transfer modelling confirms that dust obscuration can drive substantial decoupling in the spatial distributions of short and long-wavelength light. For our simulated dusty galaxies, the UV is a poor tracer of not only the amount of star formation but also its spatial extent.

Acknowledgements

Section 2 is based on work conducted in collaboration with the SHiZELS team, including Philip Best, Mark Swinbank, Edo Ibar, Ian Smail, David Sobral, and Juan

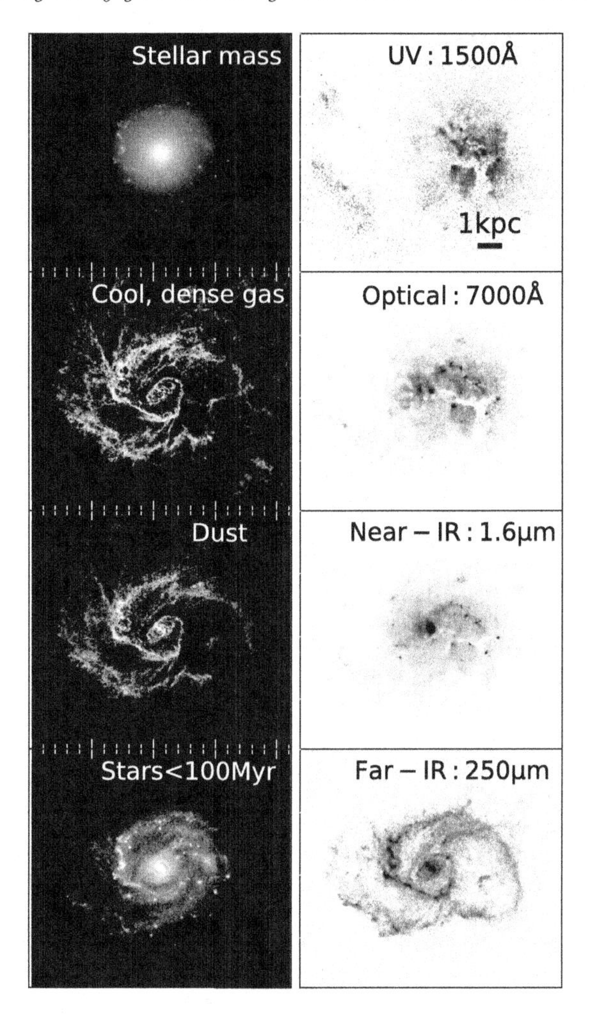

Figure 2. The wavelength-dependent morphology of a $M_* \sim 10^{11}$ galaxy in the FIRE-2 simulations at $z = 2.95$. The left-hand panels show the projected distributions of stellar mass, cold, dense gas mass, dust mass, and stars formed within 100Myr (intrinsic properties of the galaxy). The right-hand panels show the SKIRT-predicted images at different rest-frame wavelengths. The morphology is strongly dependent on the wavelength. The galaxy appears clumpy and extended in the rest-frame UV but more ordered at longer wavelengths. The UV and optical light is significantly offset from the peak of the stellar mass and SFR, appearing to trace holes in the dust. The colour scales are logarithmic and span the $70^{th} - 99^{th}$ percentiles of the flux distribution of each panel, to highlight the qualitative differences in morphology. From Cochrane *et al.* (2019).

Molina. This project will be published in a forthcoming paper. Section 3 is based on work with FIRE collaborators, Christopher Hayward, Daniel Anglés-Alcázar, Jennifer Lotz, Tyler Parsotan, Xiangcheng Ma, Dusan Kereš, Robert Feldmann, Claude-André Faucher-Giguère, and Philip Hopkins. This work was published in Cochrane *et al.* (2019).

References

Baes, M. *et al.* 2011, *ApJS*, 196, 22
Bruzual, G. & Charlot, S. 2003, *MNRAS*, 35, 1
Calistro Rivera, G. *et al.* 2018, *ApJ*, 863, 56
Camps, P. & Baes, M. 2015, *Astron. Comput.*, 9, 20

Chen, C.-C. *et al.* 2017, *ApJ*, 846, 108
Cochrane, R. K., Best, P. N., Sobral, D., Smail, I., Wake, D. A., Stott, J. P., & Geach, E. 2017, *MNRAS*, 469, 2913
Cochrane, R. K., Best, P. N., Sobral, D., Smail, I., Geach, J. E., Stott, J. P., & Wake, D. A. 2018, *MNRAS*, 475, 3730
Cochrane, R. K. *et al.* 2019, *MNRAS*, 488, 1779–1789
Draine, B. T. & Salpeter, E. E. 1979, *ApJ*, 231, 77
Dunlop, J. S. *et al.* 2017, *MNRAS*, 466, 861
Dwek, E. 1998, *ApJ*, 501, 643
Feldmann, R., Hopkins, P. F., Quataert, E., Faucher-Giguère, C.-A., & Kereš, D. 2016, *MNRAS*, 458, L14
Feldmann, R., Quataert, E., Hopkins, P. F., Faucher-Giguère, C. A., & Kereš, D. 2017, *MNRAS*, 470, 1050
Kennicutt, R. C. & Evans, N. J. 2012, *ARA&A*, 50, 531
Hodge, J. A. *et al.* 2016, *ApJ*, 833, 103
Hopkins, P. F., Kereš, D., Oñorbe, J., Faucher-Giguère, C. A., Quataert, E., Murray, N., & Bullock, J. S. 2014, *MNRAS*, 445, 581
Hopkins, P. F. *et al.* 2018, *MNRAS*, 480, 800
James, A., Dunne, L., Eales, S., & Edmunds, M. G. 2002, *MNRAS*, 335, 753
Ma, X., Hopkins, P. F., Faucher-Giguère, C. A., Zolman, N., Muratov, A. L., Kereš, D., & Quataert, E. 2016, *MNRAS*, 456, 2140
Ma, X., Hopkins, P. F., Wetzel, A. R., Kirby, E. N., Anglés-Alcázar, D., Faucher-Giguère, C.-A., Kereš, D., & Quataert, E. 2017, *MNRAS*, 467, 2430
Molina, J., Ibar, E., Swinbank, A. M., Sobral, D., Best, P. N., Smail, I., Escala, A., & Cirasuolo, M. 2017, *MNRAS*, 466, 892
Sobral, D. *et al.* 2009, *MNRAS*, 398, 75
Sobral, D., Best, P. N., Geach, J. E., Smail, I., Cirasuolo, M., Garn, T., Dalton, G. B., & Kurk, J. 2010, *MNRAS*, 1563, 1551
Sobral, D. *et al.* 2013, *ApJ*, 779, 139
Sobral, D., Best, P. N., Smail, I., Mobasher B., Stott J., & Nisbet D. 2014, *MNRAS*, 427, 3516
Sobral, D. *et al.* 2015, *MNRAS*, 451, 2303
Sparre, M., Hayward, C. C., Feldmann, R., Faucher-Giguère, C. A., Muratov, A. L., Kereš, D., & Hopkins, P. F. 2017, *MNRAS*, 466, 88
Swinbank, A. M., Sobral, D., Smail, I., Geach, J. E., Best, P. N., Mccarthy, I. G.,Crain, R. A., & Theuns, T. 2012a, *MNRAS*, 426, 935
Swinbank, A. M., Smail, I., Sobral, D., Theuns, T., Best, P. N., & Geach, J. E. 2012b, *ApJ*, 760, 13
Tielens, A. G. G. M., McKee, C. F., Seab, C. G., & Hollenbach, D. J. 1994, *ApJ*, 431, 321
Weingartner, J. C. & Draine, B. T. 2001, *ApJ*, 548, 296
van de Voort, F., Quataert, E., Hopkins, P. F., Kereš, D., Faucher-Giguère, C. A. 2015, *MNRAS*, 447, 140

Uncovering Early Galaxy Evolution in the ALMA and JWST Era
Proceedings IAU Symposium No. 352, 2019
E. da Cunha, J. Hodge, J. Afonso, L. Pentericci & D. Sobral, eds.

doi:10.1017/S1743921319009074

A sub-kiloparsec-scale view of un-lensed submillimeter galaxies

Ken-ichi Tadaki[1] **and Daisuke Iono**[1,2]

[1]National Astronomical Observatory of Japan, 2-21-1 Osawa, Mitaka, Tokyo 181-8588, Japan
email: tadaki.ken@nao.ac.jp

[2]Department of Astronomical Science, SOKENDAI (The Graduate University for Advanced Studies), Mitaka, Tokyo 181-8588, Japan

Abstract. Submillimeter galaxies at $z > 3$ building up their central cores through compact starbursts with an effective radius of 1–2 kpc. Our ALMA high-resolution observations reveal off-center gas clumps in a submillimeter galaxy at $z = 4.3$, COSMOS-AzTEC-1, as well as a rotation-dominated disk. Exploiting the kinematic properties and the spatial distribution of gas mass surface density, we find that the starburst disk is gravitationally unstable. This result is consistent with a scenario where in-situ clumps are formed through disk instability. On the other hand, we find evidence for an ex-situ clump that does not corotate with the starburst disk. The accretion of such a non-corotating clump could stimulate violent disk instability, driving gas inflows into the central regions of the galaxy. Our results suggest that compact cores are formed through an extreme starburst due to a gravitational instability, triggered by non-corotating clumps.

Keywords. galaxies: formation, galaxies: starburst, galaxies: ISM

1. Introduction

When and how did galaxies shape the Hubble sequence? Over the past three decades, high-resolution, high-quality images obtained with Hubble Space Telescope (HST) revealed the rest-optical morphologies for tens of thousands of distant galaxies. The most striking results are
1) a correlation between star-forming activity and morphology of galaxies at $0 < z < 2.5$: star-forming galaxies are typically a disk-dominated system whereas quiescent galaxies are a spheroid-dominated one (e.g., Wuyts *et al.* 2011; Bell *et al.* 2012),
2) a discovery of massive, compact quiescent galaxies at $z \sim 2$: the half-light radius of $R_{1/2} \sim 1$ kpc is a factor of 4–5 smaller than similar mass quiescent galaxies at $z = 0$ (e.g., Trujillo *et al.* 2006; van Dokkum *et al.* 2015), and
3) size evolution from $z = 3$ to $z = 0$: star-forming galaxies are always larger than quiescent galaxies and both galaxies gradually increase size over time (e.g., van der Wel *et al.* 2014).

These findings offered a paradigm for galaxy evolution: high-redshift massive galaxies evolve into giant ellipticals through two channels. In a fast channel, massive compact star-forming galaxies will become compact quiescent galaxies at $z \sim 2$ after quenching star formation (e.g., Barro *et al.* 2013). Repeating minor mergers puffs up compact quiescent galaxies, turning them into giant elliptical galaxies (e.g., Naab *et al.* 2009). In a slow channel, galaxies increase their size by in-situ star formation, and after quenching larger quenched galaxies continuously add to the quiescent population at $z = 0 - 1$ (e.g., Carrolo *et al.* 2013; Belli *et al.* 2015). This two-channel scenario explains the redshift evolution

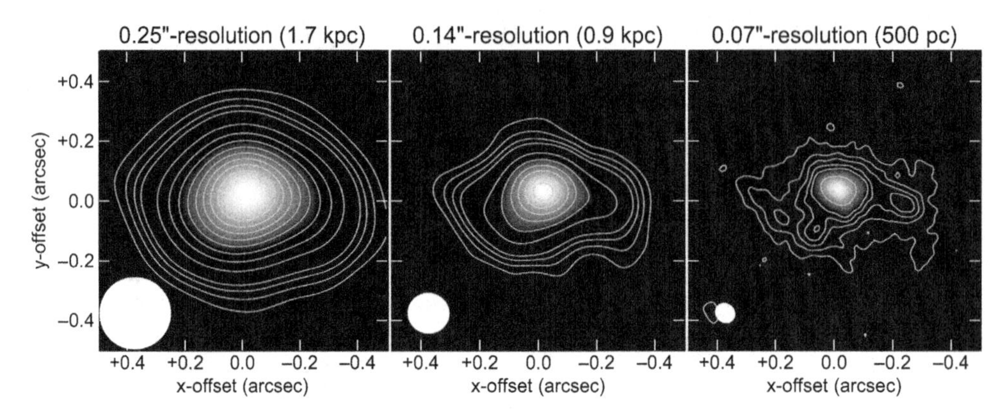

Figure 1. ALMA 870 μm continuum maps at three different spatial resolutions in COSMOS-AzTEC-1 at $z = 4.3$ (Iono *et al.* 2016). The image size is $1'' \times 1.''$Contours are plotted every 3σ from 6σ to 15σ and every 10σ from 15σ.

of both the size and the number density of star-forming/quiescent galaxies (e.g., Barro *et al.* 2013; van Dokkum *et al.* 2015). However, a big problem still remains: how did compact star-forming galaxies form at $z = 3 - 6$?

2. Overview

Submillimeter bright galaxies (SMGs) at $z > 3$ are the most likely progenitors of compact star-forming galaxies. Recent $0.''2$-resolution dust continuum observations indicate that SMGs are vigorously forming stars in the central 1-2 kpc region, supporting an evolutionary link from SMGs to compact star-forming galaxies with $R_{1/2} \sim 1$ kpc (e.g., Simpson *et al.* 2015; Ikarashi *et al.* 2015; Tadaki *et al.* 2017). What happens if we go to higher resolution? ALMA $0.''07$-resolution observations of 870 μm continuum reveal that a bright SMG at $z = 4.3$, COSMOS-AzTEC-1, is not just compact, but also has clumpy structures in the central 1-2 kpc region (Figure 1; Iono *et al.* 2016). Such off-center clumps are now commonly seen in SMGs (Hodge *et al.* 2019; Rujopakarn *et al.* 2019 although we need a better resolution than $0.''1$ to prove them (Figure 1).

An obvious next step is to investigate the kinematics in the central starburst region to understand the origin of off-center clumps. Unfortunately, dust continuum observations do not provide information of kinematics in galaxies. We therefore obtained $0.''08$-resolution observation of the CO(4-3) line in COSMOS-AzTEC-1 (Tadaki *et al.* 2018). The starburst disk is rotation-dominated with a rotation velocity-to-velocity dispersion ratio of $v/\sigma_0 \sim 3$ (Figure 2). The ordered rotation is also confirmed by [C II] and [N II] line observations (Figure 3; Tadaki *et al.* 2019a). The measured Toomre Q parameter is much below unity over the starburst disk ($Q_{\rm obs} \sim 0.2$), suggesting that off-center clumps are formed through the gravitational instability in the central 1-2 kpc region.

On the other hand, we have discovered a non-corotating clump by $0.''17$-resolution [C II] observations (Tadaki *et al.* 2019b, in prep). The [C II] clump has a large velocity offset of $\sim$200 km s^{-1} from the disk component and is located along the kinematic minor axis of disk rotation. We suggest a scenario where the ex-situ non-corotating clump develops the violent disk instability (VDI), where the disk is turbulent and highly perturbed, driving gas inflow into the central region of the galaxy (Dekel *et al.* 2009; Dekel & Burkert 2014; Danovich *et al.* 2015; Zolotov *et al.* 2015). The extreme starburst is likely to be caused by a combination of minor mergers and efficient star formation due to gravitational instability in the central 1-2 kpc region.

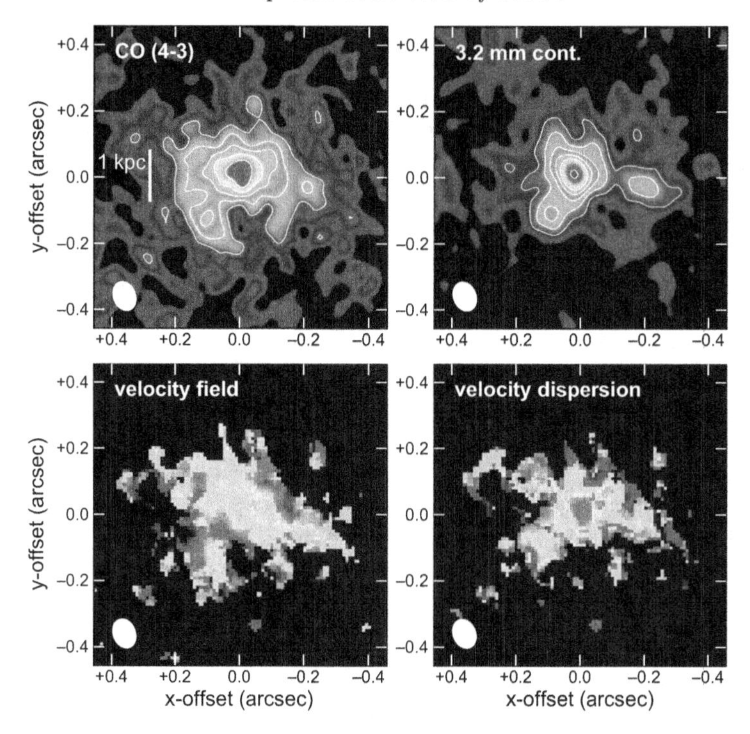

Figure 2. ALMA maps of the CO (4-3) line, 3.2 mm continuum, velocity field, velocity dispersion (Tadaki *et al.* 2018).

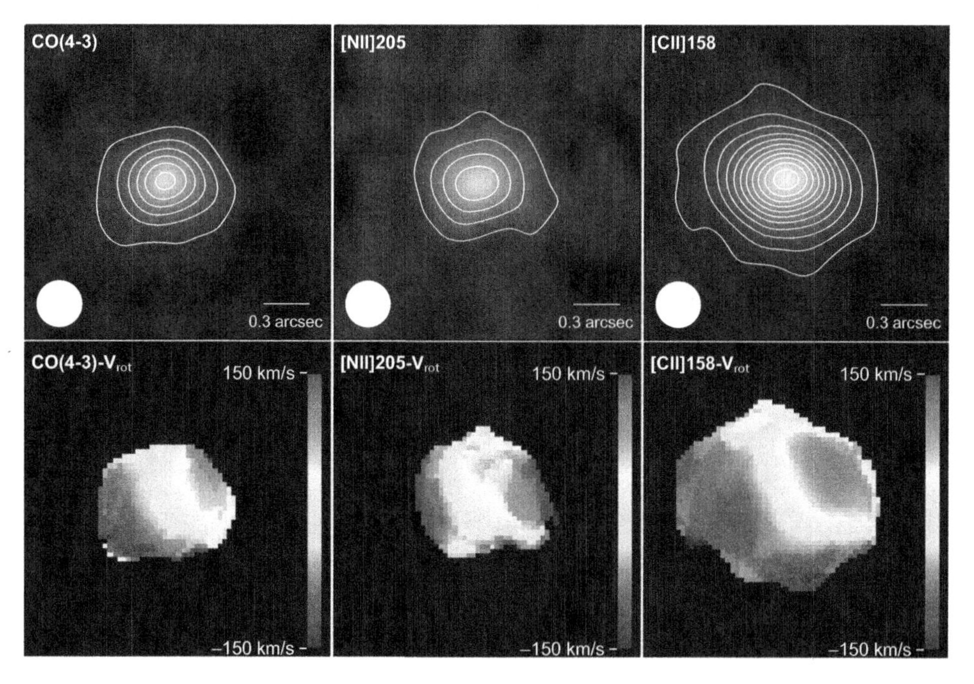

Figure 3. ALMA CO(4-3), [N II] and [C II] maps of flux and velocity field in COSMOS-AzTEC-1 (Tadaki *et al.* 2019a). The spatial resolutions are all $\sim 0.''3$. The image size is $2'' \times 2.''$Contours are plotted every 5σ from 3σ.

ALMA observations of molecular/ionized/PDR gas alone are not enough to reject the possibility of major mergers in COSMOS-AzTEC-1, because gas and new stars formed after mergers become rotation-dominated in both cases of major and minor mergers (Robertson *et al.* 2006). While COSMOS-AzTEC-1 is very faint at $< 2\ \mu$m, it is sufficiently bright at $3 - 4\mu$m with an AB magnitude of 22. Integral field spectroscopy with Near Infrared Spectrograph on James Webb Space Telescope (Dorner *et al.* 2016) will allow us to investigate the stellar kinematics through observations of stellar absorption lines in the rest-frame optical wavelengths. The ALMA-JWST synergetic observations will enable us to determine whether SMGs experienced major mergers in the past and understand the physical mechanism responsible for the extreme starburst in the early Universe.

References

Barro, G., Faber, S. M., Pérez-González, P. G., *et al.* 2013, *ApJ*, 765, 104
Bell, E. F., van der Wel, A., Papovich, C., *et al.* 2012, *ApJ*, 753, 167
Belli, S., Newman, A. B., & Ellis, R. S. 2015, *ApJ*, 799, 206
Carollo, C. M., Bschorr, T. J., Renzini, A., *et al.* 2013, *ApJ*, 773, 112
Danovich, M., Dekel, A., Hahn, O., Ceverino, D., & Primack, J. 2015, *MNRAS*, 449, 2087
Dekel, A., Birnboim, Y., Engel, G., *et al.* 2009, *Nature*, 457, 451
Dorner, B., Giardino, G., Ferruit, P., *et al.* 2016, *A&A*, 592, A113
Dekel, A. & Burkert, A. 2014, *MNRAS*, 438, 1870
Hodge, J. A., Smail, I., Walter, F., *et al.* 2019, *ApJ*, 876, 130
Ikarashi, S., Ivison, R. J., Caputi, K. I., *et al.* 2015, *ApJ*, 810, 133
Iono, D., Yun, M. S., Aretxaga, I., *et al.* 2016, *ApJL*, 829, L10
Naab, T., Johansson, P. H., & Ostriker, J. P. 2009, *ApJL*, 699, L178
Robertson, B., Bullock, J. S., Cox, T. J., *et al.* 2006, *ApJ*, 645, 986
Rujopakarn, W., Daddi, E., Rieke, G. H., *et al.* 2019, arXiv e-prints, arXiv:1904.04507
Simpson, J. M., Smail, I., Swinbank, A. M., *et al.* 2015, *ApJ*, 799, 81
Tadaki, K.-i., Genzel, R., Kodama, T., *et al.* 2017, *ApJ*, 834, 135
Tadaki, K., Iono, D., Yun, M. S., *et al.* 2018, *Nature*, 560, 613
Tadaki, K.-i., Iono, D., Hatsukade, B., *et al.* 2019, *ApJ*, 876, 1
Trujillo, I., Feulner, G., Goranova, Y., *et al.* 2006, *MNRAS*, 373, L36
van der Wel, A., Franx, M., van Dokkum, P. G., *et al.* 2014, *ApJ*, 788, 28
van Dokkum, P. G., Nelson, E. J., Franx, M., *et al.* 2015, *ApJ*, 813, 23
Wuyts, S., Förster Schreiber, N. M., van der Wel, A., *et al.* 2011, *ApJ*, 742, 96
Zolotov, A., Dekel, A., Mandelker, N., *et al.* 2015, *MNRAS*, 450, 2327

Uncovering Early Galaxy Evolution in the ALMA and JWST Era
Proceedings IAU Symposium No. 352, 2019
E. da Cunha, J. Hodge, J. Afonso, L. Pentericci & D. Sobral, eds.

doi:10.1017/S1743921319008883

Kinematics of z ∼ 4 − 6 Lyman break galaxies in ALPINE

Gareth C. Jones[1,2] and the ALPINE Collaboration

[1]Cavendish Laboratory, University of Cambridge, 19 J. J. Thomson Ave., Cambridge CB3 0HE, UK
email: gj283@cam.ac.uk

[2]Kavli Institute for Cosmology, University of Cambridge, Madingley Road, Cambridge CB3 0HA, UK

Abstract. The past century has seen massive improvements in the study of galaxy kinematics. While early work focused on single nearby galaxies, current studies with modern IFUs and interferometers (e.g., SINFONI, ALMA) allow for extension of this field to high redshift. However, the sample of galaxy observations at $z > 4$ that feature the sensitivity and resolution required for resolved dynamical characterization has been small. The **ALMA Large Program to INvestigate CII at Early times (ALPINE)** targeted 118 star-forming galaxies at $z = 4 - 6$, representing a vast increase in the sample size of potentially dynamically-characterizable sources. Using a set of diagnostic plots, we are able to characterize roughly half the sample, revealing a vast kinematic diversity and high merger rate. For the nine targets that show rotational signatures, initial tilted ring fitting with ^{3D}Barolo shows promise. With further observations (e.g., ALMA, NOEMA, MUSE), the true nature of each source will be revealed in unprecedented detail.

Keywords. galaxies: high-redshift, galaxies: interactions, galaxies: kinematics and dynamics

---

1. Introduction

Kinematics of local galaxies have been studied for more than a century (e.g., Pease 1918), and developments over this time have allowed for larger sample sizes and more detailed modelling (e.g., Förster Schreiber *et al.* 2009; de Blok *et al.* 2016; Wuyts *et al.* 2016). With the advent of modern instruments (e.g., VLT, MUSE, ALMA), it is now possible to observe high-redshift galaxies at high resolution, allowing for the precise study of their motions. However, while studies at higher redshift, including De Breuck *et al.* (2014; $z = 4.8$), Jones *et al.* (2017; $z \sim 4 - 6$), and Smit *et al.* (2018; $z \sim 6.8$), have determined the kinematics of high-redshift galaxies, a larger sample is required to make statistically significant inferences.

The need for a larger set of $z > 4$ galaxies was one of the driving goals of the ALMA Large Program to INvestigate CII at Early times (ALPINE, PI: O. Le Févre; Faisst *et al.* 2019, Le Fèvre *et al.* in prep.), which observed 118 normal, star-forming galaxies at $4.4 < z_{spec} < 5.8$ in [CII] and FIR dust continuum emission with ALMA in cycles 5 and 6. Here, we present an initial study of the kinematical diversity of the sample.

2. Methods

A preliminary examination of the 118 ALPINE [CII] cubes revealed a vast array of dynamical properties. In order to characterize each, we created several diagnostic plots (Figure 1): moment zero (integrated intensity) and moment one (velocity field) maps, position-velocity (PV) plots along two orthogonal axes, and integrated spectra.

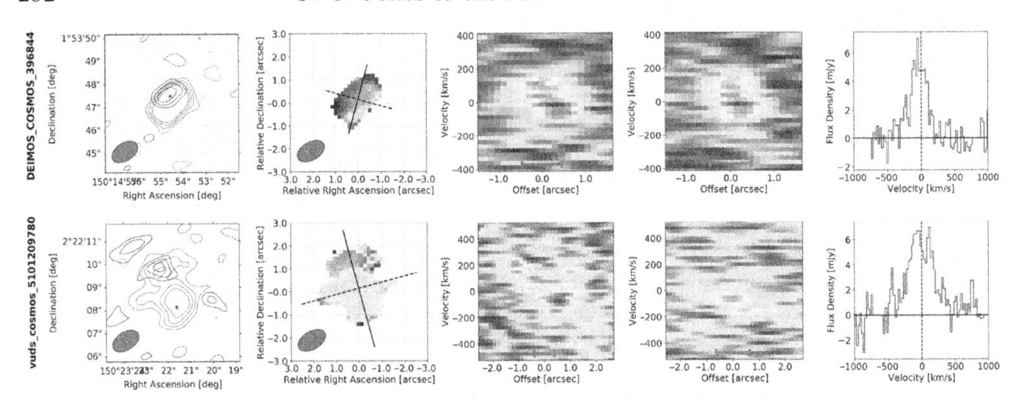

Figure 1. Examples of a rotator (top row) and merger (bottom row) from ALPINE. Column 1: HST/ACS I-band image (greyscale), [CII] emission (red contours), and FIR continuum emission (blue contours). Column 2: [CII] velocity field, with the pseudoslits for each PV diagram marked by solid (major axis) and dashed (minor axis) lines. Columns 3 and 4: position-velocity diagrams for the major and minor axes, respectively. Column 5: integrated [CII] spectrum.

Using these and ancillary multiwavelength photometry, we classified each galaxy into one of five categories: rotators, mergers, dispersion-dominated, disturbed, or weak.

Rotators show a single source in all images, a tilted PV slice along the major axis, and a straight PV slice along the minor axis. **Dispersion-Dominated Galaxies** also only show a single source, but have identical PV slices along orthogonal pseudoslits. **Mergers** show multiple sources, separated in position and/or velocity. **Disturbed Galaxies** are sources that feature strong line emission, but do not fall into one of the three above classes. **Weak Galaxies** include both those where no significant [CII] emission was detected, and those whose emission is too weak to determine its true nature.

3. Results

In this initial analysis, we find that roughly half of the ALPINE sources are weak, $\sim 20\%$ are mergers, and the remaining $\sim 30\%$ is evenly split between rotating, dispersion-dominated, and disturbed galaxies. Since this sample was intended to represent normal galaxies at high-redshift (i.e., not Malmquist biased), this suggests that the underlying sample shows a great diversity of kinematic states.

A more in-depth classification scheme has now been implemented, and will be presented in Le Fèvre *et al.* (in prep). In addition, the rotating galaxies will be further analyzed in Jones *et al.* (in prep).

References

de Blok, W. J. G., Walter, F., Smith, J.-D. T., *et al.* 2016, *AJ*, 152, 51
De Breuck, C., Williams, R. J., Swinbank, M., *et al.* 2014, *A&A*, 565, A59
Faisst, A., Béthermin, M., Capak, P., *et al.* 2019, arXiv e-prints, p. arXiv:1901.01268
Förster Schreiber, N. M., Genzel, R., Bouché, N., *et al.* 2009, *ApJ*, 706, 1364
Jones, G. C., Carilli, C. L., Shao, Y., *et al.* 2017, *ApJ*, 850, 180
Pease, F. G. 1918, Proceedings of the National Academy of Science, 4, 21
Smit, R., Bouwens, R. J., Carniani, S., *et al.* 2018, *Nature*, 553, 178
Wuyts, S., Förster Schreiber, N. M., Wisnioski, E., *et al.* 2016, *ApJ*, 831, 149

Uncovering Early Galaxy Evolution in the ALMA and JWST Era
Proceedings IAU Symposium No. 352, 2019
E. da Cunha, J. Hodge, J. Afonso, L. Pentericci & D. Sobral, eds.

doi:10.1017/S1743921319009104

What drives the [CII]/FIR deficit in submillimeter galaxies?

Matus Rybak, J. A. Hodge, G. Calistro Rivera and ALESS Collaboration

Leiden Observatory, Leiden University, Niels Bohrweg 2, 2333 CA Leiden, the Netherlands
email: mrybak@strw.leidenuniv.nl

Abstract. Submillimeter galaxies at redshift $z \geqslant 1$ show a pronounced [CII]/FIR deficit down to sub-kpc scales; however, the physical origin of this deficit remains poorly understood. We use resolved ALMA observations of the [CII], FIR and CO(3–2) emission in two $z = 3$ SMGs to distinguish between the different proposed scenarios; the thermal saturation of the [CII] emission is the most likely explanation.

Keywords. galaxies: high-redshift, galaxies: ISM, submillimeter

1. Introduction

Intrinsically bright and easy to excite, the [CII] 158-μm line has become a key probe of gas in submillimeter galaxies (SMGs) and a potentially powerful tracer of their star-formation and gas content. But the interpretation of the [CII] emission in SMGs complicated by the so-called [CII]/FIR deficit: the [CII]/FIR luminosity ratio decreases at high star-formation rate surface densities ($\Sigma_{\rm SFR}$). Indeed, recent resolved ALMA observations of $z \simeq 2-5$ SMGs have revealed a pronounced [CII]/FIR deficit ($L_{\rm [CII]}/L_{\rm FIR} = 10^{-4} - 10^{-3}$) down to sub-kpc scales (Fig. 1a).

Although several potential mechanisms for the [CII]/FIR deficit have been proposed (see below), distinguishing between them requires knowing the FUV field strength (G) and gas density ($n_{\rm H}$). While source-averaged G, $n_{\rm H}$ in SMGs have been inferred using e.g. unresolved [CII], FIR and CO observations, these can be biased as different tracers are not generally co-spatial.

2. Results and implications

We observed two z=3 SMGs – ALESS 49.1 and ALESS 57.1. – with ALMA in the [CII] and FIR continuum (Band 8, 0.15" resolution; Rybak *et al.* 2019) and CO(3–2) (Band 3, 0.6" resolution; Calistro-Rivera *et al.* 2018). Both sources show a pronounced [CII]/FIR deficit ($10^{-4} - 10^{-3}$) at 1-kpc scales (Fig. 1a), falling below the Smith *et al.* (2017) empirical trend. Concentrating on the central star-forming regions (R $\leqslant$ 2 kpc), we use the PDRToolbox photon-dominated region (PDR) models (Kaufman *et al.* 2006; Pound & Wolfire 2008) to infer G, $n_{\rm H}$ from the observed [CII]/FIR and [CII]/CO(3–2) ratios. We find $G = 10^4\ G_0$ and $n_{\rm H} = 10^4 - 10^5$ cm^{-3}, significantly higher than the source-averaged values for both $z \sim 0$ ULIRGs and high-z SMGs (Fig. 1b).

We now consider the following mechanisms for the [CII]/FIR deficit:

• **AGNs** can suppress the [CII] emission by further ionizing C^+ via soft X-rays, while boosting the FIR luminosity. However, the AGN X-ray luminosities in ALESS 49.1 and 57.1 correspond to a sphere of influence on the order of 100 pc, insufficient to explain the observed [CII]/FIR deficit over scales of few kpc.

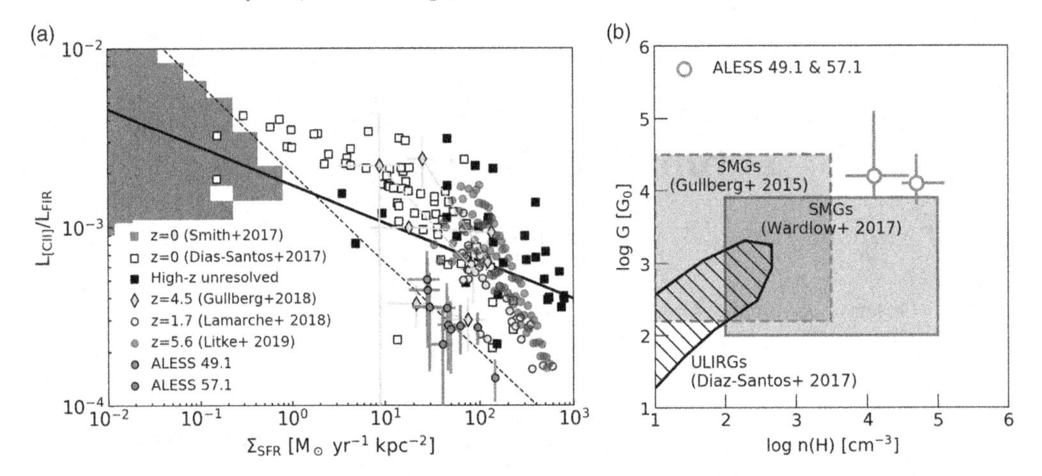

Figure 1. a): [CII]/FIR deficit in ALESS 49.1 and 57.1, compared to other $z \sim 0$ and high-redshift measurements, the empirical trend of Smith *et al.* (2017; solid line) and the thermal saturation prediction (Muñoz & Oh 2016; dashed line). **b)**: G and $n_{\rm H}$ in ALESS 49.1 and 57.1, compared to $z \sim 0$ ULIRGs and unresolved SMGs studies.

• **Positive grain charging** will reduce the photoelectric gas heating. However, although the inferred G, $n_{\rm H}$ in ALESS 49.1 and 57.1 imply substantial grain charging, the photoelectric heating is not significantly reduced.

• **Dust-bounded HII regions** where UV photons are absorbed by the dust instead of heating the gas will result in increased $L_{\rm FIR}$ and decreased $L_{\rm [CII]}$. However, the radiation pressure in ALESS 49.1 and 57.1 will expel the dust out of HII regions in $\sim 10^5$ yr, making them too short-lived to drive the [CII]/FIR deficit.

• **Thermal saturation**. At $T_{\rm gas} \gg 91$ K, the C^+ fine-structure upper-level occupancy (and $L_{\rm [CII]}$) depends only weakly on temperature (Muñoz & Oh 2016). Our PDR models imply cloud surface temperatures of $400-700$ K, indicating the [CII] emission is thermally saturated. Moreover, fitting a power-law to our data from Fig. 1a yields a best-fitting slope $\gamma = -0.5 \pm 0.1$), in agreement with the thermal saturation model ($\gamma = -0.5$).

These results imply that the pronounced [CII]/FIR deficit in SMGs is driven by the C^+ temperature saturation due to the strong FUV fields. Although limited by the sample size, this study highlights the need for resolved studies of physical conditions in SMGs, and presents a necessary stepping stone to future resolved [CII]/FIR/CO studies for representative samples of SMGs.

References

Calistro-Rivera, G. *et al.* 2019, *ApJ*, 863, 56
Díaz-Santos, T. *et al.* 2017, *ApJ*, 846, 32
Gullberg, B. *et al.* 2015, *MNRAS*, 449, 2883
Gullberg, B. *et al.* 2018, *ApJ*, 859, 12
Kaufman, M. J. *et al.* 2008, *ApJ*, 644, 283
Lamarche, C. *et al.* 2018, *ApJ*, 867, 140
Litke, K. C. *et al.* 2019, *ApJ*, 870, 80
Muñoz, J. A. & Oh, S. P. 2016, *MNRAS*, 463, 2085
Pound, M. W. & Wolfire, M. G. 2008, *ASPCS*, 394, 654
Rybak, M. *et al.* 2019, *ApJ*, 876, 112
Smith, J. D. *et al.* 2017, *ApJ*, 834, 5
Wardlow, J. *et al.* 2017, *ApJ*, 837, 12

Uncovering Early Galaxy Evolution in the ALMA and JWST Era
Proceedings IAU Symposium No. 352, 2019
E. da Cunha, J. Hodge, J. Afonso, L. Pentericci & D. Sobral, eds.

doi:10.1017/S1743921319009694

Luminous and dark matter density profiles in the inner regions of a group-scale lens at z = 0.6

Mônica Tergolina, Cristina Furlanetto and Marina Trevisan

Universidade Federal do Rio Grande do Sul, Instituto de Física
Av. Bento Gonçalves, 9500, 91501-970, Porto Alegre, Brazil
email: monica.tergolina@ufrgs.br

Abstract. Density profiles of galaxy groups can provide an insight on how large-scale structure in the Universe formed and evolved, since galaxy groups bridge the gap between individual galaxies and galaxy clusters. Studying the galaxy group that is gravitational lensing HELMS18, a submillimeter galaxy at $z = 2.39$ from the Herschel's HerMES Large Mode Survey (HELMS), we aim to probe the total density profile by combining strong gravitational lensing with kinematics of the centrally-located galaxies and kinematics of the group members. We have high-resolution data of HELMS18 obtained with the Atacama Large Millimeter/submillimeter Array (ALMA) and multi-object spectroscopic data of the group members from Gemini-GMOS. Our main goal is to match these observations to probe the DM and stellar density profiles and to establish a complete description of this galaxy group.

Keywords. gravitational lensing: strong, galaxies: groups, galaxies: general, galaxies: high-redshift

1. Introduction

Probing the mass distribution of structures in the Universe, from individual galaxies to galaxy clusters, is still an open question in modern Cosmology. Structure formation and evolution models can be constrained by studies of the internal mass distribution (Bartelmann *et al.* 2013). Of particular interest is the radial density profile of dark matter (DM), which is sensitive to baryonic processes, such as AGN and stellar feedback. As the balance of these processes vary with mass and formation history, it is essential to investigate the baryonic and DM distributions across the full mass range, from individual galaxies to clusters. Strong gravitational lensing provides a unique way to trace the mass distribution for this wide range of systems. This technique provides a measurement of the total mass inside their respective Einstein radius. Combining strong lensing (SL) with other observations enable the density profile to be measured at different radii. While baryons and DM seem to follow an isothermal density profile in early-type galaxies (ETGs), other studies found that the inner DM density profile in galaxy clusters is shallower than the Navarro-Frank-White (NFW) profile, suggesting a connection between the assembly of stars in the central galaxies and the inner halo. (Sand *et al.* 2008; Newman *et al.* 2013).

So far, very few studies focused on the measurement of the density profile in the intermediate mass range of the galaxy groups, using SL and other techniques. Thus, the main goal of this work is to probe the stellar and DM density profiles of the galaxy group that is lensing HELMS18, a dusty star-forming galaxy (DSFG) at $z = 2.39$ (Nayyeri *et al.* 2016), by combining SL with stellar kinematics of the central galaxies and kinematics of

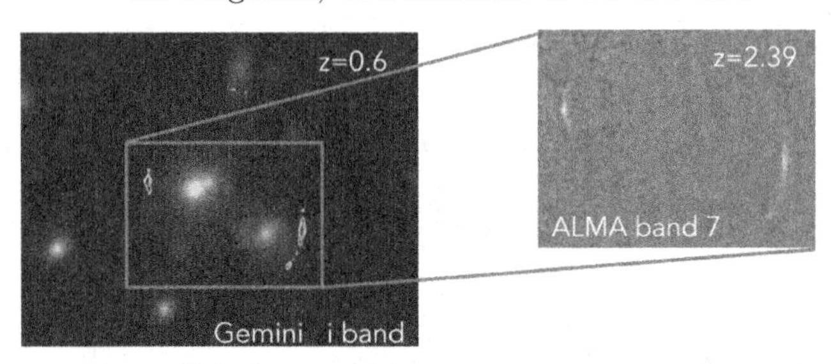

Figure 1. Superposition of Gemini and ALMA data (white contours) of the inner region of the galaxy group (*a*). ALMA band 7 data of HELMS18 (*b*).

the group members. These complementary techniques will allow us to examine the mass distribution of the group at several widely separated radii.

2. Data and Methods

We are using ALMA band 7 data, which show two gravitational arcs (Figure 1b). According to Amvrosiadis *et al.* (2018), HELMS18 system has an Einstein radius of $\sim 6.5''$. Optical images from Sloan Digital Sky Survey (SDSS) show that the lensing group has two central galaxies, an ETG and a quasar candidate (Figure 1a). Spectroscopic data from SDSS (for the ETG) indicate that the group is at $z \sim 0.6$.

We modelled the ALMA data using the semi-linear inversion method outlined in Dye *et al.* (2018), which works directly in the interferometric uv-plane on the visibility data. With the modelling we aim to: *i*) measure properties of the lens; *ii*) measure the slope of the density profile; *iii*) reconstruct the image and investigate the properties of the background source. Preliminary results on the source reconstruction indicate a main component, which is very elongated, and a second component, located ∼8 Kpc away from the main one.

As part of this project we conducted Gemini Multi-Object Spectroscopy (GMOS) observations to target both the two brightest central galaxies and the group members. Currently we are working on the data reduction. Gemini data will allow us to determine the group members, investigate stellar kinematics of the central galaxies and estimate the luminous density profile and total stellar mass. Combining SL with stellar kinematics of the centrally-located galaxies and kinematics of the group members, will allow us to measure the mass distribution of the galaxy group at different radii.

References

Amvrosiadis, A., Eales, S. A., Negrello, M., Marchetti, L., Smith, M. W. L., Bourne, N., Clements, D. L., De Zotti, G., *et al.* 2018, *MNRAS*, 475, 4939

Bartelmann, M., Limousin, M., Meneghetti, M., & Schmidt, R. 2013, *Space Science Reviews*, 177, 3

Dye, S., Furlanetto, C., Dunne, L., Eales, S. A., Negrello, M., Nayyeri, H., van der Werf, P. P., Serjeant, S., *et al.* 2018, *MNRAS*, 476, 4383

Nayyeri, H., Keele, M., Cooray, A., Riechers, D. A., Ivison, R. J., Harris, A. I., Frayer, D. T., Baker, A. J., *et al.* 2016, *ApJ* 823, 17

Newman, J. A., Cooper, M. C., Davis, M., Faber, S. M., Coil, A. L., Guhathakurta, P., Koo, D. C., Phillips, A. C., *et al.* 2013, *ApJS*, 208, 5

Sand, D. J., Treu, T., Ellis, R. S., Smith, G. P., & Kneib, J.-P. 2008, *ApJ*, 674, 711

Uncovering Early Galaxy Evolution in the ALMA and JWST Era
Proceedings IAU Symposium No. 352, 2019
E. da Cunha, J. Hodge, J. Afonso, L. Pentericci & D. Sobral, eds.

doi:10.1017/S1743921319008871

Molecular gas and dust emission in a $z = 3.63$ strongly lensed starburst merger at sub-kiloparsec scales

C. Yang[1], R. Gavazzi[2], A. Beelen[3], P. Cox[2], A. Omont[2] and M. D. Lehnert[2]

[1]European Southern Observatory, Alonso de Córdova 3107, Vitacura, Santiago, Chile
email: cyang@eso.org

[2]Institut d'Astrophysique de Paris, Sorbonne Université, CNRS (UMR 7095), 98 bis bd Arago, 75014 Paris, France

[3]Institut d'Astrophysique Spatiale, CNRS (UMR 8617), Université Paris-Sud, Université Paris-Saclay, Orsay, France

Abstract. We present $0\farcs2$–$0\farcs4$ resolution ALMA images of the submillimeter dust continuum and the CO, H_2O, and H_2O^+ line emission in a $z = 3.63$ strongly lensed dusty starburst. We construct the lens model for the system with an MCMC technique. While the average magnification for the dust continuum is about 11, the magnification of the line emission varies from 5 to 22 across the source, resolving the source down to sub-kpc scales. The ISM content reveals that it is a pre-coalescence major merger of two ultra-luminous infrared galaxies, both with a large amount of molecular gas reservoir. The approaching galaxy in the south shows no apparent kinematic structure with a half-light radius of 0.4 kpc, while the preceding one resembles a 1.2 kpc rotating disk, separated by a projected distance of 1.3 kpc. The distribution of dust and gas emission suggests a large amount of cold ISM concentrated in the interacting region.

Keywords. galaxies, high-redshift, ISM, gravitational lensing, submillimeter, molecules

1. Introduction

The strongest starbursts across cosmic time are the heavily dust-obscured submillimeter galaxies (SMGs, see e.g. Casey *et al.* 2014). With total infrared (IR) luminosities above $10^{12}\,L_\odot$, SMGs reach the limit of "maximum starbursts" (Barger *et al.* 2014) with star formation rates exceeding $1000\,M_\odot\,\mathrm{yr}^{-1}$. Such intense star formation is thought to be triggered by galaxy mergers or at least enhanced by interactions, which is consistent with ΛCDM simulations where merger rates are expected to increase with increasing redshift (e.g. Rodriguez-Gomez *et al.* 2015), although there are also pieces of evidence that some of the SMGs are rather smooth isolated clumpy disks with high gas fractions (e.g. Hodge *et al.* 2019). The nature of the SMGs remains hotly debated, in part because of the details of the structure of the molecular gas and dust content (fuel of star formation). Therefore, to better understanding the nature of SMG, we conduct a detailed study of CO and H_2O (which is a powerful diagnostic tool of the far-IR radiation fields, Yang *et al.* 2016) and dust content in a strongly lensed SMG selected from the *Herschel*-ATLAS survey (Bussmann *et al.* 2013), G09v1.97 (Yang *et al.* 2019), down to the sub-kpc scales, with the help of gain sensitivity and angular resolution from lensing magnification.

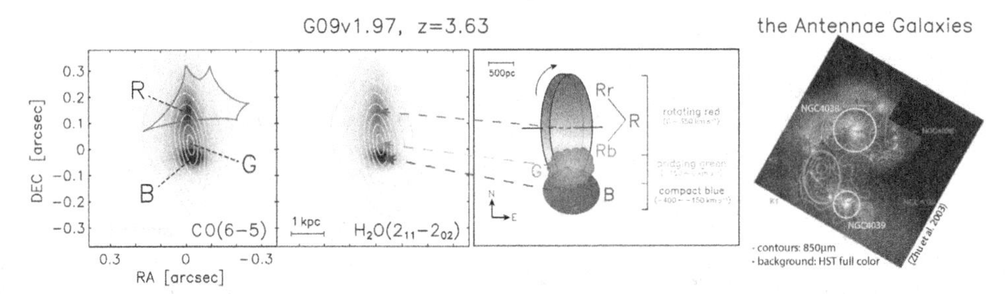

Figure 1. *The reconstructed image in the source plane, from left to right*: the moment-0 map of CO(6–5), with yellow contours of the 188 μm dust emission in the rest-frame; the moment-0 map of $H_2O(2_{11}–2_{02})$, with similar contours; the sketch of the merging scenario of G09v1.97; the optical image with white contours showing the 850 μm SCUBA image of the Antenna Galaxies (Zhu *et al.* 2003).

2. Results

We have conducted ALMA observations of the CO(6–5), $H_2O(2_{11}–2_{02})$, $H_2O^+(2_{02}–1_{11})$, $H_2O^+(2_{11}–2_{02})$ and the dust continuum emission, resulted in high signal-to-noise data-cubes with synthesis beam-sizes around 0.″2–0.″4. The dust continuum and molecular gas emission are resolved into a nearly complete ∼1.″5 diameter Einstein ring plus a weaker image in the centre, which is caused by a special dual deflector lensing configuration. The velocity structures of the three lines in the image plane are strikingly similar.

Line profile. The (image plane) spatially integrated, continuum-subtracted spectra of CO(6–5), $H_2O(2_{11}–2_{02})$, and the $J_{\rm up}=2$ H_2O^+ have the same profile, which is composite of three Gaussian components, a distinct blue (B) and two neighbouring red ones (Rr and Rb). This similarity suggests these three tracers are co-spatially located, tracing the warm dense gas that directly relates to the intense star formation.

Lens modelling. In order to derive the intrinsic properties of G09v1.97, a lens model needs to be built. Using the 0.″2 ALMA image of the dust continuum of the background source and KECK images of the two foreground deflectors, we build a double-SIE lens model with an MCMC approach. Then the lens model is applied to the line emissions per channel to reconstruct the emission in the source plane.

The nature of the SMG in the source plane. The reconstructed source-plane image of the CO(6–5) and $H_2O(2_{11}–2_{02})$ line are similarly showing a bimodality – the northern component is mainly associated with the two red-shifted Gaussian components seen in the spectrum, while the southern component is dominantly from the blue-shifted spectral components. As shown in Fig 1, the approaching galaxy B shows no apparent kinematic structure with a semi-major half-light radius of $a_{\rm s}=0.4$ kpc, while the receding galaxy R resembles an $a_{\rm s}=1.2$ kpc rotating disk. The dust emission is best modelled with two components, a compact cold dust (17 K) dust component peaking in between R and B, and an extended warmer dust (33–61 K) component dominated by the R and B galaxies. The two galaxies are separated by a projected distance of 1.3 kpc, bridged by weak line emission that is co-spatially located with the cold dust emission peak in the centre, suggesting a large amount of cold interstellar medium (ISM) in the interacting region, much like the concentration of cold dust emission found in the Antenna Galaxies.

References

Barger, A. J., Cowie, L. L., Chen, C.-C., *et al.* 2014, *ApJ*, 784, 9
Bussmann, R. S., Pérez-Fournon, I., Amber, S., *et al.* 2013, *ApJ*, 779, 25

Casey, C. M., Narayanan, D., & Cooray, A. 2014, *Phys. Rep.*, 541, 45
Hodge, J. A., Smail, I., Walter, F., *et al.* 2019, *ApJ*, 876, 130
Rodriguez-Gomez, V., Genel, S., Vogelsberger, M., *et al.* 2015, *MNRAS*, 449, 49
Tacconi, L. J., Genzel, R., Neri, R., *et al.* 2010, *Nature*, 463, 781
Yang, C., Omont, A., Beelen, A., *et al.* 2016, *A&A*, 595, A80
Yang, C., Gavazzi, R., Beelen, A., *et al.* 2019, *A&A*, 624, A138
Zhu, M., Seaquist, E. R., & Kuno, N. 2003, *ApJ*, 588, 243

SESSION 7: Lessons from local galaxies and high-z analogues

Uncovering Early Galaxy Evolution in the ALMA and JWST Era
Proceedings IAU Symposium No. 352, 2019
E. da Cunha, J. Hodge, J. Afonso, L. Pentericci & D. Sobral, eds.

doi:10.1017/S174392132000126X

Analogs of high redshift galaxies: Disentangling the complexity of the green peas

Ricardo Amorin

University of La Serena, Chile

Abstract. Young low-mass galaxies with extreme emission-line properties are ubiquitous at high redshift. However, a detailed characterisation of their physical properties, key for understanding cosmic reionisation and the early growth of galaxies, will be only possible with JWST and ELT observations. Rare lower-z analogues of these primeval galaxies provide us ideal laboratories to study in larger detail the complex physical mechanisms taking place in these extreme systems. In this talk, I will review key results from these high-z analogues, with an emphasis on lessons learned from deep spectroscopic observations of green pea galaxies at $z \lesssim 0.3$. New recent results based on high-dispersion Echelle and IFU spectroscopy of green peas will be presented. They illustrate current advantages and limitations of the chemodynamical analysis for a simultaneous study of the ionised gas kinematics, chemical enrichment and the escape of ionising photons in compact low-mass starbursts.

Uncovering Early Galaxy Evolution in the ALMA and JWST Era
Proceedings IAU Symposium No. 352, 2019
E. da Cunha, J. Hodge, J. Afonso, L. Pentericci & D. Sobral, eds.

doi:10.1017/S1743921319008962

Neutral gas and the escape of ionizing radiation: Lessons from the low-redshift Green Peas

Anne Jaskot[1,2], Jed McKinney[2], Tara Dowd[3], Sally Oey[4], Min Yun[2], Claudia Scarlata[5] and James Lowenthal[6]

[1]Astronomy Department, Williams College, Williamstown, MA 01267, USA
email: 08aej@williams.edu

[2]Department of Astronomy, University of Massachusetts, Amherst, MA 01003, USA

[3]The Chandra X-ray Center, Cambridge, MA 02138, USA

[4]Department of Astronomy, University of Michigan, Ann Arbor, MI 48109, USA

[5]Minnesota Institute for Astrophysics, University of Minnesota, Minneapolis, MN 55455, USA

[6]Department of Astronomy, Smith College, Northampton, MA 01063, USA

Abstract. How galaxies reionized the universe remains an open question, but we can gain insights from the low-redshift Green Pea galaxies, one of the only known populations of Lyman continuum (LyC) emitters. Using VLA H I 21 cm observations and *HST* UV spectra of Green Peas, we investigate how neutral gas content and geometry influence LyC and Lyα escape. Our results suggest that LyC Emitters may have high ratios of star formation rate to H I mass. Low gas covering fractions are common among the population, but not all sightlines are optically thin. Based on the observed relationship between high ionization parameters, low metallicities, and narrow Lyα profiles, we propose that weak stellar feedback at low metallicities results in a gas geometry of dense clumps within a low-density medium, which facilitates Lyα and LyC escape. We address the implications of these results for identifying LyC emitters at high redshift with *JWST* and ALMA.

Keywords. Galaxies: evolution, Galaxies: starburst, Intergalactic medium, Galaxies: ISM, Radiative transfer, Stars: massive

---

1. Introduction

One of the main goals of *JWST* is to identify the galaxies that reionized the universe. We currently do not know which galaxies dominate reionization or how much ionizing Lyman continuum (LyC) radiation escapes from galaxies into the intergalactic medium (IGM; e.g., Robertson *et al.* 2015; Madau & Haardt 2015; Finkelstein *et al.* 2019; Naidu *et al.* 2019). Because the high-redshift IGM absorbs LyC radiation, we cannot directly study LyC escape during the epoch of reionization. However, we can gain important insights by studying galaxies in the low-redshift universe and investigating the physical factors that affect the escape of ionizing radiation.

The "Green Pea" (GP) galaxies are one of the most valuable low-redshift samples for probing the mechanisms behind LyC escape. Discovered in the Sloan Digital Sky Survey (SDSS) at $z < 0.4$ Cardamone *et al.* 2009, the GPs share a number of properties with $z > 2$ galaxies, such as high [O III] λ5007 equivalent widths (200-2000 Å), low stellar masses ($M_* \sim 10^7 - 10^{10}$M$_\odot$), high specific star formation rates (sSFRs=$10^{-9} - 10^{-7}$ M$_\odot$ yr^{-1}), and low metallicities (~0.2 Z$_\odot$; e.g., Cardamone *et al.* 2009; Amorín *et al.* 2010; Izotov *et al.* 2011; 2017b; Nakajima & Ouchi 2014; Smit *et al.* 2014; Hagen *et al.* 2016). Like high-redshift galaxies, the GPs show elevated ratios of [O III] λ5007/[O II] λ3727, which indicates a highly ionized interstellar medium and which may be a diagnostic of LyC escape (e.g., Jaskot & Oey 2013; Nakajima & Ouchi 2014).

Indeed, the GPs are the only known population of low-redshift star-forming galaxies where LyC escape appears to be common. All eleven GPs probed so far show LyC escape, with escape fractions ranging from $2-72\%$ (Izotov *et al.* 2016; Izotov *et al.* 2018). Likewise, LyC emission has been detected from $z \sim 3$ galaxies with strong [O III] emission (e.g., Vanzella *et al.* 2016; Fletcher *et al.* 2019), although the average LyC escape fraction of this population may be $< 15\%$ (Rutkowski *et al.* 2017; Naidu *et al.* 2018).

In this work, we present recent analyses of the GPs' neutral gas properties using Very Large Array (VLA) 21 cm observations and *Hubble Space Telescope* (*HST*) UV spectra (Jaskot *et al.* 2017; McKinney *et al.* 2019; Jaskot *et al.* 2019). We selected a sample of 17 GPs with [O III]/[O II] ratios $\geqslant 7$ from SDSS. These galaxies are the most highly ionized star-forming galaxies at low redshift, and one galaxy in the sample, J160810+352809 (J1608), has the highest [O III]/[O II] ratio among SDSS star-forming galaxies ([O III]/[O II]= 34.9). We have obtained VLA 21 cm spectra of J1608, and all 17 GPs in the sample have far-UV spectra from the *HST* Cosmic Origins Spectrograph.

2. H I 21 cm Results: High SFR/$N_{\rm HI}$ in LyC Emitters?

The total H I content of a galaxy represents the amount of raw material available to form molecular gas and stars. Observations show that H I to stellar mass ratios increase with specific star formation rate and decrease with stellar mass (e.g., Huang *et al.* 2012b). Based on empirical scaling relations between H I masses, galaxy structure, and galaxy color (e.g., Huang *et al.* 2012b), we would expect low-mass starbursts like the GPs to have high H I gas fractions, as abundant H I fuels their intense star formation.

Instead, we find a 3σ upper limit on J1608's H I mass of $\log(M_{\rm HI}/{\rm M}_\odot)= 8.14$ (McKinney *et al.* 2019). This H I mass is not unusual for a galaxy of J1608's stellar mass ($\log(M_*/{\rm M}_\odot) = 7.04$ (Izotov *et al.* 2017b). However, it is unusual given J1608's extremely high sSFR (6.85×10^{-7} M$_\odot$ yr^{-1}) and falls below the predictions of all H I scaling relations for low-mass galaxies (McKinney *et al.* 2019). In other words, J1608 has a surprisingly high ratio of SFR to H I mass compared with other low-redshift galaxies (Figure 1a). Interestingly, two other galaxies show similarly high SFR/$M_{\rm HI}$ ratios: Tololo 1247-232 (Puschnig *et al.* 2017) and Haro 11 (Pardy *et al.* 2016), both of which are known LyC emitters. Like other GPs, J1608 has a young burst age (< 3 Myr; Izotov *et al.* 2017b; Jaskot *et al.* 2017) and likely a high ionizing photon production rate (Schaerer *et al.* 2016; Izotov *et al.* 2017a). Consequently, the standard conversion from Hα luminosity to SFR may be unreliable for J1608, and Figure 1a may instead show that LyC-emitting galaxies have higher Hα luminosities and higher ionizing fluxes per unit H I mass.

3. UV Spectra Results: Gas Geometry and Lyα and LyC Escape

The Lyα emission line and UV metal absorption lines can give us alternative perspectives on the neutral gas in GPs. Both models and observations suggest that the Lyα spectral profile width is sensitive to the H I column density, where higher column densities lead to more Lyα scattering and broader Lyα emission profiles (e.g., Verhamme *et al.* 2015; Izotov *et al.* 2018). Most, but not all, of the 17 highly ionized GPs in our sample are Lyα emitters (Jaskot *et al.* 2017; Jaskot *et al.* 2019). Two of the GPs likely do not have LyC escape along our line of sight, as they show deep Lyα absorption indicative of high H I column densities. However, most of the sample has strong, narrow Lyα emission profiles, as narrow as confirmed LyC emitters and suggestive of low H I column densities.

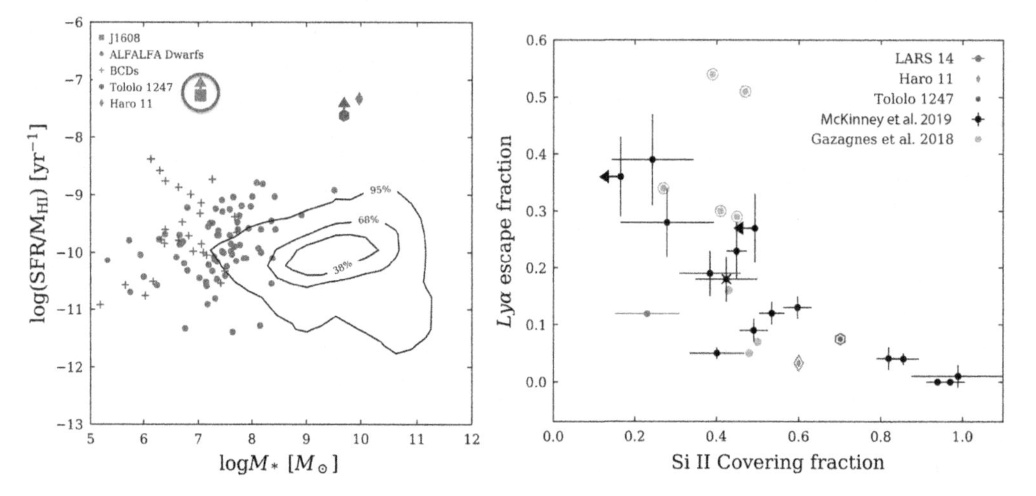

Figure 1. (a) *Left:* SFR/$M_{\rm HI}$ vs. stellar mass. The red square inside the purple circle shows the limit for J1608. The blue hexagon and red diamond represent two confirmed LyC emitters (Puschnig *et al.* 2017; Pardy *et al.* 2016). Contours show SDSS galaxies with H I masses (Huang *et al.* 2012b). Green circles and blue crosses show dwarf galaxies (Huang *et al.* 2012a; Thuan *et al.* 2016). (b) *Right:* Lyα escape fraction correlates strongly with the low-ionization gas covering fraction for our GP sample (black points) and other GPs and LyC emitters (colored points). Figures adapted from McKinney *et al.* (2019).

Some of these Lyα-emitting GPs, including J1608, show broad, underlying Lyα absorption in addition to their narrow Lyα emission. The derived H I column densities from the Lyα absorption components in the Lyα-emitting and pure Lyα-absorbing GPs range from $\log(N_{\rm HI}/{\rm cm}^{-2}) = 19.5$ to 21.5, several orders of magnitude too high for LyC escape and seemingly in contradiction to the narrow Lyα emission profiles (McKinney *et al.* 2019). These observations suggest that the neutral gas in the GPs is inhomogeneous, with optically thin sight lines coexisting with higher column density gas. From the GPs' UV metal absorption lines, we find further evidence for an inhomogeneous gas geometry. Their saturated line ratios and non-zero residual fluxes are consistent with partial gas covering fractions in the Lyα-emitting GPs (McKinney *et al.* 2019).

We find that gas covering fraction is one of the physical properties that correlates most closely with Lyα escape fraction (Figure 1b; McKinney *et al.* 2019; Jaskot *et al.* 2019). This correlation suggests that the interstellar medium (ISM) porosity regulates Lyα and LyC escape, with gaps within denser gas providing the avenues for Lyα and LyC photons to travel. However, these gaps cannot be true holes devoid of gas. If they were, Lyα photons would escape freely, without scattering, and would not form the double-peaked profiles we observe (Verhamme *et al.* 2015).

In addition to porosity, the ionization of the ISM may also affect Lyα and LyC escape. We examine the physical properties that correlate with Lyα profile width, specifically the separation of the blue and red Lyα emission peaks, $\Delta v_{\rm Ly\alpha}$. A more highly ionized ISM, as traced by line ratios such as high [O III]/[O II] and low [O I] λ6300/Hβ, is correlated with narrower Lyα profiles (Figure 2a; Jaskot *et al.* 2019). A more highly ionized ISM could be associated with lower H I column densities and hence narrow Lyα profiles. In this picture, the Lyα and LyC photons would travel through low, but not zero, column density gas channels between denser neutral clouds. Ionization would control the column density of these channels, which would then affect the Lyα profile widths.

Intriguingly, narrow $\Delta v_{\rm Ly\alpha}$ also correlates strongly with low gas metallicities (Figure 2b; Jaskot *et al.* 2019). This trend cannot be explained by dust extinction, the

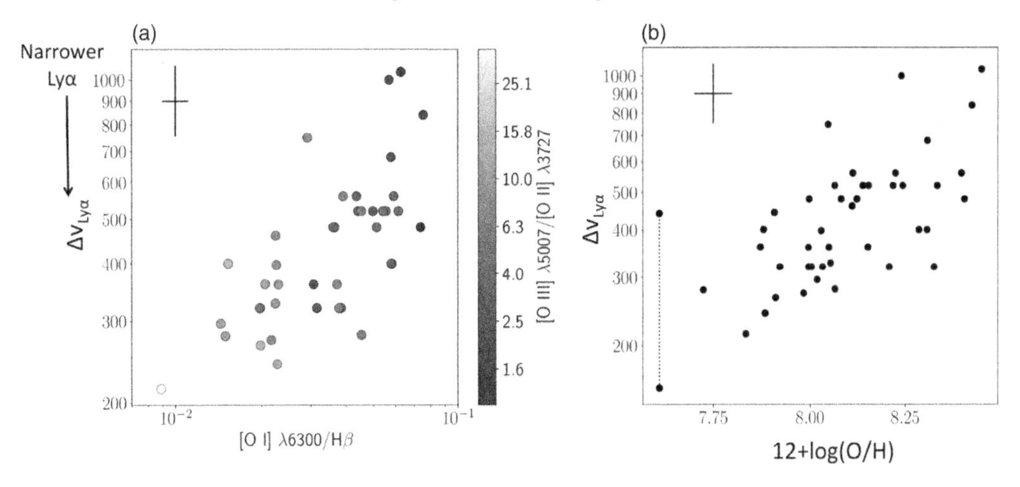

Figure 2. (a) The separation of the red and blue Lyα emission peaks, $\Delta v_{\mathrm{Ly}\alpha}$, in the GPs correlates with measures of ionization, such as [O III] λ5007/[O II] λ3727 (color scale) and [O I] λ6300/Hβ. (b) We observe an equally strong trend between $\Delta v_{\mathrm{Ly}\alpha}$ and metallicity. The dotted line connects two measures of $\Delta v_{\mathrm{Ly}\alpha}$ for a GP with a triple-peaked Lyα profile. Reproduced from Jaskot *et al.* (2019).

mass-metallicity relation, or metallicity-dependent ionizing luminosities, as none of these factors correlates strongly with $\Delta v_{\mathrm{Ly}\alpha}$. Rather, high ionization parameters appear to lie at the root of this trend, since high [O III]/[O II] ratios do correlate with low metallicities.

We propose a physical scenario that may explain the observed correlation between low metallicities, high ionization, and narrow Lyα profiles (Jaskot *et al.* 2019) and which may also explain why GPs so often show signs of LyC escape. At high metallicity, strong stellar winds push neutral gas into a relatively homogeneous superbubble shell, and LyC and Lyα photons encounter few optically thin pathways. Conversely, at low metallicity, mechanical stellar feedback is weaker (e.g., Vink *et al.* 2001; Ramachandran *et al.* 2019), and hence, low-metallicity clusters may retain more of their dense natal gas, particularly at young ages. This dense gas is likely clumpy; dense clumps near the ionizing source exhibit high ionization parameters, and low-density paths between the clumps allow Lyα and LyC photons to escape. This physical picture is consistent with the high-ionization emission, inhomogeneous geometries, young ages, and weak outflows observed in the GPs.

4. Implications

Highly ionized galaxies like the GPs frequently show direct or indirect evidence of LyC escape and may be similar to the galaxies that reionized the universe. At high redshift, *JWST* may identify candidate LyC emitters via high ionization line ratios, such as [O III]/[O II] and [O I]/[O III]. In the case of ALMA, such galaxies will show high [O III] λ88 μm/[C II] λ158 μm, which appears to be a common property of galaxies in the epoch of reionization (Inoue *et al.* 2016; Carniani *et al.* 2017; Hashimoto *et al.* 2019).

LyC emitters may show high ratios of ionizing radiation to H I mass. However, these high ratios could arise from boosted ionizing photon production and do not necessarily mean that LyC-emitting galaxies are deficient in neutral gas. In fact, we find high neutral gas columns in LyC emitter candidates. ALMA could therefore potentially detect molecular gas in LyC emitters, although their low metallicities may make such detections challenging. The significant neutral gas columns in LyC-emitter candidates are consistent with weak mechanical feedback in low-metallicity starbursts at young ages.

We have uncovered a possible link between high nebular ionization, low metallicities, and low H I column densities. Because low H I column densities lead to narrow Lyα spectral profiles, this link may help us constrain the Lyα spectral shape produced by high-redshift galaxies before its subsequent radiative transfer through the IGM. Paradoxically, *weak* feedback at low metallicities may enhance the ISM porosity, generating a low-density inter-clump medium through which ionizing photons may escape. The inhomogeneity of the neutral gas implies that LyC escape will depend on orientation, and large samples will be required to pin down the dependence of LyC escape fraction on galaxy properties. To this end, we are undertaking the Low-Redshift Lyman Continuum survey (PI Jaskot), a Cycle 26 *HST* program, which will measure escape fractions for 66 galaxies and systematically test indicators of LyC escape.

References

Amorín, R. O., Pérez-Montero, E., & Vílchez, J. M. 2010, *ApJ* (Letters), 715, L128
Cardamone, C., Schawinski, K., Sarzi, M., *et al.* 2009, *MNRAS*, 399, 1191
Carniani, S., Maiolino, R., Pallottini, A., *et al.* 2017, *A&A*, 605, A42
Finkelstein, S. L., D'Aloisio, A., Paardekooper, J.-P., *et al.* 2019, *ApJ*, 879, 36
Fletcher, T. J., Tang, M., Robertson, B. E., *et al.* 2019, *ApJ*, 878, 87
Hagen, A., Zeimann, G. R., Behrens, C., *et al.* 2016, *ApJ*, 817, 79
Hashimoto, T., Inoue, A. K., Mawatari, K., *et al.* 2019, *PASJ*, 70
Huang, S., Haynes, M. P., Giovanelli, R., *et al.* 2012, *AJ*, 143, 133
Huang, S., Haynes, M. P., Giovanelli, R., *et al.* 2012, *ApJ*, 756, 113
Inoue, A. K., Tamura, Y., Matsuo, H., *et al.* 2016, *Science*, 352, 1559
Izotov, Y. I., Guseva, N. G., & Thuan, T. X. 2011, *ApJ*, 728, 161
Izotov, Y. I., Schaerer, D., Thuan, T. X., *et al.* 2016, *MNRAS*, 461, 3683
Izotov, Y. I., Guseva, N. G., Fricke, K. J., *et al.* 2017, *MNRAS*, 467, 4118
Izotov, Y. I., Thuan, T. X., & Guseva, N. G. 2017, *MNRAS*, 471, 548
Izotov, Y. I., Worseck, G., Schaerer, D., *et al.* 2018, *MNRAS*, 478, 4851
Jaskot, A. E. & Oey, M. S. 2013, *ApJ*, 766, 91
Jaskot, A. E., Oey, M. S., Scarlata, C., *et al.* 2017, *ApJ* (Letters), 851, L9
Jaskot, A. E., Dowd, T., Oey, M. S., *et al.* 2019, *ApJ*, Submitted
Madau, P. & Haardt, F. 2015, *ApJ* (Letters), 813, L8
McKinney, J. H., Jaskot, A. E., Oey, M. S., *et al.* 2019, *ApJ*, 874, 52
Naidu, R. P., Forrest, B., Oesch, P. A., *et al.* 2018, *MNRAS*, 478, 791
Naidu, R. P., Tacchella, S., Mason, C. A., *et al.* 2019, arXiv:1907.13130
Nakajima, K. & Ouchi, M. 2014, *MNRAS*, 442, 900
Pardy, S. A., Cannon, J. M., Östlin, G., *et al.* 2016, *AJ*, 152, 178
Puschnig, J., Hayes, M., Östlin, G., *et al.* 2017, *MNRAS*, 469, 3252
Ramachandran, V., Hamann, W.-R., Oskinova, L. M., *et al.* 2019, *A&A*, 625, A104
Robertson, B. E., Ellis, R. S., Furlanetto, S. R., *et al.* 2015, *ApJ* (Letters), 802, L19
Rutkowski, M. J., Scarlata, C., Henry, A., *et al.* 2017, *ApJ* (Letters), 841, L27
Schaerer, D., Izotov, Y. I., Verhamme, A., *et al.* 2016, *A&A* (Letters), 591, L8
Smit, R., Bouwens, R. J., Labbé, I., *et al.* 2014, *ApJ*, 784, 58
Thuan, T. X., Goehring, K. M., Hibbard, J. E., *et al.* 2016, *MNRAS*, 463, 4268
Vanzella, E., de Barros, S., Vasei, K., *et al.* 2016, *ApJ*, 825, 41
Verhamme, A., Orlitová, I., Schaerer, D., *et al.* 2015, *A&A*, 578, A7
Vink, J. S., de Koter, A., & Lamers, H. J. G. L. M. 2001, *A&A*, 369, 574

Uncovering Early Galaxy Evolution in the ALMA and JWST Era
Proceedings IAU Symposium No. 352, 2019
E. da Cunha, J. Hodge, J. Afonso, L. Pentericci & D. Sobral, eds.

doi:10.1017/S1743921320000800

Local analogs of high-redshift galaxies: Metallicity calibrations at high-redshift

Fuyan Bian[1], Lisa J. Kewley[2], Brent Groves[2] and Michael A. Dopita[2]

[1]European Southern Observatory, Alonso de Córdova 3107, Casilla 19001, Vitacura, Santiago 19, Chile
email: fbian@eso.org

[2]Research School of Astronomy and Astrophysics, Australian National University, Canberra, ACT 2611, Australia

Abstract. We study the metallicity calibrations in high-redshift galaxies using a sample of local analogs of high-redshift galaxies selected from the SDSS survey. Located in the same region on the BPT diagram as star-forming galaxies at $z \sim 2$, these high-redshift analogs share the same ionized ISM conditions as high-redshift galaxies. We establish empirical metallicity calibrations between the direct gas-phase oxygen abundances and varieties of metallicity indicators in our local analogs using direct T_e method. These new metallicity calibrations are the best means to measure the metallicity in high-redshift galaxies. There exist significant offsets between these new high-redshift metallicity calibrations and local calibrations. Such offsets are mainly driven by the evolution of the ionized ISM conditions from high-z to low-z.

Keywords. galaxies: high-redshift, ISM: abundances, galaxies: abundances, galaxies: ISM

1. Introduction

The gas-phase oxygen abundance is not only modulated by the metal enrichment from massive stars, but also by the physical processes of outflow and inflow in galaxies. Studying gas-phase oxygen abundance provides great insight into the key physical processes that control the formation and evolution of galaxies. By studying the evolution of gas-phase chemical abundance depending on stellar mass (mass-metallicity relation), people can put strong constraints on galaxy build-up process and the stellar feedback models that drive galactic-scale outflows (e.g. Lilly *et al.* 2013). Therefore, it is essential to measure the chemical abundance accurately to study the metallicity evolution in galaxies over cosmic time.

Metallicity indicators based on metallicity sensitive nebular emission-line ratios are widely used to estimate the gas-phase oxygen abundance in galaxies and H II regions. These metallicity indicators are calibrated based on either photoionization models (e.g., Kewley & Dopita 2002) or empirical calibrations (Pettini & Pagel 2004). In particular, metallicity measurements in high-redshift galaxies heavily rely on the empirical calibrations, which are usually calibrated to H II regions in nearby star-forming galaxies using the metallicity derived using direct T_e method.

Photoionization models suggested the ionized ISM conditions, including the ionization parameter, the electron density, the nitrogen-to-oxygen ratio (N/O), and the spectral shape of the radiation field, can affect the metallicity sensitive strong emission-line ratios (e.g., Kewley *et al.* 2013). Observational evidence suggests that the ISM conditions change dramatically over cosmic time Steidel *et al.* 2014. For example, the ISM conditions in high-redshift star-forming galaxies have ~0.6 dex higher ionization parameters and an

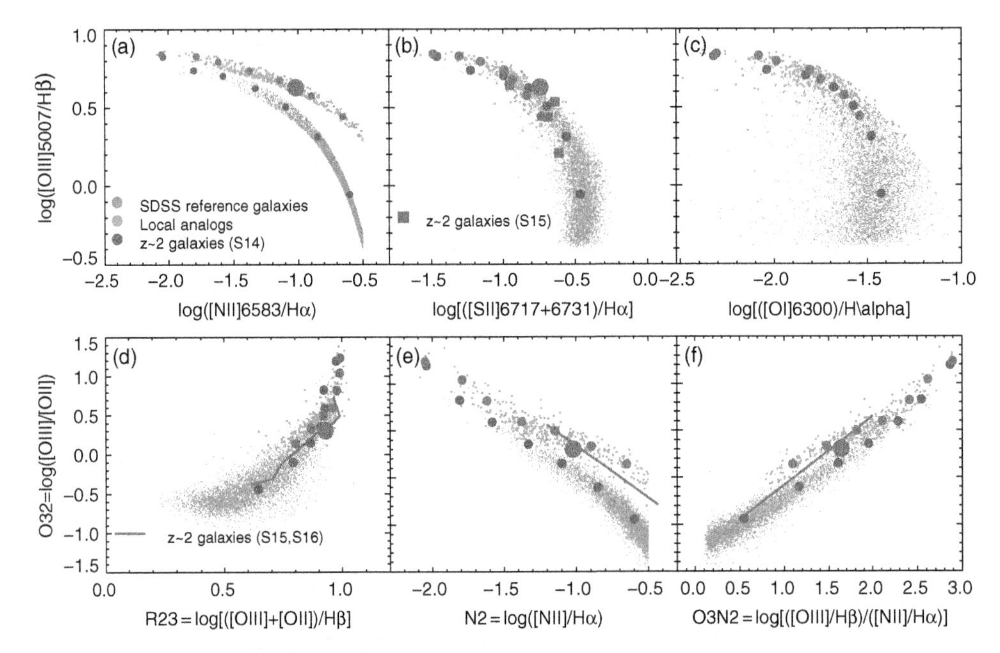

Figure 1. Optical diagnostic diagrams of local analogs of high-redshift galaxies and SDSS reference galaxies. The small blue and red points represent the individual local analogs and SDSS reference galaxies, respectively, and the large open blue and red points represent the stacked spectra of the local analogs and SDSS reference galaxies, respectively. The large purple filled circle represents the stacked spectrum of a sample of $z \sim 2$ UV-selected galaxies adopted from Steidel *et al.* 2014. The figure is adopted from Bian *et al.* 2018.

order of magnitude higher electron densities than their local counterparts (Bian *et al.* 2010).

The strong evolution of the ionized ISM raises the following question: Are the local metallicity calibrations still applicable for high redshift galaxies? We approach this issue by empirically calibrating the metallicity indicators in a sample of local analogs of high-redshift galaxies using the direct T_e method. We compare the empirical calibrations derived from the local analogs with those derived from a sample of local reference galaxies to study how metallicity calibrations change with cosmic time.

2. Sample Selection

We select local analogs of high-redshift galaxies from the SDSS MPA-JHU catalog based on their locations on the BPT diagram. These local analogs are selected in the ± 0.04 dex region of the $z \sim 2.3$ star-forming sequence defined by equation 9 in Steidel *et al.* 2014 on the BPT diagram (small blue data points in Figure 1(a)). Our studies have shown that this type of local analogs closely resembles the properties of the high-redshift galaxies (Bian *et al.* 2016, Bian *et al.* 2017). In particular, these analogs have high ionization parameters ($\log q \simeq 7.9$ cm^{-1}) and high electron densities ($n_e \simeq 120$ cm^{-3}), which are an order of magnitude higher than local star-forming galaxies. These properties are comparable to those in the $z \sim 2-3$ galaxies (e.g., Nakajima *et al.*, Sanders *et al.* 2016). These local analogs also share the same region with $z \sim 2$ star-forming galaxies in all the optical diagnostic diagrams (Figure 1).

To compare the metallicity calibrations between the local analogs and nearby galaxies, we also select local reference galaxies located in the ± 0.05 dex region of the local star-forming sequence on the BPT diagram.

3. Direct T_e Metallicity

To detect the [O III]λ4363 line, we generate composite spectra for both the local analogs and SDSS reference galaxies by stacking the individual spectra in different [N II]λ6584/Hα bins. First of all, we use the Balmer decrement to correct the dust extinction in the reduced 1D galaxy spectra from the SDSS DR9. Then we shift each of the spectra to the rest-frame wavelength based on their redshifts and resample the spectra onto a grid of wavelength from 3700Å to 7300Å with $\Delta\lambda = 1$Å. The mean flux density in the wavelength range of 4400 – 4450Å is used to normalize the spectra. The local analogs and the SDSS redshift galaxies are divided into 0.25 dex bins in [N II]λ6584/Hα from log([N II]λ6584/Hα)= x to $x + 0.25$, where $x = [-2.25, -2.00, -1.75, -1.50, -1.25, -1.00, -0.75]$. At last, we stack the spectra within each of the [N II]λ6584/Hα bins using the mean flux density at each wavelength, and a total of fourteen stacked spectra are generated: seven spectra of local analogs of high-redshift galaxies and seven spectra of SDSS reference sample.

We subtract the stellar continuum from the composite spectra using the STARLIGHT stellar population synthesis models. Then the fluxes of the [O III]λ4363 lines are measured by fitting with a Gaussian function. To get reliable direct T_e oxygen abundance measurements, we only adopt the stacked spectra with signal-to-noise ratios (S/N) of [O III]λ4363 greater than ten (S/N$>$10). After this S/N cut, six composite spectra for local analogs and four composite spectra for SDSS reference galaxies are available for our further analysis.

We use the Izotov *et al.* 2006 recipe to estimate the electron temperature in the O^{++} zone (T_e(O III)) using [O III]$\lambda\lambda$4959,5007/[O III]λ4363 ratio. The following relation between T_e(O III) and T_e(O II) is adopted from Garnett 1992: T_e(O II)=0.7T_e(O III) +3000K. The O^{++} abundance is estimated using the T_e(O III) temperature and [O III]$\lambda\lambda$4959,5007/Hβ ratio, and O^{+} abundance is estimated using the T_e(O II) temperature together with [O II]$\lambda\lambda$3726,3729/Hβ and electron density. The final oxygen abundance is the sum of the O^{++} abundance and the O^{+} abundance.

4. Results

We use the direct T_e metallicity and line ratios measured in the stacked spectra to calibrate varieties of metallicity indicators, including N2, O3N2, R23, O32, log([OIII]5007/ Hβ), and log([NeIII]3869/[OII]3727). Figure 2 shows the relations between the direct T_e oxygen abundance and the above metallicity indicators in the local analogs of high-redshift galaxies (blue squares) and the local normal star-forming galaxies (red squares). There are significant offsets between the local analogs and SDSS reference galaxies.

We use a linear equation to fit the relation between the direct oxygen abundance and the N2, O3N2,O32,and log([NeIII]3869/[OII]3727) indicators in local analogs. The results are as follows:

$$12 + \log(\mathrm{O/H}) = 8.82 + 0.49 \times \mathrm{N2}, \tag{4.1}$$

$$12 + \log(\mathrm{O/H}) = 8.97 - 0.39 \times \mathrm{O3N2}, \tag{4.2}$$

$$12 + \log(\mathrm{O/H}) = 8.54 - 0.59 \times \mathrm{O32}. \tag{4.3}$$

$$12 + \log(\mathrm{O/H}) = 7.80 - 0.63 \times \log([\mathrm{NeIII}]/[\mathrm{OII}]). \tag{4.4}$$

For the R23-Z relation, our data only cover the R23 upper branch. Therefore, a third-order polynomial is used to fit the upper branch of the R23-Z relation:

$$y = 138.0430 - 54.8284x + 7.2954x^2 - 0.32293x^3 \tag{4.5}$$

where $y = R23$ and $x = 12 + \log(\mathrm{O/H})$.

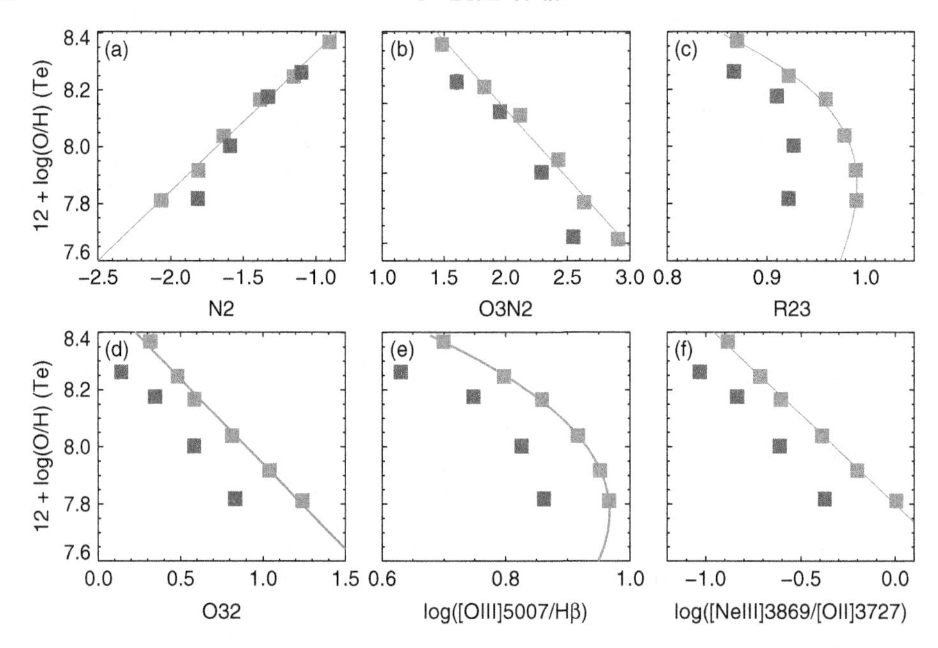

Figure 2. Metallicity calibrations of different metallicity indicators using direct T_e method in the local analogs of high-redshift galaxies (blue data points) and the SDSS reference galaxies (red data points).

For the [O III]λ5007/Hβ-Z relation, we also use a third-order polynomial to fit the data points:

$$y = 43.9836 - 21.6211x + 3.4277x^2 - 0.1747x^3 \tag{4.6}$$

where $y =$[O III]λ5007/Hβ and $x = 12 + \log(\mathrm{O/H})$.

It is worth noting that our new metallicity calibrations for high redshift galaxies are ONLY valid in the metallicity range of 7.8 to 8.4. The blue solid curves in Figure 2 show the best-fitted results for each of the metallicity indicator. Using a sample of $z \sim 2-3$ galaxies with direct oxygen abundance, Sanders *et al.* 2019 suggested that this new set of the metallicity calibrations is the best approach to measure the metallicity in high-redshift galaxies. Especially in the JWST era, these calibrations can be used to measure metallicity in star-forming galaxies at $z > 6$.

5. Discussion

The relations between the oxygen abundance and metallicity indicators for two cases are established based on the MAPPINGs photoionization models: 1. The first model with the ionization parameter of $\log q = 7.5$ and the ISM pressure is $\log(P/k) = 5.2$. This model represents the ionized conditions in nearby galaxies (Bian *et al.* 2016). 2. The second model with the ionization parameter of $\log q = 8.0$ and the ISM pressure of $\log(P/k) = 6.2$. These ISM conditions are consistent with those in $z \sim 2$ galaxies and our local analogs (Kaasinen *et al.* 2017). The relations between metallicity sensitive diagnostic line ratios and the oxygen abundance derived from the photoionization models are shown in Figure 3. Though there exist large offsets on the absolute value of the metallicity estimation between the metallicities derived from empirical calibrations and photoionization models, the metallicity calibrations between the high-redshift and low-redshift conditions derived by the photoionization models (solid lines in Figure 3) follow the same trend as those between the local analogs and the SDSS reference galaxies

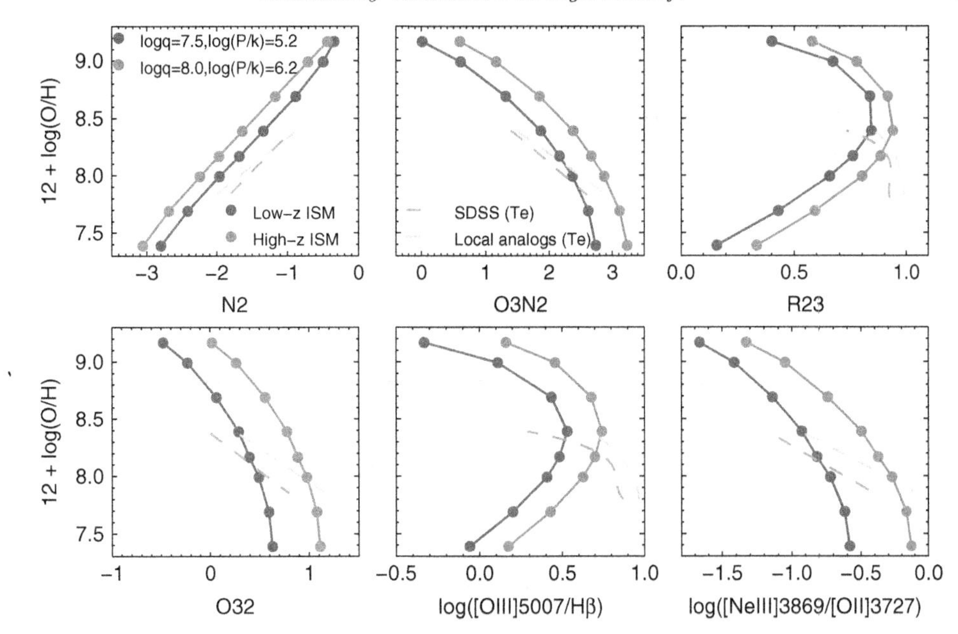

Figure 3. Relations between metallicity indicators and oxygen abundance from the MAPPINGs photoionization models. The red data points represent the photoionization models with ISM pressure $\log(P/k) = 5.2$ (electron density of $n_e \sim 10$) and photoionization parameter of $\log q = 7.5$, which are representative values found in local star-forming galaxies. The blue data points represent the photoionization models with ISM pressure $\log(P/k) = 6.2$ (electron density of $n_e \sim 100$) and photoionization parameter of $\log q = 8.0$, which are comparable to those in $z \sim 2$ star-forming galaxies and our local analogs.The dashed lines represent the metallicity calibrations derived from the direct T_e method in the SDSS reference galaxies (light red dashed line) and the local analogs of high-redshift galaxies (light blue dashed line). The figure is adopted from Bian *et al.* 2018.

(dashed lines in Figure 3). This suggests that the change of the metallicity calibration relations between the high- and low-redshift galaxies are due to the evolution of the ISM conditions.

References

Bian, F., Fan, X., Bechtold, J., *et al.* 2010, *ApJ*, 725, 1877
Bian, F., Kewley, L. J., Dopita, M. A., *et al.* 2018, *ApJ*, 859, 175
Bian, F., Kewley, L. J., Dopita, M. A., *et al.* 2017, *ApJ*, 834, 51
Bian, F., Kewley, L. J., Dopita, M. A., *et al.* 2016, *ApJ*, 822, 62
Garnett, D. R. 1992, *AJ*, 103, 1330
Izotov, Y. I., Stasińska, G., Meynet, G., *et al.* 2006, *A&A*, 448, 955
Kaasinen, M., Bian, F., Groves, B., *et al.* 2017, *MNRAS*, 465, 3220
Kewley, L. J. & Dopita, M. A. 2002, *ApJS*, 142, 35
Kewley, L. J., Maier, C., Yabe, K., *et al.* 2013, *ApJ*, 774, L10
Lilly, S. J., Carollo, C. M., Pipino, A., *et al.* 2013, *ApJ*, 772, 119
Nakajima, K. & Ouchi M. 2014, *MNRAS*, 442, 900
Pettini, M. & Pagel, B. E. J. 2004, *MNRAS*, 348, L59
Sanders, R. L., Shapley, A. E., Kriek M., *et al.* 2016, *ApJ*, 816, 23
Steidel, C. C., Rudie, G. C., Strom, A. L. *et al.* 2014, *ApJ*, 795, 165

Uncovering Early Galaxy Evolution in the ALMA and JWST Era
Proceedings IAU Symposium No. 352, 2019
E. da Cunha, J. Hodge, J. Afonso, L. Pentericci & D. Sobral, eds.

doi:10.1017/S1743921320001271

ALMA observations of local analogs of high-redshift star-forming galaxies

Thiago Gonçalves

Valongo Observatory, UFRJ, Brazil

Abstract. I will present the result of two observational projects using ALMA to investigate the properties of the molecular gas in low-redshift ($z \sim 0.2$) ultraviolet-luminous galaxies. These objects are extremely dense, highly star-forming and very metal-poor compared to other galaxies of similar stellar mass at the same redshifts, justifying their use as analogues to distant main-sequence galaxies in an attempt to understand the interplay between gas and star formation under similar conditions in the early universe. Firstly, we have observed the most metal-poor objects in our sample, in order to determine whether metallicity plays a role in CO emissivity of the molecular regions in these galaxies. Our four non-detections, with stringent upper limits, shows that CO is severely depleted, even under turbulent conditions. We have also observed one object with high spatial resolution, comparing data from CO emission and hydrogen recombination lines down to a resolution of ~ 400 pc, allowing for a detailed analysis of the conversion of gas into new stars. We are able to compare star formation laws in individual clumps and the surrounding ISM, highlighting the difference between star formation efficiencies in each environment within the galaxy. Finally, the high-resolution data offers interesting insights on the growth of supermassive black holes in these galaxies: our combined multiwavelength data shows that there must be a low-mass ($10^5 M_{\odot}$) black hole in the center of the galaxy, while bolometric luminosity in the central region is dominated by star formation activity.

Uncovering Early Galaxy Evolution in the ALMA and JWST Era
Proceedings IAU Symposium No. 352, 2019
E. da Cunha, J. Hodge, J. Afonso, L. Pentericci & D. Sobral, eds.

doi:10.1017/S1743921320001283

Lessons from the local Universe

Daniel Weisz

UC Berkeley, USA

Abstract. Resolved galaxies in the local Universe are fundamentally connected to galaxies observed at all cosmic epochs. The IMF, extinction law, distance ladder, and stellar evolution are all anchored in observations of resolved stars in the nearby Universe. In this talk, I highlight new links between resolved galaxies and those in the higher redshift Universe, and discuss how future observations of resolved stars are essential for a complete and accurate census of galaxy evolution across cosmic time.

Uncovering Early Galaxy Evolution in the ALMA and JWST Era
Proceedings IAU Symposium No. 352, 2019
E. da Cunha, J. Hodge, J. Afonso, L. Pentericci & D. Sobral, eds.

doi:10.1017/S1743921320001295

Local star-forming dwarf galaxies as windows on reionization-era stellar populations

Peter Senchyna

University of Arizona / Steward Observatory, USA

Abstract. The recent detections of high-ionization nebular line emission from species including CIV in a number of $z > 6$ galaxies have highlighted substantial deficiencies in our understanding of metal poor stars. Prominent nebular CIV has never been detected in purely star-forming systems locally, and the massive star models used to model this emission in photoionization codes have not been empirically calibrated below the metallicity of the SMC (20% solar). As a result, we are presently entirely unprepared to correctly interpret nebular emission from metal-poor stars observed with JWST and ALMA in the reionization era. We present results from a multi-pronged ongoing local ultraviolet/optical observation campaign with HST/COS, Keck/ESI, and MMT designed to address this issue by locating and characterizing stellar populations capable of powering such high-ionization emission. This work has already demonstrated that strong nebular CIV can be powered by extremely metal-poor (< 10% solar) massive stars, indicating that we may already have evidence of such low-metallicity populations in the reionization era. However, CIV at the equivalent widths detected at $z > 6$ remains elusive locally, potentially in part due to the relative paucity of known nearby galaxies at these metallicities with massive stellar populations comparable to those in $z > 6$ systems. We present a new technique to locate such nearby galaxies, and results from optical follow-up which indicate that a substantial population of highly star- forming metal-poor galaxies likely resides just below the detection limits of previous large spectroscopic surveys.

Uncovering Early Galaxy Evolution in the ALMA and JWST Era
Proceedings IAU Symposium No. 352, 2019
E. da Cunha, J. Hodge, J. Afonso, L. Pentericci & D. Sobral, eds.

doi:10.1017/S1743921320001301

DYNAMO: An upclose view of turbulent, clumpy galaxies

Deanne Fisher

Swinburne University of Technology, Australia

Abstract. Over 2/3 of all star formation in the Universe occurs in gas-rich, super-high pressure clumpy galaxies in the epoch of redshift $z \sim 1-3$. However, because these galaxies are so distant we are limited in the information available to study the properties of star formation and gas in these systems. I will present results using a sample of extremely rare, nearby galaxies (called DYNAMO) that are very well matched in gas fraction ($f_{\rm gas} \sim 20-80\%$), kinematics (rotating disks with velocity dispersions ranging $20-100$ km/s), structure (exponential disks) and morphology (clumpy star formation) to high-z main-sequence galaxies. We therefore use DYNAMO galaxies as laboratories to study the processes inside galaxies in the dominate mode of star formation in the Universe. In this talk I will report on results from our programs with HST, ALMA, Keck, and NOEMA for DYNAMO galaxies that are aimed at testing models of star formation. We have discovered of an inverse relationship between gas velocity dispersion and molecular gas depletion time. This correlation is directly predicted by theories of feedback-regulated star formation; conversely, predictions of models in which turbulence is driven by gravity only are not consistent with our data. I will also show that feedback-regulated star formation can explain the redshift evolution of galaxy star formation efficiency. I will also present results from a recently acquired map of CO(2-1) in a clumpy galaxy with resolution less than 200 pc. With maps such as these we can begin to study these super giant star forming clumps at scales that are more comparable to local surveys. I will show results for the star formation efficiency of clumps, the boundedness of clumps of molecular gas, and discuss links between star formation efficiency and formation of clumps of stellar mass. The details of clumpy systems are a direct constraint of the results of simulations, especially on the nature of feedback in the high density environments of star formation that dominate the early Universe.

Uncovering Early Galaxy Evolution in the ALMA and JWST Era
Proceedings IAU Symposium No. 352, 2019
E. da Cunha, J. Hodge, J. Afonso, L. Pentericci & D. Sobral, eds.

doi:10.1017/S1743921319008433

Diving deeper into jellyfish: The rich population of jellyfish galaxies in Abell 901/2

Fernanda Roman-Oliveira[1], Ana Chies-Santos[1] and Fabrício Ferrari[2]

[1]Departamento de Astronomia, Universidade Federal do Rio Grande do Sul, Av. Bento Gonçalves 9500, 91501-970, Porto Alegre, RS, Brazil
email: fernanda.oliveira@ufrgs.br

[2]Instituto de Matemática, Estatística e Física, Universidade Federal do Rio Grande, Rio Grande, RS, Brazil

Abstract. Jellyfish galaxies are the most extreme examples of ram pressure stripping (RPS). They represent an important path in the morphological change and quenching in galaxy clusters, however they are still not well characterised morphologically and finding them is a complex task based mainly on visual inspection. We present a study on the properties of a large sample of jellyfish candidates in the multi-cluster system A901/2. We find evidence that the multi-cluster is triggering RPS events in preferential regions in the system and that these galaxies have enhanced specific star formation rates. We also use the software Morfometryka in order to analyse the unique morphometric features in jellyfish galaxies providing a better comprehension of their physical state and future. This can help unravel the physical processes behind such extreme morphologies as well as possibly automatising the search for jellyfish galaxy candidates in large surveys in the next era of instruments.

Keywords. galaxies: evolution, galaxies: structure, galaxies: clusters: general, galaxies: intergalactic medium

---

1. Introduction

Galaxies in dense environments suffer a variety of environmental processes that accelerate their evolution. One of these mechanisms is the ram pressure stripping (RPS) and it plays a major role in the change in morphology and star formation properties of gas rich galaxies. This phenomenon occurs when a galaxy rich in gas falls into a dense environment and the interstellar medium suffers a hydrodynamic friction with the intracluster medium (Gunn & Gott (1972)), as a result it loses its cold gas in the form of tails rich in star formation. The most extreme cases of galaxies undergoing RPS are known as jellyfish galaxies. To understand how the RPS influences the evolution of galaxies in dense environments we are studying how the morphology of jellyfish galaxies is being transformed. We assess their morphometric properties with the software Morfometryka (Ferrari *et al.* (2015)).

2. Overview

As part of the OMEGA survey, we have selected a sample of 70 jellyfish galaxy candidates – the largest sample found in a single system up to date – with morphological evidences of ram pressure stripping and Hα emission in the multi-cluster system A901/2 at z$\sim$0.165. In Roman-Oliveira *et al.* (2019) we analyse their spatial distribution, velocities and their very high specific star formation rates.

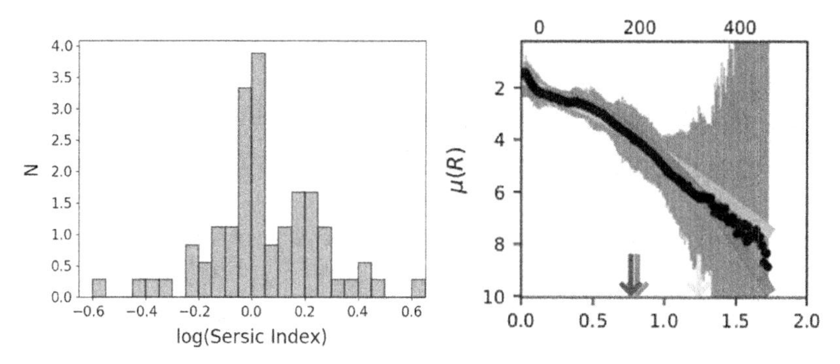

Figure 1. Left panel: The logarithmic distribution of the Sersic Index. Right panel: Brightness profile. Radius is measured in pixels (top) and in terms of the Petrosian radius (bottom), the red line is the 2D Sérsic fit.

2.1. *The multi-cluster environment*

In Ruggiero *et al.* (2019) we found that this unusually rich population of jellyfish galaxies can be explained as the result of an enhancement of the RPS on the boundaries of the four merging halos in the system. Therefore, merging systems like the A901/2 might be the best laboratories for finding and studying jellyfish galaxies and RPS effects.

2.2. *Morphological hints*

The morphology properties of jellyfish galaxies can be helpful to provide some insight on the next steps of their evolution. By performing a single Sérsic fit analysis on our sample we find a bimodal distribution as shown in the left panel of Figure 1. This could be hinting the existence of two populations of jellyfish galaxies. We are currently investigating if this is the case and what could be causing this division, e.g. the inclinations of which the galaxies are infalling.

Another interesting feature is that approximately half of the sample shows an unusual low Sérsic index, like shown in the right panel of Figure 1. Such a low concentration is unusual and often only found in dwarf galaxies.

3. Implications

We present a rich population of jellyfish galaxies in the A901/2 that shows very enhanced star formation. This large number of cases seems to be the result of a trigger in the RPS events caused by the merging multi-cluster environment. As for their morphology, we find hints for two populations of jellyfish galaxies based on their Sérsic index fit. We also find low concentrated brightness profiles in some jellyfish galaxies that resemble the morphological features of dwarf galaxies. We are now currently analysing how these properties influence the next evolutionary step of jellyfish galaxies.

References

Gunn, J. E. & Gott, J. R. 1972, *ApJ*, 176, 1

Ferrari, F., de Carvalho, R. R., & Trevisan, M. 2015, *ApJ*, 814, 15

Roman-Oliveira, F. V., Chies-Santos, A. L., Rodríguez del Pino, B., Aragón-Salamanca, A., Gray, M. E., & Bamford, S. P. 2019, *MNRAS*, 484, 892

Ruggiero, R., Machado, R. E. G., Roman-Oliveira, F. V., Chies-Santos, A. L., Lima Neto, G. B., Doubrawa, L., & Rodríguez del Pino, B. 2019, *MNRAS*, 484, 906

Uncovering Early Galaxy Evolution in the ALMA and JWST Era
Proceedings IAU Symposium No. 352, 2019
E. da Cunha, J. Hodge, J. Afonso, L. Pentericci & D. Sobral, eds.

doi:10.1017/S1743921319009505

The quest for relics: Massive compact galaxies in the local Universe

A. Schnorr-Müller[1], M. Trevisan[1], F. S. Lohmann[1], N. Mallmann[1], R. Riffel[1], A. Chies-Santos[1] and C. Furlanetto[2]

[1]Astronomy Department, Universidade Federal do Rio Grande do Sul, Brazil
[2]Physics Department, Universidade Federal do Rio Grande do Sul, Brazil

Abstract. In the local Universe there exists a rare population of compact galaxies resembling the high-redshift quiescent population in mass and size. It has been found that some of these objects have survived largely unchanged since their formation at high-z. They are called relic galaxies. With the goal of finding relic galaxies, we searched the SDSS-MaNGA DR15 release for massive compact galaxies. We find that massive compact galaxies are mostly composed of old, metal-rich and alpha enhanced stellar populations. In terms of kinematics, massive compact galaxies show ordered rotation in their velocity fields and σ_* profiles rising towards the center. They are predominantly fast rotators and show increased rotational support when compared to a mass-matched control sample of average-sized early-type galaxies. These properties are consistent with these objects being relic galaxies. However, to confirm their relic status, we need to probe larger radii ($\gtrsim 3R_e$) than probed with the current data.

Keywords. galaxies: evolution, galaxies: kinematics and dynamics

1. Introduction

Massive quiescent galaxies in the early Universe are unlike their local counterparts. In particular, they are remarkably compact (typical half-light radius $R_e \simeq 1$–2 kpc) and disk dominated (Buitrago *et al.* 2008; van der Wel *et al.* 2011). In the local Universe, there exists a rare population of compact galaxies resembling the high-redshift quiescent population in mass and size. Some of these local compact galaxies were found to be relic galaxies: objects which have survived largely unchanged since their formation at high-z (Yıldırım *et al.* 2017). The study of these relics opens a new window into early galaxy evolution.

2. Methodology

Sample selection: with the goal of finding relic galaxies, we searched the SDSS-MaNGA DR15 release for massive compact galaxies, defined as fulfilling these criteria: 1) $10^{10.5}\,M_\odot < M_* < 10^{11.5}\,M_\odot$; 2) The size of the semi-major axis of the half-light ellipse is at least $1\,\sigma$ smaller than the value predicted by the local mass-size relation for early-type galaxies. 87 galaxies satisfy these criteria.
Control sample: In order to assess if massive compact galaxies differ from average-sized quiescent galaxies in any other parameter than their size, we define two control samples: 1) a mass-matched quiescent galaxy sample and 2) a σ_e matched-control sample (where σ_e is the velocity dispersion measured inside an aperture of radius R_e). Each control sample contains 174 galaxies.
Stellar population synthesis and kinematics: To derive the stellar population properties for each spaxel in the datacubes we employed the STARLIGHT code together with

the E-MILES stellar population models (Vazdekis *et al.* 2016), fitting the stellar absorption features in 3800–7000 Å restframe wavelength range. We employed the pPXF code to measure the stellar kinematics.

3. Results

The stellar populations of massive compact galaxies: massive compact galaxies are divided into two groups: 1) old (mass-weighted age $\gtrsim 8$ Gyr), metal-rich ($Z_* \gtrsim Z_\odot$) and alpha enhanced galaxies (65% of the sample); 2) younger (mass-weighted age $\lesssim 6$ Gyr), metal rich ($Z_* \gtrsim Z_\odot$) galaxies (35%).

The rotational support of massive compact galaxies: a comparison of the V_{max}/σ_0 distribution of the massive compact galaxies sample with a mass-matched and a σ_e-matched control sample of average-sized galaxies shows that massive compact galaxies have an increased rotational support. Furthermore, massive compact galaxies are predominantly fast-rotators, in clear contrast to average-sized galaxies.

4. Conclusion

• **Clues on the formation of massive compact galaxies:** massive compact galaxies show ordered rotation in their velocity fields (except for 5 galaxies) and σ_* profiles rising towards the center. A strong anti-correlation between V_* and the Gauss-Hermite moment h_3 (which describes asymmetric deviations from a Gaussian) is observed in 80% of the sample. Simulations of major mergers with large gas fractions ($\gtrsim 30\%$) reproduce the observed kinematics, specifically this anti-correlation (Hoffman *et al.* 2009).

• **Clues on the evolution of massive compact galaxies:** for the bulk of the sample, it can be concluded that these galaxies suffered no dry major mergers, as they produce slowly rotating remnants with near-Gaussian line-of sight velocity distribution (i.e. no correlation between V_* and h_3). We can also constrain the growth of massive compact galaxies by dry minor mergers: minor mergers bring old, metal-poor and alpha-enhanced stars to the outskirts of galaxies, which is inconsistent with the high metallicities observed out to $\approx 2\,R_e$. Furthermore, frequent minor mergers decrease the rotational support of galaxies (Bournaud *et al.* 2007), which is inconsistent with the observed increased rotational support of massive compact galaxies. Thus, while we cannot a discard a modest growth through minor merging, a growth dominated by minor mergers is inconsistent with the observed properties of massive compact galaxies.

• **Are massive compact galaxies relics?** About 25% of the sample have properties consistent with a formation at $z \gtrsim 1$ and a subsequent passive evolution. However, to confirm the relic status of these objects we need to probe larger radii ($\gtrsim 3R_e$) than probed with the current data.

References

Bournaud, F., Jog, C. J., & Combes, F. 2007, *A&A*, 476, 1179

Buitrago, F., Trujillo, I., Conselice, C. J., Bouwens, R. J., Dickinson, M., & Yan, H. 2008, *ApJL*, 687, L61

Hoffman, L., Cox, T. J., Dutta, S., & Hernquist, L. 2009, *ApJ*, 705, 920

Vazdekis, A., Koleva, M., Ricciardelli, E., Röck, B., & Falcón-Barroso, J. 2016, *MNRAS*, 463, 3409

van der Wel, A., Rix, H. W., Wuyts, S., McGrath, E. J., Koekemoer, A. M., Bell, E. F., Holden, B. P., Robaina, A. R., & McIntosh, D. H. 2011, *ApJ*, 730, 38

Yıldırım, A., van den Bosch, R. C. E., van de Ven, G., Martín-Navarro, I., Walsh, J. L., Husemann, B., Gültekin, K., & Gebhardt, K. 2017, *MNRAS*, 468, 4216

SESSION 8: Synergies with other facilities & future outlook

Uncovering Early Galaxy Evolution in the ALMA and JWST Era
Proceedings IAU Symposium No. 352, 2019
E. da Cunha, J. Hodge, J. Afonso, L. Pentericci & D. Sobral, eds.

doi:10.1017/S1743921320001313

Galaxies at high z: The MUSE revolution

Roland Bacon

CRAL, University of Lyon, France

Abstract. Spectroscopic observations of galaxies at high redshift has recently been revolutionised by the Multi Unit Spectroscopic Explorer (MUSE) instrument in operation at the VLT since 2014. Thanks to its unrivalled capabilities, MUSE has been able to increase by an order of magnitude the number of spectroscopic redshifts in these fields. The most spectacular increase is at high redshift ($z > 3$), where MUSE was able to detect thousands of Lyman-alpha emitters. In the deepest exposures, MUSE is even able to goes beyond the limiting magnitude of the deepest HST exposures. These observations have led to a breakthrough in our understanding of the high redshift universe: e.g. the discovery of Lyman-alpha emission from the circumgalactic medium around individual galaxies, the role and property of low mass galaxies. In this talk I will present the latest results obtained with the MUSE observations of the Hubble deep and ultra-deep fields.

Uncovering Early Galaxy Evolution in the ALMA and JWST Era
Proceedings IAU Symposium No. 352, 2019
E. da Cunha, J. Hodge, J. Afonso, L. Pentericci & D. Sobral, eds.

doi:10.1017/S1743921319008913

Nature and physical properties of gas-mass selected galaxies using integral field spectroscopy

Leindert A. Boogaard

Leiden Observatory, Leiden University, PO Box 9513, NL-2300 RA Leiden, The Netherlands
email: boogaard@strw.leidenuniv.nl

Abstract. Mapping the molecular gas content of the universe is key to our understanding of the build-up of galaxies over cosmic time. Spectral line scans in deep fields, such as the Hubble Ultra Deep Field (HUDF), provide a unique view on the cold gas content out to high redshift. By conducting 'spectroscopy-of-everything', these flux-limited observations are sensitive to the molecular gas in galaxies without preselection, revealing the cold gas content of galaxies that would not be selected in traditional studies.

In order to capitalize on the molecular gas observations, knowledge about the physical conditions of the galaxies detected in molecular gas, such as their interstellar medium conditions, is key. Fortunately, deep surveys with integral-field spectrographs are providing an unprecedented view of the galaxy population, providing redshifts and measurements of restframe UV/optical lines for thousands of galaxies.

We present the results from the synergy between the ALMA Spectroscopic Survey of the HUDF (ASPECS), with deep integral field spectroscopy from the MUSE HUDF survey and multi-wavelength data. We discuss the nature of the galaxies detected in molecular gas without preselection and their physical properties, such as star formation rate and metallicity. We show how the combination of ALMA and MUSE integral field spectroscopy can constrain the physical properties in galaxies located around the main sequence during the peak of galaxy formation.

Keywords. galaxies: high-redshift, galaxies: formation, galaxies: ISM, techniques: spectroscopic

1. Introduction

Recent years have seen tremendous advances in the characterization of the cosmic history of star formation and it has now been established that the star formation rate (SFR) density increased with cosmic time up to a peak at $z \sim 1-3$ and then decreased until the present (for an overview, see Madau & Dickinson 2014). At each epoch, more massive star-forming galaxies are observed to have a higher star formation rate, establishing what has become known as the 'galaxy main sequence' (MS; Brinchmann *et al.* 2004; Noeske *et al.* 2007; Whitaker *et al.* 2014; Schreiber *et al.* 2015; Boogaard *et al.* 2018).

A key ingredient in our understanding of galaxy formation is the observation of the cold interstellar medium (ISM) – the 'fuel for star formation' – typically traced through carbon monoxide (^{12}CO, hereafter CO) or dust continuum emission (for a review, see Carilli & Walter 2013). Recent years have seen a significant progress in observations of CO at $z > 1$, through targeted observations of star-forming galaxies on and above the MS (e.g., Daddi *et al.* 2015; Tacconi *et al.* 2018; Silverman *et al.* 2018). This has been largely driven by the incredible sensitivity of the Atacama Large Millimeter Array (ALMA), which is now revolutionizing the field of high-redshift ISM line observations, as shown by several publications in this volume.

In order to conduct a 'complete' census of the cosmic molecular gas density, one has to probe a well defined cosmic volume without any target preselection. So called 'spectral scan surveys' are designed to do exactly that, by conducting a mosaic of observations on the sky while simultaneously scanning a large bandwidth in frequency for emission lines from the cold ISM. These are now providing the first ever constraints on the cosmic molecular gas content and hence the cosmic star formation efficiency, directly from CO observations (Decarli *et al.* 2014; Walter *et al.* 2014, 2016; Pavesi *et al.* 2018; Riechers *et al.* 2019; Decarli *et al.* 2019).

As molecular line emission can be elusive, having a large number of spectroscopic redshifts over the target field can provide essential redshift information to push spectral scans to their sensitivity limit. The Multi-Unit Spectroscopic Explorer (MUSE) on the Very Large Telescope (Bacon *et al.* 2010) is a key instrument in this context, providing 'integral field spectroscopy' over a large $1' \times 1'$ field-of-view from $4750 - 9300$ Å at a high observing efficiency.

Here, we will present the first results from the ALMA Spectroscopic Survey (ASPECS) in the Hubble Ultra Deep Field (HUDF) Large Program (Decarli *et al.* 2019), focusing specifically on the synergy between ALMA and MUSE (Boogaard *et al.* 2019). These results are part of a larger series of papers on the Band 3 data from ASPECS, further discussed in González-López *et al.* (2019); Popping *et al.* (2019); Aravena *et al.* (2019).

2. Observations: ALMA and MUSE observations in the HUDF

The ALMA Spectroscopic Survey consists of a mosaic of spectral scan observations targeting the deepest region of the HUDF, covering the complete ALMA Band 3 (84 - 115 GHz) and Band 6 (212 - 272 GHz) windows. These flux-limited observations allow one to detect both line emission from the ISM (mainly CO, [CI] and [CII]) as well as dust continuum at different redshifts, without any preselection of targets (see the top panels of Fig. 1). The HUDF is an excellent target for ASPECS, due to the wealth of multi-wavelength data available all the from the X-ray to radio. In particular, deep *HST* and *Spitzer* photometry are key to identify the host galaxies of the molecular line emission, as the stellar light from these sources is often significantly attenuated.

The MUSE data over the HUDF provides optical spectroscopy for all galaxies in the field and has revolutionized the amount of available redshift information (increasing the number of known spectroscopic redshifts by a factor $\times 10$; Bacon *et al.* 2017; Inami *et al.* 2017). Fig. 1 shows how the galaxies at different redshifts are identified by distinct spectral features that fall within the spectrograph.

Fig. 1 also illustrates how the integral field data from ALMA and MUSE can work together: On the one hand, MUSE can provide redshift information for the galaxies detected in CO emission as well as physical properties from the rest-frame UV/optical spectra (depending on the redshift). On the other hand, one can leverage the large number of MUSE redshifts to search specifically for lines in the ALMA data, as well as conduct stacking in order to detect fainter sources.

3. Results: The physical properties of the ASPECS galaxies

We search the complete ASPECS Band 3 cube for emission lines using a matched filtering approach (González-López *et al.* 2019). This reveals 16 emission line candidates at high significance (very low probability of being spurious), all of which show a counterpart in the *HST* imaging. Using the MUSE and multi-wavelength data, we identify all as rotational transitions of CO, with redshifts between $z = 1.0 - 3.6$ ($J_{\rm up} = 2 - 4$). In addition, using the MUSE redshifts as prior information, we recover two additional CO lines at signal-to-noise > 3, bringing the total sample to 18. We convert the observed CO

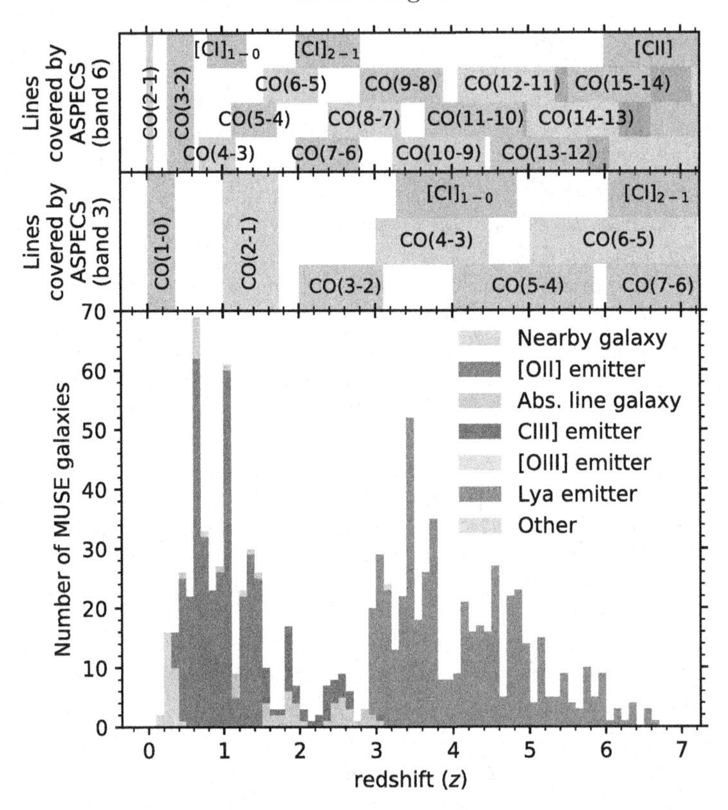

Figure 1. Top panel: Primary emission lines observable at different redshifts with the ALMA Spectroscopic Survey (ASPECS) in Band 3 and 6. **Bottom panel**: Histogram of all galaxies from the MUSE HUDF Survey (Bacon *et al.* 2017; Inami *et al.* 2017) that fall within the ASPECS field (Decarli *et al.* 2019), colored by their spectral classification. At each distinct redshift, MUSE and ALMA probe different parts of the galaxy spectra. *Figure adapted from Boogaard et al. (2019).*

luminosity to CO(1-0) using the excitation corrections derived for $z \sim 1.5$ star-forming galaxies from Daddi *et al.* (2015) and convert this to a molecular gas mass assuming an $\alpha_{\rm CO} = 3.6$ (Daddi *et al.* 2010), similar to the Galactic value (e.g., Bolatto *et al.* 2013).

For the galaxies detected in CO(2-1) up to $z = 1.5$, MUSE still covers part of the NUV spectrum (see Fig. 1), where we can use emission line ratios to probe the metallicity. We use the ratio of [O II] $\lambda 3726, 3729$/[Ne III] $\lambda 3869$ to infer the gas-phase metallicity in these galaxies, following the calibration from Maiolino *et al.* (2008), as shown in the left panel of Fig. 3. Overall, we find that the ASPECS galaxies have a (super-)solar metallicity, consistent with what would be expected from the mass-metallicity relation (e.g. Zahid *et al.* 2014), supporting the use of a Galactic $\alpha_{\rm CO}$ for these galaxies.

A key question is what kind of galaxies we detect with the flux-limited observations from ASPECS. In Fig. 2, we plot histograms of the stellar masses of the galaxies in which we detect molecular gas. What is immediately evident is that we detect molecular gas in the majority of the most massive galaxies (in terms of their stellar mass). As we move towards lower mass galaxies, the fraction of galaxies decreases, reaching a detection fraction of ~50% for galaxies with $M_* \geqslant 10^{10}$ $(10^{10.5})$ $\rm M_\odot$ at $1 < z < 2$ $(2 < z < 3)$.

In the right panel of Fig. 3, we expand the one dimensional histograms and show the ASPECS galaxies on the stellar mass - SFR plane. At $1 < z < 2$, we find that our gas-mass selected sample consists mostly of galaxies on the 'galaxy main sequence' at these

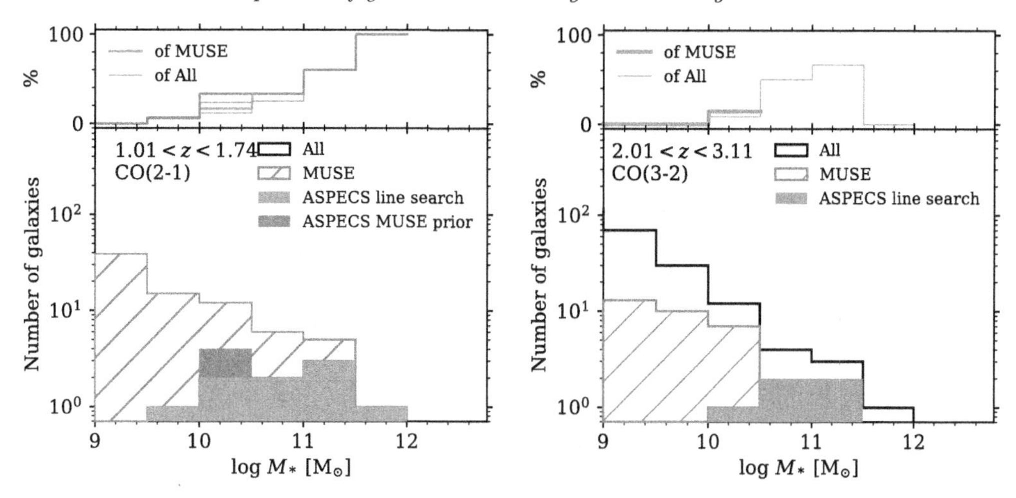

Figure 2. Detection fractions and histograms as a function of the stellar masses (M_*) of the ASPECS galaxies detected in CO(2-1) (**left panel**) and CO(3-2) (**right panel**). With a purely gas-mass selected sample we detect most of the massive galaxies, while probing down to $M_* \approx 10^{9.5}$ $M_\odot$. *Figure adapted from Boogaard et al. (2019)*

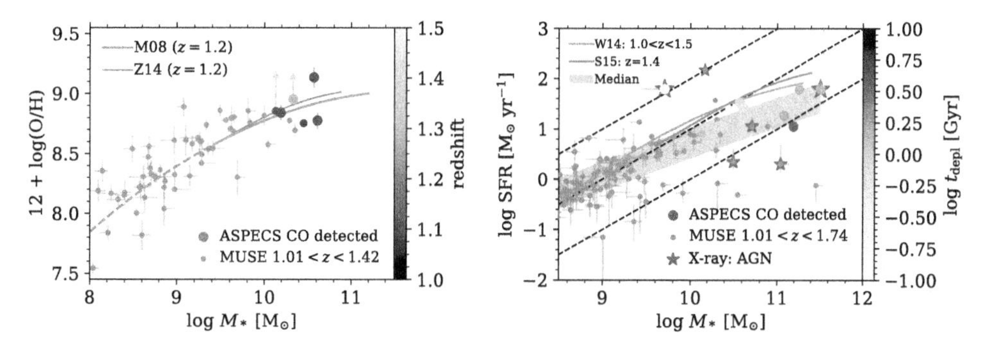

Figure 3. Left panel: Stellar mass - metallicity relation of the ASPECS galaxies detected in CO(2-1), derived from the [O II]/[Ne III] ratio (at $z < 1.42$). The ASPECS galaxies are consistent with a (super-)solar metallicity. **Right panel:** Stellar mass - star formation rate relation for the ASPECS galaxies detected in CO(2-1). With ASPECS with detect galaxies on, above and below the 'galaxy main sequence' at these redshifts. *Figures adapted from Boogaard et al. (2019)*

redshifts. Above $M_* \approx 10^{10}$ $M_\odot$, we detect almost all of the 'starburst' galaxies, defined as lying $\geqslant 0.3$ dex above the main sequence. Critically, we also detect galaxies that lie below the MS. While these galaxies have SFR that is lower than typical at their stellar mass, they still host a significant gas reservoir. Their detection in the flux-limited survey is important, as these galaxies would typically not be selected in targeted follow-up.

The detections fractions as a function of mass and SFR, together with additional CO lines found using the MUSE redshifts as a prior, suggest that there may be several galaxies for which the CO emission falls just below the individual detection threshold. The large number of systemic redshifts from MUSE can be leveraged here, in order to recover this signal through stacking. Indeed, preliminary results show that stacking the CO(2-1) undetected galaxies on the MS at $10 < \log M_* [M_\odot] < 11$ show a detection of their CO line emission. These and additional results from the stacking will be discussed in Inami *et al.* (in prep.).

4. Outlook: ASPECS and *JWST*

The combination of integral field spectroscopy with MUSE in the optical and ALMA in the (sub-)millimeter regime provides a unique tool for studying the physical properties of galaxies across cosmic time. The recently observed band 6 data from ASPECS provide additional constraints on the molecular gas content, through the deep dust continuum map at 1.2 mm and observations of atomic carbon, as well as on the CO excitation, by constraining higher-J CO lines for all sources.

The anticipated launch of the *James Webb Space Telescope* will open up a new window on these galaxies. In particular, key rest-frame UV, optical and near-infrared spectroscopy will allow further characterization of the star formation rates, metallicities and ISM conditions in these galaxies. Together with observations of the cold ISM from ALMA, these will provide critical constraints on our theory of star formation at high redshift.

Acknowledgements

The work presented here is the result of a large collaborative effort and would not have been possible without Fabian Walter, Roberto Decarli, Manuel Aravena, Jorge González-López, Chris Carilli, Paul van der Werf, Rychard Bouwens, Roland Bacon, Hanae Inami, and the other members from the ASPECS and MUSE GTO teams. The author would like to thank Paul van der Werf for providing comments on the manuscript.

References

Aravena, M., Decarli, R., Gónzalez-López, J., *et al.* 2019, *ApJ*, 882, 136
Bacon, R., Accardo, M., Adjali, L., *et al.* 2010, in Proc. SPIE, ed. I. S. McLean, S. K. Ramsay, & H. Takami, Vol. 7735, 773508
Bacon, R., Conseil, S., Mary, D., *et al.* 2017, *A&A*, 608, A1
Bolatto, A. D., Wolfire, M., & Leroy, A. K. 2013, *ARA&A*, 51, 207
Boogaard, L. A., Brinchmann, J., Bouché, N., *et al.* 2018, *A&A*, 608, A10
Boogaard, L. A., Decarli, R., González-López, J., *et al.* 2019, *ApJ*, 882, 140
Brinchmann, J., Charlot, S., White, S. D. M., *et al.* 2004, *MNRAS*, 351, 1151
Carilli, C. L., & Walter, F. 2013, *ARA&A*, 51, 1
Daddi, E., Bournaud, F., Walter, F., *et al.* 2010, *ApJ*, 713, 686
Daddi, E., Dannerbauer, H., Liu, D., *et al.* 2015, *A&A*, 577, A46
Decarli, R., Walter, F., Carilli, C., *et al.* 2014, *ApJ*, 782, 78
Decarli, R., Walter, F., Gónzalez-López, J., *et al.* 2019, *ApJ*, 882, 138
González-López, J., Decarli, R., Pavesi, R., *et al.* 2019, *ApJ*, 882, 139
Inami, H., Bacon, R., Brinchmann, J., *et al.* 2017, *A&A*, 608, A2
Madau, P., & Dickinson, M. 2014, *ARA&A*, 52, 415
Maiolino, R., Nagao, T., Grazian, A., *et al.* 2008, *A&A*, 488, 463
Noeske, K. G., Weiner, B. J., Faber, S. M., *et al.* 2007, *ApJ*, 660, L43
Pavesi, R., Sharon, C. E., Riechers, D. A., *et al.* 2018, *ApJ*, 864, 49
Popping, G., Pillepich, A., Somerville, R. S., *et al.* 2019, *ApJ*, 882, 137
Riechers, D. A., Pavesi, R., Sharon, C. E., *et al.* 2019, *ApJ*, 872, 7
Schreiber, C., Pannella, M., Elbaz, D., *et al.* 2015, *A&A*, 575, A74
Silverman, J. D., Rujopakarn, W., Daddi, E., *et al.* 2018, *ApJ*, 867, 92
Tacconi, L. J., Genzel, R., Saintonge, A., *et al.* 2018, *ApJ*, 853, 179
Walter, F., Decarli, R., Sargent, M., *et al.* 2014, *ApJ*, 782, 79
Walter, F., Decarli, R., Aravena, M., *et al.* 2016, *ApJ*, 833, 67
Whitaker, K. E., Franx, M., Leja, J., *et al.* 2014, *ApJ*, 795, 104
Zahid, H. J., Dima, G. I., Kudritzki, R.-P., *et al.* 2014, *ApJ*, 791, 130

Uncovering Early Galaxy Evolution in the ALMA and JWST Era
Proceedings IAU Symposium No. 352, 2019
E. da Cunha, J. Hodge, J. Afonso, L. Pentericci & D. Sobral, eds.

doi:10.1017/S1743921319008391

Ultra-faint Lyman Alpha Emitters with MUSE

Michael V. Maseda and the MUSE GTO Consortium

Leiden Observatory, Leiden University, Postbus 9513, NL-2300RA, Leiden, the Netherlands
email: maseda@strw.leidenuniv.nl

Abstract. Using an ultra-deep, untargeted survey with the MUSE integral field spectrograph on the ESO Very Large Telescope, we obtain spectroscopic redshifts to a depth never explored before: galaxies with observed magnitudes $m > 30-32$. Specifically, we detect objects via Lyman-α emission at $2.9 < z < 6.7$ without individual continuum counterparts in areas covered by the deepest optical/near-infrared imaging taken by the Hubble Space Telescope, the Hubble Ultra Deep Field. In total, we find more than 100 such objects in 9 square arcminutes at these redshifts, also including a number of sources that are visible only in the HST band that contains Lyman-α. Detailed HST and IRAC stacking analyses confirm the Lyman-α emission as well as the 1216 Å breaks, faint UV continua ($M_{\rm UV} \sim -15$), and optical emission lines: these objects are the faintest spectroscopically-confirmed galaxies at high-z. The blue UV continuum slopes and measurements/limits on the equivalent widths of Lyman-α, which in some cases exceeds 300 Å, are consistent with ages < 10 Myr, metallicities < 5% solar, and stellar masses < 10^{7-8} solar masses. The nature of these types of objects is intriguing as they could be the faint star-forming sources of Reionization and could represent the initial (strong) phase of stellar mass growth in galaxies.

Keywords. galaxies: dwarf, galaxies: high-redshift

---

1. Introduction

Until recently, large samples of star forming galaxies at $z > 1$ were created exclusively via color selections. These selections, such the Lyman break (Steidel *et al.* 2003), BzK (Daddi *et al.* 2004), and BX/BM selections (Steidel *et al.* 2004), suffer variously from issues of incompleteness and contamination from lower-z interlopers, as well as preferentially selecting galaxies with certain dust properties (Ly *et al.* 2011). When the selection criteria become stricter to avoid some of these issues, then the selection itself becomes more biased. For example, the Lyman break selection requires detections in several bands redward of the actual break in the spectral energy distribution (SED), potentially biasing against low-mass bursts with strong emission lines but little stellar continuum light.

To circumvent this issue, unbiased, un-targeted spectroscopic surveys are required. Previous efforts to do this utilized long-slit spectra taken over blank regions of sky (e.g. Rauch *et al.* 2008), in some cases centered on the caustics of strong gravitationally lensing systems (e.g. Ellis *et al.* (2001); Stark *et al.* (2007)). However, the area covered by these types of surveys is necessarily small ($\lesssim$ 0.2 square arcminutes), and the ability to measure line fluxes and slit loss corrections when no deep continuum imaging exists is problematic. Slitless surveys taken with prisms (e.g. Salzer *et al.* (2000)) or grisms (e.g. Momcheva *et al.* 2016) can overcome these issues, but at the cost of a higher background flux level.

The Multi-Unit Spectroscopic Explorer (Bacon *et al.* 2010), however, overcomes many of these issues. Its image slicer design provides low-background optical spectroscopy simultaneously over a large field-of-view (1 square arcminute). This makes MUSE well-suited to undertake mosaicked spectroscopic surveys as a compliment to wide-field imaging surveys: in the first deep MUSE data cubes covering the Hubble Deep Field South (HDFS), 189 individual objects have confirmed redshifts in 27 hours of spectroscopy (Bacon *et al.* 2015).

Due to the un-targeted nature of MUSE spectroscopy, 26 objects in the HDFS have detectable Lyman-α emission lines at $2.9 < z < 6.7$ with no counterpart in the *Hubble* Space Telescope (HST) WFPC2 imaging. The WFPC2 observations of the HDFS reach a 5-σ limiting AB magnitude of 29.1 in $F606W$ and 28.5 in $F814W$ (Casertano *et al.* 2000). The Hubble Ultra Deep Field (UDF), however, reaches depths more than one magnitude deeper in nine HST filters spanning the optical to near-IR (Illingworth *et al.* 2013) and is the target of deeper, 30-hour observations with MUSE (Bacon *et al.* 2017). By combining the MUSE spectroscopy with the HST imaging in the UDF, we can search for the faintest galaxies in terms of both emission line fluxes as well as continuum magnitudes.

2. Observations and Results

In order to identify emission lines without relying on a photometric detection, we use the `ORIGIN` software (Bacon *et al.* 2017, D. Mary *et al.* in prep) to find emission lines in the reduced MUSE UDF data cubes. The individual MUSE data cubes are combined into a 9 square arcminute "mosaic" with a depth of 10 hours and a single, 1 square arcminute pointing with a depth of 30 hours. All detected sources in the full dataset have redshifts determined by using a combination of automated tools and expert identifications (see Inami *et al.* 2017 for details). Lyman-α emitters (LAEs) can often be identified due to the characteristic asymmetry, resolvable at the spectral resolution of MUSE. Additionally, lower-redshift solutions can ruled out based on multiple optical emission line non-detections due MUSE's broad wavelength coverage.

We place $0.4''$ apertures in all HST bands centered on the MUSE emission line centroid. The *local* background level in each band is calculated by measuring the standard deviation of the fluxes in 250 identical apertures spread randomly within a $10''\times10''$ cutout centered on the object, while masking other sources according to the imaging segmentation maps. An object is considered "undetected" if there are no $>$5-σ detections in any of the HST bands. In total we identify 102 MUSE LAEs that have no counterpart in the HST imaging.

As these galaxies are individually undetected, we must rely on photometric stacking in order to learn about their continuum properties. We split the galaxies into three redshift bins corresponding to a single HST filter probing the restframe UV ($\sim$ 1600 Å). Figure 1 shows clear detections in the band(s) that contain Lyman-α and the (blue) UV continuum. Combined with non-detections in bands bluewards of Lyman-α, these objects fulfill the photometric criteria for the aforementioned Lyman break galaxies. This provides independent, photometric evidence of the Lyman-α identification. Based on the aperture magnitude of the UV continuum, these galaxies have $M_{\rm UV}$ values of ~ -15, more than 3 magnitudes fainter than continuum-selected samples in the field: these are the faintest spectroscopically-confirmed galaxies at high-z discovered to date (Maseda *et al.* 2018).

The detectable Lyman-α emission and extremely faint UV continuum implies that the Lyman-α equivalent widths (EWs) are extremely high, typically $>$ 300 Å (M. Maseda *et al.* in prep.). These extreme EWs can only be produced by young, low-metallicity stellar populations. Stellar population models can predict the evolution in the EW of Lyman-α as a function of age, metallicity, and the star formation history (SFH) of the galaxy. These models also predict the evolution in the observed UV continuum slopes (β). By combining measurements of the Lyman-α EW and β from the stacks, we can constrain the age-metallicity-SFH parameter space.

In Figure 2 we show how each stack's constraints on β and Lyman-α EW translates into an age, metallicity, and stellar mass (age convolved with SFH) using the models of

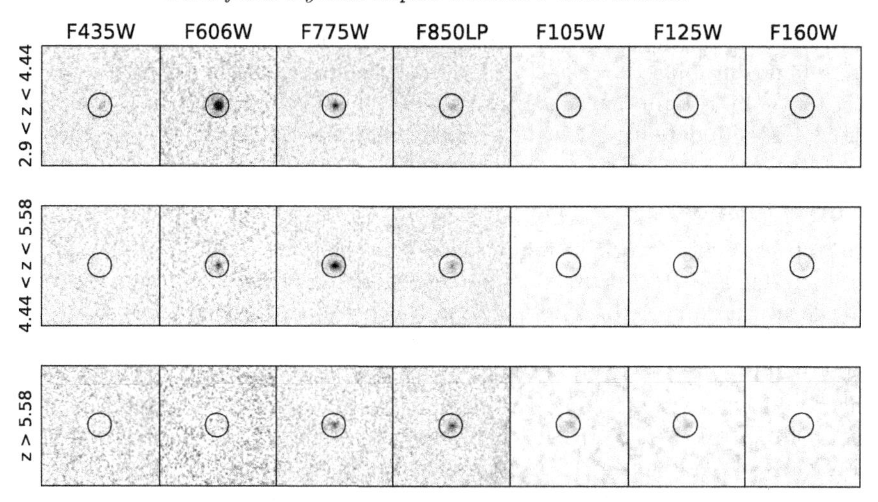

Figure 1. HST imaging stacks for the undetected LAEs in three different redshift bins. Lyman-α emission is visible in the $F606W$ (top), $F606W$ and $F775W$ (middle), and $F775W$ and $F850LP$ bands (bottom), in line with the redshift determined by MUSE. All of these photometric stacks fulfill the standard "dropout" selection criteria for LBGs at these redshifts.

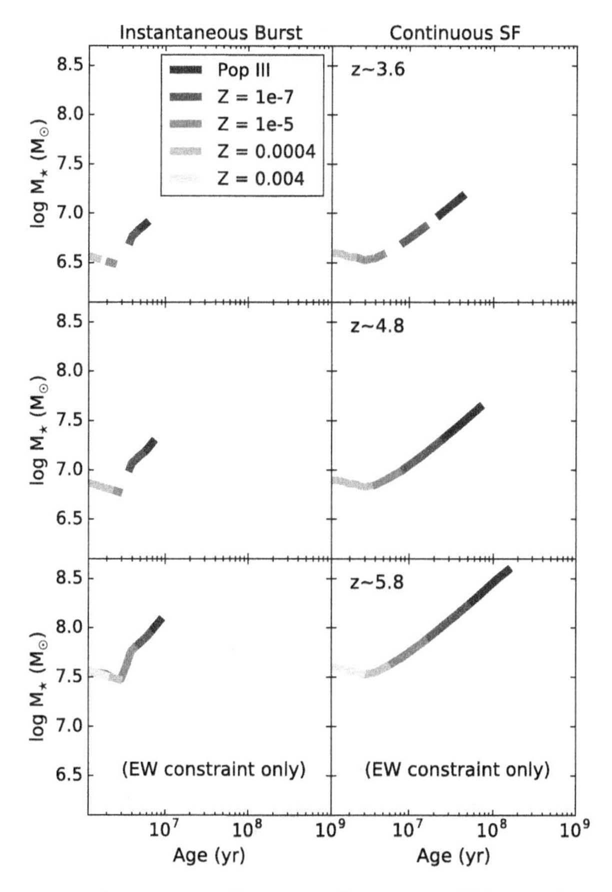

Figure 2. Stellar mass estimates as a function of age, metallicity, and star formation history (burst or continuous) based on the observed Lyman-α EWs and UV continuum slopes (β), using the models of Raiter *et al.* 2010. The highest stellar masses can be achieved with the lowest metallicities and continuous star formation, and in general the metallicity must be $<$ 0.0004 ($\sim$5% $Z_\odot$) when both the EW and β are constrained.

Raiter *et al.* 2010. Although β is not constrained for the $z > 5.58$ stack due to the lack of continuum bands probing the rest-UV at these redshifts, in general we need metallicities < 0.0004 ($\sim$5% $Z_{\odot}$), burst ages < 10 Myr, and stellar masses $< 10^{8}$ $M_{\odot}$. This implies these are galaxies undergoing their first significant episode of star formation.

3. Future Outlook

Further spectroscopic work is required to fully characterize the stellar populations in these ultra-faint LAEs. In particular, detections of restframe-optical emission lines will provide comparable information about the metallicity and age of the galaxy. At these redshifts spectroscopic observations of the restframe-optical are challenging as these features are shifted further into the near-IR and are difficult or impossible to observe from the ground. However, at certain redshifts these features are located in broadband photometry from e.g. *Spitzer*/IRAC. Since the redshifts are known *a priori* from MUSE, any photometric excess in the IRAC photometry can be attributed to e.g. Hα emission. With a measurement of Hα and Lyman-α, we can measure the ionizing photon production efficiency (ξ_{ion}), which is crucial to understand in the context of cosmic Reionization (e.g. Robertson *et al.* 2015).

At $3.829 < z < 4.955$, Hα is in IRAC's Channel 1 and Lyman-α is probed by MUSE. For a sample of 40 HST-faint LAEs, we create a stack of the IRAC Ch. 1 photometry and find a >5-σ detection within a 1.8$''$ aperture. If the Hα emission is confirmed to be real, e.g. with JWST/NIRSpec, then the implied high equivalent width makes these galaxies even more extreme than the "Extreme Emission Line Galaxies" discovered at $z \sim 1-2$ in slitless HST grism spectroscopy (e.g. Maseda *et al.* 2013, 2014), which have values of ξ_{ion} up to $\sim$25.8 (Tang *et al.* 2018 in press). An abundant population of these galaxies, as revealed with our MUSE survey, is likely to have made a significant contribution to cosmic Reionization. Further work is required to quantify this contribution.

References

Bacon, R., Accardo, M., Adjali, L., *et al.* 2010, *SPIE*, 773508
Bacon, R., Brinchmann, J., Richard, J., *et al.* 2015, *A&A*, 575, A75
Bacon, R., Conseil, S., Mary, D., *et al.* 2017, *A&A*, 608, A1
Casertano, S., de Mello, D., Dickinson, M., *et al.* 2000, *AJ*, 120, 2747
Daddi, E., Cimatti, A., Renzini, A., *et al.* 2004, *ApJ*, 617, 746
Ellis, R., Santos, M. R., Kneib, J.-P., *et al.* 2001, *ApJ*, 560, L119
Illingworth, G. D., Magee, D., Oesch, P. A., *et al.* 2013, *ApJS*, 209, 6
Inami, H., Bacon, R., Brinchmann, J., *et al.* 2017, *A&A*, 608, A2
Ly, C., Malkan, M. A., Hayashi, M., *et al.* 2011, *ApJ*, 735, 91
Maseda, M. V., van der Wel, A., da Cunha, E., *et al.* 2013, *ApJ*, 778, L22
Maseda, M. V., van der Wel, A., Rix, H.-W., *et al.* 2014, *ApJ*, 791, 17
Maseda, M. V., Bacon, R., Franx, M., *et al.* 2018, *ApJ*, 865, L1
Momcheva, I. G., Brammer, G. B., van Dokkum, P. G., *et al.* 2016, *ApJS*, 225, 27
Raiter, A., Schaerer, D., & Fosbury, R. A. E. 2010, *A&A*, 523, A64
Rauch, M., Haehnelt, M., Bunker, A., *et al.* 2008, *ApJ*, 681, 856
Robertson, B. E., Ellis, R. S., Furlanetto, S. R., *et al.* 2015, *ApJ*, 802, L19
Salzer, J. J., Gronwall, C., Lipovetsky, V. A., *et al.* 2000, *AJ*, 120, 80
Stark, D. P., Ellis, R. S., Richard, J., *et al.* 2007, *ApJ*, 663, 10
Steidel, C. C., Adelberger, K. L., Shapley, A. E., *et al.* 2003, *ApJ*, 592, 728
Steidel, C. C., Shapley, A. E., Pettini, M., *et al.* 2004, *ApJ*, 604, 534
Tang, M., Stark, D., Chevallard, J., *et al.* 2018, arXiv e-prints, arXiv:1809.09637

Discussion

S. Wuyts Based on the observed number counts of these galaxies, can you derive an estimate for the total number density of galaxies at this UV magnitude? This would be an independent measurement of (part of) the UV luminosity function at these redshifts.

Maseda We have not done this yet, but in theory we do have a well-defined selection function based on Lyman-α flux. Based on this we could estimate the volume density of these sources that have $M_{\rm UV} \sim -15$. However this is likely to be a lower limit since there should be a significant population of sources at a similar continuum magnitude that do not have such high Lyman-α EWs and hence would be unobservable in our survey. Additionally, this sample was selected for purity rather than completeness, as we manually remove any sources that are very close to HST detections even if it is clearly at a different redshift.

E. da Cunha At some level do we not expect *all* galaxies at such faint magnitudes to be undergoing some kind of bursty episode of star formation?

Maseda Bursty star formation does appear to be the norm at low masses and low metallicities in the early Universe. However, the UV magnitude is not a direct probe of the stellar mass of a galaxy, and hence for a fixed faint UV magnitude we could also expect a number of more massive, older galaxies in addition to the young starbursts. Presumably these older galaxies were once bursty in the past, but it is difficult to say how bright they were. Likewise, it is difficult to know what these high-EW LAEs will look like in a few Myrs.

Uncovering Early Galaxy Evolution in the ALMA and JWST Era
Proceedings IAU Symposium No. 352, 2019
E. da Cunha, J. Hodge, J. Afonso, L. Pentericci & D. Sobral, eds.

doi:10.1017/S1743921320001325

The role of the JWST near-infrared spectrograph NIRSpec in understanding the assembly and evolution of galaxies

Catarina Alves de Oliveira

European Space Agency, USA

Abstract. The near-infrared spectrograph NIRSpec is one of four instruments aboard the James Webb Space Telescope (JWST). It offers seven dispersers covering the wavelength range from 0.6 to 5.3 micron with resolutions from R$\sim$100 to R$\sim$2700. Using an array of micro-shutters for target selection, the multi-object spectroscopy mode of NIRSpec will be capable of obtaining spectra from a few tens to more than 200 objects simultaneously. It also features an integral field unit with a 3 by 3 arcseconds field of view, and various slits for high contrast spectroscopy of individual objects. We will provide an overview of the capabilities and performances of these three observing modes highlighting how NIRSpec will contribute to the quest to further understand the assembly and evolution of galaxies from the end of re-ionisation epoch to the present day.

Uncovering Early Galaxy Evolution in the ALMA and JWST Era
Proceedings IAU Symposium No. 352, 2019
E. da Cunha, J. Hodge, J. Afonso, L. Pentericci & D. Sobral, eds.

doi:10.1017/S1743921319008950

JWST advanced deep extragalactic survey: NIRCam imaging to $z > 10$

Marcia Rieke on behalf of JADES Collaboration

Steward Observatory, University of Arizona, 933 N. Cherry, Tucson, AZ 85721 USA
email: mrieke@as.arizona.edu

Abstract. The JWST Advanced Deep Extragalactic Survey (JADES) is a joint program of the JWST NIRCam and NIRSpec Guaranteed Time Observation (GTO) teams involving over 800 hours of observation. This paper describes the imaging portion of the program which covers nearly 200 square arc minutes divided between two well-studied fields with excellent supporting data (e.g. from Chandra, ALMA, and HST-CANDELS): GOODS North and South, including the Ultra Deep Field. NIRCam imaging will enable the study of galaxy evolution to $z \sim 10$ and higher using multi-color imaging with 9 filters covering 0.9 to 5 microns. Such data will provide photometric redshifts and a wealth of data for constructing luminosity and mass functions. A key component of the program is rapid turn around of imaging into NIRSpec target lists. Preparing for this program has benefited from the development of a mock catalog and simulated imaging to test these processes.

Keywords. galaxies: evolution, galaxies: high-redshift, surveys

---

1. Introduction

The original justification for the construction of the James Webb Space Telescope (JWST) was based on detecting the first light to be produced by stars and black holes in the Universe (Dressler 1996). This led to the nickname "First Light Machine" for JWST, but this appellation can be misleading as JWST cannot detect the first star. It will be able to detect the first galaxies and possibly structures as small as the precursors to current day globular clusters as described by Renzini at this conference. Many of JWST Guaranteed Time Observers (GTOs) have formed a collaboration to produce a legacy-scale deep survey to address the questions that underlie the questions presented at this conference. This paper describes the observing strategy adopted for the near-infrared imaging portion of the survey with the NIRSpec portion described in more detail by Bunker at this conference. In addition to providing results and insights relevant to the galaxy evolution questions posed at this conference, the JADES collaboration also hopes that this deep survey will lead to some surprises and guide new directions for study in the future. Two fields, GOODS North and GOODS South, have been chosen for observation due to the wealth of ancillary data available. The deepest observations will be centered on the UDF in GOODS South with GOODS North receiving only medium depth exposure, a level which is still sufficiently deep to detect luminous $z \sim 10$ galaxies.

This JADES program evolved from considering the need for NIRCam imaging to select targets for the NIRSpec microshutter array which provides the ability to observe 100s of objects at once. The NIRCam - NIRSpec synergy is deeper than just using NIRCam to produce finding charts for NIRSpec, however. NIRCam will detect many more sources than can be observed with NIRSpec in a program of the scale of JADES. NIRSpec data can be used to anchor relations found in the imaging data with NIRCam's larger source numbers providing improved statistics, for example. NIRCam imaging will be needed for ensuring accurate spectral energy distributions because the broad wavelength range of NIRSpec (1 - 5 microns) and fixed slit width means that wavelength dependent corrections will be needed to avoid biases introduced by diffraction. The combined NIRSpec-NIRCam data set will be more valuable than either one alone.

2. NIRCam Overview

NIRCam has been designed expressly to be efficient at executing surveys. Because NIRCam is the wavefront sensor for JWST, it must be fully redundant, and so NIRCam is comprised of two identical cameras mounted back-to-back. In survey mode, both modules can be used at the same time thus doubling the area covered. Each module has a dichroic beamsplitter which divides the incoming radiation at 2.35 microns and sends each the light into short wavelength (0.7 to 2.3 microns) and long wavelength (2.4 to 5 microns) arms that view the same area on the sky The overlap of the short and long wavelength arms is $> 96\%$. A module will always be capable of observing at two wavelengths simultaneously which further improves survey efficiency. The JDOX pages at STScI (http://www.stsci.edu/jwst/) illustrate the NIRCam field of view in detail with its $\sim 43''$ separation between modules and $\sim 4''$ gaps between the four detectors that comprise the short wavelength arms.

3. Observing Plan

Central to the JADES survey plan is observing with NIRCam and NIRSpec operated in parallel (eg. what Space Telescope Science Institute calls a coordinated parallel). This effectively doubles the amount of observing time available for the program. Figure 1 shows the layout for GOODS South. Mid-infrared imaging using MIRI will also be acquired both in prime mode on the UDF and in parallel with some of the NIRCam imaging. These data will be collected by various parts of the MIRI GTO teams. One nuance of the parallel usage of NIRCam and NIRSpec stems from the requirement that NIRSpec must formally be designated as the prime instrument because of how the microshutter array in NIRSpec is configured to match targets. This issue has been handled in our program by specifying NIRSpec as prime with central pointing coordinates chosen to yield a fully sampled NIRCam region. Additionally dither patterns suitable for both instruments also had to be selected. Last, the layouts shown in Figure 1 are for an observation date in October, and will be different if the program is executed at another time of year because of the fixed relationship between the NIRCam and NIRSpec fields of view, and the modest amount of roll control available with JWST.

Some of the NIRSpec slit assignments will be generated based on HST data. Other pointings will observe galaxies either detected only by NIRCam or shown to be particularly interesting using NIRCam data. The JADES team would like to observe NIRCam-selected objects with NIRSpec during the same four-month visibility period as when the NIRCam imaging is obtained. To fit all the imaging in the beginning of the visibility period and allow time for STScI to prepare and upload observing sequences implies that the NIRCam imaging must be processed and converted into NIRSpec target selections in at most 45 days. Williams *et al.* (2018) have created a mock catalog, JAGUAR, to simulate what the survey may see, and to provide a tool for testing analysis techniques to ensure that this rapid time scale can be met.

The NIRCam portion of the JADES survey uses seven wide ($R \sim 4$) filters supplemented with two medium width filters. Figure 2 shows the filter transmission functions overlaid on a $z \sim 5$ galaxy spectral energy distribution. Only a limited area will be observed using NIRCam's shortest W filter, F070W (0.7 microns), as JWST has the smallest advantage over HST at this wavelength, and much of the area has deep HST

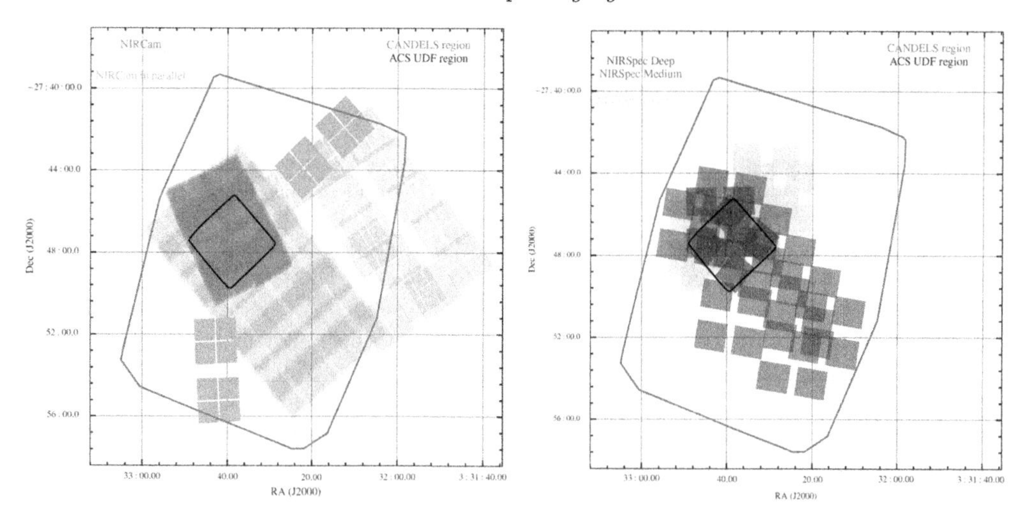

Figure 1. Footprints for the NIRCam (left), and NIRSpec (right) observations in GOODS South. The heavy outline shows the area covered by CANDELS, Grogin *et al.* (2011). The NIRCam lighter-toned areas constitute the medium survey while the deeper tones are the deep survey. The exact areas to be covered will depend on when the observations are scheduled.

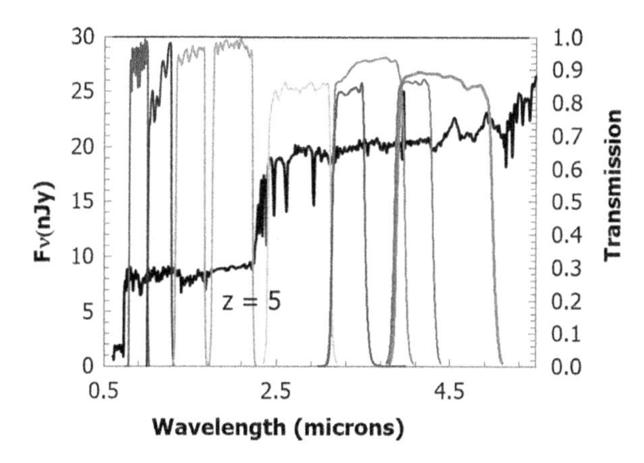

Figure 2. NIRCam filter transmission functions overlaid on a $z \sim 5$ galaxy spectral energy distribution.

imaging in similar filters. Figure 2 shows that the two medium filters that will be used subdivide nearly in half the two longest wavelength filters. By comparing fluxes in the corresponding medium and wide filters, an estimate of possible emission lines can be made. F410M (4.1 microns) will be observed on all areas because its sensitivity is nearly the same as the wider filter because the dominant zodiacal background is concentrated in the long wavelength half of F444W. The other medium filter, F335M (3.35 microns), will be used over a portion of the survey to provide additional constraints on emission lines and photometric redshifts. See Rasappu *et al.* (2016) and references therein for a discussion of how IRAC photometry has revealed strong $H\alpha$ and other emission lines in some galaxies in the range $3.5 < z < 6.0$.

4. Expected Results

JWST will of course provide the deepest view of the near- to mid-infrared universe yet and likely for the forseeable future. Current space telescopes have no observing capability

Table 1. NIRCam Imaging Sensitivity. The 10-σ depth for a point source corresponding to the average exposure times (60 ksecs for deep, 12 ksecs for medium) is tabulated.

Subsurvey	Area □′	10σ Point Source Magnitude (AB) F070W	F090W	F115W	F150W	F200W	F277W	F335M	F356W	F410M	F444W
Deep	46	—	29.5	29.8	29.9	29.9	29.5	28.8	29.4	29.0	29.1
Medium	190	28.0[a]	28.6	28.8	28.9	29.0	28.6	28.0[a]	28.6	28.1	28.3

[a] The F070W and F335M areas of the Medium survey are only 93 square arcminutes.

between HST's long wavelength end of WC3IR at 1.6 microns and IRAC on Spitzer with its 3.6 micron filter. No current facility operates beyond IRAC's 4.5 micron channel. Current groundbased telescopes lack the sensitivity needed for detecting high redshift galaxies in the near- and mid-infrared. JWST will for the first time supply data with space based sensitivities in the near-infrared. Depth is only part of the strength of JADES and related surveys. Neither IRAC nor WFC3IR have adequate spatial resolution for discerning shapes and structures in distant galaxies. NIRCam with a diffraction-limited imaging resolution of 0.12″ at 4 microns will be able to discern kiloparsec scale structure at the highest redshifts.

Table 1 presents predicted detection limits for the JADES survey. The JAGUAR mock catalog mentioned above has been constructed to match HST results as far as they go with particular attention paid to matching the redshift evolution of sizes, colors, star formation, and chemical properties based on what is known from observed galaxy populations. Williams *et al.* (2018) estimate that 1000s of galaxies at $z\sim6$, and 10s at $z\sim10$ will be detected at a level of 3.6 nJy (5σ). The exact numbers of galaxies at higher redshifts depend on what star formation histories and feedback processes are correct. Cowley *et al.* (2018) predicted numbers of high redshift galaxies to be detected at several levels of NIRCam survey depth. They also ensured that their nominal model reproduces HST UV luminosity function results. The Cowley *et al.* (2018) nominal predictions using a semi-analytical model of galaxy formation and evolution in a ΛCDM framework imply that there would be only one galaxy per NIRCam field of view at $z\sim12$ detected at 2 microns in 100,000 seconds (slightly longer than the deep JADES exposures). Adjusting the model to reproduce also the reionization redshift implied by Planck data increases the galaxy detection rate by a factor of 5. Clearly the JADES survey will provide crucial inputs to constraining what processes are at work in the early universe.

5. Summary

The JADES survey will provide a rich data set for addressing galaxy evolution questions. Depending on the exact launch data of JWST which dictates when the survey can be executed, the JADES collaboration hopes to have public releases of portions of the survey as early as consistent with ensuring that performance questions and systematics are understood adequately. These releases will include both the raw and processed data along with lessons learned. Astronomer's Proposal Tool (APT) files for the survey are already available on the STScI web site and can serve as guides to others wanting to design their own surveys.

Acknowledgements

Development of NIRCam has been supported by NASA Contract NAS5-02015 which also supports the NIRCam portion of the JADES team.

References

Cowley, W., Baugh, C. M. Cole, S., Frenk, C. S., & Lacey, C. G. 2018, *MNRAS*, 474, 2352

Dressler, A. 1996, *HST and Beyond* (Washington: Association of Universities for Research in Astronomy)

Grogin, N. A, Kocevski, D. D., Faber, S. M., Ferguson, H. C., Koekemoer, A. M., *et al.* 2011, *ApJS*, 207, 24

Rasappu, N., Smit, R., Labbe, I., Stark, D. P. Ellis, R. S., & Oesch, P. A. 2016, *MNRAS*, 461, 3886

Williams, C. C., Curtis-Lake, E., Hainline, K. N., Chevallard, J., Robertson, B. E., Charlot, S., Endsley, R., Stark, D. P., Willmer, C. N. A., Alberts, S., Amorin, R., Arribas, S., Baum, S., Bunker, A., Carniani, S., Crandall, S., Egami, E., Eisenstein, D. J., Ferruit, P., Husemann, B., Maseda, M., Maiolino, R., Rawle, T. D., Rieke, M., Smit, R., Tacchella, S., & Willott C. J. 2018, *ApJS*, 236, 33

Uncovering Early Galaxy Evolution in the ALMA and JWST Era
Proceedings IAU Symposium No. 352, 2019
E. da Cunha, J. Hodge, J. Afonso, L. Pentericci & D. Sobral, eds.

doi:10.1017/S1743921319009463

Spectroscopy with the JWST Advanced Deep Extragalactic Survey (JADES) - the NIRSpec/NIRCAM GTO galaxy evolution project

Andrew J. Bunker on behalf of the NIRSpec Instrument Science Team and the JADES collaboration

Department of Physics, University of Oxford, Keble Road, Oxford OX13RH, United Kingdom
email: andy.bunker@physics.ox.ac.uk

Abstract. I present an overview of the JWST Advanced Deep Extragalactic Survey (JADES), a joint program of the JWST/NIRCam and NIRSpec Guaranteed Time Observations (GTO) teams involving 950 hours of observation. We will target two well-studied fields with excellent supporting data (e.g., from HST-CANDELS): GOODS-North and South, including the Ultra Deep Field. The science goal of JADES is to chart galaxy evolution at $z > 2$, and potentially out to $z > 10$, using the rest-frame optical and near-IR though observations from $\approx 1 - 5\,\mu$m. Multi-colour NIRCam imaging with 9 filters will enable photometric redshifts and the application of the Lyman break technique out to unprecedented distances. NIRSpec spectroscopy (with spectral resolving powers of $R = 100$, 1000 & 2700) will measure secure spectroscopic redshifts of the photometrically-selected population, as well as stellar continuum slopes in the UV rest-frame, and hence study the role of dust, stellar population age, and other effects. Measuring emission lines can constrain the dust extinction, star formation rates, metallicity, chemical abundances, ionization and excitation mechanism in high redshift galaxies. Coupling NIRCam and NIRSpec observations will determine stellar populations (age, star formation histories, abundances) of galaxies and provide the information to correct their broad-band spectral energy distribution for likely line contamination. Potentially we can search for signatures of Population III stars such as HeII. We can address the contribution of star-forming galaxies at $z > 7$ to reionization by determining the faint end slope of the luminosity function and investigating the escape fraction of ionizing photons by comparing the UV stellar continuum with the Balmer-line fluxes.

Keywords. instrumentation: spectrographs, galaxies: evolution, formation, luminosity function

1. Introduction

In recent years, the high redshift frontier has greatly expanded thanks largely to upgraded cameras on HST, able to see further into the near-infrared, and follow-up spectroscopy with 8-10m class telescopes on the ground. Multi-wavelength imaging using broad-band filters enables redshifts to be estimated from the shape of the spectral energy distribution (SED), and one of the cleanest and best examples of these "photometric redshifts" is the Lyman break technique. This relies on the absorption at Lyman-α (1216Å) produced by individual clouds of neutral hydrogen along the line-of-sight, and this Lyman-α forest absorption is strong at high redshift, meaning that there is a sharp drop in flux at wavelengths below Lyman-α from a distant galaxy or QSO, and essentially no flux below the Lyman limit at 912Å. We can identify candidates in different redshift slices by looking at which broad-band filter a galaxy "disappears" in due to the strong redshifted spectral break. There is a danger that objects with intrinsically red colours in the optical/near-infrared, such as low-mass stars within our own galaxy, and old or dust-reddened galaxies at intermediate redshift, can mimic the Lyman break in the UV continuum of a high redshift galaxy. However, using the colours from filters longward of the putative break can remove these low-redshift interlopers to some extent.

Going beyond simply identifying candidate high redshift galaxies, we can use the luminosities (inferred from the apparent magnitudes and estimated redshifts) and the surface densities, coupled with the depth surveyed in redshift space, to determine the luminosity function. Currently, most of our knowledge of the highest redshift galaxies is restricted to the rest-frame UV (just above the 1216Å break). The UV luminosity is usually dominated by the hot, massive, short-lived OB stars, and hence is a proxy for the star formation rate, although this is subject to the stellar initial mass function and to dust obscuration.

Unfortunately, most of the UV luminosity functions at high redshift are based just on candidate Lyman Break Galaxies (LBGs) from broad-band imaging alone, with no spectroscopic confirmation. Accurate spectroscopic redshifts are critical, since the interloper fraction of lower-redshift sources is very uncertain, and also the inferred luminosity function depends strongly on the calculation of the volume sampled, which in turn depends on the selection and completeness which are strongly affected by the redshift distribution and spectral slopes of the LBGs. Current UV luminosity functions are uncertain, but for now obtaining spectroscopic redshifts at $z > 6$ is supremely challenging, since current spectroscopy is largely limited to the rest-frame UV, and the most notable emission line, Lyman-α, is often weak or absent at high redshift, perhaps through scattering of this resonant line by neutral gas in the IGM. While a minority of the $z \sim 6$ candidates have had secure spectroscopic redshifts from Lyman-α (e.g., Bunker *et al.* 2003, Stark *et al.* 2010), beyond $z \approx 7.5$ the success in confirming the photometric redshifts through spectroscopy is very low (e.g. Caruana *et al.* 2014), with some rare exceptions (e.g. Zitrin *et al.* 2015 at $z = 8.68$). Spectral breaks have been seen in low-dispersion slitless spectroscopy (e.g., Oesch *et al.* 2016 GN-z11 which may be at $z \approx 11$), confirming the broad-band photometry but not giving a definitive redshift from emission or absorption lines. While some progress has been made using other weak rest-UV lines such as CIII]1909 (Stark *et al.* 2015) and also in the far-IR with ALMA (e.g., [CII] 158 μm, Smit *et al.* 2018), the next critical step is spectroscopy of the rest-frame optical where there are many well-understood emission lines. This will be accomplished by the near-infrared spectrograph, NIRSpec, on the James Webb Space Telescope, scheduled for launch in 2021.

2. NIRSpec – the Near-Infrared Spectrograph on JWST

NIRSpec operates in the range $0.6 - 5\,\mu$m, and has three spectral resolutions: a low-dispersion prism ($R = 100$) which captures all the wavelength range with a single exposure, and medium- and high-resolution gratings ($R = 1000$ and $R = 2700$) which use 3 bands to cover the wavelength range. The unique feature of this spectrograph is its use of micro-shutter arrays, developed specifically for NIRSpec to enable multi-object spectroscopy. Each $100 \times 200\,\mu$m micro-shutter subtends $0.2'' \times 0.4''$ and can be individually commanded to open, and they are arranged in 4 arrays each containing 171×365 micro-shutters. By opening shutters on targets, we can significantly reduce the background intensity (from zodiacal light and from other astronomical sources in the field), essentially building a slitmask in space and becoming far more sensitive than slitless spectroscopy. The NIRSpec field of view covers over $3' \times 3'$ with an unvignetted area of 5.5 square arcminutes, well matched to existing deep fields and the JWST NIRCam imager. There is also a $3'' \times 3''$ integral field unit with $0.1''$ sampling for spatially-resolved spectroscopy of individual objects, along with a number of fixed slits for traditional long-slit work.

3. The JWST Advanced Deep Extragalactic Survey (JADES) GTO Observations

The NIRSpec Instrument Science Team (IST) will undertake a cohesive "wedding cake" survey in our GTO time, targetting $z > 7$ galaxies in our deepest fields (potentially out to $z \sim 20 - 30$) and $1 < z < 7$ galaxies in our Medium and Wide fields. We intend to make a genuine impact on our understanding of galaxy evolution, rather than spreading the GTO time over a number of disparate projects. Since July 2015 we have been collaborating with the NIRCam Instrument Science Team (the PI for which is Marcia Rieke at the University of Arizona – see her contribution to the proceedings of this IAU Symposium), and the NIRCam and NIRSpec ISTs have an integrated extra-galactic programme (JADES).

In collaboration with our NIRSpec-IST, the NIRCam-IST will spend 450 hours imaging our NIRSpec survey fields with 7 broad-band filters spanning $0.9 - 4.4\,\mu$m, to $10\,\sigma$ point-source depths of $AB = 29.5 - 29.8$ mag in the Deep fields, and $AB = 28.6 - 29.0$ mag in the Medium fields. We will use these images, along with the existing deep HST images at shorter wavelength (as well as other data such as from ALMA), to select targets for NIRSpec-IST spectroscopy. The Deep Tier has been allocated 150 hours of GTO time. We cover two pointings within the GOODS-South field, one of which contain the Hubble Ultra Deep Field (HUDF) region, also known as the eXtreme Deep Field (XDF). We will obtain 25 ksec using each of the three medium-dispersion ($R = 1000$) gratings, spanning $1 - 5\,\mu$m, and the high resolution F290LP/G395H mode. The spectral resolution is sufficient to resolve the [NII]6583 doublet from Hα for emission line diagnostics, and to measure line widths (for example the asymmetry of Lyman-α emission due to blue-wing absorption). Our simulations indicate that we can allocate 60–100 targets on the micro-shutter assembly (MSA), depending on target density and allowing three microshutters for each object (dithering up and down to improve background subtraction). We will also use the low-dispersion $R = 100$ prism for 100 ksec at each pointing, which captures the full wavelength range $0.8 - 5\,\mu$m and has greater sensitivity to continuum emission than $R = 1000$, and because the prism spectra are short compared to the width of the two NIRSpec 2K infrared arrays we can potentially have 3–4 targets on each spectral row, increasing our multiplex to ~200–300 targets per pointing. All these observing modes have 10σ sensitivities to line emission of $5 - 9 \times 10^{-19}\,\mathrm{erg\,cm^{-2}\,s^{-1}}$ at $\lambda > 2\,\mu$m.

The Medium Tier of the survey will use the same spectral set-ups, but with integration times 20% those of the Deep (i.e., sensitivities of about half the Deep Tier where the observations are zodiacal background limited). The Medium Tier has been allocated 200 hours of GTO time to target 12 pointings distributed over both the GOODS-North and GOODS-South fields, and we supplement this with a Wide survey over some of the rest of the GOODS/CANDELS fields, taking snapshots of 3 ksec with the R=100 prism, and 2 ksec each with the two redder $R = 2700$ gratings ($2 - 5\,\mu$m). Having three tiers of increasing area and decreasing sensitivity means that we can sample a range in the luminosity function with comparable numbers of galaxies in each of the tiers (three luminosity bins). The Wide survey will do 35 pointings for a total of 100 hours, and will focus on galaxies at intermediate redshift ($1 < z < 5$) and rare bright targets at $z > 5$. The NIRSpec-IST three-tier survey with the MSA has a total of 470 hours of GTO time, and we will spend a further 270 hours using the Integral Field Unit with the high-dispersion $R = 2700$ gratings to obtain spatially-resolved spectra of ≈50 high redshift objects, to derive kinematics (rotation curves and outflows) and metallicity gradients. Our final high redshift targets are the most distant known QSOs at $z > 6.7$, each of which we will target for 5 hours on source with the NIRSpec fixed slits and high resolution grating to determine the IGM opacity and study the Lyman-α damping wing, as a probe of reionization.

4. Emission lines from High Redshift Galaxies

NIRSpec will deliver multiple emission line measurements for high redshift galaxies in the rest-frame optical and UV (Lyα, Hα+[NII], [SII], Hβ, [OII]3727Å, [OIII]5007Å etc.), enabling us to use "BPT" diagrams (Baldwin *et al.* 1981) to address the nature of the photoionization in individual galaxies (i.e. star formation vs. AGN). Diagnostic ratios will also give the gas-phase metallicity, and we can apply the popular "R23" measure of O/H (Pagel *et al.* 1979, using [OII], [OIII] and Hβ) out to $z \approx 10$ with NIRSpec (Chevallard *et al.* 2019). At $z < 7$ with the $R = 1000$ grating, we can use the [NII] line to break the "double fork" degeneracy in the plot of R23 against metallicity. From a wider range of line ratio diagnostics we can potentially get individual abundances for C, N and O, and these abundance patterns should evolve differently with redshift due to the different timescales involved in the production of the elements.

One of the most ambitious science goals of JWST is the discovery of the first generation of stars, forming from the pristine intergalactic medium of hydrogen and helium from Big Bang nucleosynthesis, before contamination by heavier elements produced in stellar nucleosynthesis. Some simulations predict very massive stars due to a lack of metal cooling lines, with a hard spectrum that can doubly-ionize helium. An expected signature of Population III is the HeII1640 emission line, and the absence of metal lines. The search for Population III is challenging, but we can improve our chances of detecting HeII (and tighten the limits on metal lines) by stacking our JWST spectra of faint galaxies at Lyman-α.

5. Exploring Reionization with JWST

Recent results indicate the mid-point of reionization may have occurred at $z \approx 8 - 9$ (Planck Collaboration 2016). The source of necessary ionizing photons remains an open question: the number density of high redshift quasars is insufficient at $z > 6$ to achieve this (Dijkstra *et al.* 2004). There has been speculation that low-luminosity AGN might contribute more ionizing photons (Giallogno *et al.* 2015). If these are indeed numerous enough to account for reionization then they will be identified as NIRSpec targets in our Lyman break selection, and the line ratio diagnostics will reveal their AGN nature. Star-forming galaxies at high redshift are a more likely driver of reionization, but we must first determine their rest-frame UV luminosity density. Other important and poorly-constrained factors are the slope of their UV spectra (and hence the number of ionizing photons) and the escape fraction of ionizing photons from these galaxies (f_{esc}, the fraction of ionizing photons formed on the photospheres of OB stars which reach the low-density IGM).

The star-forming galaxies we see at $z \approx 6$, even to the limiting depth of the HUDF (≈ 28.5 mag in z-band, corresponding to an absolute magnitudes $M_{UV} \approx -17$), do not provide enough photons to maintain the ionization of the Universe, even if the escape fraction is 100% (Bunker *et al.* 2004). The short-fall is even more apparent around the reionization epoch at $z \approx 8$ (Lorenzoni *et al.* 2013). Undoubtedly galaxies fainter than the detection limit contribute, and current indications are that the luminosity function may have a steep faint-end slope at high redshift (perhaps a Schechter function with $\alpha \simeq -2$), but even so we need to extrapolate well beyond current observational limits to fainter luminosities for there to be sufficient photons.

Our JWST GTO programme will go $1 - 2$ mag. fainter in the UV luminosity function with LBGs from NIRCam imaging, as well as using NIRSpec spectra for accurate redshifts and better template SEDs to improve the accuracy of the UV luminosity functions at $z > 6$, which are highly uncertain at high redshift. Another huge uncertainty is the escape fraction of ionizing photons, and while some measurements exist at $z \sim 3$ the high optical depth of the IGM at $z > 6.3$ means that we may never directly observe the ionizing photons at these high redshifts. We can use the indirect method of Zackrisson *et al.* (2013) on our NIRSpec spectra to estimate the escape fraction of ionizing photons by comparing Hα emission with the extrapolated UV continuum shortward of 912Å (corrected for dust by the Balmer decrement) out to $z \sim 7$, and using Hβ out to $z \sim 10$. The Hα luminosity is tied to the number of ionizing photons which do not escape the galaxy (and are absorbed by H I). Extrapolating the UV continuum from the observed flux above 1216Å introduces some uncertainty – in many cases we can measure the UV spectral slope, β, from the $R = 100$ low-dispersion spectrum or the broad-band photometry. Indeed the blue slopes we observe at $z > 6$ in the rest-UV (Stanway, McMahon & Bunker 2005; Wilkins *et al.* 2011) could be explained through low metallicity, or a top-heavy initial mass function (IMF), which can produce between 3 and 10 times as many ionizing photons for the same 1500Å UV luminosity as a Salpeter IMF (Schaerer 2003). Binary stars can further increase this correction factor (Eldridge & Stanway 2009). However, for reionization, the important quantity is the number of escaping ionizing photons – the product of the escape fraction of ionizing photons and the emissivity of ionizing photons per UV luminosity density longward of Lyman-α (usually measured around 1500Å), and this product is more robust than the individual quantities f_{esc} and the number of ionizing photons. The ultimate goal is to measure the escape fraction (and number of escaping photons) as a function of stellar mass, star formation rate and metallicity, and also kinematics and outflows (determined by ISM line offsets from the systemic redshift). We can then address the nature of the galaxies which reionize the Universe.

References

Baldwin, J. A., Phillips, M. M. & Terlevich, R. 1981 *PASP*, 93, 5
Bunker, A. J., Stanway, E. R., Ellis, R. S., *et al.* 2003, *MNRAS*, 342L, 47
Bunker, A. J., Stanway, E. R., Ellis, R. S., *et al.* 2004, *MNRAS*, 355, 374
Caruana, J., Bunker, A. J., Wilkins, S. M., *et al.* 2014, *MNRAS*, 443, 2831
Chevallard, J., Curtis-Lake, E., Charlot, S. *et al.* 2019, *MNRAS*, 483, 2621
Dijkstra, M., Haiman, Z. & Loeb, A. 2004, *ApJ*, 601, 666
Eldridge, J. J. & Stanway, E. R. 2009, *MNRAS*, 400, 1019
Giallongo, E., Grazian, A., Fiore, F., *et al.* 2015, *A&A*, 578, 83
Lorenzoni, S., Bunker, A. J., Wilkins, S. M., *et al.* 2013, *MNRAS*, 429, 150
Oesch, P. A., Brammer, G., van Dokkum, P. G., *et al.* 2016, *ApJ*, 819, 129
Pagel, B., Edmunds, M. G., Blackwell, D. E., Chun, M. S. & Smith, G. 1979, *MNRAS* 189, 95
Planck Collaboration 2016, *A&A*, 596A, 108
Schaerer, D. 2003, *A&A*, 397, 527
Smit, R., Bouwens, R. J., Carniani, S., *et al.* 2018, *Nature*, 553, 178
Stanway, E. R., McMahon, R. G. & Bunker, A. J. 2005, *MNRAS*, 359, 1184
Stark, D. P. *et al.* 2015, *MNRAS*, 450, 1846
Stark, D. P., Ellis, R. S., Chiu, K., *et al.* 2010, *MNRAS*, 408, 1628
Wilkins, S. M. *et al.* 2011, *MNRAS*, 417, 717
Zackrisson, E., Inoue, A. K.. & Jensen, H. 2013, *ApJ*, 777, 39
Zitrin, A., Labbé, I., Belli, S., *et al.* 2015, *ApJ*, 810L, 12

Uncovering Early Galaxy Evolution in the ALMA and JWST Era
Proceedings IAU Symposium No. 352, 2019
E. da Cunha, J. Hodge, J. Afonso, L. Pentericci & D. Sobral, eds.

doi:10.1017/S1743921320001337

Toward a new understanding of disk galaxy formation

Susan Kassin

Space Telescope Science Institute, USA

Abstract. One of the most important open issues in astronomy is the assembly of galactic disks. Over the last decade this has been addressed with large surveys of the internal kinematics of galaxies spanning the last 10 billion years of the universe. I will discuss recent results from the field that show the kinematic assembly of disk galaxies since a redshift of 2.5, including recent deep 10−30 hour observations by my group with the DEIMOS spectrograph on Keck. These results strongly challenge traditional analytic models of galaxy formation and provide an important benchmark for simulations. Furthermore, I will discuss our plans for extending measurements to higher redshifts with future instruments such as the JWST's NIRSpec IFU and the E-ELT's MOSAIC and HARMONI IFUs. From mock JWST and E-ELT observations of simulated galaxies, we are learning that interpreting these observations of galaxies in the early universe, when merging is frequent, is not necessarily straightforward.

Uncovering Early Galaxy Evolution in the ALMA and JWST Era
Proceedings IAU Symposium No. 352, 2019
E. da Cunha, J. Hodge, J. Afonso, L. Pentericci & D. Sobral, eds.

doi:10.1017/S1743921320001349

Connecting observations of the first galaxies and the Epoch of Reionisation

Simon Mutch

University of Melbourne, Australia

Abstract. Dwarf galaxies are thought to be dominant contributors of ionizing photons during the Epoch of Reionisation (EoR). Our knowledge of the statistics of these high redshift galaxies is constantly improving and will take yet another important step forward with the launch of JWST. At the same time, the upper limits on the EoR 21cm power spectrum are continually falling, with a firm measurement from SKA-low being a certainty in coming years. In order to maximise what we can learn from these two complimentary observational datasets, we need to be able to model them together, self-consistently. In this talk, I will present insights into the connection between galaxy formation and the EoR gained from the DRAGONS suite of semi-analytic and hydrodynamic galaxy formation simulations. Using these we find that the steep faint end slope of the high- redshift galaxy UV luminosity function extends well beyond current observational limits, indicating that only $\sim 50\%$ of the ionising photons available for reionisation have been observed at $z < 7$. I will also discuss the relative contribution of quasars to reionisation and present constraints on ionising escape fraction models.

Uncovering Early Galaxy Evolution in the ALMA and JWST Era
Proceedings IAU Symposium No. 352, 2019
E. da Cunha, J. Hodge, J. Afonso, L. Pentericci & D. Sobral, eds.

doi:10.1017/S1743921320001350

The brightest galaxies in the dark ages: Galaxies' dust continuum emission out to the reionization era

Caitlin Casey

University of Texas Austin, USA

Abstract. Though half of cosmic starlight is absorbed by dust and reradiated at long wavelengths (3μm – 3mm), constraints on the infrared through millimeter galaxy luminosity function (the 'IRLF') are poor in comparison to the rest-frame ultraviolet and optical galaxy luminosity function, particularly at $z \geqslant 2.5$. Here we present a backward evolution model for interpreting number counts, redshift distributions, and cross-band flux density correlations in the infrared and submillimeter sky, from 70μm – 2mm, using a model for the IRLF out to the epoch of reionization. Mock submillimeter maps are generated by injecting sources according to the prescribed IRLF and flux densities drawn from model spectral energy distributions that mirror the distribution of SEDs observed in $0 < z < 5$ dusty star-forming galaxies (DSFGs). We explore two extreme hypothetical case-studies: a dust-poor early Universe model, where DSFGs contribute negligibly ($< 10\%$) to the integrated star-formation rate density at $z > 4$, and an alternate dust-rich early Universe model, where DSFGs dominate $> 90\%$ of $z > 4$ star-formation. We find that current submm/mm datasets do not clearly rule out either of these extreme models. We suggest that future surveys at 2 mm – both from ALMA and single-dish facilities – will be crucial to measuring the IRLF beyond $z > 4$.

Uncovering Early Galaxy Evolution in the ALMA and JWST Era
Proceedings IAU Symposium No. 352, 2019
E. da Cunha, J. Hodge, J. Afonso, L. Pentericci & D. Sobral, eds.

doi:10.1017/S174392131900838X

Constraining star formation timescales with molecular gas and young star clusters

Kathryn Grasha[1] and **Daniela Calzetti**[2]

[1]Research School of Astronomy and Astrophysics, Australian National University, Canberra, ACT 2611, Australia
email: kathryn.grasha@anu.edu.au

[2]Astronomy Department, University of Massachusetts, Amherst, MA 01003, USA
email: calzetti@astro.umass.edu

Abstract. Star formation provides insight into the physical processes that govern the transformation of gas into stars. A key missing piece in a predictive theory of star formation is the link between scales of individual stars and star clusters up to entire galaxies. LEGUS is now providing the information to test the overall organization and spatial evolution of star formation. We present our latest findings of using star clusters from LEGUS combined with ALMA CO observations to investigate the transition from molecular gas to star formation in local galaxies. This work paves the way for future JWST observations of the embedded phase of star formation, the last missing ingredient to connect young star clusters and their relation with gas reservoirs. Multi-wavelength studies of local galaxies and their stellar and gas components will help shed light on early phases of galaxy evolution and properties of the ISM at high-z.

Keywords. galaxies: star clusters, galaxies: stellar content, ISM: clouds, ISM: structure

1. Introduction

The process of star formation is a critical pillar in our understandings of the foundations of galaxy evolution. Despite this, the connection between the local processes at scales of individual molecular clouds, stars, and clusters with multi-kpc galactic scales remains ill-understood. Star clusters, identifiable in galaxies up to $\sim$100 Mpc, provide a direct observational signature of the star formation process.

Within this framework, the investigation of the spatial relation between young star clusters and their natal molecular gas provide an important tool to connect local scales of star formation with global scaling relations between star formation and gas reservoirs of entire galaxies (Kennicutt 1998). The path connecting the structure of galaxies and their ISM, the properties of their molecular clouds, and the characteristics of the resulting stellar clusters are virtually uncharted. The importance of our physical understanding of resolved star formation extends beyond local galaxies and will help shed light on the evolution of star formation and its subsequent impact on galactic evolution from clumpy galaxies at high redshift to the smoother disks seen today.

2. Overview

Using young star clusters from the Legacy ExtraGalactic UV Survey (LEGUS; Calzetti *et al.* 2015), an HST Treasury program of 50 nearby galaxies, and CO molecular gas, we are able to constrain the timescale for the association of star clusters and their natal molecular clouds in two nearby spiral galaxies, NGC 5194 (Figure 1; Grasha *et al.* 2019; Schinnerer *et al.* 2013) and NGC 7793 (Grasha *et al.* 2018).

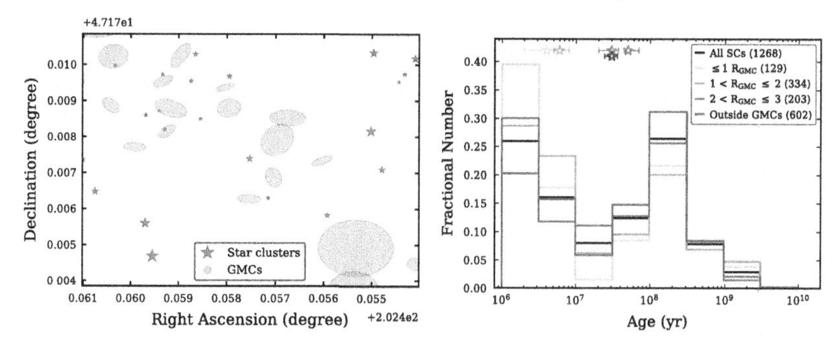

Figure 1. Left: Star clusters (pink stars) and GMCs (gray ellipses) in NGC 5194. There is a statistical tend for younger star clusters (smaller stars) to be located within the vicinity of molecular gas and older star clusters (larger stars) to be isolated. Right: Age distribution of the star clusters in bins of distance to their nearest GMC. Star clusters still located within a GMC show ages of 4 Myr (teal) whereas isolated star clusters show ages > 50 Myr (red).

We measure the distance between each star cluster to a GMC to constrain the timescale and velocity for star clusters to separate from their clouds. We find the expected trend for younger star clusters to still reside in areas where GMCs still live (Figure 1), with star clusters losing their association with their GMCs after ∼2-3 Myr in NGC 7793 and ∼4-6 Myr in NGC 5194. These timescales are a combination of stellar movement and cloud erosion via feedback and suggest that there may be environmental impacts on how long star clusters remain embedded in their natal clouds.

3. Implications

Multi-tracers of gas. The key to understanding the conversion of gas into stars likely cannot be found in a single gas phase: the transition from atomic, to molecular, to dense gas, and to stars, and the underlying mechanisms that maintain the balance between the different gas phases are ultimately what drives and controls the star formation cycle.

A pivotal, ill-studied component in this transition is the dense gas phase of the ISM, highlighting the importance for incorporating the critical role of dense gas in spatially linking gas to regions of newborn stars with facilities such as the Large Millimeter Telescope and ALMA. The different timescales for CO—cluster association versus that of dense gas are crucial for a quantitative description of the evolution of young stellar populations and the co-evolution of star formation, stellar populations, and gas.

Mechanisms that drive the star cluster–molecular gas timescale. Star formation is understood as a consequence of dense gas concentrated within giant molecular clouds (GMCs) that undergo fragmentation and other feedback processes that act to suppress star formation (Krumholz 2014). The short timescales we derive indicate that feedback and stellar wind dominate over other effects such as supernovae (Dale *et al.* 2014). The exact timescale for the expulsion of gas has broad ranging implications from constraining multiple stellar generations in globular clusters (Krause *et al.* 2013) to the reionization of the Universe (Ma *et al.* 2015).

References

Calzetti, D. *et al.* 2015, *AJ*, 149, 51
Dalc, J. *ct al.* 2014, *MNRAS*, 442, 694
Grasha, K. *et al.* 2018, *MNRAS*, 481, 1016
Grasha, K. *et al.* 2019, *MNRAS*, 483, 4707
Kennicutt, R. C. 1998, *ApJ*, 498, 541

Krause, M. *et al.* 2013, *A&A*, 550, A49
Krumholz, M. 2014, *Phys. Rep.*, 539, 49
Ma, X. *et al.* 2015, *MNRAS*, 453, 960
Schinnerer, E. *et al.* 2013, *ApJ*, 779, 42

Uncovering Early Galaxy Evolution in the ALMA and JWST Era
Proceedings IAU Symposium No. 352, 2019
E. da Cunha, J. Hodge, J. Afonso, L. Pentericci & D. Sobral, eds.

doi:10.1017/S1743921319009359

Towards the first radio galaxies

Israel Matute[1,2], Jose Afonso[1,2], Luca Bizzocchi[3], Cirino Pappalardo[1,2], Hugo Messias[4] and Stergios Amarantidis[1,2]

[1]Institute of Astrophysics and Space Sciences, Lisbon, PT
email: imatute@oal.ul.pt

[2]Faculty of Sciences, Univ. of Lisbon, Lisbon, PT

[3]Max Planck Institute for Extraterrestrial Physics (MPE), Garching, DE

[4]Joint ALMA Observatory, Santiago de Chile, CHL

Abstract. Powerful AGN have been detected up to very high redshifts ($z \sim 6-8$), well within the Epoch of Reionization (EoR), but the lack of powerful radio-galaxies among such sources strongly disagrees with the expectations based on the known radio population up to $z \sim 5$. Our group has been pursuing a detailed analysis of the faintest population of radio sources detected in the deepest fields searching for clues of these first radio galaxies. This paper describes our strategy and presents a highly confident candidate. The results, once follow-up of all candidates is completed, will have significant implications for the upcoming generation of all-sky deep radio surveys such as ASKAP-EMU, Westerbork-WODAN, and SKA itself.

Keywords. galaxies: active, galaxies: high-redshift, galaxies: distances and redshifts, galaxies: evolution, radio continuum: galaxies, submillimeter

---

1. Overview

One of the most fundamental objectives of extragalactic astronomy is the understanding of the birth of the first objects in the Universe, and how this led an almost completely neutral Universe, at redshifts $z \gtrsim 11$, to an almost completely ionised one at redshifts of around 6−7. Although theory has advanced significantly over the last few years (Santos *et al.* 2011), direct observations of sources in the Epoch of Reionisation (EoR) are still only barely manageable: distant powerful quasars, at $z \sim 6-8$, and star-forming galaxies selected from Lyman-break techniques *possibly* reaching $z \sim 8-11$ (Oesch *et al.* 2016); and the occasional detection of elusive gamma-ray burst hosts.

Furthermore, the statistical detection of neutral hydrogen will be performed with LOFAR and MWA beyond $z \sim 6$ while in the next decade, the SKA will be able to directly study the HI 21cm forest against a bright Radio Galaxy (RG), placed well within the EoR - this relies, however, in **finding even more distant powerful RGs**.

But, even though *i*) current radio telescopes are sensible enough to detect equivalent sources to local powerful RGs (like 3C273 and Cygnus A) up to $z \sim 7$; *ii*) QSOs have been detected up to $z \sim 7.5$ (Bañados *et al.* 2018); *iii*) cosmological simulations show that even earlier activity is possible within the ΛCDM framework through hierarchical, gas-rich mergers and *iv*) the extrapolation of radio LFs predict a significant number of RGs above $z > 6$ (30 radio-loud AGN per deg^2 at 100μJy flux level; S^3 simulations, Wilman *et al.* (2008)), only a handful have been detected so far at $z \sim 6$ (Fig. 1).

2. Methodology and current status

Our group has been exploring the possibility that many of these expected high-z RGs are hidden within the faint, unidentified fraction of radio sources in the deep cosmological

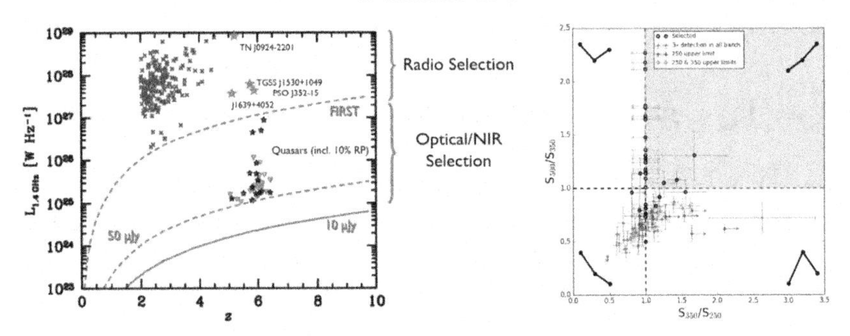

Figure 1. *Left*) Compilation of the highest-z known Radio-Galaxies . *Right*) Herschel colors for radio candidates in the COSMOS field were used as proxies for their high-z nature.

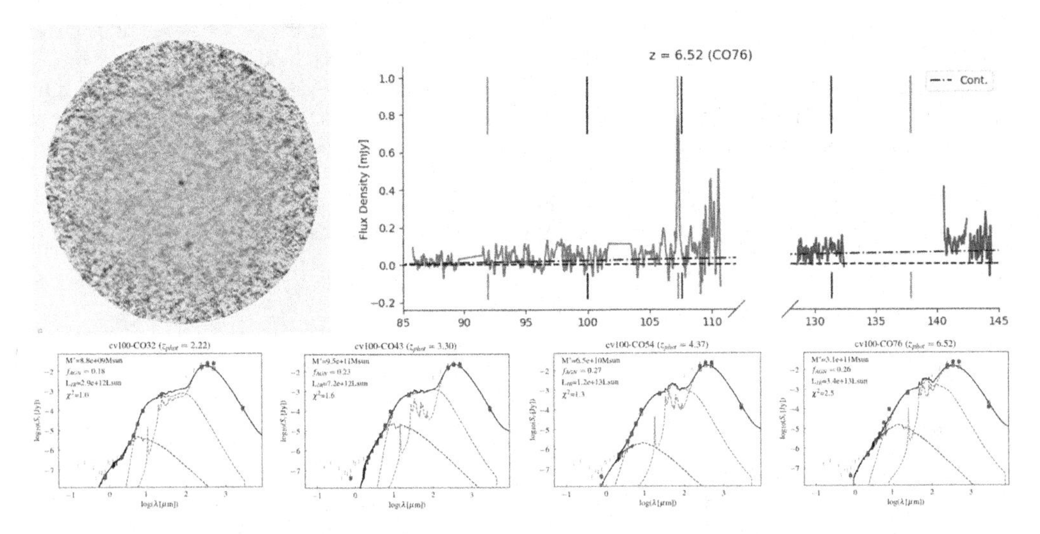

Figure 2. *Top-left*) ALMA 3mm continuum image of CVLA100. *Top-right*) Combined band3 + band4 spectrum of CVLA100 showing only one single line at the 10-σ level. The continuum level, indicated by the dot-dashed line, spectral index is compatible con thermal emission in the Rayleigh-Jeans regime ($\alpha \sim 1.9 \pm 0.2$). *Bottom*) The best fits are found at the redshift permitted by the line detection and have been computed using SED3Fit Berta *et al.* (2013).

fields. We use the FIR colors (Fig. 1) as a proxy of their high-z nature (Dowell *et al.* 2014) and measured the *Herschel* SPIRE fluxes using the full timeline information at radio position (Bendo et al. 2013). The candidates are selected after a detailed analysis of the photo-z based on their UV-to-radio SED. A handful of candidates have been found in the COSMOS + ECDFS fields while analysis is ongoing in other fields (UDS, AEGIS, GAMA, Stripe82).

One of the most promising candidates is CVLA100. Its overall SED is shown in Fig. 2 as well as its mm spectrum after follow-up with IRAM and ALMA. Although the redshift of the source is still not constrained due to the detection of a single line in the ALMA spectrum, the possibilities are severely limited and are currently being extensively evaluated by a detailed analysis with energy balance algorithms such as SED3FIT and CIGALE. The conclusions of this work, whether the high-z nature of this (or any other) candidate source is confirmed or not, will impact the definition of future radio surveys with SKA.

References

Banados, E., Venemans, B. P., & Mazzucchelli, C. 2018, *Nature*, 553, 473B
Bendo, G. *et al.* 2013, *MNRAS*, 433, 3062
Berta, S., Lutz, D., Santini, P., *et al.* 2013, *A&A*, 551, 100
Dowell, C. D., Conley, A., Glenn, J., *et al.* 2014, *ApJ*, 780, 75
Oesch, P. A., Brammer, G., van Dokkum, P. G., *et al.* 2016, *ApJ*, 819, 2
Santos, M. G., Silva, M. B., Pritchard, J. R., Cen, R., & Cooray, A. 2011, *A&A*, 527, A93
Wilman, R. J., Miller, L., Jarvis, M. J., *et al.* 2008, *MNRAS*, 388, 1335

Uncovering Early Galaxy Evolution in the ALMA and JWST Era
Proceedings IAU Symposium No. 352, 2019
E. da Cunha, J. Hodge, J. Afonso, L. Pentericci & D. Sobral, eds.

doi:10.1017/S1743921319008408

Guitarra, a Simulator for the JWST/NIRCam

Christopher N. A. Willmer[1], Kevin N. Hainline[1], Emma Curtis-Lake[2] and JADES Team

[1]Steward Observatory, University of Arizona
933 N Cherry Avenue, Tucson, AZ 85721, USA
emails: cnaw@as.arizona.edu, kevinhainline@email.arizona.edu

[2]Kavli Institute for Cosmology Cambrigde, University of Cambridge, Madingley Road, Cambridge CB3 0HA, UK
email: curtis@iap.fr

Abstract. We present an overview of *Guitarra*, a simulator for the Near Infrared Camera that creates scenes from catalogues of mock or real sources using the current best estimates of the instrument characteristics and the pattern on the sky of the observations.

Keywords. techniques: image processing, methods: numerical

1. Introduction

The (*James Webb Advanced Deep Extragalactic Survey* JADES, Rieke, this volume) will observe both GOODS fields taking data simultaneously with the JWST NIRSpec and NIRCam instruments or NIRCam and MIRI (Alberts, this volume) as part of the Guaranteed Time Observations. Most JADES observations will be carried out during JWST Cycle 1, and one of the challenges faced by the team will be providing a vetted list of galaxies for NIRSpec follow-up in the time span of about 45 days from the downlink of raw data. This poster presents a brief description of *Guitarra*, a suite of programmes that generates scenes designed to test the pipeline that will ultimately provide the spectroscopic sample.

2. Description

Guitarra† is a collection of *fortran* routines with *perl* and *python* wrapper scripts that creates astronomical scenes. These use the instrument footprint measured during ground-based cryogenic tests at Goddard Space Flight Center, point spread function models calculated by WebbPSF (Perrin *et al.* 2014), and the JWST positions on the sky output by the Astronomer's Proposal Tool (APT) used to prepare the observations. The catalogues used to create scenes contain positions, magnitudes in the NIRCam and MIRI bands and shapes described by Sersic parameters. For the realisation shown in Fig. 1 we used a combination of the mock catalogue of Williams *et al.* (2018) with a subsample of CANDELS data (van der Wel *et al.* 2012), processed using *BEAGLE* (Chevallard & Charlot (2016) to generate NIRCam fluxes (Curtis-Lake *et al.* in preparation). The images are then processed as the real data will be and mosaiced. The detection of sources on these mosaics uses *Sextractor* (Bertin & Arnouts 1996) and the fluxes are used by Hainline *et al.* (in preparation), to calculate photometric redshifts using *EAZY* (Brammer *et al.*

† Available at https://github.com/cnaw/guitarra

Figure 1. Part of a scene created by *Guitarra* using a merger of the CANDELS van der Wel *et al.* (2012) and JAGUAR Williams *et al.* (2018) catalogues. The field simulates the JADES deep observations in GOODS-S in the F115W, F200W and F356W NIRCam filters.

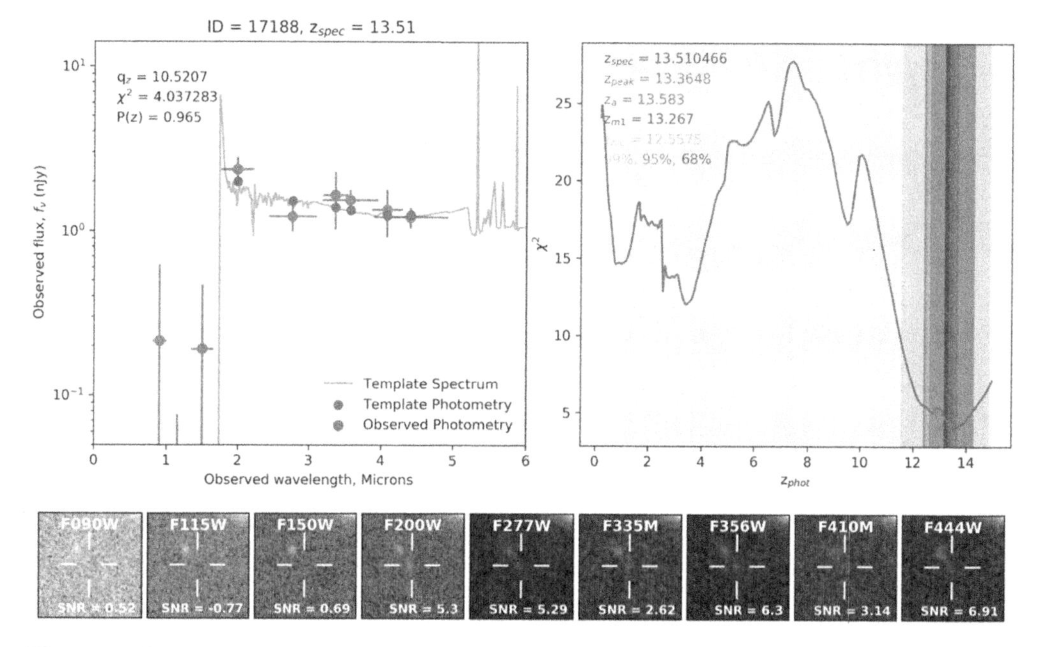

Figure 2. Reconstructed spectrum of a simulated high redshift galaxy detected in the simulated GOODS field. *Guitarra* allows testing the object detection and photometric redshift calculation algorithms (Hainline *et al.* in preparation).

2008) and *BEAGLE* (Chevallard & Charlot 2016). Fig. 2 shows a very high redshift galaxy detected in the simulated deep GOODS-S field. In addition to testing the pipeline from processing raw data to providing source catalogues, the simulations generated by *Guitarra* will be used to identify and quantify the completeness and selection functions that will affect the JADES survey, an essential step for all statistical analyses of these data (e.g., Bouwens *et al.* 2015; Finkelstein *et al.* 2015).

We gratefully acknowledge funding from the JWST/NIRCam contract to the University of Arizona, NAS5-02015.

References

Bertin, E. & Arnouts, S. 1996, *AAS*, 117, 393
Brammer, G. B., van Dokkum, P. G., & Copi, P. 2008, *ApJ*, 686, 1503
Bouwens, R. J., Illingworth, G. D., Oesch., P. A., *et al.* 2015, *ApJ*, 803, 34

Finkelstein, S. L., Ryan, R. E., Papovich, C., *et al.* 2015, *ApJ* 810, 71
Perrin, M. D., Sivaramakrishnan, A., Lajoie, C.-P., *et al.* 2014, *Proceedings SPIE* 9143, 91433X
van der Wel, A., Bell, E. F., Häussler, B., *et al.* 2012, *ApJS* 203, 24
Williams, C. C., Curtis-Lake, E., Hainline, K. N. *et al.* 2012, *ApJS* 236, 33

Author index

Printed in the United States
By Bookmasters